Lecture Notes in Computer Science   5182

Commenced Publication in 1973
Founding and Former Series Editors:
Gerhard Goos, Juris Hartmanis, and Jan van Leeuwen

Il-Yeol Song   Johann Eder
Tho Manh Nguyen (Eds.)

Data Warehousing and Knowledge Discovery

10th International Conference, DaWaK 2008
Turin, Italy, September 1-5, 2008
Proceedings

 Springer

Volume Editors

Il-Yeol Song
Drexel University
College of Information Science and Technology
Philadelphia, PA 19104, USA
E-mail: song@drexel.edu

Johann Eder
University of Klagenfurt
Information and Communication Systems
Universitätsstraße 65-67, 9020 Klagenfurt, Austria
E-mail: johann.eder@univie.ac.at

Tho Manh Nguyen
Vienna University of Technology
Institute of Software Technology and Interactive Systems
Favoritenstraße 9-11/188, 1040 Wien, Austria
E-mail: tho@ifs.tuwien.ac.at

Library of Congress Control Number: Applied for

CR Subject Classification (1998): H.2, H.3, H.4, C.2, H.2.8, H.5, I.2, J.1

LNCS Sublibrary: SL 3 – Information Systems and Application, incl. Internet/Web and HCI

ISSN	0302-9743
ISBN-10	3-540-85835-0 Springer Berlin Heidelberg New York
ISBN-13	978-3-540-85835-5 Springer Berlin Heidelberg New York

Springer is a part of Springer Science+Business Media

springer.com

© Springer-Verlag Berlin Heidelberg 2008
Printed in Germany

Typesetting: Camera-ready by author, data conversion by Scientific Publishing Services, Chennai, India
Printed on acid-free paper   SPIN: 12466535   06/3180   5 4 3 2 1 0

Preface

Data Warehousing and Knowledge Discovery have been widely accepted as key technologies for enterprises and organizations as a means of improving their abilities in data analysis, decision support, and the automatic extraction of knowledge from data. With the exponentially growing amount of information to be included in the decision making process, the data to be processed is becoming more and more complex in both structure and semantics. Consequently, the process of retrieval and knowledge discovery from this huge amount of heterogeneous complex data constitutes the reality check for research in the area.

During the past few years, the International Conference on Data Warehousing and Knowledge Discovery (DaWaK) has become one of the most important international scientific events to bring together researchers, developers and practitioners. The DaWaK conferences serve as a prominent forum for discussing the latest research issues and experiences in developing and deploying data warehousing and knowledge discovery systems, applications, and solutions. This year's conference, the 10th International Conference on Data Warehousing and Knowledge Discovery (DaWaK 2008), continued the tradition of facilitating the cross-disciplinary exchange of ideas, experience and potential research directions. DaWaK 2008 sought to disseminate innovative principles, methods, algorithms and solutions to challenging problems faced in the development of data warehousing, knowledge discovery and data mining applications.

The papers presented covered a number of broad research areas on both theoretical and practical aspects of data warehousing and knowledge discovery. In the areas of data warehousing, the topics covered included advanced techniques in OLAP and multidimensional designing and modeling, advanced OLAP and Cube Processing, innovation of ETL processes and data integration problems, materialized view optimization, distributed and parallel processing in data warehousing, data warehouses and data mining applications integration, multidimensional analysis of text documents, and data warehousing for real-world applications such as medical applications, spatial applications, and bioinformatics data warehouses. In the areas of data mining and knowledge discovery, the topics covered included text mining and taxonomy, web information discovery, stream data analysis and mining, ontology-based data and non-standard data mining techniques, machine learning, constraint-based mining, and traditional data mining topics such as mining frequent item sets, clustering, association, classification, ranking, and applications of data mining technologies to real-world problems. It is especially notable to see that some papers covered emerging real-world applications such as bioinformatics, geophysics, and terrorist networks, as well as integration of multiple technologies such as conceptual modeling of knowledge discovery process and results, integration of semantic web into data warehousing and OLAP technologies, OLAP mining, and imprecise or fuzzy OLAP. All these papers show that data warehousing and knowledge discovery technologies are maturing and making an impact on real-world applications.

From 143 submitted abstracts, we received 121 papers from 37 countries. The Program Committee finally selected 40 papers, making an acceptance rate of 33.05% of submitted papers.

We would like to express our gratitude to all the Program Committee members and the external reviewers, who reviewed the papers very profoundly and in a timely manner. Due to the high number of submissions and the high quality of the submitted papers, the reviewing, voting and discussion process was an extraordinarily challenging task. We would also like to thank all the authors who submitted their papers to DaWaK 2008, as their contributions formed the basis of this year's excellent technical program.

Many thanks go to Gabriela Wagner for providing a great deal of assistance in administering the DaWaK management issues as well as to Mr. Amin Andjomshoaa for the conference management software.

September 2008

Il Yeol Song
Johann Eder
Tho Manh Nguyen

Organization

Conference Program Chairpersons

Il-Yeol Song	Drexel University, USA
Johann Eder	University of Klagenfurt, Austria
Tho Manh Nguyen	Vienna University of Technology, Austria

Program Committee Members

Alberto Abello	Universitat Politecnica de Catalunya, Spain
Jose Balcazar	Universitat Politecnica de Catalunya, Spain
Elena Baralis	Politecnico di Torino, Italy
Ladjel Bellatreche	LISI/ENSMA, France
Bettina Berendt	Humboldt University Berlin, Germany
Petr Berka	University of Economics Prage, Czech Republic
Jorge Bernardino	ISEC, Polytechnic Institute of Coimbra, Portugal
Sourav Bhowmick	Nanyang Technological University, Singapore
Francesco Bonchi	ISTI - C.N.R., Italy
Henrik Bostrom	University of Skövde, Sweden
Jean Francois Boulicaut	INSA Lyon, France
Stephane Bressan	National University of Singapore, Singapore
Peter Brezany	University of Vienna, Austria
Luca Cabibbo	Università Roma Tre, Italy
Tru Hoang Cao	Ho Chi Minh City University of Technology, Vietnam
Jesus Cerquides	University of Barcelona, Spain
Rada Chirkova	NC State University, USA
Sunil Choenni	University of Twente and Dutch Ministry of Justice, Netherlands
Ezeife Christie	University of Windsor, Canada
Frans Coenen	The University of Liverpool, UK
Bruno Cremilleux	Université de Caen, France
Alfredo Cuzzocrea	University of Calabria, Italy
Agnieszka Dardzinska	Bialystok Technical University, Poland
Karen Davis	University of Cincinnati, USA
Kevin Desouza	University of Washington, USA
Draheim Dirk	Software Competence Center Hagenberg, Austria
Johann Eder	University of Klagenfurt, Austria
Tapio Elomaa	Tampere University of Technology, Finland
Roberto Esposito	University of Turin, Italy
Vladimir Estivill-Castro	Griffith University, Australia
Ling Feng	Tsinghua University, China

Nikos Mamoulis	University of Hong Kong, Hong Kong
Giuseppe Manco	ICAR-CNR, National Research Council, Italy
Damiani Maria-Luisa	Università degli Studi di Milano, Italy
Michael May	Fraunhofer Institut für Autonome Intelligente Systeme, Germany
Rosa Meo	University of Turin, Italy
Mukesh Mohania	I.B.M. India Research Lab, India
Yang-Sae Moon	Dept. of Computer Science, Kangwon National University, Korea
Eduardo Morales	INAOE, Mexico
Alexandros Nanopoulos	Aristotle University of Thessaloniki, Greece
Wee-Keong Ng	Nanyang Tech. University, Singapore
Tho Manh Nguyen	Vienna University of Technology, Austria
Richard Nock	Université des Antilles-Guyane, Martinique, France
Arlindo Oliveira	IST/INESC-ID, Portugal
Themis Palpanas	University of Trento, Italy
Byung-Kwon Park	Dong-A University, Korea
Torben Bach Pedersen	Aalborg University, Denmark
Dino Pedreschi	University of Pisa, Italy
Jaakko Peltonen	Helsinki University of Technology, Finland
Clara Pizzuti	ICAR-CNR, Italy
Lubomir Popelinsky	Masaryk University, Czech Republic
David Powers	The Flinders University of South Australia, Australia
Jan Ramon	Katholieke Universiteit Leuven, Belgium
Zbigniew Ras	University of North Carolina, Charlotte, USA
Andrew Rau-Chaplin	Dalhousie University, Canada
Mirek Riedewald	Cornell University, USA
Gilbert Ritschard	University of Geneva, Switzerland
Stefano Rizzi	University of Bologna, Italy
Domenico Sacca	University of Calabria & ICAR-CNR, Italy
Monica Scannapieco	University of Rome, Italy
Josef Schiefer	Vienna University of Technology, Austria
Markus Schneider	University of Florida, USA
Michael Schrefl	University of Linz, Austria
Giovanni Semeraro	University of Bari, Italy
Alkis Simitsis	Stanford University, USA
Dan Simovici	University of Massachusetts at Boston, USA
Andrzej Skowron	Warsaw University Banacha 2, Poland
Carlos Soares	University of Porto, Portugal
Il-Yeol Song	Drexel University, USA
Min Song	New Jersey Institute of Technology, USA
Jerzy Stefanowski	Poznan University of Technology, Poland
Olga Stepankova	Czech Technical University, Czech Republic
Reinhard Stolle	BMW Car IT, Germany
Jan Struyf	Katholieke Universiteit Leuven, Belgium
Mika Sulkava	Helsinki University of Technology, Finland
Ah-Hwee Tan	Nanyang Technological University, Singapore

Table of Contents

Data Warehouse and Data Mining

Clustering I

Clustering II

Mining Data Streams

Classification

Text Mining and Taxonomy I

Text Mining and Taxonomy II

Machine Learning Techniques

Data Mining Applications

UML-Based Modeling for What-If Analysis

Matteo Golfarelli and Stefano Rizzi

DEIS, University of Bologna, Viale Risorgimento 2, Bologna, Italy
golfare@csr.unibo.it, stefano.rizzi@unibo.it

Abstract. In order to be able to evaluate beforehand the impact of a strategical or tactical move, decision makers need reliable previsional systems. What-if analysis satisfies this need by enabling users to simulate and inspect the behavior of a complex system under some given hypotheses. A crucial issue in the design of what-if applications in the context of business intelligence is to find an adequate formalism to conceptually express the underlying simulation model. In this experience paper we report on how this can be accomplished by extending UML 2 with a set of stereotypes. Our proposal is centered on the use of activity diagrams enriched with object flows, aimed at expressing functional, dynamic, and static aspects in an integrated fashion. The paper is completed by examples taken from a real case study in the commercial area.

Keywords: What-if analysis, data warehouse, UML, simulation.

1 Introduction

An increasing number of enterprises feel the need for obtaining relevant information about their future business, aimed at planning optimal strategies to reach their goals. In particular, in order to be able to evaluate beforehand the impact of a strategical or tactical move, decision makers need reliable previsional systems. Data warehouses, that have been playing a lead role within business intelligence (BI) platforms in supporting the decision process over the last decade, are aimed at enabling analysis of past data, and are not capable of giving anticipations of future trends. That's where what-if analysis comes into play.

What-if analysis can be described as a data-intensive simulation whose goal is to inspect the behavior of a complex system (i.e., the enterprise business or a part of it) under some given hypotheses (called *scenarios*). More pragmatically, what-if analysis measures how changes in a set of independent variables impact on a set of dependent variables with reference to a *simulation model* offering a simplified representation of the business, designed to display significant features of the business and tuned according to the historical enterprise data [7].

Example 1. A simple example of what-if query in the marketing domain is: *How would my profits change if I run a 3 × 2 (pay 2 – take 3) promotion for one week on all audio products on sale?* Answering this query requires a simulation model to be built. This model, that must be capable of expressing the complex

I.-Y. Song, J. Eder, and T.M. Nguyen (Eds.): DaWaK 2008, LNCS 5182, pp. 1–12, 2008.
© Springer-Verlag Berlin Heidelberg 2008

relationships between the business variables that determine the impact of promotions on product sales, is then run against the historical sale data in order to determine a reliable forecast for future sales.

Surprisingly, though a few commercial tools are already capable of performing forecasting and what-if analysis, very few attempts have been made so far outside the simulation community to address methodological and modeling issues in this field (e.g., see [6]). On the other hand, facing a what-if project without the support of a methodology is very time-consuming, and does not adequately protect the designer and her customers against the risk of failure.

From this point of view, a crucial issue is to find an adequate formalism to conceptually express the simulation model. Such formalism, by providing a set of diagrams that can be discussed and agreed upon with the users, could facilitate the transition from the requirements informally expressed by users to their implementation on the chosen platform. Besides, as stated by [3], it could positively affect the accuracy in formulating the simulation problem and help the designer to detect errors as early as possible in the life-cycle of the project. Unfortunately, no suggestion to this end is given in the literature, and commercial tools do not offer any general modeling support.

In this paper we show how, in the light of our experience with real case studies, an effective conceptual description of the simulation model for a what-if application in the context of BI can be accomplished by extending UML 2 with a set of stereotypes. As concerns static aspects we adopt as a reference the multidimensional model, used to describe both the source historical data and the prediction; the YAM^2 [1] UML extension for modeling multidimensional cubes is adopted to this end. From the functional and dynamic point of view, our proposal is centered on the use of activity diagrams enriched with object flows. In particular, while the control flow allows sequential, concurrent, and alternative computations to be effectively represented, the object flow is used to describe how business variables and cubes are transformed during simulation.

The paper is structured as follows. In Section 2 we survey the literature on modeling and design for what-if analysis. In Section 3 we outline the methodological framework that provides the context for our proposal. Section 4 discusses how we employed UML 2 for simulation modeling; it proposes some examples taken from a case study concerning branch profitability and explains how we built the simulation model. Finally, Section 7 draws the conclusion.

2   Related Literature

In the literature about simulation, different formalisms for describing simulation models are used, ranging from colored Petri nets [10] to event graphs [8] and flow charts [2]. The common trait of these formalisms is that they mainly represent the dynamic aspects of the simulation, almost completely neglecting the functional (how are data transformed during the simulation?) and static (what data are involved and how are they structured?) aspects that are so relevant for data-intensive simulations like those at the core of what-if analysis in BI.

A few related works can be found in the database literature. [5] uses constraint formulae to create hypothetical scenarios for what-if analysis, while [9] explores the relationships between what-if analysis and multidimensional modeling. [4] presents the SESAME system for formulating and efficiently evaluating what-if queries on data warehouses; here, scenarios are defined as ordered sets of hypothetical modifications on multidimensional data. In all cases, no emphasis is placed on modeling and design issues.

In the context of data warehousing, there are relevant similarities between simulation modeling for what-if analysis and the modeling of ETL (Extraction, Transformation and Loading) applications; in fact, both ETL and what-if analysis can be seen as a combination of elemental processes each transforming an input data flow into an output. [15] proposes an ad hoc graphical formalism for conceptual modeling of ETL processes. While such proposal is not based on any standard formalisms, other proposals extend UML by explicitly modeling the typical ETL mechanisms. For example, [14] represents ETL processes by a class diagram where each operation (e.g., conversion, log, loader, merge) is modeled as a stereotyped class. All these proposals cannot be considered as feasible alternatives to ours, since the expressiveness they introduce is specifically oriented to ETL modeling. On the other hand, they strengthen our claim that extending UML is a promising direction for achieving a better support of the design activities in the area of BI.

Finally, we mention two relevant approaches for UML-based multidimensional modeling [1,12]. Both define a UML profile for multidimensional modeling based on a set of specific stereotypes, and represent cubes at three different abstraction levels. On the other hand, [1] is preferred to [12] for the purpose of this work since it allows for easily modeling different aggregation levels over the base cube, which is essential in simulation modeling for what-if analysis.

3   Methodological Framework

A what-if application is centered on a *simulation model*. The simulation model establishes a set of complex relationships between some *business variables* corresponding to significant entities in to the business domain (e.g., products, branches, customers, costs, revenues, etc.). In order to simplify the specification of the simulation model and encourage its understanding by users, we functionally decompose it into *scenarios*, each describing one or more alternative ways to construct a *prediction* of interest for the user. The prediction typically takes the form of a multidimensional cube, meant as a set of cells of a given type, whose dimensions and measures correspond to business variables, to be interactively explored by the user by means of any OLAP front-end. A scenario is characterized by a subset of business variables, called *source variables*, and by a set of additional parameters, called *scenario parameters*, that the user has to value in order to execute the model and obtain the prediction. While business variables are related to the business domain, scenario parameters convey information technically related to the simulation, such as the type of regression adopted for forecasting and the number of past years to be considered for regression. Distinguishing source variables among business variables is important since it enables the user to understand which are the "levers" that she can independently adjust to drive the simulation; also non-source business variables are involved in scenarios, where they are used to store simulation results. Each scenario may give rise to different simulations, one for each assignment of values to the source variables and of the scenario parameters.

Example 2. In the promotion domain of Example 1, the source variables for the scenario are the type of promotion, its duration, and the product category it is applied to; possible scenario parameters are the forecasting algorithm and its tuning parameters. The specific simulation expressed by the what-if query reported in the text is determined by giving values "3 × 2", "one week" and "audio", respectively, to the three source variables. The prediction is a Sales cube with dimensions week and product and measures revenue, cost and profit, which the user could effectively analyze by means of any OLAP front-end.

Designing a what-if application requires a methodological framework; the one we consider, presented by [6], relies on the seven phases sketched in the following:

1. *Goal analysis* aims at determining which business phenomena are to be simulated, and how they will be characterized. The goals are expressed by (i) identifying the set of business variables the user wants to monitor and their granularity; and (ii) outlining the relevant scenarios in terms of source variables the user wants to control.
2. *Business modeling* builds a simplified model of the application domain in order to help the designer understand the business phenomenon, enable her to refine scenarios, and give her some preliminary indications about which aspects can be neglected or simplified for simulation.
3. *Data source analysis* aims at understanding what information is available to drive the simulation, how it is structured and how it has been physically deployed, with particular regard to the cube(s) that store historical data.
4. *Multidimensional modeling* structurally describes the prediction by taking into account the static part of the business model produced at phase 2 and respecting the requirements expressed at phase 1. Very often, the structure of the prediction is a coarse-grain view of the historical cube(s).
5. *Simulation modeling* defines, based on the business model, the simulation model allowing the prediction to be constructed, for each given scenario, from the source data available.
6. *Data design and implementation*, during which the cube type of the prediction and the simulation model are implemented on the chosen platform, to create a prototype for testing.
7. *Validation* evaluates, together with the users, how faithful the simulation model is to the real business model and how reliable the prediction is. If the approximation introduced by the simulation model is considered to be unacceptable, phases 4 to 7 are iterated to produce a new prototype.

The five analysis/modeling phases (1 to 5) require a supporting formalism. Standard UML can be used for phases 1 (use case diagrams), 2 (a class diagram coupled with activity and state diagrams) and 3 (class and deployment diagrams), while any formalism for conceptual modeling of multidimensional databases can be effectively adopted for phase 4 (e.g., [1] or [12]). On the other hand, finding in the literature a suitable formalism to give broad conceptual support to phase 5 is much harder.

4   A Wish List

Phase 5, simulation modeling, is the core phase of design. In the light of our experience with real case studies of what-if analysis in the BI context, we enunciate a wish list for a conceptual formalism to support it:

♯1 The formalism should be capable of coherently expressing the simulation model according to three perspectives: functional, that describes how business variables are transformed and derived from each other during simulation; dynamic, required to define the simulation workflow in terms of sequential, concurrent and alternative tasks; static, to explicitly represent how business variables are aggregated during simulation.

♯2 It should provide constructs for expressing the specific concepts of what-if analysis, such as business variables, scenario parameters, predictions, etc.

♯3 It should support hierarchical decomposition, in order to provide multiple views of the simulation model at different levels of abstraction.

♯4 It should be extensible so that the designer can effectively model the peculiarities of the specific application domain she is describing.

♯5 It should be easy to understand, to encourage the dialogue between the designer and the final users.

♯6 It should rely on some standard notation to minimize the learning effort.

UML perfectly fits requirements ♯4 and ♯6, and requirement ♯5 to some extent. In particular, it is well known that the stereotyping mechanism allows UML to be easily extended. As to requirements ♯1 and ♯3, the UML diagrams that best achieve integration of functional, dynamic and static aspects while allowing hierarchical decomposition are *activity diagrams*. Within UML 2, activity diagrams take a new semantics inspired by Petri nets, which makes them more flexible and precise than in UML 1 [13]. Their most relevant features for the purpose of this work are summarized below:

– An *activity* is a graph of *activity nodes* (that can be action, control or object nodes) connected by *activity edges* (either control flows or object flows).

– An *action node* represents a task within an activity; it can be decorated by the rake symbol to denote that the action is described by a more detailed activity diagram.

– A *control node* manages the control flow within an activity; control nodes for modeling decision points, fork and synchronization points are provided.

- An *object node* denotes that one or more instances of a given class are available within an activity, possibly in a given state. Input and output objects for activities are denoted by overlapping them to the activity borderline. A datastore stereotype can be used to represent an object node that stores non-transient information.
- *Control flows* connect action nodes and control nodes; they are used to denote the flow of control within the activity.
- *Object flows* connect action nodes to object nodes and vice versa; they are used to denote that objects are produced or consumed by tasks.
- *Selection* and *transformation* behaviors can be applied to object nodes and flows to express selection and projection queries on object flows.

Though activity diagrams are a nice starting point for simulation modeling since they already support advanced functional and dynamic modeling, some extensions are required in order to attain the desired expressiveness as suggested by requirement ♯2. In particular, it is necessary to define an extension allowing basic multidimensional modeling of objects in order to express how simulation activities are performed on data at different levels of aggregation.

5   Expressing Simulation Models in UML 2

In our proposal, the core of simulation modeling is a set of UML 2 diagrams organized as follows:

1. A use case diagram that reports a what-if analysis use case including one or more scenario use cases.
2. One or more class diagrams that statically represent scenarios and multidimensional cubes. A scenario is a class whose attributes are scenario parameters; it is related via an aggregation to the business variables that act as source variables for the scenario. Cubes are represented in terms of their dimensions, levels and measures.
3. An activity diagram (called *scenario diagram*) for each scenario use case. Each scenario diagram is hierarchically exploded into activity diagrams at increasing level of detail. All activity diagrams represent, as object nodes, the business variables, the scenario parameters and the cubes that are produced and consumed by tasks.

Representation of cubes is supported by YAM² [1], a UML extension for conceptual multidimensional modeling. YAM² models concepts at three different detail levels: *upper*, *intermediate*, and *lower*. At the upper level, *stars* are described in terms of *facts* and *dimensions*. At the intermediate level, a fact is exploded into *cells* at different aggregation granularities, and the aggregation *levels* for each dimension are shown. Finally, at the lower level, *measures* of cells and *descriptors* of levels are represented.

In our approach, the intermediate and lower levels are considered. The intermediate level is used to model, through the cell stereotype, the aggregation granularities at which data are processed by activities, and to show the combinations of dimension levels (level stereotype) that define those granularities. The lower level allows single measures of cells to be described as attributes of cells, and their type to be separately modeled through the KindOfMeasure stereotype.

In order to effectively use YAM^2 for simulation modeling, three additional stereotypes are introduced for modeling, respectively, scenarios, business variables and scenario parameters:

> *name*: scenario
> *base class*: class
> *description*: classes of this stereotype represent scenarios
> *constraints*: a scenario class is an aggregation of business variable classes (that represent its source variables)

> *name*: business variable
> *base class*: class
> *description*: classes of this stereotype represent business variables
> *tagged values*:   · isNumerical (type Boolean, indicates whether the business variable can be used as a measure)
>   · isDiscrete (type Boolean, indicates whether the business variable can be used as a dimensions)

> *name*: scenario parameter
> *icon*: SP
> *base class*: attribute
> *description*: attributes of this stereotype represent parameters that model user settings concerning scenarios
> *constraints*: a scenario parameter attribute belongs to a scenario class

---

Besides these basic stereotypes, and considering the characteristics of each specific application domain, the designer may define some ad hoc activity and dependency stereotypes to model, respectively, recurrent types of activities and specific roles of object flows within such activities. In Section 6, in the context of the case study, we will see some examples of ad hoc stereotyping.

6   A Case Study

Orogel S.p.A. is a large Italian company in the area of deep-frozen food. It has a number of branches scattered on the national territory, each typically entrusted with selling and/or distribution of products. Its data warehouse includes a number of data marts, one of which dedicated to commercial analysis and centered on a Sales cube with dimensions Month, Product, Customer, and Branch.

The managers of Orogel are willing to carry out an in-depth analysis on the *profitability* (i.e., the net revenue) of branches. More precisely, they wish to know if, and to what extent, it is convenient for a given branch to invest on either selling or distribution, with particular regard to the possibility of taking new

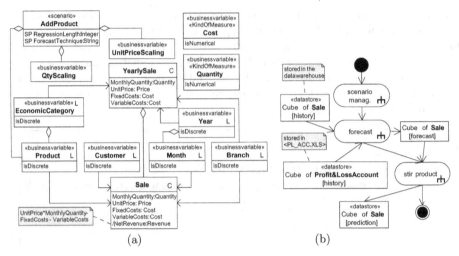

Fig. 1. Class diagram (a) and scenario diagram (b)

customers or new products. Thus, the four scenarios chosen for prototyping are: (i) analyze profitability during next 12 months in case one or more new products were taken/dropped by a branch; and (ii) analyze profitability during next 12 months in case one or more new customers were taken/dropped by a branch. Decision makers ask for analyzing profitability at different levels of detail; the finest granularity required for the prediction is the same of the Sales cube.

The main issue in simulation modeling is to achieve a good compromise between reliability and complexity. To this end, in constructing the simulation model we adopted a two-step approach that consists in first forecasting past data, then "stirring" the forecasted data according to the events (new product or new customer) expressed by the scenarios. We mainly adopted statistical techniques for both the forecasting and the stirring steps; in particular, linear regression is employed to forecast unit prices, quantities and costs starting from historical data taken from the commercial data mart and from the profit and loss account during a past period taken as a reference. Based on the decision makers' experience, and aimed at avoiding irrelevant statistical fluctuations while capturing significant trends, we adopted different granularities for forecasting the different measures of the prediction cube [6].

6.1 Representing the Simulation Model

The four what-if use cases (one for each scenario) are part of a use case diagram – not reported here for space reasons– that, as suggested by [11], expresses how the different organization roles take advantage of BI in the context of sales analysis.

As to static aspects, the class diagram shown in Figure 1.a gives a (partial) specification of the multidimensional structure of the cubes involved. Sale is the base cell; its measures are MonthlyQuantity, UnitPrice, FixedCosts, VariableCosts and NetRevenue (the latter is derived from the others), while the dimensions are Product, Customer, Month and Branch. Aggregations within dimensions represent roll-up hierarchies (e.g., products roll up to economic categories). Both dimension levels and measure types are further stereotyped as business variables. YearlySale is a cell derived from Sale by aggregation on EconomicCategory, Year and Branch. Finally, the top section of the diagram statically represents the AddProduct scenario in terms of its parameters (RegressionLength and Forecast-Technique) and source variables (UnitPriceScaling, QtyScaling and Product).

As to dynamic aspects, the add product use case is expanded in the scenario diagram reported in Figure 1.b, that provides a high-level overview of the whole simulation process. The rake symbol denotes the activities that will be further detailed in subdiagrams. Object nodes whose instances are cubes of cells of class <Sale> are named as Cube of <Sale>. The state in the object node is used to express the current state of the objects being processed (e.g., [forecast]).

The activity nodes of the context diagram are exploded into a set of hierarchical activity diagrams whose level of abstraction may be pushed down to describing tasks that can be regarded as atomic. Some of them are reported here in a simplified form and briefly discussed below:

- Activity forecast (Figure 2.a) is aimed at extrapolating sale data for the next twelve months. This is done separately for the single measures. In particular, forecasting general costs requires to extrapolate the future fixed and variable costs from the past profit and loss accounts, and scale variable costs based on the forecasted quantities. Input and output objects for forecast are emphasized by placing them on the activity borderline. The transformation stereotype expresses which measure(s) are selected from an object flow.
- The quantity forecast for next year (Figure 2.b) can be done, depending on the value taken by parameter ForecastTechnique, either by judgement (the total quantities for next year are directly specified by the user) or by regression (based on the total quantities sold during the last RegressionLength years); in both cases, the total quantity is then apportioned on the single months, products and customers proportionally to the quantities sold during the last 12 months. The selection stereotype expresses which objects are selected from an object flow. Note also the use of dependency stereotypes to specify the roles taken by objects flows within standard activities such as regression and apportion; for instance, with reference to a regression activity, length and history denote the input flows that provide, respectively, the temporal interval and the historical data to be used as a basis for regression.
- Finally, Figure 2.c explodes the stir product activity, that simulates the effects of adding a new product of a given type by reproducing the sales events related to a representative product of the same type in the same branch. First, the past sales of the reference product are scaled according to the user-specified percentages stored in QtyScaling and UnitPriceScaling. Then, cannibalization[1] on forecasted sales for the other products is simulated by

---

[1] *Cannibalization* is the process by which a new product gains sales by diverting sales from existing products, which may deeply impact the overall profitability.

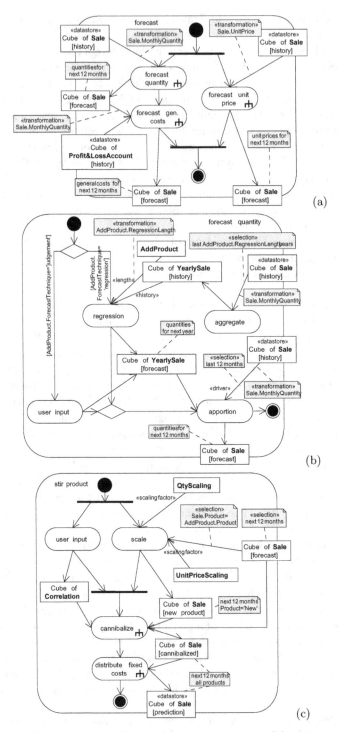

Fig. 2. Activity diagrams for forecast (a), forecast quantity (b) and stir product (c)

applying a product correlation matrix built by judgmental techniques. Finally, fixed costs are properly redistributed on the single forecasted sales.

6.2   Building the Simulation Model

In this section we give an overview of the approach we pursued to build the UML simulation model for Orogel. The starting points are the use case diagram, the business model and the multidimensional model obtained, respectively, from phases 1, 2 and 4 of the methodology outlined in Section 3. For simplicity, we assume that the multidimensional model is already coded in YAM^2.

1. The class diagram is created first, by extending the multidimensional model that describes the prediction with the statical specification of scenarios, source variables and scenario parameters.
2. For each scenario reported in the use case diagram, a high-level scenario diagram is created. This diagram should show the macro-phases of simulation, the main data sources and the prediction. The object nodes should be named consistently with the classes diagram.
3. Each activity in each scenario diagram is iteratively exploded and refined into additional activity diagrams. As new, more detailed activities emerge, business variables and scenario parameters from the class diagram may be included in activity diagrams. Relevant aggregation levels for processing business variables within activities may be identified, in which case they are increasingly reported on the class diagram. Refinement goes on until the activities are found that are elemental enough to be understood by an executive designer/programmer.

7   Conclusion

To sum up, our approach to simulation modeling fulfills the wish list proposed in Section 4 as follows: ($\sharp$1) Static, functional and dynamic aspects are modeled in an integrated fashion by combining use case, class and activity diagrams; ($\sharp$2) Specific constructs of what-if analysis are modeled through the UML stereotyping mechanism; ($\sharp$3) Multiple levels of abstraction are provided by both activity diagrams, through hierarchical decomposition, and class diagrams, through the three detail levels provided by YAM^2; ($\sharp$4) Extensibility is provided by applying the stereotyping mechanism; ($\sharp$5) Though completely understanding the implications of a UML diagram is not always easy for business users, the precision and methodological rigor encouraged by UML let them more fruitfully interact with designers, thus allowing solutions to simulation problems to emerge easily and clearly during analysis even when, in the beginning, users have little or no idea about how the basic laws that rule their business world should be coded; ($\sharp$6) UML is a standard. In the practice, the approach proved successful in making the design process fast, well-structured and transparent.

   A critical evaluation of the proposed approach against its possible alternatives unveils that the decisive factor is the choice of adopting UML as the modeling language rather than devising an ad hoc formalism. Indeed, adopting UML poses some constraints in the syntax of diagrams (for instance, the difficulty of directly showing on activity diagrams the aggregation level at which cells are processed); on the other hand it brings some undoubted advantages to the designer, namely, the fact of relying on a standard and widespread formalism. Besides, using hierarchical decomposition of activity diagrams to break down the complexity of modeling increases the scalability of the approach.

We remark that the proposed formalism is oriented to support simulation modeling at the *conceptual* level, which in our opinion will play a crucial role in reducing the overall effort for design and in simplifying its reuse and maintenance. Devising a formalism capable of adequately expressing the simulation model at the *logical* level, so that it can be directly translated into an implementation, is a subject for our future work.

References

1. Abelló, A., Samos, J., Saltor, F.: YAM²: a multidimensional conceptual model extending UML. Information Systems 31(6), 541–567 (2006)
2. Atkinson, W.D., Shorrocks, B.: Competition on a divided and ephemeral resource: A simulation model. Journal of Animal Ecology 50, 461–471 (1981)
3. Balci, O.: Principles and techniques of simulation validation, verification, and testing. In: Proc. Winter Simulation Conf., Arlington, VA, pp. 147–154 (1995)
4. Balmin, A., Papadimitriou, T., Papakonstantinou, Y.: Hypothetical queries in an OLAP environment. In: Proc. VLDB, Cairo, Egypt, pp. 220–231 (2000)
5. Dang, L., Embury, S.M.: What-if analysis with constraint databases. In: Proc. British National Conf. on Databases, Edinburgh, Scotland, pp. 307–320 (2004)
6. Golfarelli, M., Rizzi, S., Proli, A.: Designing what-if analysis: Towards a methodology. In: Proc. DOLAP, pp. 51–58 (2006)
7. Kellner, M., Madachy, R., Raffo, D.: Software process simulation modeling: Why? what? how? Journal of Systems and Software 46(2-3), 91–105 (1999)
8. Kotz, D., Toh, S.B., Radhakrishnan, S.: A detailed simulation model of the HP 97560 disk drive. Technical report, Dartmouth College, Hanover, NH, USA (1994)
9. Koutsoukis, N.-S., Mitra, G., Lucas, C.: Adapting on-line analytical processing for decision modelling: the interaction of information and decision technologies. Decision Support Systems 26(1), 1–30 (1999)
10. Lee, C., Huang, H.C., Liu, B., Xu, Z.: Development of timed colour Petri net simulation models for air cargo terminal operations. Computers and Industrial Engineering 51(1), 102–110 (2006)
11. List, B., Schiefer, J., Tjoa, A.M.: Process-oriented requirement analysis supporting the data warehouse design process - a use case driven approach. In: Ibrahim, M., Küng, J., Revell, N. (eds.) DEXA 2000. LNCS, vol. 1873, pp. 593–603. Springer, Heidelberg (2000)
12. Luján-Mora, S., Trujillo, J., Song, I.-Y.: A UML profile for multidimensional modeling in data warehouses. Data & Knowledge Engineering 59(3), 725–769 (2006)
13. OMG. UML: Superstructure, version 2.0 (2007), www.omg.org
14. Trujillo, J., Luján-Mora, S.: A UML based approach for modelling ETL processes in data warehouses. In: Song, I.-Y., Liddle, S.W., Ling, T.-W., Scheuermann, P. (eds.) ER 2003. LNCS, vol. 2813, pp. 307–320. Springer, Heidelberg (2003)
15. Vassiliadis, P., Simitsis, A., Skiadopoulos, S.: Conceptual modeling for ETL processes. In: Proc. DOLAP, McLean, VA, pp. 14–21 (2002)

Model-Driven Metadata for OLAP Cubes from the Conceptual Modelling of Data Warehouses*

Jesús Pardillo, Jose-Norberto Mazón, and Juan Trujillo

Lucentia Research Group,
Department of Software and Computing Systems,
University of Alicante, Spain
{jesuspv,jnmazon,jtrujillo}@dlsi.ua.es

Abstract. The development of a data warehouse is based on the definition of a conceptual multidimensional model. This model is used to obtain the required database metadata for implementing the data-warehouse repository. Surprisingly, current approaches for multidimensional modelling overlook the necessity of generating additional data-cube metadata to allow end-user tools to query the data warehouse. To overcome this situation, we propose an approach based on "model-driven engineering" techniques in order to automatically derive both kinds of metadata in a systematic and integrated way. Specifically, in this paper, we describe how to obtain the data-cube metadata for "on-line analytical processing" (OLAP) tools. As a proof of concept, our approach has been implemented in the ECLIPSE development platform, thus showing its feasibility.

Keywords: OLAP, data warehouse, customisation, conceptual modelling, multidimensional modelling, model-driven engineering.

1 Introduction

A *data warehouse* is an integrated database that provides adequate information in a proper way to support decision making. The components of a data warehouse are usually depicted as a multilayer architecture (data sources, ETL processes, data repository, data customisation, and end-user tools) in which data from one layer is derived from data of the previous layer [1]. Therefore, data warehouses are heterogeneous systems and its development claims for the need of managing metadata by assuring integration and interoperability between the different layers. Importantly, data repository and customisation layers are closely linked each other, since the former defines the adequate database metadata to store the data, whereas the latter defines data-cube metadata that end-user tools employ for properly accessing the data warehouse.

---

* Supported by the TIN2007-67078 project from the Spanish Ministry of Education and Science and FPU grants AP2006-00332 and AP2005-1360, and by the PAC08-0157-0668 project from the Castilla-La Mancha Ministry of Education and Science.

I.-Y. Song, J. Eder, and T.M. Nguyen (Eds.): DaWaK 2008, LNCS 5182, pp. 13–22, 2008.

The development of a data warehouse is based on the *multidimensional modelling* [2,3] which arranges data into intuitive analysis structures, *i.e.*, facts to be analysed together with a set of dimensions that contextualize them. This modelling paradigm resembles the traditional database methods [4] because it is divided into a variety of steps during which a conceptual design phase is carried out, whose results are transformed into logical data models as the basis for the implementation. Actually, once a conceptual multidimensional model is defined, two kinds of logical models must be derived: (i) a model of the data repository which determines the required database metadata to structure the data warehouse, and (ii) a model of the data cubes which contains the necessary metadata to allow end-user tools to query the data warehouse in a suitable format. Therefore, an approach for data-warehouse development must provide mechanisms for enriching database metadata for end-tool support in a systematic and automatic way as pointed out in [5].

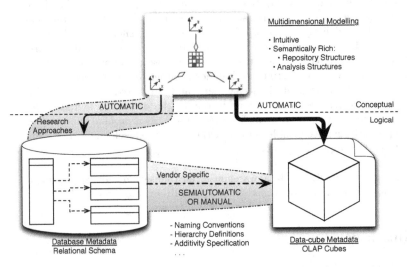

Fig. 1. Our approach for model-driven development of OLAP metadata

Surprisingly, current approaches based on multidimensional modelling (see the left-hand side of Fig. 1), only focus on deriving the database metadata from a conceptual multidimensional model [4,6,7,8], thus overlooking the derivation of the necessary data-cube metadata. Moreover, current commercial data-warehouse development platforms derive the data-cube metadata from logical models on an *ad-hoc* basis (lower part of Fig. 1). For instance, a common data-warehousing solution in the ORACLE platform[1] consists of using the WAREHOUSE BUILDER for designing the data warehouse, and then using the DISCOVERER to analyse it. In spite of the fact that ORACLE provides designers with mechanisms to integrate both tools, they are vendor-specific and require certain tedious post-processing (*e.g.*, renaming data entities or defining aggregation hierarchies) to

---

[1] URL: http://www.oracle.com/technology/products/index.html (March 2008)

configure a suitable end-user catalogue of the underlying data repository. This way of proceeding is prone-to-fail and requires great amount of effort, since the data-cube metadata is manually defined from the database metadata. Therefore, the global cost of the data-warehouse development is increased.

The generation of data-cube metadata then poses some interesting research challenges, because such metadata should be derived together with database metadata in an integrated way without any reference to a specific software platform or technology. For this aim, we advocate in [9] the use of "model-driven engineering" [10] techniques for deriving both metadata from conceptual multidimensional models. Now, in this work, we focus on designing the required mappings to automatically derive metadata for "on-line analytical processing" (OLAP) tools [11], as we shown in the right-hand side of Fig. 1; whereas database metadata derivation has been previously addressed in [12].

The main contribution of our proposal is that the semantic gap between an expressive conceptual multidimensional model and its implementation is bridged via the integrated generation of OLAP and database metadata, thus allowing the full potential analysis of any OLAP tool over the deployed database structures. Furthermore, this generation is done automatically which ameliorates the tedious task of metadata definition and prevents the emergence of human errors.

The rest of this paper is structured as follows: the next section discusses the related work. Section 3 describes the involved mapping between conceptual models and logical OLAP metadata. Then, Section 4 presents the development architecture to implement the proposed transformations. In the last section, we expound the conclusions of this research and sketch the ongoing work.

2   Related Work

One of the most popular approaches for designing data warehouses is described in [6]. In this work, the development of the data warehouse consists of three phases: (i) extracting data from distributed operational sources, (ii) organising and integrating data consistently into the data warehouse, and (iii) accessing the integrated data in an efficient and flexible fashion. The authors focus on the second phase, by proposing a general methodological framework for data-warehouse design based on the "dimensional-fact model", a particular notation for the conceptual multidimensional design. Once a dimensional schema is defined, it can be mapped on a relational model by defining the corresponding database structures, *i.e.*, tables, columns, and so on. The third phase is aimed to define data cubes, but even although the authors point out that this phase is highly important to be able to properly use data-analysis tools, they do not provide any concrete solution. In the same way, other interesting multidimensional approaches have been investigated in [4,7,8]. Unfortunately, they only focus on obtaining database metadata from conceptual multidimensional models without considering data-cube metadata.

To the best of our knowledge, the only work that addresses the generation of data-cube metadata is proposed in [13], where it is explained how to derive database schemata together with configurations for OLAP tools from a conceptual multidimensional model. However, this approach directly generates vendor-specific metadata by adapting the conceptual model to the target system. Our approach solves this drawback, since we take advantage of a vendor-neutral development architecture. In this architecture, we employ an intermediate metadata-interchange standard [14] that enables us to abstract the underlying software technologies whilst the conceptual multidimensional modelling provides semantically rich and intuitive primitives for the data-warehouse design. Both are managed by applying "model-driven engineering" [10] techniques for deriving both database and data-cube metadata in a systematic and automatic way.

3   Automatic Generation of OLAP Metadata

Our approach employs a three-layer model-transformation architecture based on conceptual, logical, and physical design phases, thus resembling the traditional database design process. Firstly, we specify the information requirements into a conceptual multidimensional model [15]. From this model, as we stated in [9], the logical model of the data-warehouse repository [12] and OLAP data cubes are derived, both represented in a vendor-independent format [14]. Finally, the physical implementation for a specific vendor platform is obtained. Specifically, in this paper, we discuss the necessary model mappings between our conceptual modelling framework [15] and the logical OLAP model.

Running Example. To exemplify our approach through the paper, consider the following running example, which is inspired by an example of [?]. We assume that the multidimensional conceptual model for the automobile-sales domain shown in Fig. 2 has been specified according to our approach defined in [15]. Herein, we focus on the *autosale* fact (represented as ▦) to analyse the automobile sales by means of the following measures (FA): the *quantity* sold, the *price* for sale, and their *total* amount. They are analysed around several dimensions (☲) that describe a sale: *auto*, *time* (tagged as a temporal dimension, {*isTime*}), *dealership*, *salesperson*, and *customer*. Each dimension contains several aggregation hierarchies to enable *roll-up* and *drill-down* OLAP operations [11] on sale data cubes. For instance, a customer can be viewed from different aggregation levels (☷): from *customer data*, analysts can aggregate (☺) automobile sales into *cities*, these into *regions*, and regions into *states*[2]. In order to simplify the entire aggregation path, sales can be also aggregated by means of two alternative paths, *i.e.*, into regions, and city into states. For representing dimensions themselves in an OLAP analysis, any dimension exposes many *dimension attributes* (DA) such as customer *born date* or even identifiers (D) such as their own *names*.

---

[2] In our models, an entire aggregation path is identified by a given name and each of its ends is labelled with '*r*' ('*d*') for specifying the roll-up (drill-down) direction.

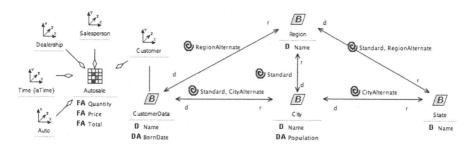

Fig. 2. The conceptual multidimensional model of automobile sales

3.1   Model Mappings for Deriving OLAP Metadata

In this work, the specification of logical OLAP metadata is carried out by means of the "common warehouse metamodel" (CWM) [14], an industry standard for information management, specifically oriented to the metadata interchange. Therefore, CWM provides the *OLAP* package that allows us to represent OLAP metadata in a vendor-neutral way. In this package, OLAP metadata are stored in *schemata*. Each schema is organised by means of *cubes* and *dimensions*. Each cube also has *cube regions* and *member selection groups* that configure its granularity and dimensions. Moreover, cubes have *cube dimension associations* to relate them to the required dimensions. On the other hand, every OLAP dimension also has aggregation *levels* that define aggregation *hierarchies* which are related each other by means of *hierarchy level associations*. Models conformed with this OLAP package can be later directly implemented into a CWM-compliant OLAP tool or even additionally translated into a particular vendor solution. In the following paragraphs, we explain the mappings between our conceptual framework for multidimensional modelling [15] and the OLAP metadata represented in CWM by using our running example.

Mapping multidimensional models and OLAP schemata. An entire (conceptual) multidimensional model m is mapped into an OLAP SCHEMA s^3. Hence, the *Autosales* model of Fig. 2 would be mapped into an *Autosales* schema. From this mapping, the multidimensional-modelling elements contained in m are then mapped into their OLAP counterparts in s.

Mapping facts and measures into OLAP cubes. Given the previous mapping, every conceptual fact in m is mapped into a CUBE in s. Moreover, mapping facts implies another mapping for establishing the cube UNIQUEKEYS that identify the data cells based on the mapped dimensions D_o. For instance, in the *Autosale* cube derived from the *Autosale* fact (see Fig. 2), where $D_o = \{Auto, Time,$

---

[3] Through these mappings, we assume $V = \{v_1, v_2, \ldots, v_n\}$ for any symbol. In addition, when the element names of the two involved domains crash, we use v_c for the conceptual source element and v_o for the OLAP target.

Dealership, Salesperson, Customer}, an *AutosaleKey* unique key would be defined with the following attributes: {*AutoKey, TimeKey, DealershipKey, SalespersonKey, CustomerKey*}. Concerning measure mappings, these cube ATTRIBUTEs are created in CWM: *Quantity, Price*, and *Total*. However, OLAP cubes also require specifying their CUBEREGIONs that define D_o and MEMBERS-ELECTIONGROUPs defining their granularity. Therefore, a *LowestLevels* region would be mapped by selecting the lowest aggregation level of each $d_o \in D_o$, e.g., the *CustomerData* level in the *Customer* dimension of Fig. 2. Finally, each dimension employed in a cube is linked by means of the corresponding CUBED-IMENSIONASSOCIATION.

Mapping dimensions and aggregation levels. For each conceptual dimension d_c in m, its OLAP counterpart d_o is mapped into s. Also, every conceptual aggregation level l_c in d_c is mapped into its corresponding OLAP level l_o in d_o. On the one hand, every dimension attribute a in l_c is mapped into a CWM ATTRIBUTE belonging to l_o and into another attribute related to d_o. On the other hand, every identifier i_c in l_c is also mapped into an attribute i_o within unique keys for both l_o and d_o. For instance, the *Customer* OLAP dimension would contain a *CustomerKey* defined with the {*CustomerDataName, CityName, RegionName, StateName*} attributes, whereas *CustomerDataBornDate* and *CityPopulation* would remain as regular attributes. In addition, for temporal dimensions, this mapping also establishes the ISTIME attribute in CWM in order to facilitate their management by OLAP tools.

Mapping aggregation hierarchies. Every of the conceptual hierarchy is mapped into a LEVELBASEDHIERARCHY where all attributes $A_c \cup I_c$ of each aggregation level l_o is also mapped as the same manner as d_c (together with their corresponding unique keys from I_c). Moreover, OLAP metadata store a HIERARCHYLEVE-LASSOCIATION hla that links and sorts every l_o into the OLAP hierarchy. Each hla also include $A_c \cup I_c$ and a unique key from the related l_o. Focusing on the *Customer* hierarchies (see Fig. 2), we would have the following aggregation paths, which would be mapped into the OLAP metadata shown in Fig. 3: *Standard* = (*CustomerData, City, Region, State*), *RegionAlternate* = (*CustomerData, Region, State*), *CityAlternate* = (*CustomerData, City, State*).

Mapping dimension attributes into level-based attributes. Dimension-related attributes $A_c \cup I_c$ that are involved through the previous mapping also require an explicit mapping into their OLAP level-based counterparts $A_o \cup I_o$. Specifically, a dimension attribute a_c is mapped into a CWM ATTRIBUTE a_o for the corresponding OLAP container, *i.e.*, levels, hierarchies, and so on. In addition, a level identifier i_c is mapped into a i_o within the unique key of the related container in order to identify its entries. Due to these mappings are spread through many others, in Table 1, we show an example correspondence for each of the involved CWM types for the customer dimension. Finally, with regards to data-type mappings, each conceptual data type is directly mapped into its CWM counterpart (note that Fig. 2 omits them for the sake of clarity).

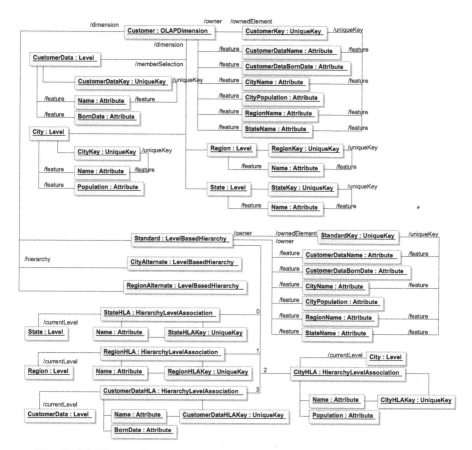

Fig. 3. OLAP metadata mapped in CWM to specify the customer dimension

Table 1. Mapping the *Name* attribute of *Region* and *State* levels of *Customer*s

Logical OLAP Metadata			Conceptual
Attribute	Container	CWM Type	Level (*Customer* dim.)
RegionName	*Customer*	DIMENSION	*Region*
StateName	*Customer*	DIMENSION	*State*
RegionName	*RegionAlternate*	LEVELBASEDHIERARCHY	*Region*
StateName	*RegionAlternate*	LEVELBASEDHIERARCHY	*State*
Name	*RegionHLA*	HIERARCHYLEVELASSOC.	*Region*
Name	*StateHLA*	HIERARCHYLEVELASSOC.	*State*
Name	*Region*	LEVEL	*Region*
Name	*State*	LEVEL	*State*

In short, with the designed mappings, we can obtain the OLAP metadata to query the database structures [12] with an integrated and vendor-neutral approach [9]. It is worth noting that OLAP metadata need to refer to database metadata, such as columns for a relational schema. However, for the sake of simplicity, we assume that there is an implicit mapping between these metadata based on matching element names, *e.g.*, in a relational schema, there is an *Autosale* table that deploy the *Autosale* OLAP cube. Interestingly, we can easily assume it, because both kinds of metadata, are generated from the same conceptual multidimensional model.

4   Implementation

The formal specification and deployment of the previous mappings is based on one of the best-known initiatives for "model-driven engineering" [10], namely the "model-driven architecture" (MDA) [14]. Concerning MDA, the proposed mappings relate two models with regards to the underneath platform: a "platform-independent model" (PIM) that represents a multidimensional model at the conceptual level, and two "platform-specific models" (PSMs) for representing both data-warehouse and OLAP metadata at the logical level with regards to a particular software technology but in a vendor-independent manner. For the PIM, we extend the "unified modelling language" (UML) [14] by means of our UML profile for multidimensional modelling of data warehouses [15]. On the other hand, our PSM are represented in CWM [14] as we stated in the previous discussion: the repository structures with the *Relational* package [12] and the analysis structures with the *OLAP* package. Therefore, we employ a vendor-neutral representation, *i.e.*, independently of any tool, thus focusing only on the problems involved in the conceptual–logical mapping. For implementing this one, we use the "query/view/transformation" (QVT) [14] language that contains a declarative part to easily design *model-to-model relations*.

The proposed model-transformation architecture has been implemented in the ECLIPSE development platform[4] as we show in Fig. 4. Due to its modular design, there are several developments to support the MDA standards (*e.g.*, the MODEL DEVELOPMENT TOOLS (MDT) for modelling with UML, MEDINI QVT for specifying declarative QVT relations, or SMARTQVT for imperative ones), which we have combined and adapted in order to provide an "integrated development environment" (IDE) to manage data-warehousing projects in our model-driven approach. With it, we have implemented the running example as a proof of concept (see Fig. 4). Specifically, in the left-hand side of the figure, the *Autosales* conceptual model is implemented in order to automatically transform it into the corresponding logical OLAP metadata (shown on the right-hand side), by means of applying the QVT relations at the centre of the figure. Due to the space constraints, we omit here an explanation of the transformation trace that originates the target metadata.

---

[4] URL: http://www.eclipse.org (March 2008)

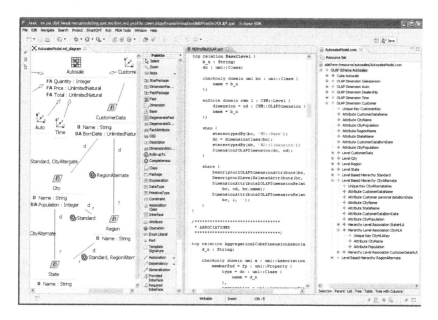

Fig. 4. Our IDE based on ECLIPSE of the case study concerning automobile sales

5   Conclusions

Hitherto, no methodological approach has been proposed for the integrated development of data warehouses, considering both database and data-cube vendor-neutral metadata. Moreover, no one has taken advantage of employing a model-driven approach to automatically derive these kinds of metadata from conceptual multidimensional models in an integrated vendor-neutral way. It would allow data-warehouse developers to focus on the high-level description of the system instead of low-level and tool-dependent details. Therefore, we propose to extend the model-transformation architecture presented in [12] to consider the automatic generation of OLAP metadata from the same conceptual multidimensional model. In this work, we discuss the involved conceptual–logical mappings and present a set of QVT relations to formalise the mapping between our conceptual modelling framework for data warehouses [15] and the analysis structures of the OLAP metadata represented in CWM [14], a vendor-neutral standard for metadata interchange. As a proof of concept, our proposal has been implemented in the ECLIPSE open-source platform, showing the feasibility of the automatic metadata generation for OLAP tool from conceptual multidimensional models.

We plan to extend this approach to consider advanced multidimensional properties such as degenerated facts and dimensions, also investigating how their logical OLAP counterparts are implemented. We would also investigate the derivation of specific OLAP metadata for non-traditional OLAP tools such as in mobile environments [17,18], as well as their enrichment with advanced customisation issues such as OLAP preferences [19]. Apart from OLAP metadata, other analysis techniques, such as data mining [20], should be also considered. In addition, we will further investigate the existing mappings between data-cube and database structures.

References

1. Jarke, M., Lenzerini, M., Vassiliou, Y., Vassiliadis, P.: Fundamentals of Data Warehouses. Springer, Heidelberg (2000)
2. Inmon, W.H.: Building the Data Warehouse. Wiley, Chichester (2005)
3. Kimball, R., Ross, M.: The Data Warehouse Toolkit: The Complete Guide to Dimensional Modeling. Wiley, Chichester (2002)
4. Hüsemann, B., Lechtenbörger, J., Vossen, G.: Conceptual data warehouse modeling. In: DMDW, p. 6 (2000)
5. Rizzi, S., Abelló, A., Lechtenbörger, J., Trujillo, J.: Research in data warehouse modeling and design: dead or alive?. In: DOLAP, pp. 3–10 (2006)
6. Golfarelli, M., Maio, D., Rizzi, S.: The Dimensional Fact Model: A Conceptual Model for Data Warehouses. Int. J. Cooperative Inf. Syst. 7(2-3), 215–247 (1998)
7. Abelló, A., Samos, J., Saltor, F.: YAM^2: a multidimensional conceptual model extending UML. Inf. Syst. 31(6), 541–567 (2006)
8. Prat, N., Akoka, J., Comyn-Wattiau, I.: A UML-based data warehouse design method. Decis. Support Syst. 42(3), 1449–1473 (2006)
9. Pardillo, J., Mazón, J.N., Trujillo, J.: Towards the Automatic Generation of Analytical End-user Tool Metadata for Data Warehouses. In: BNCOD, pp. 203–206 (2008)
10. Bézivin, J.: Model Driven Engineering: An Emerging Technical Space. In: GTTSE, pp. 36–64 (2006)
11. Chaudhuri, S., Dayal, U.: An Overview of Data Warehousing and OLAP Technology. SIGMOD Record 26(1), 65–74 (1997)
12. Mazón, J.N., Trujillo, J.: An MDA approach for the development of data warehouses. Decision Support Systems 45(1), 41–58 (2008)
13. Hahn, K., Sapia, C., Blaschka, M.: Automatically Generating OLAP Schemata from Conceptual Graphical Models. In: DOLAP, pp. 9–16 (2000)
14. Object Management Group (OMG): Catalog of OMG Specifications (March 2008), http://www.omg.org/technology/documents/spec_catalog.htm
15. Luján-Mora, S., Trujillo, J., Song, I.Y.: A UML profile for multidimensional modeling in data warehouses. Data Knowl. Eng. 59(3), 725–769 (2006)
16. Giovinazzo, W.A.: Object-Oriented Data Warehouse Design: Building A Star Schema. Prentice Hall, Englewood Cliffs (2000)
17. Cuzzocrea, A., Furfaro, F., Saccà, D.: Hand-OLAP: A System for Delivering OLAP Services on Handheld Devices. In: ISADS, pp. 80–87 (2003)
18. Maniatis, A.S.: The Case for Mobile OLAP. In: Lindner, W., Mesiti, M., Türker, C., Tzitzikas, Y., Vakali, A.I. (eds.) EDBT 2004. LNCS, vol. 3268, pp. 405–414. Springer, Heidelberg (2004)
19. Rizzi, S.: OLAP preferences: a research agenda. In: DOLAP, pp. 99–100 (2007)
20. Zubcoff, J.J., Pardillo, J., Trujillo, J.: Integrating Clustering Data Mining into the Multidimensional Modeling of Data Warehouses with UML Profiles. In: Song, I.-Y., Eder, J., Nguyen, T.M. (eds.) DaWaK 2007. LNCS, vol. 4654, pp. 199–208. Springer, Heidelberg (2007)

An MDA Approach for the Development of Spatial Data Warehouses

Octavio Glorio and Juan Trujillo

Department of Software and Computing Systems,
University of Alicante, Spain
{oglorio,jtrujillo}@dlsi.ua.es

Abstract. In the past few years, several conceptual approaches have been proposed for the specification of the main multidimensional (MD) properties of the spatial data warehouses (SDW). However, these approaches often fail in providing mechanisms to univocally and automatically derive a logical representation. Even more, the spatial data often generates complex hierarchies (i.e., many-to-many) that have to be mapped to large and non-intuitive logical structures (i.e., *bridge tables*). To overcome these limitations, we implement a Model Driven Architecture (MDA) approach for spatial data warehouse development. In this paper, we present a spatial extension for the MD model to embed spatiality on it. Then, we formally define a set of Query/ View/ Transformation (QVT) transformation rules which allow us, to obtain a logical representation in an automatic way. Finally, we show how to implement the MDA approach in our Eclipse-based tool.

Keywords: Spatial Data Warehouse, Model Driven Architecture.

1 Introduction

Data warehouse (DW) systems provide companies with many years of historical information for the success of the decision-making process. Nowadays, it is widely accepted that the basis for designing the DW repository is the multidimensional (MD) modeling [1,2]. Even more, the multidimensional schema is used in the development of other parts of the DW, such the OLAP server or the ETL module. On the other hand, SDWs integrate spatial data in the DW structure, coming in the form of spatial MD hierarchies levels and measures. The spatial data represent geometric characteristics of a real world phenomena; to this aim these data have an absolute and relative position, an associated geometry and some descriptive attributes.

Various approaches to conceptual modeling for SDWs are proposed in [3,4], where multidimensional models are extended with spatial elements. However, these approaches are lacking in formal mechanisms to automatically obtain the logical representation of the conceptual model and the development time and cost is increased. Even more, the spatial data often generates complex structures that had to be mapped to non-intuitive logical structures.

I.-Y. Song, J. Eder, and T.M. Nguyen (Eds.): DaWaK 2008, LNCS 5182, pp. 23–32, 2008.
© Springer-Verlag Berlin Heidelberg 2008

In order to overcome a similar limitation, but without taking account of spatial data, we have previously described a MDA framework [5] to accomplish the development of DW and we formally present a set of Query/View/Transformation (QVT) [6] transformations for MD modeling of the DW [7]. In this paper, we introduce spatial data on this approach to accomplish the development of SDWs using MDA. Therefore, we focus on (i) extending the conceptual level with spatial elements, (ii) defining the main MDA artifacts for modeling spatial data on a MD view, (iii) formally establishing a set of QVT transformation rules to automatically obtain a logical representation tailored to a relational spatial database (SDB) technology, and (iv) applying the defined QVT transformation rules by using our MDA tool (based on Eclipse development platform [8]), thus obtaining the final implementation of the SDW in a specific SDB technology (PostgreSQL [9] with the spatial extension PostGIS [10]).

The main advantages of our approach are (i) productivity is improved and development time and cost decreased, (ii) assure high quality SDWs, (iv) easy to adapt to new platform technologies, and (v) assure correct and intuitive design and analysis.It is important to notice that, the spatial multidimensional schema of the DW is used here to generate the repository and to enrich the OLAP service. But also the others cornerstone of the DW system have to take advantage of the conceptual spatiality added, and this is out of the scope of this work.

The remainder of this paper is structured as follows. A brief overview of MDA and QVT is given in section 2. Section 3 presents the related work. Section 4 describes our MDA approach for development of SDWs. An example is provided in section 5 to show how to apply MDA and QVT transformation rules by using a Eclipse-based tool and to obtain implementable code for a specif SDB engine. Finally, section 6 points out our conclusions and future works.

2   Overview of MDA and QVT

Model Driven Architecture (MDA) is an Object Management Group (OMG) standard [11] that addresses the complete life cycle of designing, deploying, integrating, and managing applications by using models in software development. MDA encourages specifying a Platform Independent Model (PIM) which contains no specific information of the platform or the technology that is used to realize it. Then, this PIM can be transformed into a Platform Specific Model (PSM) in order to include information about the specific technology that is used in the realisation of it on a specific platform. Later, each PSM is transformed into code to be executed on each platform. PIMs and PSMs can be specified by using any modeling language, but typically MOF-compliant languages, as the Unified Modeling Language (UML) [12], are used since they are standard modeling languages for general purpose and, at the same time, they can be extended to define specialized languages for certain domains (i.e. by metamodel extensibility or profiles).

Nowadays, the most crucial issue in MDA is the transformation between PIM and PSM. Thus, OMG defines QVT [6], an approach for expressing these MDA transformations. This is a standard for defining transformation rules between MOF-compliant models.

3   Related Work

Various approaches for the conceptual and logical design of SDWs have been proposed in the last few years [3,4]. To this aim, several authors define conceptual elements of SDWs, i.e., spatial measures and dimensions. For example, Stefanovic [13] propose three types of spatial dimensions based on the spatial references of the hierarchy members: non-spatial (a usual thematic hierarchy), spatial-to-non-spatial (a level has a spatial representation that rolls-up to a non-spatial representation), and fully spatial (all hierarchy levels are spatial).

Regarding measures, Stefanovic [13] distinguish numerical and spatial measures; the latter represent the collection of pointers to spatial objects. Rivest [14] extend the definition of spatial measures and include measures represented as spatial objects or calculated using spatial metric or topological operators.

However,these approaches are presenting the following drawbacks: (i) they do not define a standard notation for conceptual modeling (e.g. UML), (ii) some of the transformations to generate logical and physical models from a conceptual model are not totally automatic in a formal way, (iii) they are based on a specific implementation (e.g. star schema in relational databases), and (iv) use to generate additional complex structures (i.e., *bridge tables* [1]) to support spatial data decreasing the simplicity and intuitiveness of the model. Even more, to the best of our knowledge, there is no effort developed for aligning the design of SDWs with the general MDA paradigm. In order to overcome these pitfalls, we developed the present approach for the development of SDWs.

4   A MDA Implementation for Spatial Data Warehouses Development

In this section, the MD modeling of a SDW is aligned with MDA. We show how to define (i) the main MDA artifacts for spatial MD modeling, and (ii) a set of QVT transformation rules between these models. In Fig. 1, we show a symbolic diagram of our approach: from the PIM (spatial MD model), several PSMs (logical representations) can be obtained by applying several QVT transformations.

4.1   PIM Definition

This PIM is developed following our UML profile presented in [15]. This profile contains the necessary stereotypes in order to elegantly represent main MD properties at the conceptual level by means of a UML class diagram in which the information is clearly organised into facts and dimensions. These facts and dimensions are modeled by *Fact* (represented as ▦) and *Dimension* (⊯) stereotypes, respectively. Facts and dimensions are related by shared aggregation relationships (the *Association* UML metaclass) in class diagrams. While a fact is

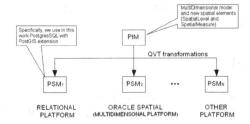

Fig. 1. Overview of our MDA approach for MD modeling of SDW repository

composed of measures or fact attributes (*FactAttribute*, **FA**), with respect to dimensions, each aggregation level of a hierarchy is specified by classes stereotyped as *Base* (*B*). Every *Base* class can contain several dimension attributes (*DimensionAttribute*, **DA**) and must also contain a *Descriptor* attribute (*Descriptor*, **D**). An association stereotyped as *Rolls-upTo* (*Rolls-upTo*,℗) between *Base* classes specifies the relationship between two levels of a classification hierarchy. Within it, role *R* represents the direction in which the hierarchy rolls up, whereas role *D* represents the direction in which the hierarchy drills down.

Nevertheless, this profile lacks from spatial expressivity. Therefore, we enrich the MD model with the minimum required description for the correct integration of spatial data, coming in spatial levels and spatial measures. Then, we implement this spatial elements in the base MD UML profile [15]. The new spatial elements are a generalization of the MD levels and the MD measures. In our previous profile these elements correspond to the *Base* and *FactAttribute* stereotypes, respectively. We named the new stereotypes *SpatialLevel*(*⊞*) and *SpatialMeasure*(✳). Finally, we add a property to these new stereotypes in order to geometrically describe them. All the allowed geometric primitives use to describe elements are group in a enumeration element named *GeometricTypes*. These primitives are included on ISO [16] and OCG [17] SQL spatial standards, in this way we ensure the final mapping from PSM to platform code. The complete profile can be seen on Fig. 2.

4.2 PSM Definition

In a MD modeling, platforms specific means that the PSM is specially designed for a kind of a specific database. According to Kimball [1], the most common representation for MD models is relational, thus, we assume that our PSM is a relational specific one.

Our PSM is modeled using the relational metamodel from CWM (Common Warehouse Metamodel) [18], since it is a standard to represent the structure of data resources in a relational database. Besides it is SQL:1999 compliant, so we will be able to obtain SQL code in an easy and straightforward way from relational CWM models. Furthermore, CWM metamodels can all be used as source or target for MDA transformations, since they are MOF-compliant and QVT can be applied [19]. For the sake of clarity, we use the part of the relational metamodel, shown in figure 3.

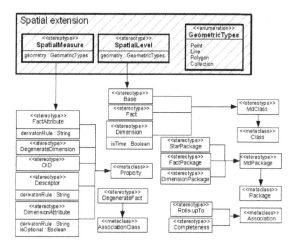

Fig. 2. UML Profile for conceptual MD modeling with spatial elements in order to support spatial data integration

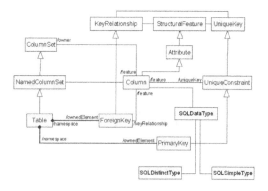

Fig. 3. Part of the Relational CWM metamodel

4.3   From PIM to PSM

In this paper, transformations are given following the declarative approach of QVT [6], thus relations between elements of the metamodels are used for constructing the transformation between models. We use the diagrammatic of the declarative part of QVT to develop each transformation. Each of these transformations contains the following elements:

- Two or more domains: each domain identifies one candidate model (i.e. PIM or PSM metamodels)
- A relation domain: it specifies the kind of relation between domains, since it can be marked as checkonly (labeled as C) or as enforced (labeled as E).
- When and Where clause: it specifies the pre and post conditions that must be satisfied to carry out the relation.

In [7], we have developed every transformation to obtain a relational PSM from its corresponding MD PIM by using the diagrammatic (i.e. transformation diagrams) and textual notation of QVT. Now, we add some spatial features to this PIM in order to obtain a PSM for relational spatial database platforms.

SpatialLevel2Column

In Fig. 4 the *SpatialLevel2Column* transformation is shown. A spatial level is matching with a column on a precise table corresponding to the dimension which the level belongs to. This table was previously generated by the *Dimension2Table* transformation defined on [7].In previous approach, we only map the level attributes to columns, the level itself was not mapped. Now, things are different, a column is created by every spatial level. This new column of spatial type keeps the geometry of the level. Then, this geometry can be used to compute precise OLAP analysis across complex hierarchies without using additional complex and non-intuitive structures. In the where clause, we define a prefix named *geom* to use in the names of the spatial columns generated.

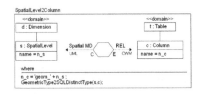

Fig. 4. Transforming SpatialLevel into a column on a dimension table

GeometricType2SQLDistinctType

This transformation takes the spatial type (i.e., polygon, line, point, etc.) of the level or measure and fixes the type of the new column created by *Spatial-Level2Column* or *SpatialMeasure2Column*. The key of this transformation is the mapping between the value of the *GeometricType* to a specific *SQLDistinctType*. The use of this CWM type is to represent proprietary and complex types. Indeed, we use for the last reason, to represent a complex type, a geometric type. In Fig. 5 we show the *GeometricType2SQLDistinctType* transformation: a spatial level was previously mapped into a column and now this column fixes the geometric type using the spatial description of the level. In analogous way, a column created by a spatial measure fixes it geometric type using the spatiality of the conceptual element.

SpatialMeasure2Column

SpatialMeasure2Column transformation is similar to *SpatialLevel2Column* (see In Fig. 4). Now, a spatial measure matches with a column on a precise table previously generated by the *Fact2Table* transformation defined on [7] and corresponding to the fact which the measure belongs to.

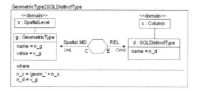

Fig. 5. Transforming GeometricType of a SpatialLevel into the type of the respective column

5   Case Study

In this section we provide an example about a olive grove study to show how to apply the transformations defined in section 4.3 in order to obtain the corresponding PSM from a PIM. Even more, we obtain a simple implementation able to support complex structures (i.e., many-to-many hierarchies) without additional structures. To build the conceptual schema of the example, we use our Eclipse-based diagram editor wich automatic reconfigures the palette and the semantic using an UML Profile, in this case we use the PIM defined in section 4.1. In Fig. 6 we show the complete model inside a window of this tool.

We focus on the *Olive Grove* fact. This fact (represented as ⊞) contains a measure named *area* (spatial measure stereotyped represent as ✳). On the other hand, we also consider the following dimensions (⚏) as contexts to be analyse: *Time* and *Location*. We focus on the *Location* dimension, with the following hierarchy spatial levels(☞): *UTM Coordinate* and *Province*. *UTM coordinate* corresponds to the higher level of the *Location* dimension, then *Province* corresponds to a lower spatial level that represents the provincial division. All the spatial levels and measures involve have a polygon shape, therefore respective geometric attributes are of *polygon GeometricType*.

Due to space constraints, we focus on the *Location* dimension transformation. The transformation result of each spatial level involverd is given by *SpatialLevel2Column* and *GeometricType2SQLDistinctType* (see Fig. 7): the spatial level is transformed into elements from CWM Relational metamodel according to these transformations. Basically, the *SpatialLevel2Column* creates a column in the dimension table in order to describe the spatiality of the level. Then, the geometric type of this new column is fixed by *GeometricType2SQLDistinctType*.

Notice in this example the many-to-many hierarchy between the spatial levels. This situation do not allow precise and intuitive SDW design and analysis. Malinowski [20] proposes to model these hierarchies using *bridge tables* [1]. However, the traditional mapping must additionally include information about measure distribution and this information is keep on complex structures that had to be created by meaning a programming effort and a complexity added to the model. With our implementation we can compute a precise aggregation between complex hierarchies only using the spatial column that describe levels involved as is shown in the following query:

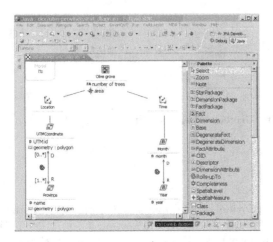

Fig. 6. Conceptual schema of Case Study

```
SELECT SUM(trees_number*AREA(INTERSECTION(utm_coordinate_geometry,
province_geometry))/AREA(utm_coordinate_geometry)) AS spatial_agg
FROM location_dimension INNER JOIN olive_groves_fact_table ON
location_dimension.PK_utm_id = olive_groves_fact_table.FK_utm_id
GROUP BY province_name
```

In the spatial query above, see the factor inside the SQL aggregation function, $AREA(INTERSECTION(utm\_coordinate\_geometry, province\_geometry))/ AREA(utm\_coordinate\_geometry)$. This factor ponderate the number of trees of a UTM coordinate while aggregating in provinces. The spatiality added to the MD conceptual model finish in the enrichment of the classical aggregation query to produce precise results.

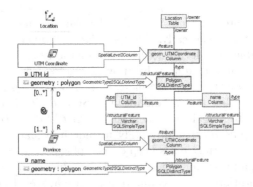

Fig. 7. Applying SpatialLevel2Column and GeometricType2SQLDistinctType transformations

6   Conclusion and Future Work

In this paper, we have enrich our previous MDA-oriented framework [7] for the development of DWs to integrate spatial data and develops SDWs. This framework addresses the design of the whole SDW system by aligning every component of the DW with the different MDA viewpoints (CIM, PIM, and PSM). We have focused on describing one part of our framework: an MDA approach for the development of the SDW repository and OLAP service based on the MD modeling, since it is the cornerstone of any DW system. We have defined the corresponding MDA artifacts: our spatial MD modeling profile has been used as a PIM and the CWM relational package as a PSM, while the transformations are formally and clearly established by using the QVT language. Finally, some examples has been presented in order to show the feasibility of our proposal and the simplicity of the logical implementation generated.

Our short-term intentions include improving our MDA approach for the development of SDWs by adding other PSMs according to several platforms. Furthermore, we plan to extend the spatial elements presented to some complex levels and measures (i.e., temporal measures, time dimensions).

Acknowledgements

This work has been partially supported by the ESPIA project (TIN2007-67078) from the Spanish Ministry of Education and Science, and by the QUASIMODO project (PAC08-0157-0668) from the Castilla-La Mancha Ministry of Education and Science (Spain). Octavio Glorio is funded by the University of Alicante under the 11th Latin American grant program.

References

1. Kimball, R., Ross, M.: The Data Warehouse Toolkit, 2nd edn. Wiley, Chichester (2002)
2. Inmon, W.: Building the Data Warehouse. Wiley Sons, Chichester (2002)
3. Bimonte, S., Tchounikine, A., Miquel, M.: Towards a spatial multidimensional model. In: DOLAP 2005: Proceedings of the 8th ACM international workshop on Data warehousing and OLAP, pp. 39–46. ACM, New York (2005)
4. Malinowski, E., Zimányi, E.: Representing spatiality in a conceptual multidimensional model. In: GIS 2004: Proceedings of the 12th annual ACM international workshop on Geographic information systems, pp. 12–22. ACM, New York (2004)
5. Mazón, J.-N., Trujillo, J., Serrano, M., Piattini, M.: Applying mda to the development of data warehouses. In: DOLAP 2005: Proceedings of the 8th ACM international workshop on Data warehousing and OLAP, pp. 57–66. ACM, New York (2005)
6. OMG 2nd Revised Submission: MOF 2.0 Query/Views/Transformations (2002), http://www.omg.org/cgi-bin/doc?ad/05-03-02
7. Mazón, J.-N., Trujillo, J.: An MDA approach for the development of data warehouses. Decision Support Systems 45(1), 41–58 (2008)

8. The Eclipse Foundation: Eclipse, version 3.3.1.1 (Visited January 2008), http://www.eclipse.org

9. PostgreSQL Global Development Group: PostgreSQL, version 8.3 (February 2008), http://www.postgresql.org

10. Refractions Research : PostGIS, version 1.3.2 (December 2007), http://postgis.refractions.net/

11. OMG, Object Management Group: Model Driven Architecture (MDA) (2004), http://www.omg.org/mda/

12. Object Management Group: Unified Modeling Language (UML), version 2.1.1 (February 2007), http://www.omg.org/technology/documents/formal/uml.htm

13. Stefanovic, N., Han, J., Koperski, K.: Object-based selective materialization for efficient implementation of spatial data cubes. IEEE Transactions on Knowledge and Data Engineering 12(6), 938–958 (2000)

14. Rivest, S., Bèdard, Y., Marchand, P.: Toward better support for spatial decision making: Defining the characteristics of spatial on-line analytical processing. Geomatica 55(4), 539–555 (2001)

15. Luján-Mora, S., Trujillo, J., Song, I.Y.: A UML profile for multidimensional modeling in data warehouses. Data Knowl. Eng. 59(3), 725–769 (2006)

16. International Organization for Standardization, http://www.iso.org

17. Open Geospatial Consortium: Simple Features for SQL, http://www.opengeospatial.org

18. OMG, Object Management Group: Common Warehouse Metamodel (CWM) Specification 1.0.1 (2002), http://www.omg.org/cgi-bin/doc?formal/03-03-02

19. Kleppe, A., Warmer, J., Bast, W.: MDA Explained. The Practice and Promise of The Model Driven Architecture. Addison Wiley, Chichester (2003)

20. Malinowski, E., Zimányi, E.: Hierarchies in a multidimensional model: from conceptual modeling to logical representation. Data Knowl. Eng. 59(2), 348–377 (2006)

Built-In Indicators to Discover Interesting Drill Paths in a Cube

Véronique Cariou[2], Jérôme Cubillé[1], Christian Derquenne[1], Sabine Goutier[1],
Françoise Guisnel[1], and Henri Klajnmic[1]

[1] EDF Research and Development
1 avenue du Général de Gaulle, 92140 Clamart, France
`firstname.lastname@edf.fr`
[2] ENITIAA
Rue de la Géraudière, BP 82225, 44322 Nantes CEDEX 3, France
`veronique.cariou@enitiaa-nantes.fr`

Abstract. OLAP applications are widely used by business analysts as a decision support tool. While exploring the cube, end-users are rapidly confronted by analyzing a huge number of drill paths according to the different dimensions. Generally, analysts are only interested in a small part of them which corresponds to either high statistical associations between dimensions or atypical cell values. This paper fits in the scope of discovery-driven dynamic exploration. It presents a method coupling OLAP technologies and mining techniques to facilitate the whole process of exploration of the data cube by identifying the most relevant dimensions to expand. At each step of the process, a built-in rank on dimensions is restituted to the users. It is performed through indicators computed on the fly according to the user-defined data selection. A proof of the implementation of this concept on the ORACLE 10G system is described at the end of the paper.

Keywords: OLAP, Data cube, Olap Mining, Data Mining, ORACLE 10G.

1 Introduction

Most of large companies resort to business intelligence applications to analyze data contained in their Data Warehouses. In particular, On-Line Analytical Processing (OLAP) applications are powerful decision support tools for business analysis. They offer a multidimensional view of data, by calculating and displaying indicators either in detail or with a certain level of aggregation. This view of data is easily understandable by business analysts who can become direct end-users of the corresponding software. Exploration of the data is performed thanks to interactive navigation operators.

Nevertheless, as soon as multidimensional databases become very large, the number of cells of the cube dramatically increases. Business analysts are rapidly confronted to a difficult and tedious task in order to analyze such multidimensional databases with multiple criteria. They would like to answer such types of questions:

I.-Y. Song, J. Eder, and T.M. Nguyen (Eds.): DaWaK 2008, LNCS 5182, pp. 33–44, 2008.
© Springer-Verlag Berlin Heidelberg 2008

- How to quickly find which cell of the multidimensional table contains a value very different from expected?
- How to rapidly discover the dimensions which are the most correlated or associated one to another?

Commercial OLAP products do not provide such intelligent analysis operators.

This paper describes a method to perform a dynamic analysis of a multidimensional cube in order to guide end-users' navigation. The aim is to identify the most interesting dimensions to expand, given a current slice under study. The built-in indicators, we developed, help the user by proposing him or her at each step of the process the dimensions to expand first. They relate to the most interesting statistical associations between the dimensions and reveal the most atypical cells values.

The originality of this work is the adaptation of Data Mining methods to multidimensional environment where detailed data are no more available and only aggregated ones can be used. Our solution is based on a tight coupling between OLAP tools and statistical methods. Detection of interesting dimensions to expand is performed thanks to built-in indicators computed instantaneously during exploration without any pre-computation. They constitute an add-in OLAP tool providing a support to decision making.

The paper is organized as follows. In section 2, we outline relevant works in the context of discovery-driven exploration. In section 3, the built-in indicators to discover relevant dimensions are detailed. Finally, section 4 presents the implementation on the ORACLE 10G system.

2   Related Works

This paper fits in the scope of discovery-driven exploration of a data cube. Analyzing manually multidimensional data with a huge number of dimensions rapidly becomes a tedious and difficult task for the analyst. Discovery-driven exploration cares about a partial automation of this task in order to find rapidly interesting parts of the cube. First, Sarawagi et al. [7] and Chen [4] introduced specific operators for both the discovery of exceptions in a cube and the detection of interesting drill paths. These operators are based on the notion of degree of exception of a cell, defined as the standardized residual between the aggregated cell value and the value predicted with statistical models. In [7], predicted values are based on log-linear saturated models for *count* measures while Caron et al. [3] have adopted a similar approach with multi-way ANOVA for *quantitative* measures. Going on this approach, Sarawagi proposed other discovery-driven operators such as DIFF [8], INFORM [9] and RELAX [10].

This paper presents the next stage of interactive analysis of OLAP data cube first proposed in Cariou et al. [2]. Here, we address the guided navigation problem. Starting at the top most level where all dimensions are aggregated to a single value, it results in ranking the most relevant dimensions to expand. Our work is closely related to the framework proposed by Sarawagi. Nevertheless, we stress two main contributions compared to previous studies. In [7], the dimension indicators are based on the computation of the most exceptional cell value in a dimension. In our framework, the ranking dimension restitutes the amount of information the new dimension brings to the actual analysis. In a sense, our work is closer to user adaptive exploration of multidimensional data [9] and to rank deviations [6]. These two latter methods aim at computing a global measure of dimension interestingness comparing anticipated cell values against actual cells. While the authors only consider the crossing of two dimensions, our work addresses the global problem that is a top-down expanding of several dimensions in a data cube. The other contribution of this paper lies in the way indicators are built. In [7], discovery-driven exploration is computationally expensive since it requires the computation of both the entire data cube and the indicators of exceptions. In the proposed method, we take advantages of OLAP functionalities so that indicators can be computed instantaneously according to the dimensions interactively chosen to structure the slice. It offers a compromise between computation performance and model complexity.

3   Finding an Interesting Path

3.1   Introduction

The framework of this paper is in the scope of a dynamic analysis of a cube. We propose a method to guide end-users navigation through the multidimensional data by proposing them, at each step of the process, the dimensions to expand first. They relate to the most interesting statistical associations and reveal the most atypical cells values (in terms of comparison to the assumption of the independence model as presented in previous works [2]).

Data-driven built-in indicators, calculated on the fly, measure the degree of interest of each expandable dimension. At each step, the adopted approach for guided navigation includes two distinct tasks:
- first, the sorting of the expandable dimensions by decreasing order of interest,
- then, once a dimension has been chosen and expanded by the user, the identification and the display of the most atypical cells of the current slice.

Insofar as it is supposed that, at each step, the user interprets a new table, the method aims to only emphasize new information conditionally to the information already highlighted in the slice displayed at the previous step.

In this work, we suppose that the dimensions have simple hierarchies (two levels: the first one with only the position ALL and the second one with all the categories).

Two methods are developed, one for a *count* measure (such as the number of customers) and the other for a *quantitative* measure (such as electrical consumption). They differ from the kind of built-in indicators used to sort the expandable dimensions. In the case of a *count* measure, the method is based on the deviation of the observed cells values against the assumption of the independence model. Deviation is computed for each possible dimension with a kind of Tschuprow's T indicator [1]. In the case of a *quantitative* measure, the method is based on the calculation of the variation among the cells means which is given by the sum of squares among cells (denoted SSB in classical analysis of variance [5]). In this paper, we only detail the method developped in the case of a *count* measure.

The restitution is done by a color scale applied on each dimension which proposes to the end-user the dimensions rank.

3.2 Working Example

We illustrate our approach with an example based on a database issued from a marketing context. In this example, we consider a simplified detailed table with the following relation scheme:

CUSTOMER(#CUST, TARIFF, DWELLING, HEATING, OCCUPATION, WATER HEATING, OLDNESS OF HOUSE, DATE OF CONTRACT SUBSCRIPTION, TYPE OF DWELLING, CONSUMPTION).

Table 1. Example of a customers detailed table

#CUST	TARIFF	DWELLING	HEATING	...	CONSUMPTION
1	Tariff 1	House	Electrical		212
2	Tariff 2	Flat	Gas		153
3	Tariff 1	Flat	Electrical		225
4	Tariff 2	Flat	Gas		142
5	Tariff 3	House	Fuel Oil		172
...	...	...	...	...	...

One tuple of Table. 1 is related to one customer and will be called a *statistical unit* in this article. From this detailed table, the corresponding data cube is structured by 8 dimensions: TYPE OF contract, TYPE OF DWELLING, TYPE OF HEATING, and by the following measures: number of customers and mean of electrical CONSUMPTION. On our snapshots, the category 'TOT' stands for the common value 'ANY' or 'ALL' which represents the total aggregated level of a dimension.

3.3 Forward vs. Global Approach

To carry out a guided dynamic analysis of a multidimensional cube, two types of approaches can be planned:

- A *global solution* will seek, at each step k of the process, the top-k dimensions (chosen among all the dimensions of the data cube) which partition the selected data while revealing the most homogeneous groups. Such an approach forgets which axes have been chosen at a previous step and optimizes a problem with k dimensions. It can occur that the set of k dimensions proposed at step k does not have or have a little overlapping with the set of $k - 1$ dimensions of the previous step. This method does not help the end-user discovering the data in more details because each step of the process is independent from the previous ones. Such a solution requires for the end-user to analyze this new presentation since it can reveal the data from a quite different point of view.

- A *forward approach* (step by step) will go through in the data cube without forgetting the previous choices made by the end-user. At each step k of the process, the method will seek among the dimensions not already expanded, the one that gives the best partition of the selected data when added to the $k-1$ dimensions already chosen at previous stages. This method helps the end-user discovering the data just by adding at each step only one new characteristic in order to focus more precisely on the different trends.

Our approach is a step by step approach. At the kth step, the algorithm ranks the dimensions not already expanded at previous steps, by their degree of interest. Let us recall that each slice of a data cube partitions the entire fact table into groups. More precisely, groups (which also correspond to cells) result of the cartesian product of the dimensions values associated with the slice. Let us suppose that Π_k cells are displayed on the slice at the kth step and let us denote d_r an expandable dimension not already considered in the analysis. The unfolding of d_r at the $(k+1)^{th}$ step leads to a finer embedded partition with $\Pi_{k+1} > \Pi_k$ cells. The degree of interest of dimension d_r is based on the amount of information provided by the obtained partition when adding this dimension. To measure this degree of interest, two different built-in indicators are defined depending on the kind of the measure: a *count* measure or a *quantitative* one (not described in this paper).

In the following paragraph, we describe the main algorithm's principles based on our motivating example. Each cell of the data cube reports the number of customers. This measure is structured by the dimensions "TARPUI" (TARIFF), "CONTRAT" (DATE OF CONTRACT SUBSCRIPTION), "TYPRES" (TYPE OF DWELLING), "TYPHAB" (DWELLING), "CONSTRUCT" (OLDNESS OF HOUSE), "STOC" (OCCUPATION), "CHFP" (HEATING), "EAUCHA" (WATER HEATING).

3.4   General Idea of the Algorithm

Initialization: Choice of the Studied Population. The process starts from the most aggregated level, i.e. all the available dimensions are still in page axes and none of them have been expanded to structure a table. Let us note D=all structuring dimensions. In our example, D={TARPUI, CONTRAT, TYPRES, TYPHAB, CONSTRUCT, STOC, CHFP, EAUCHA}.

First of all, the end-user can define a data selection (also called subpopulation in the statistical formalism) on which he wants to perform his analysis. The method allows him to select positions on a set of dimensions called D_{popu} in order to restrict the subpopulation of study. After this initialization step, no further selection will be possible during the navigation process and those axes will not be expandable anymore. This potentially defines a new set D^* of expandable dimensions. If no subpopulation is selected $D^* = D$. That is to say D^*={potential expandable dimensions}=$D \backslash D_{popu}$.

Step 1: Choice of the First Dimension. When the process is executed for the first time, it calculates an indicator for each expandable dimension. The value taken by this indicator for each dimension of D^* enables to classify those

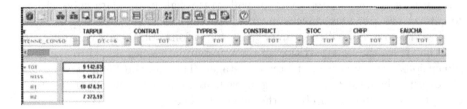

Fig. 1. Application snapshot at the end of the first step

dimensions depending on their degree of interest. The choice of the first dimension to unfold remains to the user who can decide to choose a dimension having a weaker one but which seems more relevant according to his data knowledge and the purpose of his analysis. This first dimension will then be noted $d_{(1)}$ and the data are displayed to the user according to this dimension. The two subsets of dimensions are updated: $D^* = D^* \backslash d_{(1)}$ and $D_1 = \{d(1)\}$. In the example, $d_{(1)} =$ TYPHAB involving $D^* = \{$CONTRAT, TYPRES, CONSTRUCT, STOC, CHFP, EAUCHA$\}$ and $D_1 = \{$TYPHAB$\}$.

Step 2: Choice of the Second Dimension. At the second step, the problem is to find which dimension of D^* will provide the best partition when crossed with $d_{(1)}$. The quality of the resulting partition is measured by a second built-in indicator calculated for each dimension of D^*. The value of this indicator allows to order the dimensions of D^* relating to their degree of interest. As at the first step, the choice of the dimension to unfold remains to the user who can decide to expand a dimension having a weaker capacity of discrimination but which seems more relevant according to the analysis he wants to perform. This dimension will then be noted $d_{(2)}$ and the data are displayed to the user according to the two dimensions $d_{(1)}$ and $d_{(2)}$. At the end of this step, $D^* = D^* \backslash d_{(2)}$ and $D_2 = \{d_{(1)}, d_{(2)}\}$. In the example, $d_{(2)} =$ EAUCHA, so we have $D^* = \{$CONTRAT, TYPRES, CONSTRUCT, STOC, CHFP$\}$ and $D_2 = \{$TYPHAB, EAUCHA$\}$.

Step s: Choice of the sth Dimension ($s > 2$). For all the dimensions expanded until the $s - 1^{th}$ step ($D_{s-1} = \{d_{(1)}, d_{(2)}, \ldots, d_{(s-1)}\}$), the idea is to build a virtual dimension $d_{(v)}$ resulting from the crossing of all dimensions in D_{s-1}. So, the problem to solve at the sth step is a simple generalization of the second step: how to identify the dimension of D^* that will give the best partition when crossed with the virtual dimension $d_{(v)}$? The algorithm of step 2 is repeated. In the following paragraphs, we will present the notations before describing the built-in indicators for a *count* measure.

3.5   Common Notations

Let us note $\widetilde{X}$ one measure of the cube, structured by several dimensions (in our experiments, more than 10 dimensions may occur). As already said, with the dimensions values selected on the page axes, we define a data selection (also called subpopulation) E of the whole cube (also called population) P.

If no selection is made on the page axes, $E = P$. Once E is fixed, we call t the slice resulting from the crossing of the chosen dimensions. Slice t is displayed as a summarized table corresponding to dimensions selected in lines and columns. A line or a column of the table where the value is TOT, is called a margin of the table (it corresponds to the ALL position in literature). The margins of the table are not included in this slice t.

Let us assume that p is the number of expandable dimensions (denoted D^*) belonging to the sub-population E.

The slice t is defined by the crossing of k_t dimensions $D_t = \{d_1, \dots, d_r, \dots, d_{k_t}\}$ and the choice of a measure $\widetilde{X}$. Let us note:
- n_i the size of the cell $i \in t$ as the number of statistical units in the cell i,
- n_E the size of the sub-population E, we notice that $\forall t, n_E = \sum_{i \in t} n_i$,
- M_{d_r} the number of distinct values on the dimension d_r ('ALL' is excluded)
- $x_j^{(i)}$ the value of $\widetilde{X}$ for the statistical unit j in the cell i of the slice t.

3.6   Count Measure

The problem we are interested in is to compute an indicator that expresses the level of interest for each expandable dimension according to a *count* measure. Given a *count* measure, $x_j^{(i)} = 1$ for all statistical units j belonging to the cell i of the slice t. Let $n_i = n_{i_r i_k}$ be the size of the cell i defined at the crossing of the position i_r of the dimension d_r and the position i_k of the dimension d_k.

Step 1: Choice of the First Dimension. *Construction of the indicator.* The aim of the algorithm is to order the dimensions in D^* (where $card(D^*) = p$) by decreasing order of relevancy regarding their ability to create good partitions of the subpopulation E. To perform this, we define the level of interest of a dimension $d_r \in D^*$ as its average association measure with the other dimensions. It is measured by the following built-in indicator $\overline{T}(d_r)$:

$$\overline{T}(d_r) = \frac{1}{p-1} \sum_{\substack{k=1 \\ k \neq r}}^{p} T(d_r, d_k)$$

We use the expression $T(d_r, d_k)$ of the Tschuprow's T for each $d_r \in D^*$ and $d_k \in D^* \backslash d_r$:

$$T(d_r, d_k) = \sqrt{\frac{\chi^2(d_r, d_k)}{n_E \sqrt{(M_{d_r} - 1)(M_{d_k} - 1)}}}$$

where

$$\chi^2(d_r, d_k) = \sum_{i_r=1}^{M_{d_r}} \sum_{i_k=1}^{M_{d_k}} \frac{\left(n_{i_r i_k} - \frac{n_{i_r \bullet} n_{\bullet i_k}}{n_E}\right)^2}{\frac{n_{i_r \bullet} n_{\bullet i_k}}{n_E}}$$

and $n_{i_r \bullet} = \sum_{i_k=1}^{M_{d_k}} n_{i_r i_k}$ and $n_{\bullet i_k} = \sum_{i_r=1}^{M_{d_r}} n_{i_r i_k}$.

General interpretation of the indicator. We have defined this indicator based on the Tschuprow's T because, among the multiple indicators measuring the statistical link between two characters dimensions, the Tschuprow's T:
- measures the gap between the displayed table and the expected table in the case of independence of the two crossed dimensions. By definition $0 \leq T(d_r, d_k) < 1$.

- $T(d_r, d_k) = 0$ corresponds to the independence data distribution; i.e. the observed values for d_r do not depend on the observed values for d_k.
- The more $T(d_r, d_k)$ is close to 1, the farthest the distribution of the statistical units is from the situation of independence.

- is bounded what facilitates its restitution by a system of color-coding.
- takes into account the number of positions of the dimensions, in penalizing the dimensions with a high number of positions. By that way, slices with a large number of cells are penalized which guarantees a better readable result.

The user is then free to choose or not the dimension with the highest value of the indicator. Let us note $d_{(1)}$ the dimension chosen at the end of the first stage. Then D^* is updated as follows $D^* = D^* \backslash d_{(1)}$.

We precise that the built-in indicator $\overline{T}(d_r)$ cannot be assimilated to Tschuprow's T, nor interpreted like one. This indicator only enables to rank the dimensions by descending values of $\overline{T}(d_r)$. This indicator is symmetric: $T(d_k, d_r) = T(d_r, d_k)$. We take advantage of this property in order to reduce computations tasks when implemented within OLAP management system.

▶ *Example 1*: Let us recall that, in our motivating example, the number of customers is defined as a *count* measure. Applied on this measure, the following results are obtained:

The algorithm sorts the dimensions by descending values of $\overline{T}(d_r)$ and restitutes to the user a scale. In the example shown on Fig. 2, dimension EAUCHA ("WATER_HEATING") has the highest value of the indicator. We will suppose that, at this first stage, the user chooses the best axis proposed by the algorithm, namely $d_{(1)} = $ EAUCHA.   ◀

Step 2: Choice of the Second Dimension. At the second step, the order of the expandable dimensions of D^* is based on the calculations already carried

	TYPRES	CONSTRUCT	TARPUI	CONTRAT	TYPHAB	STOC	CHFP	EAUCHA	$\overline{T}(d_r)$
TYPRES		0,0869	0,0820	0,1014	0,1833	0,1930	0,1882	0,1353	0,1386
CONSTRUCT	0,0869		0,0885	0,1759	0,1241	0,1876	0,1802	0,2397	0,1547
TARPUI	0,0820	0,0885		0,1831	0,1651	0,1064	0,2757	0,2796	0,1686
CONTRAT	0,1014	0,1759	0,1831		0,1455	0,2436	0,1269	0,2744	0,1787
TYPHAB	0,1833	0,1241	0,1651	0,1455		0,2928	0,2293	0,2101	0,1929
STOC	0,1930	0,1876	0,1064	0,2436	0,2928		0,2264	0,3541	0,2291
CHFP	0,1882	0,1802	0,2757	0,1269	0,2293	0,2264		0,5336	0,2515
EAUCHA	0,1353	0,2397	0,2796	0,2744	0,2101	0,3541	0,5336		0,2895

Fig. 2. Tschuprow's T matrix and $\overline{T}(d_r)$

out at the 1st step. Thus, the Tschuprow's T calculated at the previous step conditionally to the choice of $d_{(1)}$, i.e. $T(d_r, d_{(1)})$ is ordered. D^* is updated as follows $D^* = D^* \backslash d_{(2)}$.

▶ *Example 2*: In the previous example, $d_{(1)}$ =EAUCHA and the values of $T(d_r, d_{(1)})$ are retained on Fig. 3. We can suppose that, at this stage, the user chooses also the best axis proposed by the algorithm, namely $d_{(2)}$ =CHFP. Thus D_2 ={ EAUCHA, CHFP }.   ◀

	CHFP	STOC	TARPUI	CONTRAT	CONSTRUCT	TYPHAB	TYPRES
EAUCHA	0,5336	0,3541	0,2796	0,2744	0,2397	0,2101	0,1353

Fig. 3. Values of $T(d_r,$EAUCHA$)$ for the choice of the second dimension

Step s: Choice of the sth Dimension ($s > 2$). We suppose at this stage that s-1 dimensions $d_{(1)}, d_{(2)}, \ldots, d_{(s-1)}$ were already expanded. We generalize Tschuprow's T to choose the sth dimension. The idea is to consider a virtual dimension $d_{(v)}$ resulting from the crossing of all the dimensions $d_{(1)}, d_{(2)}, \ldots, d_{(s-1)}$ already expanded. This virtual dimension has at the most $M_{d_{(1)}} \times M_{d_{(2)}} \times \ldots \times M_{d_{(s-1)}}$ positions. Indeed, only active cells of the crossing are considered i.e. non empty cells. By this way, the classical criterion based on a two-way contingency table is generalized to any s crossing dimensions. We calculate for each dimension $d_r \in D^*$:

$$T(d_r, d_{(v)}) = \sqrt{\frac{\chi^2(d_r, d_{(v)})}{n_E \sqrt{(M_{d_r} - 1)(\Pi_{s-1} - 1)}}}$$

with:
- Π_{s-1} the number of active cells defined by the crossing of dimensions in $D_{s-1} = \{d_{(1)}, d_{(2)}, \ldots, d_{(s-1)}\}$ i.e. the number of $d_{(v)}$ categories.
- $\chi^2(d_r, d_{(v)}) = \sum_{i_r=1}^{M_{d_r}} \sum_{i_v=1}^{\Pi_{s-1}} \frac{(n_{i_r i_v} - \frac{n_{i_r \bullet} n_{\bullet i_v}}{n_E})^2}{\frac{n_{i_r \bullet} n_{\bullet i_v}}{n_E}}$
- $n_{i_r \bullet} = \sum_{i_v=1}^{\Pi_{s-1}} n_{i_r i_v}$ and $n_{\bullet i_v} = \sum_{i_r=1}^{M_{d_r}} n_{i_r i_v}$

The dimensions' order will be established starting from the values of this indicator.

3.7   How to Display the Dimensions' Rank to the End-User?

At each step of the process, the algorithm establishes a rank on the dimensions based on the values of the built-in indicators. This rank has been displayed to the end-user with a scale of color. As far as we are interested in the variation of the criteria, a logarithmic scale has been adopted. The bounds of the scale have been determined empirically based on experiments on various datasets.

4   Experiment

In this paragraph, the aim is to show, as a proof of concept, the easiness of the implementation on an existing OLAP Management System. The experiment has been performed on real data issued from a marketing context, presented previously in our motivating example. Currently, OLAP databases have been built with ORACLE 10G which also offers a web restitution of the report. It enables to build a Web application with different reports and graphs on a specific part of the company activity in order to facilitate end-users data understanding.

We have implemented the two presented built-in indicators in the case of a *quantitative* measure and a *count* measure. We have also linked this work with another algorithm of detection of interesting cells [2]. When the business analyst has chosen the dimensions of interest, the display of the table with colored cells is instantaneous. Nevertheless, two factors can influence the processing time. As the system is based on a client-server architecture, computation is dependent on network performance. Furthermore, the number of expandable dimensions influences the system response time but it never exceeds in our experiments 5 s.

The principle of the implementation is based on different *Analytical Workspaces* (OLAP data cube in ORACLE 10G):
- one *Technical Analytic Workspace* contains all the generic built-in indicators algorithms,
- different *Data Analytic Workspaces* contain the specific data (sales, marketing, human resources). They can query the *Technical Analytical Workspace* on demand.

No pre-computation is required in the *Data Analytical Workspace* since all the computations are performed on the fly. Our algorithms are implemented in a data independent way and can be applied for hyper cubes with a great number of dimensions (in our motivating example, eight dimensions are studied). The performances only depend on the number of the displayed cells whatever the number of tuples in the fact table is. The implementation consists in a set of programs implemented in the Oracle L4G (Language 4th generation) language which also allows a web restitution.

When a user connects itself on the multidimensional application, the following page is displayed (see Fig. 4). After choosing (eventually) a sub-population, the extra-button *Navigation* should be pressed to execute the first step of the

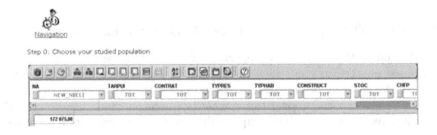

Fig. 4. Navigation initialization

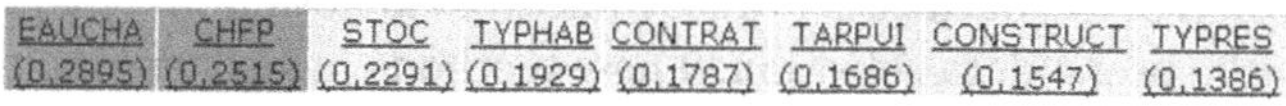

EAUCHA	CHFP	STOC	TYPHAB	CONTRAT	TARPUI	CONSTRUCT	TYPRES
(0.2895)	(0.2515)	(0.2291)	(0.1929)	(0.1787)	(0.1686)	(0.1547)	(0.1386)

Fig. 5. Built-in rank on dimensions

drill-through process. A first program is executed. Given a slice, this program detects if the displayed measure is a *count* or a *quantitative* measure and launches the appropriate program. Each appropriate program performs built-in indicators for all expandable dimensions and presents the results, on the color's scale with hypertext links (see Fig. 5).

Helped by this tool, the analyst makes its own dimension choice in order to analyze the data:
- At step 1, no dimension has already been selected. The dimension chosen by the end-user will be displayed as a line axis.
- At step 2, one dimension has already been expanded. The dimension chosen by the analyst will be displayed as a column axis.
- At step s $(s > 2)$ of the process, s-1 dimensions have already been expanded and the dimension $d_{(s-1)}$ chosen at the previous step s-1 will be moved from column axis to row axis. The chosen dimension $d_{(s)}$ will then be put as a column axis.

When one axis is chosen and the new slice is displayed, the extra-button Navigation is again available. The drill-through process can be re-run for a further step, and so on. If we want to perform these algorithms on a new Data *Analytical Workspace*, as our programs are data-independent, only two tasks have to be done:
- to define the link between the *Data Analytic Workspace* and the *Technical Analytic Workspace*
- to tag the measures of interest. We have indeed to specify if a given measure in the *Data Analytic Workspace* is a *count* measure or a *quantitative* measure.

Furthermore, to simplify the maintenance of the system, we have defined a referential in the *Technical Analytic Workspace* to link each measure with its type. It is possible to imagine that as it exists functions as *mean, median, forecast*, our algorithms could be integrated as add-in functions in different OLAP Management Systems.

5 Conclusion and Future Works

In this paper, we describe an original work to facilitate business analysts' interactive multidimensional data exploration. The approach adopted herein can be applied within the framework of a dynamic analysis of the cube. The proposed method is based on the calculation of indicators which enables to rank expandable dimensions by degree of interest. We consider two cases depending on the kind of measure. For a *quantitative* measure, the indicator corresponds to the relative amount of variation of the square of the correlation ratio. For a *count* measure, a generalized Tschuprow's T is proposed. In both cases, without any pre-computation of the data cube, those indicators are evaluated instantaneously according to the users' data selection. These methods have been implemented, in a data independent way, as proof of concept under ORACLE 10G system. Further investigations are needed in order to take into account hierarchical dimensions. Indeed, several strategies can be considered to expand a hierarchical dimension at a current step, depending on whether drill-down is restricted to the directly lower level or not. Moreover, the way a hierarchical level is unfolded (globally or partially) may impact the way built-in indicators are defined.

References

1. Agresti, A.: Categorical Data Analysis. Wiley, New York (1990)
2. Cariou, V., Cubillé, J., Derquenne, C., Goutier, S., Guisnel, F., Klajnmic, H.: Built-in Indicators to Automatically Detect Interesting Cells in a Cube. In: Song, I.-Y., Eder, J., Nguyen, T.M. (eds.) DaWaK 2007. LNCS, vol. 4654, pp. 123–134. Springer, Heidelberg (2007)
3. Caron, E.A.M., Daniels, H.A.M.: Explanation of exceptional values in multi-dimensional business databases. European Journal of Operational Research 188(3), 884–897 (2008)
4. Chen, Q.: Mining exceptions and quantitative association rules in OLAP data cube. PhD Thesis, School of Computing Science, Simon Fraser University, British Columbia, Canada (1999)
5. Jobson, J.D.: Applied Multivariate Data Analysis. Regression and Experimental Design, vol. I. Springer, New York (1991)
6. Palpanas, T., Koudas, N.: Using Datacube aggregates for approximate querying and deviation detection. IEEE Trans. on Knowledge and Data Engineering 17(11), 1–11 (2005)
7. Sarawagi, S., Agrawal, R., Megiddo, N.: Discovery-driven exploration of OLAP data cubes. Technical report, IBM Almaden Research Center, San Jose, USA (1998)
8. Sarawagi, S.: Explaining differences in multidimensional aggregates. In: Proceedings of the 25th International Conference On Very Large Databases (VLDB 1999) (1999)
9. Sarawagi, S.: User-adaptative exploration of multidimensional data. In: Proceedings of the 26th International Conference On Very Large Databases (VLDB 2000) (2000)
10. Sathe, G., Sarawagi, S.: Intelligent Rollups in Multidimensional OLAP Data. In: Proceedings of the 27th International Conference On Very Large Databases (VLDB 2001) (2001)

Upper Borders for Emerging Cubes

Sébastien Nedjar, Alain Casali, Rosine Cicchetti, and Lotfi Lakhal

Laboratoire d'Informatique Fondamentale de Marseille (LIF),
Aix-Marseille Université - CNRS
Case 901, 163 Avenue de Luminy, 13288 Marseille Cedex 9, France
lastname@lif.univ-mrs.fr

Abstract. The emerging cube computed from two relations r_1, r_2 of categorical attributes gather the tuples for which the measure value strongly increases from r_1 to r_2. In this paper, we are interested in borders for emerging cubes which optimize both storage space and computation time. Such borders also provide classification and cube navigation tools. Firstly we study the condensed representation through the classical borders *Lower / Upper*, then we propose the borders *Upper* / Upper* more reduced than the previous ones. We soundly state the connexion between the two representations by using cube transversals. Finally, we perform experiments about the size of the introduced representations. The results are convincing and reinforce our idea that the proposed borders are relevant candidates to be the smallest condensed representation of emerging cubes and thus can be really interesting for trend analysis in OLAP databases.

1 Introduction and Motivations

Recently research work in data mining and data warehousing aims to state comparisons between trends, *i.e.* to capture important variations of measures within a cube [1,2] or trend reversals between two transaction databases [3] and two categorical database relations [4]. This kind of knowledge is of great interest in many applications. For instance, if two database relations are collected in two different geographical areas, for various samples of population or for several classes of individuals, it is possible to exhibit behavior modifications between two populations, contrasts between their characteristics or deviations from a check sample. In other typical applications, the chronological dimension makes it possible to distinguish two relations. For example, when managing a data warehouse, a first relation can be constituted by the whole data accumulated until now while the second one results from refreshment data.

By tackling such an issue, we have proposed the concept of emerging cube [4] to capture trend modifications in OLAP databases. Provided with two database relations of categorical attributes or dimensions, the emerging cube encompasses all the tuples satisfying a *twofold emergence constraint*: their measure value is weak in a relation (constraint C_1) and significant in the other (constraint C_2).

In this paper we focus on succinct representations (borders) of the Emerging Cube. Such borders have several advantages. Firstly, they optimize both storage space and computation time. They also provide a relevant structure for an easy and fast classification (let us notice that in a binary context, emerging patterns have proved to be efficient

I.-Y. Song, J. Eder, and T.M. Nguyen (Eds.): DaWaK 2008, LNCS 5182, pp. 45–54, 2008.

classifiers [3]). Moreover, the user can focus on the most interesting tuples by making use of the borders as a starting point for the navigation within the cube. Finally, borders can be used as a new constraint on OLAP dimensions [5]. More precisely, we make the following contributions:

- We study the borders *Lower / Upper* of the emerging cube which are classical when the solution space is convex [6]. They are made of the minimal tuples (L) and the maximal ones (U) satisfying the twofold emergence constraints C_1 and C_2.
- We propose the new borders *Upper* / Upper* encompassing on one hand the previous maximal tuples (U) and on the other hand the maximal tuples which do not satisfy C_1 but one of their generalizing tuple has to verify C_2 (U^*).
- We characterize the link between the two couples of borders by making use of the concept of cube transversal [7] and propose a comparative discussion.
- We give two different algorithmic methods to compute the borders: on one hand by using existing data mining algorithms and on the other hand with a database approach and SQL queries.
- For comparing the size of the borders L and U^*, we perform experiments by using data sets classical in data mining and data warehousing. The yielded results are specially interesting: when compared to L, U^* brings a systematic gain with a factor varying from 1.35 to 275.

The paper is organized as follows. In section 2, we resume the characterization of the Emerging Cube. Section 3 is devoted to the two representations through borders of the Emerging Cube. The experiments are described in section 4. Finally, some conclusions and future work are given in section 5.

2   Emerging Cubes

The emerging cube characterization fits in the framework of the cube lattice of a categorical database relation r: $CL(r)$ [7]. The latter is a search space organizing the tuples, possible solutions of the problem, according to a generalization / specialization order, denoted by $\preceq_g$, capturing a similar semantics than ROLL-UP/DRILL-DOWN [8]. These tuples share the same structure than the tuples of r but attributes dimensions can be provided with the value ALL [8]. Moreover, we append to these tuples a virtual tuple which only encompasses empty values in order to close the structure. Any tuple of the cube lattice generalizes the tuple of empty values. For handling the tuples of $CL(r)$, the operator + is defined. Provided with a couple of tuples, it yields the most specific tuple which generalizes the two operands.

Example 1. Let us consider the relation DOCUMENT$_1$ (*cf.* Table 1) giving the quantities of books sold by Type, City, Publisher and Language. In $CL($DOCUMENT$_1)$, let us consider the sales of Novels in Marseilles, what ever be the publisher and language, *i.e* the tuple (Novel, Marseilles, ALL, ALL). This tuple is specialized by the two following tuples of the relation: (Novel, Marseilles, Collins, French) and (Novel, Marseilles, Hachette, English). Furthermore, (Novel, Marseilles, ALL, ALL) $\preceq_g$ (Novel, Marseilles, Collins, French) exemplifies the generalization order between tuples.

Table 1. Relation example DOCUMENT$_1$

Type	City	Publisher	Language	Quantity
Novel	Marseilles	Collins	French	1000
Novel	Marseilles	Hachette	English	1000
Textbook	Paris	Hachette	French	1000
Essay	Paris	Hachette	French	6000
Textbook	Marseilles	Hachette	English	1000

Table 2. Relation example DOCUMENT$_2$

Type	City	Publisher	Language	Quantity
Textbook	Marseilles	Collins	English	3000
Textbook	Paris	Hachette	English	3000
Textbook	Marseilles	Hachette	French	3000
Novel	Marseilles	Collins	French	3000
Essay	Paris	Hachette	French	2000
Essay	Paris	Collins	French	2000
Essay	Marseilles	Hachette	French	1000

In the remainder of the paper, we only consider the aggregative functions COUNT and SUM. Furthermore to preserve the antimonotone property of SUM, we assume that the measure values are strictly positive.

Definition 1 (Measure function associated with an aggregative function). *Let f be an aggregative function, r a categorical database relation and t a multidimensional tuple. We denote by $f_{val}(t, r)$ the value of the function f for the tuple t in r.*

Example 2. If we consider the Novel sales in Marseilles, for any Publisher and Language, *i.e.* the tuple (Novel, Marseilles, ALL, ALL) of CL(DOCUMENT$_1$) we have: SUM_{val}((Novel, Marseilles, ALL, ALL), DOCUMENT$_1$) = 2000.

Definition 2 (Emerging Tuple). *A tuple $t \in CL(r_1 \cup r_2)$ is said emerging from r_1 to r_2 if and only if it satisfies the two following constraints C_1 and C_2:*

$$\begin{cases} f_{val}(t, r_1) < MinThreshold_1 \ (C_1) \\ f_{val}(t, r_2) \geq MinThreshold_2 \ (C_2) \end{cases}$$

Example 3. Let $MinThreshold_1 = 2000$ be the threshold for the relation DOCUMENT$_1$ (*cf.* Table 1) and $MinThreshold_2 = 2000$ the threshold for DOCUMENT$_2$ (*cf.* Table 2), the tuple t_1 =(Textbook, Marseilles, ALL, ALL) is emerging from DOCUMENT$_1$ to DOCUMENT$_2$ because SUM_{val}(t_1, DOCUMENT$_1$) = 1000 ($<$ MinThreshold$_1$) and SUM_{val}(t_1, DOCUMENT$_2$) = 6000 ($\geq$ MinThreshold$_2$). In contrast, the tuple t$_2$ = (Essay, Marseilles, ALL, ALL) is not emerging because SUM_{val}(t$_2$, DOCUMENT$_2$) = 1000.

Definition 3 (Emergence Rate). *Let r_1 and r_2 be two unicompatible relations, $t \in CL(r_1 \cup r_2)$ a tuple and f an additive function. The emergence rate of t from r_1 to r_2, noted $ER(t)$, is defined by:*

$$ER(t) = \begin{cases} 0 \text{ if } f_{val}(t, r_1) = 0 \text{ and } f_{val}(t, r_2) = 0 \\ \infty \text{ if } f_{val}(t, r_1) = 0 \text{ and } f_{val}(t, r_2) \neq 0 \\ \dfrac{f_{val}(t, r_2)}{f_{val}(t, r_1)} \text{ otherwise.} \end{cases}$$

We observe that when the emergence rate is greater than 1, it characterizes trends significant in r_2 and not so clear-cut in r_1. In contrast, when the rate is lower than 1, it highlights immersing trends, relevant in r_1 and not in r_2.

Example 4. From the two relations DOCUMENT$_1$ and DOCUMENT$_2$, we compute $ER($ (Textbook, Marseilles, ALL, ALL)) $= 6000/1000$. Of course the more the emergence rate is high, the more the trend is distinctive. Therefore, the quoted tuple means a jump for the Textbook sales in Marseilles between DOCUMENT$_1$ and DOCUMENT$_2$.

Definition 4 (Emerging Cube). *We call Emerging Cube the set of all the tuples of $CL(r_1 \cup r_2)$ emerging from r_1 to r_2. The emerging cube, noted $EmergingCube(r_1, r_2)$, is a convex cube [6] with the hybrid constraint "t is emerging from r_1 to r_2". Thus it is defined by: $EmergingCube(r_1, r_2) = \{t \in CL(r_1 \cup r_2) \mid C_1(t) \wedge C_2(t)\}$, with $C_1(t) = f_{val}(t, r_1) < MinThreshold_1$ and $C_2(t) = f_{val}(t, r_2) \geq MinThreshold_2$.*

Let us underline that all the tuples of the emerging cube are provided with an emerging rate greater than the ratio $\frac{MinThreshold_2}{MinThreshold_1}$.

3 Borders for Emerging Cubes

The emerging cube is a convex cube or more generally a convex space, thus it can be represented through the classical borders *Lower / Upper* [6]. In this section we recall the definition of the borders *Lower / Upper* and introduce a new representation: the borders *Upper* / Upper*. The latter make it possible to decide whether a tuple is emerging or not. Finally we soundly state the relationship between the two proposed couples of borders by using the concept of cube transversal.

3.1 Lower / Upper Borders

In this section, we recall the definition of the borders *Lower / Upper* then, through Proposition 1, we explain how to decide whether a tuple is emerging or not.

Definition 5 (Lower / Upper Borders). *The emerging tuples can be represented by the borders: U which encompasses the emerging maximal tuples and L which contains all the emerging minimal tuples according to the generalization order $\preceq_g$.*

$$\begin{cases} L = \min_{\preceq_g}(\{t \in CL(r_1 \cup r_2) \mid C_1(t) \wedge C_2(t)\}) \\ U = \max_{\preceq_g}(\{t \in CL(r_1 \cup r_2) \mid C_1(t) \wedge C_2(t)\}) \end{cases}$$

Proposition 1. *The borders Lower / Upper are a representation for the Emerging Cube:* $\forall\, t \in CL(r_1 \cup r_2)$, *t is emerging from r_1 to r_2 if and only if $\exists (l, u) \in (L, U)$ such that $l \preceq_g t \preceq_g u$. In other words, t is emerging if and only if it belongs to the range $[L; U]$.*

Example 5. With our relations examples DOCUMENT$_1$ and DOCUMENT$_2$, table 3 gives the borders *Lower / Upper* for the Emerging Cube, by preserving a similar value for the thresholds ($MinThreshold_1 = 2000$, $MinThreshold_2 = 2000$). Provided with the borders, we know that the tuple (Novel, ALL, Collins, French) is emerging because it specializes the tuple (Novel, ALL, Collins, ALL) which belongs to the border L while generalizing the tuple (Novel, Marseilles, Collins, French) of the border U. Furthermore, the tuple (ALL, Marseilles, Hachette, ALL) is not emerging. Even if it generalizes the tuple (Textbook, Marseilles, Hachette, French) which belongs to the border U, it does not specialize any tuple of the border L.

Table 3. Borders *Lower / Upper* of the Emerging Cube

U	(Novel, Marseilles, Collins, French)	(Textbook, Marseilles, Collins, English)
	(Textbook, Paris, Hachette, English)	(Textbook, Marseilles, Hachette, French)
	(Essay, Paris, Collins, French)	
L	(Novel, ALL, Collins, ALL)	(Novel, ALL, ALL, French)
	(Textbook, Marseilles, ALL, ALL)	(Textbook, Paris, ALL, ALL)
	(Textbook, ALL, Collins, ALL)	(Textbook, ALL, ALL, French)
	(Textbook, ALL, ALL, English)	(Essay, ALL, Collins, ALL)
	(ALL, Marseilles, Collins, ALL)	(ALL, Marseilles, ALL, French)
	(ALL, Paris, Collins, ALL)	(ALL, Paris, ALL, English)
	(ALL, ALL, Collins, French)	(ALL, ALL, Collins, English)

3.2 Upper* / Upper Borders

In this section, we introduce a new representation: the borders *Upper* / Upper*. This representation is based on the maximal tuples satisfying the constraint C_2 (they are significant in r_2) without verifying C_1 (thus they are also significant in r_1 and are not considered as emerging). Moreover we provide an optimization of our search space because we no longer consider the cube lattice of $r_1 \cup r_2$, but instead the cube lattice of r_2. Actually, by its very definition (Cf. constraint C_2) any emerging tuple is necessarily a tuple of $CL(r_2)$.

Definition 6 (Upper* / Upper Borders). *The Emerging Cube can be represented through two borders: U which is the set of maximal emerging tuples and U^* encompassing all the maximal tuples not satisfying the constraint C_1 but belonging to the cube lattice of r_2 and sharing at least a common value with a tuple of U. In a more formal way, we have:*

$$
\begin{cases}
U^* = \max_{\preceq_g}(\{t \in CL(r_2) \text{ such that } \neg C_1(t) \\
\quad \text{and } \exists\, u \in U : t + u \neq (ALL, ..., ALL)\}) \\
U = \max_{\preceq_g}(\{t \in CL(r_2) \text{ such that } C_1(t) \wedge C_2(t)\})
\end{cases}
$$

By considering the search space $CL(r_2)$, we need to assess the twofold following constraint for the tuples of U^*: they do not satisfy C_1 but one of their generalizing tuple has to verify C_2. The latter condition is enforced in our definition by: $t + u \neq (ALL, ..., ALL)$ which means that t and u have a common value for at least a dimension. Thus it exists a tuple generalizing both t and u and sharing their common value for the considered dimension (all the other dimensions can be provided with ALL).

Example 6. With our relations examples DOCUMENT$_1$ and DOCUMENT$_2$, table 4 gives the borders *Upper* / *Upper* of the cube emerging from DOCUMENT$_1$ to DOCUMENT$_2$. For instance, the tuple (Novel, Marseilles, ALL, ALL) belongs to U^* because the number of novels sold in Marseilles is greater than the given thresholds not only in r_2 (C_2) but also in r_1 ($\neg C_1$).

With the following proposition, we are provided with a simple mechanism to know whether a tuple is emerging or not using the borders *Upper* / *Upper*.

Proposition 2. *The borders Upper* / Upper are a representation for the Emerging Cube:* $\forall t \in CL(r_1 \cup r_2)$, *t is emerging from r_1 to r_2 if and only if* $\forall l \in U^*$, $l \npreceq_g t$ *and* $\exists u \in U$ *such that* $t \preceq_g u$. *Thus t is emerging if and only if it belongs to the range* $]U^*; U]$.

Table 4. Borders *Upper* of the Emerging Cube

U^*	(Novel, Marseilles, ALL, ALL)	(Textbook, ALL, Hachette, ALL)
	(ALL, Marseilles, Hachette, English)	(Essay, Paris, Hachette, French)

Example 7. By resuming the couple of tuples exemplifying proposition 1, we obtain a similar result with these new borders. The tuple (Novel, ALL, Collins, French) is emerging because it generalizes the tuple (Novel, Marseilles, Collins, French) which belongs to the border U and does not generalize any tuple of the border U^*. Moreover the tuple (ALL, Marseilles, Hachette, ALL) is not emerging because it generalizes the tuple (ALL, Marseilles, Hachette, English) of the border U^*.

3.3   Relationship between Borders

In the framework of frequent pattern mining, Mannila *et al.* [9] state the link between the positive and negative borders. Experiments on the size of previous borders are reported in [10]. In a similar spirit, we characterize the relationship between the two couples of borders *Lower* / *Upper* and *Upper* / *Upper* through L and U^*. The characterization is based on the concept of cube transversal [7] and aims to give a sound foundation for the proposed borders along with experimental results and take benefit of existing and proved to be efficient minimal transversal algorithms [11,7] adapted for computing L using U^*. We apply this principle for performing our experiments (*cf.* Section 4).

Definition 7 (Cube Transversal). *Let T be a set of tuples ($T \subseteq CL(r)$) and $t \in CL(r)$ be a tuple. t is a cube transversal of T over $CL(r)$ if and only if $\forall t' \in T$, $t \npreceq_g t'$. t is a minimal cube transversal if and only if t is a cube transversal and $\forall t' \in$*

$CL(r), t'$ is a cube transversal and $t' \preceq_g t \Rightarrow t = t'$. The set of minimal cube transversals of T are denoted by $cTr(T)$ and defined as follows:

$$cTr(T) = \min_{\preceq_g}(\{t \in CL(r) \mid \forall t' \in T, t \npreceq_g t'\})$$

Let $\mathbb{A}$ be an antichain of $CL(r)$ ($\mathbb{A}$ is a set of tuples such that all the tuples of $\mathbb{A}$ are not comparable using $\preceq_g$). We can constrain the set of minimal cube transversals of r using $\mathbb{A}$ by enforcing each minimal cube transversal t to be more general than at least one tuple u of the antichain $\mathbb{A}$. The new related definition is the following:

$$cTr(T, \mathbb{A}) = \{t \in cTr(T) \mid \exists u \in \mathbb{A} : t \preceq_g u\}$$

Through the following proposition, we state the link between the *Lower* border and the *Upper** border.

Proposition 3. $L = cTr(U^*, U)$. Furthermore, $\forall\, t \in CL(r_1 \cup r_2)$, t is emerging if and only if t is a cube transversal of U^* and $\exists\, u \in U$ such that $t \preceq_g u$.

3.4 How to Compute Borders of Emerging Cubes

There are two different ways to compute borders of emerging cubes without developing a new algorithm: on one hand by using existing data mining algorithms and on the other hand with a database approach and SQL queries.

In a binary data mining context, various algorithms are specially efficient for computing the upper borders (GenMax [12] or Mafia [13]). By using proposition 3, computing the lower border can be performed from the cube transversals of the $Upper^*$ border. This approach is less costly than using classical algorithms computing constrained sets because the quoted algorithms are particularly efficient. Thus the computation of the borders *Upper** / *Upper* is a preliminary step when computing the classical borders *Lower / Upper*.

Using a database approach is less efficient because it does not exist embedded operator for computing the maximal or minimal (w.r.t generalization). Nevertheless, it provides a very common method which can be easily enforced to achieve the expected borders.

3.5 Discussion

In data mining, the borders frequently used to represent a solution space provided with a convex space structure are the borders *Lower / Upper* [9]. According to our knowledge, the borders *Upper** / *Upper* have never been proposed nor experimented in the literature. However they have undeniable advantages:

– The experimental study conducted to validate our new representation (*cf.* Section 4) shows a very significant gain for the size of the border U^* when compared to L ;
– the efficient computation of the classical borders is based on an algorithm computing the maximal tuples (GenMax [12] or Mafia [13]). Thus achieving the border U^* is a preliminary step for the computation of the border L.

Thus, by using the borders *Upper** / *Upper*, we save not only memory (because of the size of U^*) but also the computation time of the cube transversals for computing L once U^* is obtained (*cf.* proposition 3).

4   Experimental Evaluations

In order to validate the representation based on the borders *Lower* / *Upper* (*cf.* Section 3.1) and *Upper** / *Upper* (*cf.* Section 3.2), we perform experiments to evaluate the size of the borders: *Lower* (L), *Upper** (U^*). For this purpose, we use seven classical data sets [1]. We choose real and synthetic data sets. Their characteristics are reported in table 5. They are:

- the data set of census PUMSB extracted from "PUMSB sample file",
- the real data set SEP85L containing weather conditions at various weather stations or lands from December 1981 to November 1991. This weather data set has been frequently used in calibrating various cube algorithms,
- the synthetic data sets T10I4D100K and T40I10D100K, built up from sale data.

In our experiments, for computing the *Upper** border (U^*), we choose the algorithm MAFIA [13] because of its efficiency and availability. In order to evaluate the size of the *Lower* border (L), we use the algorithm MCTR [7]. The $MinThreshold_2$ threshold has a value slightly lower than the lowest value of $MinThreshold_1$. Moreover the latter varies in a range appropriated to each data set. By this way, we obtain significant results (neither too low nor too high). We propose a presentation both graphic and quantitative for the evaluation of the number of elements in the borders.

Table 5. Data Sets

Name	#Tuples	#Attributes	#Values
PUMSB	49 046	74	2 113
SEP85L	1 015 367	9	7 175
T10I4D100K	100 000	10	1 000
T40I10D100K	100 000	40	1 000

In any case, the border U^* is significantly more reduced than L. For instance, for real data sets, in the most critical cases when the threshold is very low and borders are large, the gain varies according to a factor 4.2 (SEP85L) or 5.4 (PUMSB). By increasing the threshold, the border size obviously decreases. We establish that the gain factor provided by U^* when compared to L is even more important. This factor varies from 4.3 (SEP85L) to 10.4 (PUMSB). The experiments ran on synthetic data confirm and enlarge these results with a gain factor varying from 76 to 275 (T40I10D100K) or 25 to 55 (T40I10D100K) for U^*.

---

[1] http://fimi.cs.helsinki.fi/ and
    http://cdiac.ornl.gov/ftp/ndp026b/SEP85L.DAT.Z

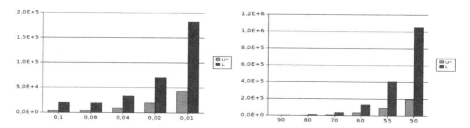

Fig. 1. Experimental Results for SEP85L and PUMSB

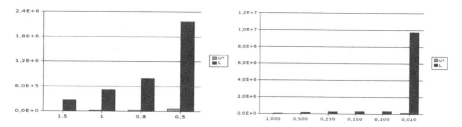

Fig. 2. Experimental Results for T40I10D100K and T10I4D100K

5   Conclusion

The Emerging Cube makes it possible to compare trends in the context of OLAP databases. It is a convex cube [6] which can be represented through the borders *Lower / Upper*. The borders *Lower / Upper* are classical when the solution space is convex. We have proposed and characterized a novel representation: the borders *Upper* / Upper*. We state the link between the borders *Lower* and *Upper** by using the concept of cube transversal. Experimental evaluations establish for various data sets that the size of this new border is really lower than the one of the border *Lower / Upper*.

Since the borders *Upper* / Upper* and *Lower / Upper* do not make possible to retrieve the measures of any emerging tuple, a possible research perspective is to investigate a novel compact representation, the emerging closed cube. This representation could be based on the cube closure [7].

Another continuation of the presented work, is to enlarge the " *Cubegrade* " problem [2] to couples of categorical attribute relations r_1 and r_2 and thus to extract couples of tuples from r_1 and r_2 for which the measures change in a significant way. The considered tuple couples are related by generalization/specialization or mutation (a different value for each dimension).

References

1. Sarawagi, S.: Explaining differences in multidimensional aggregates. In: Proceedings of the 25th International Conference on Very Large Data Bases, VLDB, pp. 42–53 (1999)
2. Imielinski, T., Khachiyan, L., Abdulghani, A.: Cubegrades: Generalizing association rules. Data Mining and Knowledge Discovery, DMKD 6(3), 219–257 (2002)

3. Dong, G., Li, J.: Mining border descriptions of emerging patterns from dataset pairs. Knowledge Information System 8(2), 178–202 (2005)

4. Nedjar, S., Casali, A., Cicchetti, R., Lakhal, L.: Emerging cubes for trends analysis in OLAP databases. In: Song, I.-Y., Eder, J., Nguyen, T.M. (eds.) DaWaK 2007. LNCS, vol. 4654, pp. 135–144. Springer, Heidelberg (2007)

5. Hurtado, C.A., Mendelzon, A.O.: Olap dimension constraints. In: Popa, L. (ed.) Proceedings of the Twenty-first ACM SIGACT-SIGMOD-SIGART Symposium on Principles of Database Systems, Madison, Wisconsin, USA, June 3-5, pp. 169–179. ACM, New York (2002)

6. Casali, A., Nejar, S., Cicchetti, R., Lakhal, L.: Convex cube: Towards a unified structure for multidimensional databases. In: Wagner, R., Revell, N., Pernul, G. (eds.) DEXA 2007. LNCS, vol. 4653, pp. 572–581. Springer, Heidelberg (2007)

7. Casali, A., Cicchetti, R., Lakhal, L.: Extracting semantics from datacubes using cube transversals and closures. In: Proceedings of the 9th ACM SIGKDD International Conference on Knowledge Discovery and Data Mining, KDD, pp. 69–78 (2003)

8. Gray, J., Chaudhuri, S., Bosworth, A., Layman, A., Reichart, D., Venkatrao, M., Pellow, F., Pirahesh, H.: Data cube: A relational aggregation operator generalizing group-by, cross-tab, and sub-totals. Data Mining and Knowledge Discovery 1(1), 29–53 (1997)

9. Mannila, H., Toivonen, H.: Levelwise Search and Borders of Theories in Knowledge Discovery. Data Mining and Knowledge Discovery 1(3), 241–258 (1997)

10. Flouvat, F., Marchi, F.D., Petit, J.M.: A thorough experimental study of datasets for frequent itemsets. In: Proceedings of the 5th International Conference on Data Mining, ICDM, pp. 162–169 (2005)

11. Gunopulos, D., Mannila, H., Khardon, R., Toivonen, H.: Data mining, hypergraph transversals, and machine learning. In: Proceedings of the 16th Symposium on Principles of Database Systems, PODS, pp. 209–216 (1997)

12. Gouda, K., Zaki, M.: Efficiently Mining Maximal Frequent Itemsets. In: Proceedings of the 1st IEEE International Conference on Data Mining, ICDM, pp. 3163–3170 (2001)

13. Burdick, D., Calimlim, M., Gehrke, J.: MAFIA: A maximal frequent itemset algorithm for transactional databases. In: Proceedings of the 17th International Conference on Data Engineering, ICDE, pp. 443–452 (2001)

Top_Keyword: An Aggregation Function for Textual Document OLAP

Franck Ravat, Olivier Teste, Ronan Tournier, and Gilles Zurfluh

IRIT (UMR5505), Université de Toulouse, 118 route de Narbonne
F-31062 Toulouse Cedex 9, France
{ravat,teste,tournier,zurfluh}@irit.fr

Abstract. For more than a decade, researches on OLAP and multidimensional databases have generated methodologies, tools and resource management systems for the analysis of numeric data. With the growing availability of digital documents, there is a need for incorporating text-rich documents within multidimensional databases as well as an adapted framework for their analysis. This paper presents a new aggregation function that aggregates textual data in an OLAP environment. The Top_Keyword function (Top_Kw for short) represents a set of documents by their most significant terms using a weighing function from information retrieval: *tf.idf*.

Keywords: OLAP, Aggregation function, Data warehouse, Textual measure.

1 Introduction

OLAP (On-Line Analytical Processing) systems allow analysts to improve the decision making process. These systems are based on multidimensional modelling of decisional data [6]. In OLAP systems, data is analysed according to different levels of detail and aggregation functions (e.g. sum, average, minimum…) are used to provide a synthetic view. Drilling operations (i.e. drill-down and roll-up), which are frequently used by decision makers, make intensive use of aggregation functions. For example, in Fig. 1, a decision maker analyses the number of keywords used within publications by authors and by month. To get a more global vision the decision maker changes the analysis detail level and rolls up from the month detail level of the TIME analysis axis to the year level. As a consequence, monthly values are aggregated into yearly values, applying the selected aggregation function (COUNT).

Analysis based on numeric centric multidimensional database is a well mastered technique [15]. However, according to a recent study only 20% of the data that spreads over an information system is numeric centric [17]. The remaining 80%, namely "digital paperwork," mainly composed of textual data, stays out of reach of OLAP due to the lack of tools and adapted processing. Not taking into account these data leads inevitably to the omission of relevant information during an analysis process and may end in producing erroneous analyses [17].

I.-Y. Song, J. Eder, and T.M. Nguyen (Eds.): DaWaK 2008, LNCS 5182, pp. 55–64, 2008.
© Springer-Verlag Berlin Heidelberg 2008

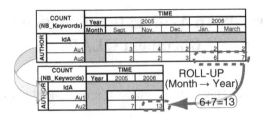

Fig. 1. An analysis with a Roll-up process

Recently, XML[1] technology has provided a wide framework for sharing documents within corporate networks or over the Web. Textual data in XML format is now a conceivable data source for OLAP systems.

By *multidimensional document analysis* throughout this paper we mean to analyse in an OLAP framework text-rich document data sources, e.g. conference proceedings, quality control reports, e-books…

The problem may be summarised as follows: during an analysis of data extracted from text-rich XML documents, textual analysis indicators are used. However, how one may aggregate textual data when all available aggregation functions of the OLAP environment are based on arithmetic functions (sum, average…)?

1.1 Related Works

In order to analyse data extracted from XML documents, three types of propositions have been made. Firstly, were proposed the multidimensional analysis of documents within a standard OLAP framework [8, 9, 5, 17]. Nevertheless, all these propositions limit themselves to numerical indicators and do not allow content analysis of these documents. Secondly, some works have detailed adaptations of aggregation functions for XML data. An aggregation function for XML structures, XAggregation, has been proposed [18, 19]. And it has recently been followed by an adaptation of the Cube function for XML data [20]. These new functions allow the analysis of XML documents. But these operators are not text-oriented, thus, these operators do not allow the analysis of the textual content of XML documents. Thirdly, in order to answer more precisely to the problem of the analysis of the contents of text-rich XML documents, in [11], the authors describe a set of aggregation functions adapted to textual data inspired by text mining techniques. Unfortunately, the authors provide no detailed description, no formal specification and no implementation guidelines for their functions. Moreover, the authors describe their framework with an associated model based on an xFACT structure [10] but they do not specify how to handle textual indicators within the framework.

In conclusion, the integration of methods and analysis tools adapted to text-rich XML documents within the OLAP environment is still a promising issue.

1.2 Objectives and Contributions

Our goal is to provide an OLAP framework adapted for the analysis of text-rich XML document contents. We require a new approach for aggregating textual data. Our

---

[1] XML, Extended Markup Language, from http://www.w3.org/XML/.

contribution for multidimensional document analysis is to provide an adapted framework composed of textual analysis indicators associated to compatible aggregation functions [12]. We thus revise measure concept to take into consideration textual measures. To allow analyses based on textual measures, textual data must be aggregated by an adapted aggregation function. Inspired by MAXIMUM$_k$ that aggregates a set of numbers into the k highest values, we define a function that aggregates a set of text fragments into the k most representative terms of those fragments.

The rest of the paper is organised as follows: the following section defines our adapted constellation multidimensional model; section 3 specifies the adapted aggregation function and finally section 4 describes the implementation.

2 Multidimensional Conceptual Model

Traditional multidimensional models (see [16, 13] for recent surveys) allow the analysis of numerical indicators according to analysis axes. However, these models are not sufficient for performing OLAP analyses on text-rich documents. Our proposition is to extend these traditional models with textual analysis indicators. The model is a constellation [6, 13] composed of dimensions and facts.

Dimensions model analysis axes and are composed of a set of parameters which are organised into one or more hierarchies. Each *hierarchy* represents an analysis perspective along the axis. The *parameters* represent different levels according to which analysis data may be observed. The subject of analysis, namely a *fact*, is a conceptual grouping of measures. Traditionally these measures are numeric, thus this concept must be extended in order to cope with our issue of textual data.

2.1 Measures in the OLAP Environment

In order to consider specificities of documents the concept of measure is extended.

Definition: A *measure M* is defined by $M = (m, type, f_{AGG})$ where:
 − m is the measure name;
 − *type* is the measure type (numerical additive, non additive,....);
 − f_{AGG} is the list of available aggregation functions for the measure.

---

f_{AGG} is used by manipulation languages (see for example [13]) to ensure multidimensional query consistency. Note that F_{AGG} is the list of all available aggregation functions of the OLAP system. The measure type conditions the list of compatible aggregation functions (F^T_{AGG}), thus for a measure type T ($F^T_{AGG} \subseteq F_{AGG}$). Amongst the compatible functions, the designer selects the aggregation functions that will be available and that will constitute the list f_{AGG} ($f_{AGG} \subseteq F^T_{AGG}$).

Different Types of Measures. Two types of measures are distinguished:
Numerical measures are exclusively composed of numbers and are either additive or semi-additive [4, 6]. With additive numerical measures, all classical aggregation

functions operate. Semi-additive measures represent instant values, i.e. snapshots such as stock or temperature values and the *sum* function does not operate.

Textual measures are composed of textual data that are non numeric and non additive [4, 6]. Contents of textual measures may be words, bags of words, paragraphs or even complete documents. According to these different possibilities the aggregation process may differ, thus several types of textual measures are distinguished:

- A *raw textual measure* is a measure whose content corresponds to the textual content of a document for a document fragment (e.g. the content of a scientific article in XML format stripped of all the XML tags that structure it).
- An *elaborated textual measure* is a measure whose content is taken from a raw textual measure and has undergone a certain amount of processing. A textual measure such as a *keyword* measure is an elaborated textual measure. This kind of measure is obtained after applying processes on a raw textual measure such as withdrawing stop words and keeping the most significant ones regarding the document's context.

OLAP Aggregation Functions. According to the measure type, not all aggregation functions may be used for specifying analyses. Table 1 lists the compatibility between measure types (T) and available aggregation functions (F^T_{AGG}). Within OLAP environments several arithmetic aggregation functions are available:

- Additive function: SUM (the sum of values that have to be aggregated);
- Semi-additive functions: AVG (the average of the values), MIN and MAX (the minimal or the maximal values).
- Generic functions: COUNT (counting the number of instances to be aggregated) and LIST (the identity function that list all the values).

Table 1. The different types of measures and the possible aggregation functions

Measure Type	Applicable Functions	Example
Numeric, Additive	Additive, Semi-Additive, Generic	A quantity of articles
Numeric, Semi-Additive	Semi-Additive, Generic	Temperature values
Textual, Raw	Generic	Content of an article
Textual, Keyword	Generic	Keywords of a fragment of a document

2.2 Example

A decision maker analyses a collection of scientific articles published by authors at a certain date (see Fig. 2 where graphic notations are inspired by [2]). The fact *ARTICLES* (subject) has three analysis indicators (measures): a numerical measure (*Accept_Rate*, the acceptance rate of the article), a raw textual measure (*Text*, the whole textual content of the article) and a keyword elaborated textual measure (*Keyword*). The fact *ARTICLES* is linked to 2 dimensions: *AUTHORS* and *TIME*.

Within this example, the measure *Accept_Rate* is associated to AVG, MIN, MAX and LIST aggregation functions. The two other measures are associated only to the two generic aggregation functions (COUNT and LIST)

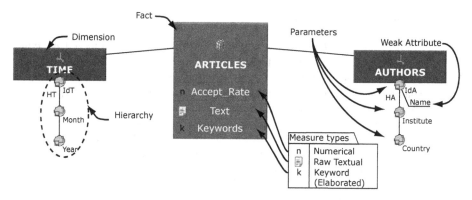

Fig. 2. Example of a constellation with textual measures for document content anlaysis

3 Top_Kw Aggregation Function

Our objective is to provide adapted aggregation functions for textual measures (raw and elaborated). However, only generic aggregation functions operate on such measures, which is not sufficient for multidimensional analysis. Thus, we propose to extend the list of available functions dedicated to textual measures. In order to handle elaborated textual measures, more precisely keyword measures, we have previously defined an aggregation function inspired by the AVERAGE function. This function aggregates a set of keywords into a smaller set of more general keywords [12].

In the rest of this paper we focus on the aggregation of raw textual measures, e.g. the full text scientific articles in XML format.

Preliminary remark: note that raw textual measures are stripped from stop words (e.g. words like: a, an, the, these...).

3.1 Formal Specification

In order to summarise data from a raw textual measure, there is a need for aggregating its textual data. The objective of the aggregation function TOP_Kw_k is to extract from a textual measure (composed of n words) a set of k most representative terms.

As in the arithmetic aggregation function MAX_k which extracts the k highest values of a set of numbers, this function extracts the k terms that have the most representativeness from a text represented through a set of n words.

$$Top\_Kw_k : \quad W^n \quad \longrightarrow \quad T^k$$
$$\{w_1, \ldots, w_n\} \quad \mapsto \quad \langle t_1, \ldots, t_k \rangle$$

Eq. 1

With as **input**: a text represented through a set of n words and as **output** an ordered subset of W^n composed of the k most representative terms. $T^k \subseteq W^n$, $\forall\ t_i \in T^k$, $t_i \in W^n$. Concretely, the function orders all the words that represent documents according to a weight assigned corresponding to its representativeness regarding the document and returns k first weighed terms.

3.2 Ordering Terms of Documents

In order to determine the k most representative terms, we adapted to the OLAP context well-known and well-mastered techniques from Information Retrieval (IR) [1]. IR uses functions to assign weights to terms within documents. Following this example, we shall use and adapt a function that weighs terms according to their representativness and thus allowing their ordering according to this representativness.

In information retrieval, it is necessary to know the representativness of a term compared to the whole collection of documents that contain this term. In the OLAP context, it is not necessary to know the representativness of terms according to the whole collection but rather according to the whole set of textual fragments that are to be aggregated with the function. The issue is to have to operate on a variable list that changes at each OLAP manipulation, instead of a fixed list (the collection).

3.3 Aggregation and Displaying within a Multidimensional Table

Context. The restitution of an OLAP analysis is done through multidimensional tables (see Fig. 1). Values are placed in cells c_{ij} that are at the intersection of ith line and the jth column. Each cell contains the aggregated values of the analysed measures.

Note that for each cell of a multidimensional table, the aggregation function is applied. Each cell represents a certain number of documents or fragments of documents. To each cell c_{ij} corresponds:

- A set of documents (or fragments of documents) D_{ij} composed of d_{ij} documents;
- A total number of terms n_{ij} that are in each of the d_{ij} documents.

Weight Calculus. Within each cell, weights are assigned to each one of the n_{ij} terms. In order to "rank" these terms we use a weighing function. We chose the *tf.idf* function that corresponds to the product of the representativness of a term within the document (*tf*: term frequency) with the inverse of its representativness within all available documents of the collection (*idf*: inverse document frequency). We have adapted this function to our context, i.e. the *idf* is calculated only for the documents of the cell (not the whole collection). For each cell c_{ij} and each term t corresponds:

- A number of occurrences $n_{ij}(t)$ of the term t in the document of c_{ij}, i.e. D_{ij};
- A number of documents $d_{ij}(t)$ that contain the term t amongst the documents of c_{ij} $(d_{ij}(t) \leq d_{ij})$.

This gives us, for each term of the cell c_{ij} an adapted *tf.idf* function:

$$tf_{ij}(t) = \frac{n_{ij}(t)}{n_{ij}} \text{ and } idf_{ij}(t) = \log\frac{d_{ij}+1}{d_{ij}(t)} \qquad \text{Eq. 2}$$

Thus the weight of the t term is: $w_{ij}(t) = tf_{ij}(t) \times idf_{ij}(t)$ \qquad Eq. 3

In the previous equations (see Eq. 2), *tf(t)* is the number of times when the term t is in a fragment of text normalised by the total number of terms of that fragment of text. This reduces the bias introduced by very long fragments of text compared to small ones. The quantity *idf(t)* is the inverse of the number of documents that contain the term t compared to the number of documents contained in the cell c_{ij}. The 1 added is to avoid obtaining a null *idf* if all the documents contain the term t. The use of a logarithm function is useful to smooth the curve thus reducing the consequences of large values compared to small ones. We invite the reader to consult [14] for a recent detailed analysis of the *tf.idf* function.

Restitution. The application of the *tf.idf* function (see Eq. 3) to a set of documents whose words have been extracted, allows obtaining a list of weighed terms.

Words are extracted from each document of the set D_{ij} and weights are assigned to each by applying the adapted *tf.idf* function. If a weighed term appears in several documents of D_{ij}, the weights are calculated separately and then added in order to increase the repsentativness of the term. Weighed terms are then ordered according to their weights: $L_{ij} = <t_1,\dots, t_n> \mid w_{ij}(t_1)>w_{ij}(t_2)>\dots> w_{ij}(t_n)$.

The aggregation function TOP_KW_n finally extracts the first n terms of the list L_{ij} with the highest weights and displays them into the corresponding c_{ij} cell of the multidimensional table.

3.4 Example

In the following example, the function TOP_KW is set to return the 2 most representative terms ($k = 2$). The analysis consists in the major keywords of scientific articles analysed according to their author and the year of publication (see Fig. 3). For this analysis, the aggregation function is applied to four groups of documents: one for each cell of the multidimensional table that correspond to each of the following braces: ($Au1$, 2005), ($Au1$, 2006), ($Au2$, 2005) and ($Au2$, 2006). Note that only 4 terms are represented.

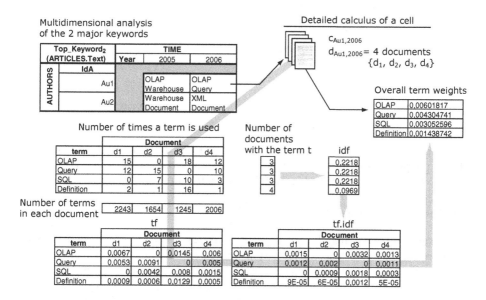

Fig. 3. Example of a detailed analysis using the function TOP_KW_2

4 Implementation

The aggregation function has been implemented within an existing prototype [13]. This prototype is based on a relational environment (ROLAP). The multidimensional database is in an Oracle $10g2$ RDBMS (with XMLDB to handle the document contents) and the high-end interface is a Java client.

4.1 Pre-processing of Document Contents

The more a term is used in a document, the higher the weight of that term given by *tf.idf* function is. The problem is that very common words (named *stop words* in information retrieval) will have a very high weight although they are not the most representative terms. On a raw measure, the solution consists in removing from the aggregation process all stop words (e.g. a, an, the, as, at…).

4.2 Processing for the Aggregation

The TOP_KW$_k$ aggregation function rests on the following algorithm. This algorithm takes as input the set of documents D_{ij} of a cell c_{ij}. Weighed_Term_List is a list of terms extracted from the D_{ij} documents and associated to a weight that correspond to the representativity of each term regarding the whole D_{ij} document set.

```
For Each c_ij cell Do
  Create a new empty Weighed_Term_List;
  For each document d of D_ij Do
    For each word of d Do
      If word is a Stop Word Then drop the word;
      Else Do
        Process weight of word;
        If word not in Weighed_Term_List Then
          Insert word and Insert weight
        Else
          Add new weight to already existing weight;
        End_If
      End_If
    End_For
  End_For
  Order Terms in Weighed_Term_List by decreasing weights
  Extract the first k terms to the cell c_ij
End_For
```

Note that some of the instructions of this algorithm may be removed if they are handled as pre-processing steps. Due to lack of space only some of the numerous optimisations will be briefly presented in the following subsection.

4.3 Performance Issues

Aggregation functions may be optimised more or less easily according to their type (distributive, algebraic and holistic). Holistic aggregation functions are difficult to optimise [3, 7] as the aggregation process may not rely on intermediate result for processing the final result. Unfortunately, the modified *tf.idf* function used in the aggregation function renders it holistic. Nevertheless, a simple solution consists in using numerous materialised views for execution for speeding up runtime.

Another solution would consist in using a precalculated specific view that would provide intermediate results in order to speed up the computation of the term weights. For example, it would be possible to precalculate the *tf* part of the weighing formula for each term and document as this only requires parsing the document. However, the computation of the *idf* part would have to be done at runtime.

5 Conclusion

Within this paper, we have defined a new aggregation function for a text-rich document OLAP framework. It is possible to have a synthetic vision of document sets by extracting the k most representative keywords of each set. The aggregation function TOP_KEYWORD (TOP_KW$_k$ for short) rests on an adapted *tf.idf* weighing function. This weighing function allows ordering words of document sets or sets of fragments of documents and the aggregation function selects the first k words.

Several future works may be undertaken. Firstly, in information retrieval, there exists the notion of "relevance feedback" that consists in adding terms to a query to increase the reliability of the terms used in a query. In a similar way, relevance feedback could be used to add terms to those returned by the aggregation function. The final result would then correspond to a more precise set of terms. Secondly, there exists several variants and even complete alternatives of the *tf.idf* function [14]. We think that a comparative study of these different weighing functions should be done in order to optimise the implementation of our aggregation function. Finally, in order to enrich our environment we are considering to define and implement other aggregation functions for textual data such as, for example, those stated in [11].

References

1. Baeza-Yates, R., Ribeiro-Neto, B.: Modern Information Retrieval. Addison Wesley, Reading (1999)
2. Golfarelli, M., Maio, D., Rizzi, S.: The Dimensional Fact Model: A Conceptual Model for Data Warehouses. IJCIS 7(2-3), 215–247 (1998)
3. Gray, J., Bosworth, A., Layman, A., Pirahesh, H.: Data Cube: A Relational Aggregation Operator Generalizing Group-By, Cross-Tab, and Sub-Total. In: ICDE, pp. 152–159 (1996)
4. Horner, J., Song, I.-Y., Chen, P.P.: An analysis of additivity in OLAP systems. In: DOLAP, pp. 83–91 (2004)
5. Keith, S., Kaser, O., Lemire, D.: Analyzing Large Collections of Electronic Text Using OLAP. In: APICS, Conf. in Mathematics, Statistics and Computer Science, pp. 17–26 (2005)
6. Kimball, R.: The data warehouse toolkit, 1996, 2nd edn. John Wiley and Sons, Chichester (2003)
7. Lenz, H.J., Thalheim, B.: OLAP Databases and Aggregation Functions. In: SSDBM 2001, pp. 91–100 (2001)
8. McCabe, C., Lee, J., Chowdhury, A., Grossman, D.A., Frieder, O.: On the design and evaluation of a multi-dimensional approach to information retrieval. In: SIGIR, pp. 363–365 (2000)

9. Mothe, J., Chrisment, C., Dousset, B., Alau, J.: DocCube: Multi-dimensional visualisation and exploration of large document sets. JASIST 54(7), 650–659 (2003)
10. Nassis, V., Rajugan, R., Dillon, T.S., Wenny, R.J.: Conceptual Design of XML Document Warehouses. In: Kambayashi, Y., Mohania, M., Wöß, W. (eds.) DaWaK 2004. LNCS, vol. 3181, pp. 1–14. Springer, Heidelberg (2004)
11. Park, B.K., Han, H., Song, I.Y.: XML-OLAP: A Multidimensional Analysis Framework for XML Warehouses. In: Tjoa, A.M., Trujillo, J. (eds.) DaWaK 2005. LNCS, vol. 3589, pp. 32–42. Springer, Heidelberg (2005)
12. Ravat, F., Teste, O., Tournier, R.: OLAP Aggregation Function for Textual Data Warehouse. In: ICEIS 2007, vol. DISI, pp. 151–156 (2007)
13. Ravat, F., Teste, O., Tournier, R., Zurfluh, G.: Algebraic and graphic languages for OLAP manipulations. ijDWM 4(1), 17–46 (2007)
14. Robertson, S.: Understainding Inverse Document Frequency: On theoretical arguments for IDF. Journal of Documentation 60(5), 503–520 (2004)
15. Sullivan, D.: Document Warehousing and Text Mining. Wiley John & Sons, Chichester (2001)
16. Torlone, R.: Conceptual Multidimensional Models. In: Rafanelli, M. (ed.) Multidimensional Databases: Problems and Solutions, ch.3, pp. 69–90. Idea Group Inc. (2003)
17. Tseng, F.S.C., Chou, A.Y.H.: The concept of document warehousing for multi-dimensional modeling of textual-based business intelligence. J. DSS 42(2), 727–744 (2006)
18. Wang, H., Li, J., He, Z., Gao, H.: Xaggregation: Flexible Aggregation of XML Data. In: Dong, G., Tang, C.-j., Wang, W. (eds.) WAIM 2003. LNCS, vol. 2762, pp. 104–115. Springer, Heidelberg (2003)
19. Wang, H., Li, J., He, Z., Gao, H.: OLAP for XML Data. In: CIT, pp. 233–237 (2005)
20. Wiwatwattana, N., Jagadish, H.V., Lakshmanan, L.V.S., Srivastava, D.: X3: A Cube Operator for XML OLAP. In: ICDE, pp. 916–925 (2007)

Summarizing Distributed Data Streams for Storage in Data Warehouses

Raja Chiky and Georges Hébrail

TELECOM ParisTech
LTCI-UMR 5141 CNRS
Paris, France
`firstname.lastname@telecom-paristech.fr`

Abstract. Data warehouses are increasingly supplied with data produced by a large number of distributed sensors in many applications: medicine, military, road traffic, weather forecast, utilities like electric power suppliers etc. Such data is widely distributed and produced continuously as data streams. The rate at which data is collected at each sensor node affects the communication resources, the bandwidth and/or the computational load at the central server. In this paper, we propose a generic tool for summarizing distributed data streams where the amount of data being collected from each sensor adapts to data characteristics. Experiments done on electric power consumption real data are reported and show the efficiency of the proposed approach.

1 Introduction

A growing number of applications in areas like utilities, retail industry and sensor networks deal with a challenging type of data: data elements arrive in multiple, continuous, rapid and time-varying data streams. A key requirement of such applications is to continuously monitor and react to interesting phenomena occurring in the input streams. For instance, electric power utilities may want to alert their customers when their consumption is going to exceed a threshold during a period with expensive electric power.

Many solutions have been developed recently to process data streams that arrive from a single or from multiple locations. The concept of Data Stream Management System (DSMS) has been introduced and several systems (including commercial ones) have been developed [4,7,17,18,21]. These systems enable to process data streams on-the-fly and issue alarms or compute aggregates in real-time without storing all data on disk. But in many applications there is a need to keep track of such data in a data warehouse for further analyses. For instance electric power utilities (or telecommunication companies) need to keep track of the way their customers use electric power (telecommunication services) upon time: applications concern power (telecommunication) consumption forecast, definition of new price policies, ... It is inconceivable to load all such data in a data warehouse due to its volume, arrival rate and possible spatial distribution.

This paper proposes an approach to summarize the contents of a large number of distributed data streams describing uni-dimensional numerical data, where each data source corresponds to a sensor which periodically records measures in a specific field (electric power consumption, use of telecommunication services, ...). The approach is based on

I.-Y. Song, J. Eder, and T.M. Nguyen (Eds.): DaWaK 2008, LNCS 5182, pp. 65–74, 2008.

optimized temporal sampling of data sources in order to produce summaries of good quality with respect to computing, storage and network constraints. The computation of the summaries is decentralized and some sensor computing capacities are used.

The basic scheme is to choose the best policy of summarizing data over each temporal window t taking into account data of the previous window $t - 1$. This assumes that data is somewhat periodic and that the window length is chosen to be compatible with this period. This assumption is realistic in many applications and in particular in the one studied in this paper concerning sensors which are communicating electric power meters.

The paper is organized as follows. Section 2 describes related work in this area. In Section 3, we present the proposed summarizing system. Section 4 evaluates the approach on real data and Section 5 gives an outlook upon our ongoing and future research in this area.

2   Related Work

Summarizing online data streams has been largely studied with several techniques like sketching ([8,9]), sampling and building histograms ([2,10]). Authors in [15] experimentally compare existing summarizing techniques and use the best one in terms of reconstruction accuracy; they also introduce a user-defined model to reduce data quality when data gets older. However, most work concentrates on summarizing a single stream and without exploiting possible bi-directional communications between a central server and its sensors nor computing facilities within sensors.

In the context of DSMS's, load shedding techniques ([1,19]) drop randomly data stream records when the load of the system increases beyond what it can handle. But these approaches do not scale to a large number of streams issued by remote sensors.

In the context of sensor networks, several adaptive sampling techniques have been developed to manage limited resources. The aim is to preserve network bandwidth (or the energy used for wireless communications) and storage memory by filtering out data that may not be relevant in the current context. Data collection rates thus become dynamic and adapted to the environment.

An adaptive sampling scheme which adjusts data collection rates in response to the contents of the stream was proposed in [11]. A Kalman filter is used at each sensor to make predictions of future values based on those already seen. The sampling interval SI is adjusted based on the prediction error. If the needed sampling interval for a sensor exceeds that is allowed by a specified Sampling Interval range, a new SI is requested to the server. The central server delivers new SIs according to available bandwidth, network congestion and streaming source priority. Again a probabilistic model is used in [5]: a replicated dynamic probabilistic model minimizes communications between sensor nodes and the central server. The model is constructed by exploiting intelligently historical data and spatial correlations across sensors. Correlations between sensors can be computed incrementally as described in [16].

A backcasting approach to reduce communication and energy consumption while maintaining high accuracy in a wireless sensor network was proposed in [20]. Backcasting operates by first activating only a small subset of sensor nodes which communicate their information to a fusion center. This provides an estimate of the environment being sensed, indicating some sensors may not need to be activated to achieve a desired level of accuracy. The fusion center then backcasts information based on the estimate to the network and selectively activates additional sensors to obtain a target error level. In this approach, adaptive sampling can save energy by only activating a fraction of the available sensors.

In [14], authors present a feedback control mechanism which makes dynamic and adaptable the frequency of measurements in each sensor. Sampled data are compared against a model representing the environment. An error value is calculated on the basis of the comparison. If the error value is more than a predefined threshold, then a sensor node collects data at a higher sampling rate; otherwise, the sampling rate is decreased. Sensor nodes are completely autonomous in adapting their sampling rate.

In [13], authors present a method to prevent sensor nodes to send redundant information; this is predicted by a sink node using an ARIMA prediction model. Energy efficiency is achieved by suppressing the transmission of some samples, whose ARIMA based prediction values are within a predefined tolerance value with respect to their actual values. A similar approach is proposed by Cormode and Garofalakis [6]. Their results show that reduced communication between sensors and the central server can be sufficient by using an appropriate prediction model. A wide range of queries (including heavy hitters, wavelets and multi-dimensional histograms) can be answered by the central server using approximate sketches.

3   Architecture of the Summarizing System

We consider a distributed computing environment, describing a collection of n remote *sensors* and a designated *central server*. Sensors communicate with the central server which is responsible for loading into the data warehouse measurements done by sensors. We assume that communications between the central server and sensors are bi-directional and that there is no communication between sensors. Such architecture is representative of a large class of applications. In the particular case of electric power metering, several million meters feed a unique data warehouse but using intermediate network nodes which play the role of the central server of our model. There are typically several hundreds of meters directly connected to an intermediate node.

We assume that sensor measures are generated regularly at the same rate for every sensor. We cut out all streams into periods (also called temporal windows) of the same duration. A typical period corresponds to a duration of one day. A temporal window is composed of p elements for each sensor (the number of elements is the same for all sensors). Elements issued by a sensor during a period is also referred as a *curve* in the following. For each sensor, the number of elements gathered by the central server within a temporal window will vary between m (fixed parameter) and p. Let us define s as the maximum number of elements which can be received by the central server from the n sensors during a temporal window ($s < n * p$).

From the observation of what is happening during a period $t - 1$, we determine a data collection model to apply to the n sensors for period t. The problem consists of finding the best "policy of summarizing" each sensor, respecting the constraints of the maximum number of data collected by the central server (parameter s) and the minimum number of data taken from each sensor (parameter m).

The data collection model should preserve as much as possible the detailed information by reducing summarizing errors. Several measures can be used to measure errors. We use in this paper the Sum of Square Errors SSE, which is the square of the L_2 distance between the original curve $C = \{c_1, c_2, ..., c_p\}$ and an estimation from the summarized curve $\hat{C} = \{\hat{c}_1, \hat{c}_2, ..., \hat{c}_p\}$. SSE is computed as follows: $SSE(C, \hat{C}) = \sum_{i=1}^{p}(c_i - \hat{c}_i)^2$ where p is the number of elements in C. Other error measures can be defined depending on the application. SSE is well adapted to electric power consumption. Note that error measures cannot be defined arbitrarily and some properties (not studied in this paper) are necessary for the optimization problem to be solved in reasonable time.

During a period t, sensors send data to the server at the scheduled sampling rate along with SSE's values for different assumptions of sampling rates corresponding to sending between m and p values. Then the server solves an optimization problem to find the best sampling rate for each sensor during period $t + 1$. Actually, we assume that the number of sensors connected to a server is large enough to distribute uniformly with time all data exchanges so that there is no temporary overload of the server.

3.1   Sensor Side Module

Several sampling schemes are considered by each sensor to summarize its curve during a period. Almost all methods of curve compression can be used depending on the handled data. For the needs of our application, we concentrated on three methods: regular sampling, curve segmentation and wavelet compression.

Regular sampling: Curves are sampled using a step j chosen between 1 and $m' = \lfloor \frac{p}{m} \rfloor$ ($\lfloor x \rfloor$ is the floor function). $j = 1$ means that all elements are selected, $j = 2$ means that every two elements are selected and so on. This technique is described as *regular* because selected points are temporally equidistant, a jump j is made between two sampled elements. As for the estimation of the original curve from the sampled elements, two interpolation methods are used: "Linear Interpolation" and "Constant Interpolation". Given one interpolation method, SSE's for the different values of j can be computed by the sensor and sent to the server.

Curve segmentation: Segmentation approximates a curve by using a *piecewise representation*. The curve is split into a given number of segments and each segment is associated with a constant value. Segmentation has been considered in many areas and there exists an algorithm from Bellman [3] which computes exactly the segment bounds which minimize the SSE between the curve and its piecewise representation given the number of segments. Other error measures can be used and lead to other optimization algorithms (see [12]). The number of segments applicable to a sensor varies between $\lfloor \frac{m}{2} \rfloor$ and p. Indeed, the number of elements of each segment is also needed to construct the piecewise representation. If a k-segmentation is applied to a curve C, the central server needs $\{(\hat{c}_1, n_1), (\hat{c}_2, n_2), ..., (\hat{c}_k, n_k)\}$ to construct the piecewise representation $\hat{C}$, where $\hat{c}_l$ is the average value of segment l composed of n_l elements. Thus, we can assign an integer value j to a curve from 1 (all elements are collected) to m' (optimally the curve is segmented in $\left\lfloor \frac{p}{2*m'} \right\rfloor$ segments). Here again each sensor performs segmentation locally with different numbers of segments, computes the corresponding SSE's and can send SSE's corresponding to different summarizing levels indexed by a value j varying between 1 and m'.

Curve compression: Wavelet coefficients are projections of the curve onto an orthogonal set of basis vectors. The choice of basis vectors determines the type of wavelets. Haar wavelets are often used for their ease of computation and are adapted to electric power consumption curves. The estimation of a curve from its summarized version is a linear combination of a number of basis vectors chosen from $\{1, 2^1, 2^2, ...2^J\}$, ($j = 1$ means the whole curve is collected). J is the biggest integer such that $2^J \leq m'$. The number of data points in a curve must be a power of 2 in the case of the Haar wavelet. If it is not the case zeros are padded at the right end of the curve. Again, each sensor computes the wavelet decomposition with different levels indexed by j varying from 1 to m' and send the corresponding SSE's to the server.

At the end of every temporal window, sensors compute the SSE's corresponding to each summarizing procedure (regular sampling, curve segmentation and curve compression) and send for each value of index j the minimum of errors between these three methods to the central server.

3.2 Central Server Module

At the end of each period, the central server receives from each sensor different values of SSE's corresponding to different levels of summarization, indexed by j varying between 1 and m'. These SSE's are stored in a matrix $W_{n*m'}$ of n lines and m' columns (n is the number of sensors connected to the server). Element w_{ij} of the matrix corresponds to the SSE obtained when collecting data from sensor i with a summarizing level j.

Instead of distributing equally the summarizing levels between all sensors for period $t + 1$, an optimization is performed to assign different sampling rates to sensors from SSE's computed during period t. The central server solves the following optimization problem minimizing the sum of SSE's on sensors:

$$\text{Minimize } \sum_{i=1}^{n} \sum_{j=1}^{m'} (W_{ij} \times X_{ij})$$

subject to:

$$\begin{cases} X_{ij} = 0 \text{ or } 1 \\ \sum_{j=1}^{m'} X_{ij} = 1 & i \text{ from 1 to n} \\ \sum_{i=1}^{n} \sum_{j=1}^{m'} \left(\left\lfloor \frac{p}{j} \right\rfloor \times X_{ij} \right) \leq s \text{ i from 1 to n} \end{cases}$$

This is a problem of assignment of summarizing levels to the curves respecting the constraints above. A variable $X_{ij} = 1$ means that the central server assigns summarizing level j to the curve of index i. The second constraint $\sum_{j=1}^{m'} X_{ij} = 1$ imposes only one value of j per curve. Lastly, the third constraint means that the number of data to be communicated to the central server should not exceed the threshold s. Once the optimization problem is solved, the j_i are sent to each sensor. The sensor uses the summarizing method corresponding to the smallest SSE value. At the end of the period, sensors send the SSEs to update the matrix of errors as explained above. This process continues as long as data streams arrive on line. This optimization problem can be solved using standard integer linear programming techniques. We used a solver based on the simplex method and Branch-and-bound methods. This approach solves in reasonable time(less than one minute using a standard PC 2GHZ with 1.5 GB) problems with less than 30000 variables, which is quite enough in our context (up to 1000 sensors connected to each intermediate node).

4 Experimental Validation

The approach described above has been assessed empirically on real data against a standard non adaptive approach called the *fixed* approach. In this approach, the three summarizing methods have been considered but the number of selected elements is equally distributed between sensors depending on parameter s.

Experiments were carried out on two real data sets issued from electric meters. Each meter produces a curve representing the electric power consumption of one customer. The first data set describes 1000 meters each producing one measure every 10 minutes during one day (144 measurements per meter per day). This data set is used to assess the efficiency of the approach in the case of a *stationary* consumption for each customer: the optimized sampling rates are assessed on the same curve which is used to choose them. This experiment is referred as "stationary window" below.

The second data set describes 140 meters each producing one measure every 30 minutes but during one year (365 days). It is used to assess the efficiency on a nonstationary consumption for each customer: the optimized sampling rates are assessed on the curve of a different day than the one used for choosing them. This second experiment will be referred as "jumping window" below.

These data sets are those available today. They do not show very high rates of elements. In the next future with the generalization of automatic communicating meters, each meter will possibly generate a measure every one second.

4.1 Stationary Window Results

In these experiments, we used $m = 7$ ($m' = \lfloor \frac{144}{7} \rfloor = 20$). This means that sampled curves consist of at least 7 values if they are built regularly, of 3 values if built with segmentation and of 9 with wavelet compression. We point out that a curve in our example consists of 144 points. Several values of threshold s have been tested: the results are presented in Table 1. Each column of the table corresponds to a threshold value s. $s = 144000 = 144 * 1000$ means that all elements are collected for all curves. Experimented thresholds are equal to 144000 divided by one of $\{1, 2, 3, 5, 7, 10, 15\}$. For instance, $s = 48000$ is the third of total number of data ($\frac{144000}{3}$).

For each method the Sum of Square Errors have been computed. The standard deviation is also presented to illustrate the dispersion of number of data computed by optimization around the average which is equal to $\frac{s}{1000}$ ($n = 1000$).

Table 1. Experimental Results. SSE: Sum of Square Error. opt: sampling steps (#segments or #wavelet coefficients) obtained by optimization. avg: Average and std: Standard Deviation. Samp CI: Regular Sampling with Constant Interpolation, Samp LI: Regular Sampling with Linear Interpolation, seg: Segmentation and wav: Wavelet Compression.

Threshold s	144000	72000	48000	28800	20571	14400	9600
SSE opt Samp CI	0	696.36	1571.79	3037.02	4265.46	5839.95	8767.05
SSE opt Samp LI	0	434.35	996.34	1969.67	2781.54	3883.9	5676.6
SSE opt seg	0	161.62	385.13	874.1	1346.58	2027.17	3432.16
SSE opt wav	0	381.15	916.22	1881.66	2724.08	3888.6	6118.8
SSE opt all	0	139.29	346.05	801.91	1240.62	1859.6	3026.24
SSE fixed CI	0	2770.8	4329	8040	9313	13157	16048
SSE fixed LI	0	1956	2821	4245	5501	7024	9087
SSE fixed seg	0	505.8	860.8	1535	2042.6	2761.5	4677.1
SSE fixed wav	0	1385.32	2725.35	4766.09	4766.09	7179.5	7179.5
avg data/meter	144	72	48	28.8	20.57	14.4	9.6
std data CI	0	50.3	44.8	32.05	24.48	15.62	4.4
std data LI	0	51	45.45	33.6	24.7	16.5	4.9
std data seg	0	43	35.73	25.32	17.64	10.6	3.9
std data wav	0	48.9	42	30	21.8	13.5	3

As can be seen in Table 1, the optimization performs well for all summarizing methods, it considerably reduces the Sum of Square Errors compared to the "fixed" approach. For instance, in the case of regular sampling with Linear Interpolation, the optimization reduces the SSE by about 77% for threshold $s = 72000$ compared to a "fixed regular sampling". The transmission cost of SSE values is not included in the reported experiments. This impacts only on the comparison between the 'fixed' and 'optimized' approaches, since SSE's transmission is not needed in the 'fixed' approach. However, the number of SSE values to transmit is bounded and can be considered as negligible compared to the number of elements p in the window.

In our experiments, segmentation always gives better results than the two other summarizing methods. We also experimented with the case where meters choose themselves the summarizing method to adopt according to the affected j. We notice that the Sum of Square Errors is again reduced thanks to the automatic selection of summarizing method (line 'SSE opt all' of table 1).

We observe high values of standard deviations mainly for high values of threshold s. The sampling rates significantly deviate from the average. This confirms the importance of selecting sampling steps by optimization. The same observation is true for all summarizing methods.

4.2 Jumping Window Results

Previous experiments evaluate the error made over a period from an optimization carried out over the same period. The global approach consists of summarizing curves over a period based on optimization results computed over a previous period. We present here the method adopted for a streaming environment as well as first experiments completed on 140 load curves available over one year period.

Optimization Using d-1. We used the second dataset of 140 curves measured every 30 minutes over one year to validate the efficiency of the proposed data collection scheme. Table 2 gives the percentage of periods (temporal windows) for which the SSE using optimized summarizing is lower than the SSE using "fixed" summarizing. The length of the temporal window used for these experiments is one day. The optimization stage uses data of the previous day and resulting steps j are applied to data of the current day. We experimented thresholds equal to the total number 6720 divided by each of $\{1, 2, 3, 5, 7, 10, 15\}$.

Table 2. Percentage of days whose SSE with optimization is better than "fixed" methods. opt-1: Optimization at day $d - 1$ and sampling at d. %d < 10% SSE corresponds to the percentage of days whose SSE opt-1 is lower by 10 % than the SSE obtained by "fixed" methods.

Threshold s	6720	3360	2240	1344	960	672	448
%d< 10% SSE LI opt-1	-	84%	84%	84%	85%	85%	85%
%d < 10% SSE CI opt-1	-	84%	84%	85%	86%	97%	100%
%d < 10% SSE seg opt-1	-	83%	83%	84%	84%	84%	84%
%d < 10% SSE wav opt-1	-	84%	85%	94%	89%	100%	100%

Results in Table 2 show that the approach is very accurate. In fact, the optimized summarizing methods decrease SSE by at least 10% compared to the fixed methods on more than 83% of the days and for all tested threshold values. Note that the more threshold decreases the more performances of optimized summarizing methods increase.

Optimization Using d-7. Previous experiments were done on data of day $d - 1$ to summarize data of day d. Fig. 1 shows the evolution of SSEs during one month (30 days) for threshold $s = 1344$. We observe that errors are not distributed regularly with time. This is due to the weekly cycle of electric power consumption. In fact the consumption remains generally stable during 5 working days and the consumption level changes on

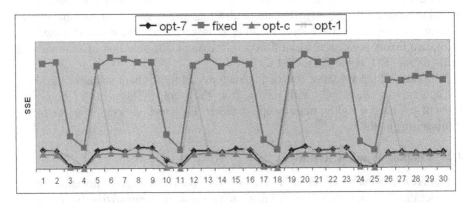

Fig. 1. Evolution of SSEs during 30 days. opt-7: Optimization at day d-7 and sampling at d. opt-1: Optimization at day d-1 and sampling at d. opt-c: Optimization and sampling at day d (current day). fixed: Fixed segmentation.

weekends. Here sampling rates for mondays are determined with data of sundays and those of saturdays based on data of fridays. Additional experiments were carried out with a sampling scheme for a day d based on sampling rates computed from day $d - 7$. Results are illustrated in curve opt-7 of Fig. 1 (curve opt-7). We note that this *smoothes* the SSE evolution and thus decreases errors due to the periodicity of electric power consumption.

5   Conclusion and Future Work

We have proposed a new generic framework for summarizing distributed data streams to be loaded into a data warehouse. The proposed approach limits the load of the central server hosting the data warehouse and optimizes the sampling rates assigned to each input stream by minimizing the sum of square errors. This optimization is dynamic and adapts continuously with the contents of the streams. Moreover, the compression technique used for each stream adapts continuously to its contents and is selected among several approaches.

Experiments have been done on real data describing measures of electric power consumption generated by customer meters. These experiments show the efficiency of the approach. Other experiments are currently being done on larger data sets describing more meters on a longer period.

This work suggests several developments which are under study:

- Testing the approach on more chaotic data with other summarizing techniques.
- Study how error measures different from SSE can be used, for instance the L_1 distance or L_∞ to minimize the maximum of errors instead of the mean. This impacts on the optimization algorithm.
- Extending the approach to sensors producing multidimensional numerical data.

References

1. Babcock, B., Datar, M., Motwani, R.: Load shedding for aggregation queries over data streams. In: 20th International Conference on Data Engineering (2004)
2. Babcock, B., Datar, M., Motwani, R.: Sampling From a Moving Window Over Streaming Data. In: 13th Annual ACM-SIAM Symposium on Discrete Algorithms (SODA 2002), pp. 633–634 (2002)
3. Bellman, R.: On the approximation of curves by line segments using dynamic programming. Communications of the ACM 4(6), 284 (1961)
4. Carney, D., Cetinternel, U., Cherniack, M., Convey, C., Lee, S., Seidman, G., Stonebraker, M., Tatbul, N., Zdonik, S.: Monitoring streams - A New Class of Data Management Applications. In: Proc. Int. Conf. on Very Large Data Bases, pp. 215–226 (2002)
5. Chu, D., Deshpande, A., Hellerstein, J.M., Hong, W.: Approximate Data Collection in Sensor Networks using Probabilistic Models. In: Proceedings of the 22nd international Conference on Data Engineering. ICDE 2006 (2006)
6. Cormode, G., Garofalakis, M.: Approximate Continuous Querying over Distributed Streams. ACM Transactions on Database Systems 33(2) (June 2008)
7. Demers, A., Gehrke, J., Panda, B., Riedewald, M., Sharma, V., White, W.M.: Cayuga: A general purpose event monitoring system. In: CIDR, pp. 412–422 (2007)

8. Dobra, A., Garofalakis, M., Gehrke, J., Rastogi, R.: Processing complex aggregate queries over data streams. In: Proceedings of the 2002 SIGMOD Conference, pp. 61–72 (2002)
9. Gehrke, J., Korn, F., Srivastava, D.: On computing correlated aggregates over continual data streams. In: Proceedings of the 2001 SIGMOD Conference, pp. 13–24 (2001)
10. Guha, S., Koudas, N., Shim, K.: Data-streams and histograms. In: Proceedings of the 2001 STOC Conference, pp. 471–475 (2001)
11. Jain, A., Chang, E.Y.: Adaptive sampling for sensor networks. In: Proceeedings of the 1st international Workshop on Data Management For Sensor Networks: in Conjunction with VLDB 2004, Toronto, Canada (2004)
12. Keogh, E., Chu, S., Hart, D., Pazzani, M.: An Online Algorithm for Segmenting Time Series. In: Proceedings of IEEE International Conference on Data Mining, pp. 289–296 (2001)
13. Liu, C., Wu, K., Tsao, M.: Energy Efficient Information Collection with the ARIMA model in Wireless Sensor Networks. In: Proceedings of IEEE GlobeCom 2005, St. Louis, Missouri (November 2005)
14. Marbini, A.D., Sacks, L.E.: Adaptive sampling mechanisms in sensor networks. In: London Communications Symposium, London, UK (2003)
15. Palpanas, T., Vlachos, M., Keogh, E., Gunopulos, D.: Streaming Time Series Summarization Using User-Defined Amnesic Functions. IEEE Transactions on Knowledge and Data Engineering (2008)
16. Sakurai, Y., Papadimitriou, S., Faloutsos, C.: Automatic Discovery of Lag Correlations in Stream Data. In: ICDE 2005, Tokyo, Japan (2005)
17. Stanford Stream Data Management (STREAM) Project,
    http://www-db.stanford.edu/stream
18. StreamBase Systems, Inc., http://www.streambase.com/
19. Tatbul, N., Cetintemel, U., Zdonik, S., Cherniack, M., Stonebraker, M.: Load shedding in a data stream manager. In: Proc. of the 29th Intl. Conf. on Very Large Databases (VLDB 2003) (2003)
20. Willett, R., Martin, A., Nowak, R.: Backcasting: adaptive sampling for sensor networks. In: Proceedings of the Third international Symposium on information Processing in Sensor Networks, Berkeley, California (2004)
21. Zdonik, S., Stonebraker, M., Cherniack, M., Cetintemel, U., Balazinska, M., Balakrishnan, H.: The Aurora and Medusa Projects. In: Bulletin of the Technical Committee on Data Engineering, March 2003, pp. 3–10. IEEE Computer Society, Los Alamitos (2003)

Efficient Data Distribution for DWS

Raquel Almeida[1], Jorge Vieira[2], Marco Vieira[1], Henrique Madeira[1],
and Jorge Bernardino[3]

[1] CISUC, Dept. of Informatics Engineering, Univ. of Coimbra, Coimbra, Portugal
[2] CISUC, Critical Software SA, Coimbra, Portugal
[3] CISUC, ISEC, Coimbra, Portugal

Abstract. The DWS (Data Warehouse Striping) technique is a data partitioning approach especially designed for distributed data warehousing environments. In DWS the fact tables are distributed by an arbitrary number of low-cost computers and the queries are executed in parallel by all the computers, guarantying a nearly optimal speed up and scale up. Data loading in data warehouses is typically a heavy process that gets even more complex when considering distributed environments. Data partitioning brings the need for new loading algorithms that conciliate a balanced distribution of data among nodes with an efficient data allocation (vital to achieve low and uniform response times and, consequently, high performance during the execution of queries). This paper evaluates several alternative algorithms and proposes a generic approach for the evaluation of data distribution algorithms in the context of DWS. The experimental results show that the effective loading of the nodes in a DWS system must consider complementary effects, minimizing the number of distinct keys of any large dimension in the fact tables in each node, as well as splitting correlated rows among the nodes.

Keywords: Data warehousing, Data striping, Data distribution.

1 Introduction

A data warehouse (DW) is an integrated and centralized repository that offers high capabilities for data analysis and manipulation [9]. Typical data warehouses are periodically loaded with new data that represents the activity of the business since the last load. This is part of the typical life-cycle of data warehouses and includes three key steps (also known as ETL): Extraction, Transformation, and Loading. In practice, the raw data is extracted from several sources and it is necessary to introduce some transformations to assure data consistency, before loading that data into the DW.

In order to properly handle large volumes of data, allowing to perform complex data manipulation operations, enterprises normally use high performance systems to host their data warehouses. The most common choice consists of systems that offer massive parallel processing capabilities [1], [11], as Massive Parallel Processing (MPP) systems or Symmetric MultiProcessing (SMP) systems. Due to the high price of this type of systems, some less expensive

I.-Y. Song, J. Eder, and T.M. Nguyen (Eds.): DaWaK 2008, LNCS 5182, pp. 75–86, 2008.

alternatives have already been proposed and implemented [7], [10], [8]. One of those alternatives is the Data Warehouse Stripping (DWS) technique [3], [6].

In the DWS technique the data of each star schema [3], [4] of a data warehouse is distributed over an arbitrary number of nodes having the same star schema (which is equal to the schema of the equivalent centralized version). The data of the dimension tables is replicated in each node of the cluster (i.e., each dimension has exactly the same rows in all the nodes) and the data of the fact tables is distributed over the fact tables of the several nodes. It is important to emphasize that the replication of dimension tables does not represent a serious overhead because usually the dimensions only represent between 1% and 5% of the space occupied by all database [9]. The DWS technique allows enterprises to build large data warehouses at low cost. DWS can be built using inexpensive hardware and software (e.g., low cost open source database management systems) and still achieve very high performance. In fact, DWS data partitioning for star schemas balances the workload by all computers in the cluster, supporting parallel query processing as well as load balancing for disks and processors. The experimental results presented in [4] show that a DWS cluster can provide an almost linear speedup and scale up.

A major problem faced by DWS is the distribution of data to the cluster nodes. In fact, DWS brings the need for distribution algorithms that conciliate a balanced distribution of data among nodes with an efficient data allocation. Obviously, efficient data allocation is a major challenge as the goal is to place the data in such way that guarantees low and uniform response times from all cluster nodes and, consequently, high performance during the execution of queries.

This paper proposes a generic methodology to evaluate and compare data distribution algorithms in the context of DWS. The approach is based on a set of metrics that characterize the efficiency of the algorithms, considering three key aspects: data distribution time, coefficient of variation of the number of rows placed in each node, and queries response time. The paper studies three alternative data distribution algorithms that can be used in DWS clusters: round-robin, random, and hash-based.

The structure of the paper is as follows: section 2 presents the data distribution algorithms in the context of DWS; section 3 discusses the methodology for the evaluation of data distribution algorithms; section 4 presents the experimental evaluation and Section 5 concludes the paper.

2   Data Distribution in DWS Nodes

In a DWS cluster OLAP (On-Line Analytical Processing) queries are executed in parallel by all the nodes available and the results are merged by the DWS middleware (i.e., middleware that allows client applications to connect to the DWS system without knowing the cluster implementation details). Thus, if a node of the cluster presents a response time higher than the others, all the system is affected, as the final results can only be obtained when all individual results become available.

In a DWS installation, the extraction and transformation steps of the ETL process are similar to the ones performed in typical data warehouses (i.e., DWS does not require any adaptation on these steps). It is in the loading step that the nodes data distribution takes place. Loading the DWS dimensions is a process similar to classical data warehouses; the only difference is that they must be replicated in all nodes available. The key difficulty is that the large fact tables have to be distributed by all nodes.

The loading of the facts data in the DWS nodes occurs in two stages. First, all data is prepared in a DWS Data Staging Area (DSA). This DSA has a data schema equal to the DWS nodes, with one exception: fact tables contain one extra column, which will register the destination node of each facts row. The data in the fact tables is chronologically ordered and the chosen algorithm is executed to determine the destination node of each row in each fact table. In the second stage, the fact rows are effectively copied to the node assigned. Three key algorithms can be considered for data distribution:

- **Random data distribution:** The destination node of each row is randomly assigned. The expected result of such an algorithm is to have an evenly mixed distribution, with a balanced number of rows in each of the nodes but without any sort of data correlation (i.e. no significant clusters of correlated data are expected in a particular node).
- **Round Robin data distribution:** The rows are processed sequentially and a particular predefined number of rows, called a window, is assigned to the first node. After that, the next window of rows is assigned to the second node, and so on. For this algorithm several window sizes can be considered, for example: 1, 10, 100, 1000 and 10000 rows (window sizes used in our experiments). Considering that the data is chronologically ordered from the start, some effects of using different window sizes are expected. For example, for a round-robin using size 1 window, rows end up chronologically scattered between the nodes, and so particular date frames are bound to appear evenly in each node, being the number of rows in each node the most balanced possible. As the size of the window increases, chronological grouping may become significant, and the unbalance of total number of facts rows between the nodes increases.
- **Hash-based data distribution:** In this algorithm, the destination node is computed by applying a hash function [5] over the value of the key attribute (or set of attributes) of each row. The resulting data distribution is somewhat similar to using a random approach, except that this one is reproducible, meaning that each particular row is always assigned to the same node.

3   Evaluating Data Distribution Algorithms

Characterizing data distribution algorithms in the context of DWS requires the use of a set of metrics. These metrics should be easy to understand and be derived directly from experimentation. We believe that data distribution algorithms can be effectively characterized using three key metrics:

- **Data distribution time (DT):** The amount of time (in seconds) a given algorithm requires for distributing a given quantity of data in a cluster with a certain number of nodes. Algorithms should take the minimum time possible for data distribution. This is especially important for periodical data loads that should be very fast in order to make the data available as soon as possible and have a small impact on the data warehouse normal operation.
- **Coefficient of variation of the amount of data stored in each node (CV):** Characterizes the differences in the amount of fact rows stored in each node. CV is the standard deviation divided by the mean (in percentage) and is particularly relevant when homogenous nodes are used and the storage space needs to be efficiently used. It is also very important to achieve uniform response times from all nodes.
- **Queries response time (QT):** Characterizes the efficiency of the data distribution in terms of the performance of the system when executing user queries. A good data distribution algorithm should place the data in such way that allows low response times for the queries issued by the users. As query response time is always determined by the slowest node in the DWS cluster, data distribution algorithms should assure well balanced response times at node level. QT represents the sum of the individual response times of a predefined set of queries (in seconds).

To assess these metrics we need representative data and a realistic set of queries to explore that data. We used the recently proposed TPC Benchmark DS (TPC-DS) [12], as it models a typical decision support system (a multichannel retailer), thus adjusting to the type of systems that are implemented using the DWS technique.

Evaluating the effectiveness of a given data distribution algorithm is thus a four step process:

1. **Define the experimental setup** by selecting the software to be used (in special the DBMS), the number of nodes in the cluster, and the TPC-DS scale factor.
2. **Generate the data** using the "dbgen2" utility (Data Generator) of TPC-DS to generate the data and the "qgen2" utility (Query generator) to transform the query templates into executable SQL for the target DBMS.
3. **Load the data** into the cluster nodes and measure the data distribution time and the coefficient of variation of the amount of data stored in each node. Due to the obvious non-determinism of the data loading process, this step should be executed (i.e., repeated) at least three times. Ideally, to achieve some statistical representativeness it should be executed a much larger number of times; however, as it is a quite heavy step, this may not be practical or even possible. The data distribution time and the CV are calculated as the average of the times and CVs obtained in each execution.
4. **Execute queries** to evaluate the effectiveness of the data placing in terms of the performance of the user queries. TPC-DS queries should be run one at a time and the state of the system should be restarted between consecutive

executions (e.g, by performing a cache flush between executions) to obtain execution times for each query that are independent from the queries run before. Due to the non-determinism of the execution time, each query should be executed at least three times. The response time for a given query is the average of the response times obtained for each of the three individual executions.

4 Experimental Results and Analysis

In this section we present an experimental evaluation of the algorithms discussed in Section 2 using the approach proposed in Section 3.

4.1 Setup and Experiments

The basic platform used consist of six Intel Pentium IV servers with 2Gb of memory, a 120Gb SATA hard disk, and running PostgreSQL 8.2 database engine over the Debian Linux Etch operating system. The following configuration parameters were used for PostgreSQL 8.2 database engine in each of the nodes: 950 Mb for shared_buffers, 50 Mb for work_mem and 700 Mb for effective_cache_size.

The servers were connected through a dedicated fast-Ethernet network. Five of them were used as nodes of the DWS cluster, being the other the coordinating node, which runs the middleware that allows client applications to connect to the system, receives queries from the clients, creates and submits the sub queries to the nodes of the cluster, receives the partial results from the nodes and constructs the final result that is sent to the client application.

Two TPC-DS scaling factors were used, 1 and 10, representing initial data warehouse sizes of 1Gb and 10Gb, respectively. These small factors were used due to the limited characteristics of the cluster used (i.e., very low cost nodes) and the short amount of time available to perform the experiments. However, it is important to emphasize, that even with small datasets it is possible to assess the performance of data distribution algorithms (as we show further on).

4.2 Data Distribution Time

The evaluation of the data distribution algorithms started by generating the facts data in the DWS Data Staging Area (DSA), located in the coordinating node. Afterwards, each algorithm was used to compute the destination node for each facts row. Finally, facts rows were distributed to the corresponding nodes. Table 1 presents the time needed to perform the data distribution using each of the algorithms considered.

As we can see, the algorithm using a hash function to determine the destination node for each row of the fact tables is clearly the less advantageous. For the 1Gb DW, all other algorithms tested took approximately the same time to populate the star schemas in all nodes of the cluster, with a slight advantage to round-robin 100 (although the small difference in the results does not allow us to draw any general conclusions). For the 10 Gb DW, the fastest way

Table 1. Time (in the format hours:minutes:seconds) to copy the replicated dimension tables and to distribute facts data across the five node DWS system

	Distribution time	
Algorithm	1 Gb	10 Gb
Random	0:33:16	6:13:31
Round-robin 1	0:32:09	6:07:15
Round-robin 10	0:32:31	6:12:52
Round-robin 100	0:31:44	6:13:21
Round-robin 1000	0:32:14	6:16:35
Round-robin 10000	0:32:26	6:22:51
Hash-based	0:40:00	10:05:43

to distribute the data was using round-robin 1, with an increasing distribution time as a larger window for round-robin is considered. Nevertheless, round-robin 10000, the slowest approach, took only more 936 seconds than round-robin 1 (the fastest), which represents less than 5% extra time.

4.3 Coefficient of Variation of the Number of Rows

Table 2 displays the coefficient of variation of the number of rows sent to each of the five nodes, for each fact table of the TPC-DS schema.

Table 2. CV(%) of number of rows in the fact tables in each node

Facts table	Random		RR1		RR10		RR100		RR1000		RR10000		Hash-based	
	1Gb	10Gb	1Gb	10Gb	1Gb	10Gb	1Gb	10Gb	1Gb	10Gb	1Gb	10Gb	1Gb	10Gb
c_returns	0,70	0,21	0,00	0,00	0,02	0,00	0,18	0,01	1,21	0,15	8,96	1,51	0,64	0,07
c_sales	0,15	0,04	0,00	0,00	0,00	0,00	0,00	0,00	0,00	0,02	1,55	0,10	0,24	0,07
inventory	0,06	0,02	0,00	0,00	0,00	0,00	0,00	0,00	0,00	0,00	0,10	0,02	0,00	0,00
s_returns	0,18	0,08	0,00	0,00	0,01	0,00	0,08	0,00	0,87	0,04	7,53	0,94	0,22	0,12
s_sales	0,11	0,05	0,00	0,00	0,00	0,00	0,01	0,00	0,01	0,01	0,94	0,06	0,14	0,08
w_returns	0,84	0,18	0,00	0,00	0,03	0,00	0,34	0,03	3,61	0,30	35,73	3,64	0,99	0,20
w_sales	0,35	0,12	0,00	0,00	0,00	0,00	0,02	0,00	0,02	0,00	3,79	0,00	0,15	0,01

For both the data warehouses with 1Gb and 10 Gb, the best equilibrium amongst the different nodes in terms of number of rows in each fact table was achieved using round-robin 1. The results obtained for the random and hash-based distributions were similar, particularly for the 1Gb data warehouse.

The values for the CV are slightly lower for 10Gb than when a 1Gb DSA was used, which would be expected considering that the maximum difference in number of rows was maintained but the total number of rows increased considerably.

As the total number of rows in each fact table increases, the coefficient of variation of the number of rows that is sent to each node decreases. If the number of rows to be distributed is considerably small, a larger window for the round-robin distribution will result in a poorer balance of total number of facts rows among the nodes. Random and hash-based distributions also yield a better equilibrium of total facts rows in each node if the number of facts rows to distribute is larger.

4.4 Queries Response Time

To assess the performance of the DWS system during query execution, 27 queries from the TPC Benchmark DS (TPC-DS) were run. The queries were selected based on their intrinsic characteristics and taking into account the changes needed for the queries to be supported by the PostgreSQL DBMS. Note that, as the goal is to evaluate the data distribution algorithms and not to compare the performance of the system with other systems, the subset of queries used is sufficient. The complete set of TPC-DS queries used in the experiments can be found in [2].

Data Warehouse of 1Gb. Figure 1 shows the results obtained for five of the TPC-DS queries. As we can see, for some queries the execution time is highly dependent on the data distribution algorithm, while for some other queries the execution time seems to be relatively independent from the data distribution algorithm used to populate each node. The execution times for all the queries used in the experiments can be found at [2].

As a first step to understand the results for each query, we analyzed the execution times of the queries in the individual nodes of the cluster. Due to space reasons we only focus on the results of queries 24 and 25. These results are listed in Table 3, along with the mean execution time and the coefficient of variation of the execution times of all nodes.

By comparing the partial execution times for query 25 (see Table 3) to its overall execution time (displayed in Figure 1), it is apparent that the greater the unbalance of each node's execution time, the longer the overall execution time of the query. The opposite, though, is observed for query 24: the distribution with the largest unbalance of the cluster nodes' execution times is also the fastest. In fact, although in this case round-robin 10000 presents one clearly slower node, it is still faster than the slowest node for any of the other distributions, resulting in a faster overall execution time for the query.

The analysis of the execution plan for query 24 showed that the steps that decisively contribute for the total execution time are three distinct index scans (of the indexes on the primary keys of dimension tables *customer*, *customer_address*, and *item*), executed after retrieving the fact rows from table *web_sales* that comply with a given date constraint (year 2000 and quarter of year 2). Also for query 25, the first step of the execution is retrieving the fact rows from table *catalog_returns* that correspond to year 2001 and month 12, after which four index scans are executed (of the indexes on the primary keys of dimension tables *customer*, *customer_address*, *household_demographics*, and *customer_demographics*).

In both cases, the number of eligible rows (i.e., rows from the fact table that comply with the date constraint) determines the number of times each index is scanned. Table 4 depicts the number of rows in table *web_sales* and in table

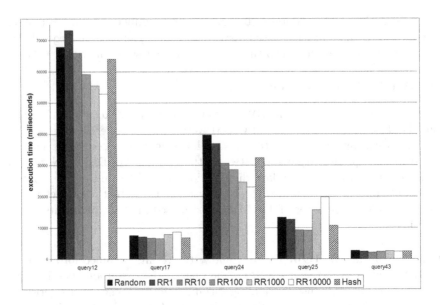

Fig. 1. Execution times for each data distribution of a 1Gb data warehouse

Table 3. Execution times in each node of the cluster (DW of 1Gb)

Query	Node	execution times (ms)						
		Random	RR1	RR10	RR100	RR1000	RR10000	Hash-based
24	1	31391	30893	28702	24617	19761	20893	27881
	2	28089	30743	29730	24812	20284	3314	27465
	3	38539	35741	29296	23301	20202	3077	29500
	4	31288	29704	29683	24794	23530	6533	31595
	5	35783	33625	28733	27765	21144	21976	30782
cv(%)		12,49%	7,72%	1,70%	6,54%	7,19%	85,02%	6,07%
25	1	8336	8519	7775	7426	8293	1798	7603
	2	7073	9094	8338	7794	13763	1457	7109
	3	12349	11620	7523	7885	14584	6011	9022
	4	8882	8428	7175	8117	2927	1533	9732
	5	8782	8666	7561	7457	1881	19034	8621
cv(%)		21,60%	14,47%	5,59%	3,79%	71,22%	126,57%	12,62%

catalog_returns in each node, for each distribution, that correspond to the date constraints being applied for queries 24 and 25.

As we can observe, the coefficient of variation of the number of eligible facts rows in each node increases as we move from round-robin 1 to round-robin 10000, being similar for random and hash-based distributions. This is a consequence of distributing increasingly larger groups of sequential facts rows from a chronologically ordered set of data to the same node: with the increase of the round-robin "window", more facts rows with the same value for the date key will end up in

Table 4. Number of facts rows that comply with the date constraints of queries 24 (table *web_sales*) and 25 (table *catalog_returns*)

Fact table	Node	# of facts rows						
		Random	RR1	RR10	RR100	RR1000	RR10000	Hash-based
web_sales	1	4061	4055	4054	4105	3994	9529	4076
	2	3990	4055	4053	4052	4283	19	3999
	3	4101	4056	4055	4002	3999	7	4044
	4	4042	4056	4053	4002	3998	740	4139
	5	4083	4055	4062	4116	4003	9982	4019
cv(%)		1,06%	0,01%	0,09%	1,34%	3,14%	128,58%	1,35%
catalog_returns	1	489	477	487	499	404	16	483
	2	475	477	470	495	982	6	462
	3	490	477	472	511	960	227	470
	4	468	477	479	457	28	10	484
	5	464	478	478	424	12	2127	487
cv(%)		2,49%	0,09%	1,40%	7,54%	100,03%	194,26%	2,24%

the same node, resulting in an increasingly uneven distribution (in what concerns the values for that key). In this case, whenever the query being run applies a restriction on the date, the number of eligible rows in each node will be dramatically different among the nodes for a round-robin 10000 data distribution (which results in some nodes having to do much more processing to obtain a result than others), but more balanced for random or round-robin 1 or 10 distributions.

Nevertheless, this alone does not account for the results obtained. If that was the case, round-robin 10000 would be the distribution with the poorer performance for both queries 24 and 25, as there would be a significant unbalance of the workload among the nodes, resulting in a longer overall execution time. The data in Table 5 sheds some light on why this data distribution yielded a good performance for query 24, but not for query 25: it displays the average time to perform two different index scans (the index scan on the index of the primary key of the dimension table *customer*, executed while running query 24, and the index scan on the index of the primary key of the dimension table *customer_demographics*, executed while running query 25) as well as the total number of distinct foreign keys (corresponding to distinct rows in the dimension table) present in the queried fact table, in each node of the system, for round-robin 1 and round-robin 10000 distributions.

In both cases, the average time to perform the index scan on the index over the primary key of the dimension table in each of the nodes was very similar for round-robin 1, but quite variable for round-robin 10000. In fact, during the execution of query 24, the index scan on the index over the primary key of the table *customer* was quite fast in nodes 1 and 5 for the round-robin 10000 distribution and, in spite of having the largest number of eligible rows in those nodes, they ended up executing faster than all the nodes for the round-robin 1 distribution. Although there seems to be some preparation time for the execution of an index scan, independently of the number of rows that are afterwards looked

Table 5. Average time to perform an index scan on dimension table *customer* (query 24) and on dimension table *customer_demographics* (query 25)

Query	Algorithm	Node	execution time (ms)	index scan on dimension		total # of diff. values of foreign key in f. table
				avg time (ms)	# of times perf.	
24	round-robin 1	1	30893	3.027	4055	42475
		2	30743	3.074	4055	42518
		3	35741	3.696	4056	42414
		4	29704	2.858	4056	42458
		5	33625	3.363	4055	42419
	round-robin 10000	1	20893	0.777	9529	12766
		2	3314	17.000	19	12953
		3	3077	19.370	7	12422
		4	6533	1.596	740	12280
		5	21976	0.725	9982	12447
25	round-robin 1	1	8519	0.019	38	28006
		2	9094	0.019	32	27983
		3	11620	0.019	22	28035
		4	8428	0.019	32	28034
		5	8666	0.017	39	28035
	round-robin 10000	1	1798	66.523	1	29231
		2	1457	73.488	1	29077
		3	6011	34.751	2	29163
		4	1533	-	0	29068
		5	19034	19.696	146	23454

for in the index (which accounts for the higher average time for nodes 2, 3 and 4), carefully looking at the data on Table 5 allows us to conclude that the time needed to perform the index scan in the different nodes decreases when the number of distinct primary key values of the dimension that are present in the fact table scanned also decreases.

This way, the relation between the number of distinct values for the foreign keys and the execution time in each node seams to be quite clear: the less distinct keys there are to look for in the indexes, the shorter is the execution time of the query in the node (mostly because the less distinct rows of the dimension that need to be looked for, the less pages need to be fetched from disk, which dramatically lowers I/O time). This explains why query 24 runs faster in a round-robin 10000 data distribution: each node had fewer distinct values of the foreign key in the queried fact table. For query 25, as the total different values of foreign key in the queried fact table in each node was very similar, the predominant effect was the unbalance of eligible rows, and round-robin 10000 data distribution resulted in a poorer performance.

These results ended up revealing an crucial aspect: some amount of clustering of fact tables rows, concerning each of the foreign keys, seems to result in an improvement of performance (as happened for query 24), but too much clustering, when specific filters are applied to the values of that keys, result in a decrease of performance (as happened for query 25).

Data Warehouse of 10Gb. The same kind of results were obtained for a DWS system equivalent to a 10Gb data warehouse, and the 3 previously identified behaviours were also found: queries whose execution times do not depend on the distribution, queries that run faster on round-robin 10000, and queries that run faster on the random distribution (and consistently slower on round-robin 10000). In this case, as the amount of data was significantly higher, the random distribution caused better spreading of the data than the round-robin 10 and 100 caused in the 1Gb distribution. But even though the best distribution was not the same, the reason for it is similar: eligible rows for queries were better distributed among the nodes and lower number of distinct primary keys values of the dimension on the fact tables determined the differences.

5 Conclusion and Future Work

This work analyzes three data distribution algorithms for the loading of the nodes of a data warehouse using the DWS technique: random, round-robin and a hash-based algorithm. Overall, the most important aspects we were able to draw from the experiments were concerning two values: 1) the number of distinct values of a particular dimension within a queried fact table and 2) the number of rows that are retrieved after applying a particular filter in each node.

As a way to understand these aspects, consider, for instance, the existence of a data warehouse with a single fact table and a single dimension, constituted by 10000 facts corresponding to 100 different dimension values (100 rows for each dimension value). Consider, also, that we have the data ordered by the dimension column and that there are 5 nodes. There are two opposing distributions possible, which distribute evenly the rows among the five nodes (resulting 2000 rows in each node): a typical round-robin 1 distribution that copies one row to each node at a time, and a simpler one that copies the first 2000 rows to the first node, the next 2000 to the second, and so on.

In the first case, all 100 different dimension values end up in the fact table of each node, while, in the second case, the 2000 rows in each node have only 20 of the distinct dimension values. As consequence, a query execution on the first distribution may imply the loading of 100% of the dimension table in all of the nodes, while on the second distribution a maximum of 20% of the dimension table will have to be loaded in each node, because each node has only 20% of all the possible distinct values of the dimension.

If the query run retrieves a large number of rows, regardless of their location on the nodes, the second distribution would result in a better performance, as fewer dimension rows would need to be read and processed in each node. On the other hand, if the query has a very restrictive filter, selecting only a few different values of the dimension, then the first distribution will yield a better execution time, because these different values will be more evenly distributed among the nodes, resulting in a more distributed processing time, thus lowering the overall execution time for the query.

The aforementioned effects suggest an optimal solution to the problem of the loading of the DWS. As a first step, this loading algorithm would classify all the dimensions in the data warehouse as large dimensions and small dimensions. Exactly how this classification would be done depends on the business considered (i.e., on the queries performed) and must also account the fact that this classification might be affected by subsequent data loadings. The effective loading of the nodes must then consider complementary effects: it should minimize the number of distinct keys of any large dimension in the fact tables of each node, minimizing the disk reading on the nodes and, at the same time, it should try to split correlated rows among the nodes, avoiding that eligible rows of typical filters used in the queries end up grouped in a few of them.

However, to accomplish that, it appears to be impossible to decide beforehand a specific loading strategy to use without taking the business into consideration. The suggestion here would be to analyze the types of queries and filters mostly used in order to decide what would be the best solution for each case.

References

1. Agosta, L.: Data Warehousing Lessons Learned: SMP or MPP for Data Warehousing. DM Review Magazine (2002)
2. Almeida, R., Vieira, M.: Selected TPC-DS queries and execution times, http://eden.dei.uc.pt/~mvieira/
3. Bernardino, J., Madeira, H.: A New Technique to Speedup Queries in Data Warehousing. In: Symp. on Advances in DB and Information Systems, Prague (2001)
4. Bernardino, J., Madeira, H.: Experimental Evaluation of a New Distributed Partitioning Technique for Data Warehouses. In: International Symp. on Database Engineering and Applications, IDEAS 2001, Grenoble, France (2001)
5. Jenkins, B.: "Hash Functions", "Algorithm Alley". Dr. Dobb's Journal (September 1997)
6. Critical Software SA, "DWS", www.criticalsoftware.com
7. DATAllegro, "DATAllegro v3", www.datallegro.com
8. ExtenDB, ExtenDB Parallel Server for Data Warehousing, http://www.extendb.com
9. Kimball, R., Ross, M.: The Data Warehouse Toolkit: The Complete Guide to Dimensional Modeling, 2nd edn. J. Wiley & Sons, Inc., Chichester (2002)
10. Netezza: The Netezza Performance Server DW Appliance, http://www.netezza.com
11. Sun Microsystems, Data Warehousing Performance with SMP and MPP Architectures, White Paper (1998)
12. Transaction Processing Performance Council, TPC BenchmarkTM DS (Decision Support) Standard Specification, Draft Version 32 (2007), http://www.tpc.org/tpcds

Data Partitioning in Data Warehouses: Hardness Study, Heuristics and ORACLE Validation

Ladjel Bellatreche, Kamel Boukhalfa, and Pascal Richard

Poitiers University - LISI/ENSMA France
{bellatreche,boukhalk,richardp}@ensma.fr

Abstract. Horizontal data partitioning is a non redundant optimization technique used in designing data warehouses. Most of today's commercial database systems offer native data definition language support for defining horizontal partitions of a table. Two types of horizontal partitioning are available: primary and derived horizontal fragmentations. In the first type, a table is decomposed into a set of fragments based on its own attributes, whereas in the second type, a table is fragmented based on partitioning schemes of other tables. In this paper, we first show hardness to select an optimal partitioning schema of a relational data warehouse. Due to its high complexity, we develop a hill climbing algorithm to select a near optimal solution. Finally, we conduct extensive experimental studies to compare the proposed algorithm with the existing ones using a mathematical cost model. The generated fragmentation schemes by these algorithms are validated on Oracle 10g using data set of APB1 benchmark.

1 Introduction

Data warehouse applications store large amounts of data from various sources. Usually, these applications are modeled using relational schemes like star and snow flake schemes. A star schema is usually queried in various combinations involving fact and dimension tables. These queries are called star join queries characterized by: (i) the presence of *restriction operations* on dimension tables and (ii) the fact table participates in every join operation. Note that joins are well known to be expensive operations, especially when the relations involved are substantially larger than main memory which is usually the case of data warehouse applications.

Horizontal partitioning (HP) has been largely used in designing distributed and parallel databases in last decades [1]. Recently, it becomes a crucial part of physical database design [2, 3, 4], where most of today's commercial database systems offer native data definition language support for defining horizontal partitions of a table [2]. HP allows tables, materialized views and indexes to be partitioned into disjoint sets of rows physically stored and usually accessed separately. It does not replicate data, thereby reducing storage requirement and minimizing maintenance overhead. There are two versions of HP [1]: *primary* and *derived*. Primary HP of a relation is performed using attributes defined on

I.-Y. Song, J. Eder, and T.M. Nguyen (Eds.): DaWaK 2008, LNCS 5182, pp. 87–96, 2008.

that relation. Derived HP, on the other hand, is the fragmentation of a relation using attribute(s) defined on another relation(s). In other words, derived HP of a table is based on the fragmentation schema of another table (the fragmentation schema is the result of the partitioning process of a given table). The derived partitioning of a table R according a fragmentation schema of S is feasible if and only if there is a join link between R and S (R contains a foreigner key of S).

In the context of relational data warehouses, we have proposed a methodology to partition a star schema [5] that we can summarize as follows: (1) partition some/all dimension tables using the primary HP and then (2) partition the facts table using the fragmentation schemes of the fragmented dimension tables. This methodology may generate an important number of horizontal fragments of the fact table, denoted by N: $N = \prod_{i=1}^{g} m_i$, where m_i and g are the number of fragments of the dimension table D_i and the number of dimension tables participating in the fragmentation process, respectively.

To avoid the explosion of this number, we have formalized the problem of selecting horizontal partitioning schema as an optimization problem: Given a representative workload Q defined on a data warehouse schema $\{D_1, ..., D_n, F\}$, and a constraint (maintenance bound W) representing the number of fact fragments[1], identify dimension tables that can be used to derived horizontal partition the fact table F, such that $\sum_{Q_i \in Q} f_{Q_i} \times Cost(Q_i)$ is minimized and $N \leq W$.

In this paper, we propose a proof of NP-hardness of this problem and novel algorithm for selecting partitioning schemes of dimension and fact tables.

This paper is divided in 7 sections: Section 2 gives a deep state of the art on the problem of selecting HP schema of a relational database. Section 3 gives different concepts to understand HP selection problem. Section 4 presents the hardness proof of our problem. Section 5 proposes a hill climbing algorithm for selecting near optimal schema. Section 6 gives the experimental results using an adaptation of the APB-1 benchmark and a validation of our findings on Oracle10g. Section 7 concludes the paper by summarizing the main results and suggests future work.

2   Related Work

HP has been widely studied in traditional databases [6], distributed and parallel databases [1, 7], where several algorithms for selecting optimal and near optimal fragmentation schema were proposed. Many academic and industrial studies showed its importance for physical design phase [2, 3, 4]. Recently, most of commercial systems advocate data partitioning, where a spectacular evolution in commercial DBMSs is identified. To show this evolution, we focus on Oracle DBMS, since it offers several partitioning modes.

The first horizontal partitioning mode supported by Oracle was *Range* partitioning (in Oracle 8). Oracle 9 and 9i added others modes like *Hash* and *List*

---

[1] The number can be given by the database administrator (DBA).

and *Composite* (Range-Hash, Range-List). Hash mode decomposes the data according to a hash function (provided by the system) applied to the values of the partitioning columns. List partitioning splits a table according list values of a column. Composite partitioning is supposed by using PARTITION - SUB-PARTITION statement [2]. Recently, Oracle 11g offers a great evolution of HP, where several fragmentation methods were supported. For instance, the Composite partitioning methods have been enriched to include all possible combinations of basic methods (Range, Hash and List): List-List, Range-Range, List-Range, List-Hash, etc. Two other interesting modes were supported: (1) *Column partitioning*, where a table is decomposed using a virtual attribute defined by an expression, using one or more existing columns of a table, and storing this expression as meta data only and (2) *Referential partitioning* that allows to partition a table by leveraging an existing parent-child relationship [4]. This partitioning is similar to derived horizontal partitioning. Unfortunately, a table may be partitioned using method using *only one table*. This evolution shows the important role of HP in designing databases and data warehouses. It also motivates us to study in deep this problem.

By exploring the main studies done by academic and industrial communities, we figure out that most of approaches for selecting horizontal partitioning schema of a database suppose a decomposition of domain values of attributes participating in the fragmentation process. These approaches may be classified into two main categories: (1) *user-driven approaches* and (2) *query-driven approaches*. In the first category, user decomposes domain values of each partitioning attribute based on his/her knowledge of applications (queries) and *a priori imposes* the number of generated horizontal fragments. The main drawbacks of this category are: (i) absence of metric guarantying the efficiency of the obtained partitioning schema, (ii) difficulty to choose partitioning attributes among attributes candidate for the fragmentation process and (iii) difficulty of identifying the best manner in decomposing each domain attribute. In query-driven partitioning, the domain values of fragmentation attributes are splitted based on queries defined on the database schema. In this category, several algorithms were proposed in traditional databases that we can classify into three classes: *predicate-based approach* [1, 8], *affinity-based approach* [9] and *cost-based approach* [6]. The main drawback of this category is that users do not control the number of generated fragments. Based on this analysis, developing HP algorithms reducing query processing cost and offering users the possibility to set the number of generated fragments becomes a crucial performance issue that should be addressed.

3   Background

As for redundant structure (materialized views, indexes) selection problems, HP schema selection is done based a set of most frequently queries, where each query an access frequency. Each query Q_j has a set of selection predicates SP^j, where

---

[2] These modes are also supported by other commercial databases like SQL Server, Sybase, DB2, etc.

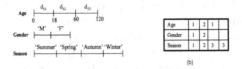

Fig. 1. (a) An Example of a Decomposition of Domain and (b) its Coding

each predicate $p_i^{D_k}$ is defined by $A_i^k \; \theta \; Value$, such as: A_i^k is an attribute of dimension table D_k, $\theta \in \{=, <, >, \le, \ge\}$, and Value $\in Domain(A_i^k)$ [8]. These predicates are essential for the HP process. To partition dimension tables, we first decompose domains of their fragmentation attributes into sub domains. To illustrate this approach, we consider three fragmentation attributes: *Age* and *Gender* of dimension table *CUSTOMER* and *Season* of dimension table *TIME*, where each one has a domain:

Dom(Age) =]0, 120], Dom(Gender)={'M', 'F'} and Dom(Season)={ "Summer", "Spring", "Autumn", "Winter"}. Each domain is partitioned by a user into sub domains as follows: Dom(Age) $= d_{11} \cup d_{12} \cup d_{13}$, with $d_{11} =$]0, 18], $d_{12} =$]18, 60[, $d_{13} =$ [60, 120]. Dom(Gender) $= d_{21} \cup d_{22}$, with $d_{21} = \{$'M'$\}$, $d_{22} = \{$'F'$\}$. Dom(Season) $= d_{31} \cup d_{32} \cup d_{33} \cup d_{34}$, where $d_{31} = \{$ "Summer"$\}$, $d_{32} = \{$ "Spring"$\}$, $d_{33} = \{$ "Autumn"$\}$, and $d_{34} = \{$ "Winter"$\}$. These sub domains are given in Figure 1a.

The decomposition of fragmentation attributes domains may be represented by multidimensional arrays, where each array represents the domain partitioning of a fragmentation attribute. The value of each cell of a given array representing an attribute $A_i^{D_k}$ belongs to $[1..n_i]$, where n_i represents the number of sub domains of attribute $A_i^{D_k}$. This multidimensional representation gives the partitioning schema of each dimension table. Fragmentation schemes depend on the values of cells: (1) If all cells of a given attribute have the different values this means that all sub domains will be considered in partitioning process of corresponding dimension table. (2) If all cells for a given attribute have the same value this means that the attribute will not participate in the fragmentation process. (3) If some cells of a given attribute have the same value then their corresponding sub domains will be merged into one. Figure 1b gives an example of coding of a fragmentation schema based on three attributes Gender, Season and Age. Once dimension tables are fragmented, we can easily derived partition the fact table.

Our coding may suffer from multi-instantiation. To solve this problem, we use Restricted Growth Functions [10].

4   Hardness Study

We consider the HP of data warehouse through a simplified decision problem that considers the derived HP of the fact table based on partitioning schema of one dimension table using only one attribute. The corresponding optimization problem consists in computing a partition of the fact table so that the number of partitions is bounded by a constant B and the maximum number of Input/Output operations is minimized. We state the decision problem as follows:

Problem: One-Domain Horizontal Partitioning

- **Instance:**
  - a set D of disjoint sub domains $\{d_1, \cdots, d_n\}$ of the fragmentation attribute of the partitioned dimension table and the number of Input/Output operations in order to read data corresponding to the sub domain d_i in the fact table, denoted $l(d_i), 1 \leq i \leq n$.
  - a set of queries $q_1, \cdots, q_m$ and for each query q_j the list $f(q_j) \subseteq D$ of used sub domains until the query completion: $\{d_{j1}, \cdots, d_{jn_j}\}$, where n_j is the number of sub domains used in the fact table to run q_j.
  - two positive integers K and L, where K is the maximum number of partitions that can be created and L is the maximum number of Input/Output operations allowed for each query, $L \geq \sum_{d \in f(q)} l(d)$.

Question: Can D be partitioned in at most K subsets, $D_1, \cdots, D_K$ such that every query requires at most L Input/Output operations.

The optimal number of Input/Output operations required by a query q is: $l(q) = \sum_{d \in f(q)} l(d)$. It assumes that only required data are loaded in memory to run q. According to a given partition, the number of Input/Output operations increases since all data of a partition are loaded when used by a given query, even if that query do not requires all data of the partition (i.e., a subset of domains in the partition). Thus, the number of Inputs Outputs operations required by a query after partitioning do not depend on the used sub domains but only on used partitions. The number of Input/Output operations while loading a partition D is defined by: $l(D) = \sum_{d \in D} l(d)$. As a consequence the number of Input/Output operations required by running a query can be defined as: $l(q) = \sum_{D \in F(q)} l(D)$, where $F(q)$ is the list of partitions used by a query q.

The objective is to perform a derived HP of the fact table such that the number of partitions is limited to K and the number of Input/Output operations is bounded by L for every query. Obviously, if $K \geq n$, the optimal HP is achieved by defining exactly one partition to every $d_i \in D$. In that way, every query only loads required data during its execution. We shall see that our simplified decision problem becomes hard when $K < n$. We also assume $L \geq \sum_{d \in f(q)} l(d)$ since otherwise the answer of the One-Domain Horizontal Partitioning is always *false*.

4.1 NP-Completeness

Theorem 1. *One-Domain HP is NP-Complete in the strong sense.*

Proof: One-Domain HP clearly belongs to NP since if one guesses a partition of D, then a polynomial time algorithm can check that at most K partitions are used and that every query requires at most L Input/Output operations. We now prove that One-Domain HP is NP-Complete in the strong sense. We shall use 3-Partition that is strongly NP-Complete [11].

Problem: 3-Partition
Instance: Set A of $3m$ elements, a bound $B \in Z^+$, and a size $s(a) \in Z^+$ for each $a \in A$ such that $B/4 < s(a) < B/2$ and such that $\sum_{a \in A} s(a) = mB$.
Question: Can A be partitioned into m disjoint sets $A_1, \cdots, A_m$ such that, for $1 \leq i \leq m$, $\sum_{a \in A_i} s(a) = B$ (note that each A_i must therefore contain exactly three elements from A)?

To prove the NP-Completeness of One-Domain HP, we reduce from 3-Partition. To every 3-Partition instance, an instance of One-Domain HP is defined as follows: (i) to every $a_i \in A$, a subdomain d_i is created so that $l(d_i) = s(a_i), 1 \leq i \leq 3m$, (ii) $3m$ queries are created such that every query uses exactly one subdomain: $f(q_i) = \{d_i\}, 1 \leq i \leq 3m$ and $K = L = B$.

Clearly the transformation is performed in polynomial time since it consists in a one-to-one mapping of 3-partition elements into sub domains and queries. We now prove that we have a solution to the 3-partition instance if, and only if, we have a solution to the One-Domain HP instance.

(Necessary condition) Assume that we have a solution of the One-Domain HP, then it satisfies the following conditions:

• Since $B/4 < l(d) < B/2$, every subset of D will be define with exactly 3 sub domains (as in every 3-Partition instance).
• Since we have a feasible solution of the One-Domain Horizontal Partitioning, then no query requires more than B Input/Output operations. By construction we verify that: $\sum_{d \in D} l(d) = mB$. As a consequence, every query requires exactly B Input/Output in the fact tables (otherwise it is not a solution). Using a one-to-one mapping of sub domains into elements of 3-Partition, a feasible solution of the 3-partition instance is obtained.

(Sufficient condition) Assume that we have a solution to the 3-Partition instance. Then, every subset A_i as a total size of B and is composed of exactly 3 elements of A. Starting from A_1, we define a sub domain partition using subdomains with same indexes of elements belonging to A_1. Since every query is associated to exactly one sub domain and three subdomains are grouped in every partition, then exactly three queries use a given partition. As a consequence, the number of Input/Output associated to these 3 corresponding queries is exactly B. Repeating this process for every remaining subset A_i, then a feasible solution of the One-Dimension HP Problem is obtained.

5   Hill Climbing Heuristic

In this section, we propose a hill climbing heuristic for selecting a HP schema. It consists of the following two steps: (1) find an initial solution and (2) iteratively improve the initial HP solution by using hill climbing heuristics until no further reduction in total query processing time can be achieved and the maintenance bound B is satisfied. This is done by applying a couple of operations on the initial data partitioning schema.

The initial solution may be generated by assigning cell values randomly. In this study, we avoid the random generation. We adapt our affinity algorithm

[6]. It consists in grouping the most frequently used sub domains by assigning them the same value. In the second step, we apply two operations on the initial solution to reduce the total query processing cost, namely, *Merge* and *Split*. The merge function defined as $Merge(SD_i^k, SD_j^k, A_k, FS)$ combines two sub-domains SD_i^k and SD_j^k of an the attribute A_k of the multidimensional representation of a partitioning schema FS into one sub domain. Therefore, there can be a maximum $(\frac{n_i \times (n_i - 1)}{2})$ merge operations that may be applied during each iteration. This operation is used when the number of generated fragments is greater than the maintenance constraint B.

The split function defined as $Split(SD_i^k, A_k, FS)$ is the dual of the merge function. It splits the sub-domain SD_i^k (already merged) of the attribute A_k into two sub-domains.

The application of these operations is controlled by a cost model computing the number of inputs and outputs required for executing a set of queries on the selected HP schema. A particularity of this model is that it considers buffer size when executing a query. For lack of space it cannot be presented in this paper. For more details, refer to our technical report at: http://wwww.lisi.ensma.fr/.

6   Experimental Studies and ORACLE Validation

We have conducted many experimental studies to compare the hill climbing algorithm (HC) with genetic (GA) and simulated annealing (SA) algorithms. Before conducting these studies, we have modified GA and SA proposed in [5] by incorporating Restricted Growth Functions to avoid multi-instantiation. This comparison is done using a mathematical cost model.

Dataset: We use the dataset from the APB1 benchmark [12]. The star schema of this benchmark has one fact table Actvars and four dimension tables: Actvars(24 786 000 tuples), Prodlevel (9 000 tuples), Custlevel (900 tuples), Timelevel (24 tuples) and Chanlevel (9 tuples).

Workload: We have considered a workload of 60 single block queries (i.e., no nested subqueries) with 40 selection predicates defined on 12 different attributes (ClassLevel, GroupLevel, FamilyLevel, LineLevel, DivisionLevel, YearLevel, MonthLevel, QuarterLevel, RetailerLevel, CityLevel, GenderLevel, AllLevel). The domains of these attributes are split into 4, 2, 5, 2, 4, 2, 12, 4, 4, 4, 2 et 5 sub domains, respectively. Our algorithms have been implemented using Visual C++ performed under an Intel Centrino with a memory of 1 Gb and ORACLE 10g.

We have conducted many experimental studies to set up different parameters for each heuristic: number of generations, mutation rate, cross over rate for GA, temperature for SA). Due to the stochastic nature of our heuristics, we have executed several times each heuristic we consider the average.

Figure 2 compares the performance of the proposed algorithms: GA, hill climbing (HC) and SA by varying the threshold from 100 to 500 and using 40 selection predicates. SA outperforms GA and HC. HC explores only one solution and may suffer from optimum local solution phenomena. Increasing the threshold increases

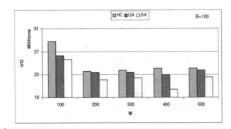
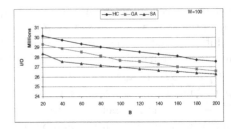

Fig. 2. Effect of threshold on IOs   **Fig. 3.** Effect of Buffer Size on IOs

performance of HP, since more attributes may participate in the fragmentation process.

In Figure 3, we study the effect of buffer size on performance of queries. To do so, we vary the buffer size from 20 to 200 pages and we execute the three algorithms for each value. This experiment shows the impact of buffer on query performance, especially, when we increase its size.

To show the effect of the number of predicates on the total performance, we vary the number of predicates contained in the 60 queries. We create four classes of queries, where each class has a different number of predicates (10, 20, 30 and 40). The three algorithms are executed from each class, where the threshold is equal 100 (W=100). The results show that the number of predicates used in the workload influences the total performance. When the number is small, several queries do not get benefit from the partitioning, since they access a large number of fragments. In this case, several union operations are performed to generate the final result (Figure 4). When the number of predicates increases, the number of relevant fragments decreases, especially if these predicates are defined on fragmentation attributes.

Figure 5 gives the running time required by each heuristic, where the threshold value is set to 100 and 40 selection predicates are used. GA and SA are time consuming compare to HC which uses simple operations (merge and split). If DBA privileges the quality of the obtained fragmentation schema s/he may choose SA. In the case, where the execution time of the partitioning algorithm is the most important criteria, s/he may choose HC.

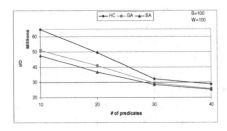
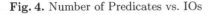
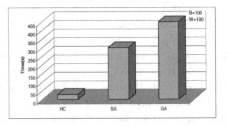

Fig. 4. Number of Predicates vs. IOs   **Fig. 5.** Execution Time per Algorithm

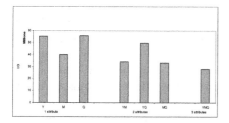

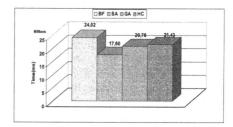

Fig. 6. The Choice of Attributes   **Fig. 7.** Validation on ORACLE10G

In Figure 6, we study the effect of choosing attributes of dimension tables for partitioning the fact table. In this experiment, we consider the dimension table TimeLevel. We have considered all possible combinations of the three attributes (Monthlevel(M), Yearlevel(Y) and Quarterlevel(Q). This experiment shows the importance of the attribute Monthlevel in partitioning of TimeLevel. Whenever it is used an interesting reduction of query processing cost is identified. This is because it is mostly used by the queries. But when all attributes are used a significant reduction is also identified since our queries use simultaneously these attributes in their selection predicates.

In order to identify the real quality of three algorithms, we conducted an implementation of the generated solutions obtained our algorithms in ORA-CLE10G. We have considered the following scenario: GA and SA generate a fragmentation schema with 80 sub star schemas. GA uses 5 attributes and dimension tables (Timelevel, Custlevel and Chanlevel) to decompose the warehouse. SA uses 5 attributs and all dimension tables. HC generates 96 sub schemas using 4 attributes and 3 dimension tables (Prodlevel, Timelevel et Chanlevel). But to do this experiments, we are faced to a serious problem: (1) no support of multiple partitioning: most of DBMSs support composite partitioning (range-hash and range-list) using partition and sub partition statements and since our algorithms may generate partitioning schemas with many attributes and (2) no support of derived HP when several tables are used. To deal with these two problems, we proposed an implementation support for multiple partitioning and derived HP. A rewriting program that takes any global query (defined on no partitioned schema) and generates another query rewritten on fragments. We execute our queries in each fragmented schema and we calculate the cost of executing each query using ORACLE tool. Figure 7 shows the following results: (i) HP is crucial for reducing query performance and (ii) the choice of fragmentation algorithm is important. SA outperforms GA and HC which confirms the theoretical results.

7 Conclusion

HP has been largely studied by academic community and advocated by most of commercial database systems, where they offer native data definition language support for defining horizontal partitions of a table using several modes. In this paper, we show the need of HP in data warehouse. We derive the complexity of problem of selecting an optimal partitioning schema and we study its hardness. To the best of our knowledge, we are the first to study in details this complexity. We propose a hill climbing algorithm for selecting near optimal HP schema. We present an extensive experimental evaluation of three HP algorithms: GA, SA and HC using mathematical cost model and Oracle10g. We are working on the problem of selecting a mixed fragmentation schema, since actual data warehouse applications manage star schemas with huge dimension tables.

References

1. Özsu, M.T., Valduriez, P.: Principles of Distributed Database Systems, 2nd edn. Prentice Hall, Englewood Cliffs (1999)
2. Sanjay, A., Narasayya, V.R., Yang, B.: Integrating vertical and horizontal partitioning into automated physical database design. In: Sigmod 2004, pp. 359–370 (June 2004)
3. Papadomanolakis, S., Ailamaki, A.: AutoPart: Automating Schema Design for Large Scientific Databases Using Data Partitioning. In: Proceedings of the 16th International Conference on Scientific and Statistical Database Management (SS-DBM 2004), June 2004, pp. 383–392 (2004)
4. Eadon, G., Chong, E.I., Shankar, S., Raghavan, A., Srinivasan, J., Das, S.: Supporting table partitioning by reference in oracle. In: SIGMOD 2008 (to appear, 2008)
5. Bellatreche, L., Boukhalfa, K., Abdalla, H.I.: Saga: A combination of genetic and simulated annealing algorithms for physical data warehouse design. In: 23rd British National Conference on Databases, July 2006, pp. 212–219 (2006)
6. Bellatreche, L., Karlapalem, K., Simonet, A.: Algorithms and support for horizontal class partitioning in object-oriented databases. The Distributed and Parallel Databases Journal 8(2), 155–179 (2000)
7. Navathe, S., Karlapalem, K., Ra, M.: A mixed partitioning methodology for distributed database design. Journal of Computer and Software Engineering 3(4), 395–426 (1995)
8. Ceri, S., Negri, M., Pelagatti, G.: Horizontal data partitioning in database design. In: Proceedings of the ACM SIGMOD International Conference on Management of Data. SIGPLAN Notices, pp. 128–136 (1982)
9. Karlapalem, K., Navathe, S.B., Ammar, M.: Optimal redesign policies to support dynamic processing of applications on a distributed database system. Information Systems 21(4), 353–367 (1996)
10. Tucker, A., Crampton, J., Swift, S.: Rgfga: An efficient representation and crossover for grouping genetic algorithms. Evol. Comput. 13(4), 477–499 (2005)
11. Garey, M.R., Johnson, D.S.: Computers and Intractability; A Guide to the Theory of NP-Completeness. W. H. Freeman & Co., New York (1990)
12. OLAP Council, Apb-1 olap benchmark, release ii (1998), http://www.olapcouncil.org/research/resrchly.htm

A Robust Sampling-Based Framework
for Privacy Preserving OLAP

Alfredo Cuzzocrea, Vincenzo Russo, and Domenico Saccà

ICAR Institute and University of Calabria
I-87036 Cosenza, Italy
{cuzzocrea,russo,sacca}@si.deis.unical.it

Abstract. A robust sampling-based framework for privacy preserving OLAP is introduced and experimentally assessed in this paper. The most distinctive characteristic of the proposed framework consists in adopting an innovative *privacy OLAP notion*, which deals with the problem of preserving the privacy of OLAP aggregations rather than the one of data cube cells, like in conventional perturbation-based privacy preserving OLAP techniques. This results in a greater theoretical soundness, and lower computational overheads due to processing massive-in-size data cubes. Also, the performance of our privacy preserving OLAP technique is compared with the one of the method *Zero-Sum*, the state-of-the-art privacy preserving OLAP perturbation-based technique, under several perspectives of analysis. The derived experimental results confirm to us the benefits deriving from adopting our proposed framework for the goal of preserving the privacy of OLAP data cubes.

1 Introduction

The problem of ensuring the *privacy* of OLAP data cubes [16] arises in several fields ranging from advanced *Data Warehousing* (DW) and *Business Intelligence* (BI) systems to sophisticated *Data Mining* (DM) tools. In DW and BI systems, decision making analysts wish to avoid that *malicious users* access perceptive ranges of multidimensional data in order to infer *sensitive knowledge* by means of *inference techniques* [31,32], or *attack* corporate data cubes via violating user roles, grants and revokes. In DM tools, domain experts wish to avoid that malicious users infer *critical-for-the-task knowledge* from authoritative DM results such as frequent item sets, patterns and regularities, clusters, discovered association rules, and mining models over multidimensional data cubes. In both application scenarios, *privacy preserving OLAP issues* are significant research challenges that are more and more attracting the attention of large communities of academic and industrial researchers. This claim is also confirmed by the large number of papers focusing on privacy preserving OLAP appeared in literature recently (e.g., [3,18,20,21,22,24,29,31,32,34]).

Privacy preservation of data cubes refers to the problem of *ensuring the privacy of data cube cells*, and, in turn, the one of queries defined over collections of cells [29]. The goal of such an approach is to hide sensitive information and knowledge during data management activities, according to the general guidelines drawn by Sweeney in

I.-Y. Song, J. Eder, and T.M. Nguyen (Eds.): DaWaK 2008, LNCS 5182, pp. 97–114, 2008.
© Springer-Verlag Berlin Heidelberg 2008

her seminal paper [30] that focuses on the privacy of online-published relational databases. During last years, this problem has indeed became of great interest for the Data Warehousing and Databases research communities, due to its exciting theoretical challenges as well as its relevance and practical impact in modern real-life OLAP systems and applications. On a more conceptual plane, theoretical aspects are mainly devoted to study how *probability* and *statistics schemes* as well as rule-based models can be applied in order to obtain *privacy preserving data cubes*. On a more practical plane, researchers and practitioners aim at integrating convenient privacy preserving solutions within the core layers of commercial OLAP server platforms.

Basically, in order to tackle privacy preservation challenges in OLAP, researchers have proposed models and algorithms within a broader context encompassing two main classes: *restriction-based techniques*, and *perturbation-based techniques*. First ones propose limiting the number of query kinds that can be posed against the target OLAP server. Second ones propose perturbing data cells by adding random noise at various levels of the target database, ranging from schemas, like in [26], to query answers, like in [4]. As we motivate in Sect. 2, while both these techniques have proved to be reliable in the context of privacy preserving databases, they are quite inadequate and inefficient when OLAP data cubes are considered.

Access control techniques for data cubes are another research challenge related to the privacy preserving OLAP scientific field. Basically, these techniques refer to the problem of ensuring the *security* of data cube cells, i.e. *restricting the access of unauthorized users to specific sub-domains of the target data cube*, according to well-known concepts studied and assessed in the context of relational database systems security (e.g., [17,21,25]). Although this exciting topic is orthogonal to ours, in this research we focus the attention on privacy preserving OLAP issues, and remand the reader to active literature on access control issues in OLAP.

Beyond effectiveness and efficiency limitations, actual proposals also lack of a rigorous theoretical foundation, as they do not consider any so-called *privacy OLAP notion*, which is a novel concept we introduce in this paper. In other words, while these proposals focus the attention on the privacy of data cells, they completely neglect to introduce a rigorous notion of privacy in their research. It is a matter of fact to note that, contrary to the trend of actual privacy preserving OLAP proposals, similar initiatives in the context of privacy preserving Databases [30] and Data Mining [2] instead introduce a proper notion of privacy for their data processing goals, i.e. *they formally define what privacy means in their research*. A positive side-effect of such an approach is represented by the amenity of devising properties and theorems on top of the theoretical framework founding on the privacy notion. According to our vision, the lack of a proper privacy OLAP notion is a relevant limitation for privacy preserving OLAP research, which we fulfill in this work.

Starting from these considerations, in this paper we present an innovative proposal that addresses the privacy preserving OLAP problem from a completely-original perspective, which improves the results achieved in past research efforts. We propose a *robust sampling-based framework for computing privacy preserving data cubes at a provable computational cost*. Contrary to state-of-the-art initiatives, in our framework we introduce a meaningful privacy OLAP notion that considers the privacy of OLAP aggregates, and, by adopting this notion as theoretical baseline, we devise a theoretical framework that allows us to nicely treat privacy preservation of data cubes. In particular, due to typical OLAP data cube processing requirements, which usually involve tight computational bounds dictated by enormous sizes and high dimension number (e.g., see [10]), in our reference scenario we also consider a space bound constraint B that imposes us to adopt the well-known *data cube compression paradigm* (e.g., [10,19]). Basically, techniques adhering to this paradigm, called *approximate query answering techniques*, propose to compute compressed representations of data cubes in order to mitigate computational overheads deriving from evaluating resource-intensive OLAP queries. According to the proposed methodology, input queries are issued against the compressed data cube instead of the original one, thus obtaining *approximate answers* whose introduced query error is perfectly tolerable for OLAP analysis goals (e.g., see [10]).

Given an input data cube A and the space bound B, the main goal of our proposed framework is that of computing the *sampling-based synopsis data cube A', whose aggregations are obtained via satisfying the so-called privacy constraint*. In particular, the privacy constraint requires that approximate answers over the synopsis data cube embed a certain *degree of privacy*, which is measured by means of a meaningful *privacy metrics*, and is bounded by a given *privacy threshold* determined by application-oriented requirements. The metric-based approach for handling privacy of data is well-established in the community (e.g., [14]), due to its flexibility and nice theoretical properties. To compute the synopsis data cube A', we introduce the so-called *privacy grid*, a *grid-based partition* of A. The privacy grid allows us to meaningfully exploit the multi-resolution nature of OLAP data, and hence obtain an *effective information gain* during the computation of A'. In this respect, in our framework the *granularity* of the privacy grid (i.e., the size of its elementary cell) is meaningfully chosen as an *adequately-small fraction* of the *selectivity* of queries populating typical *query-workloads* posed against A. Note that this selectivity can be easily gathered thanks to popular active/monitoring components that one can find in conventional OLAP server platforms. Also, since a compression is introduced (due to sampling), in our experimental assessment we also carefully test the *degree of approximation* of retrieved answers, as supporting privacy of answers without considering the accuracy of answers is useless. The results of our experimental assessment clearly show that our framework is able to provide privacy preserving answers that simultaneously retain a good degree of approximation. Therefore, we can claim that the underlying *accuracy constraint* is also satisfied, beyond the (main) privacy constraint.

It should be noted that the so-delineated constraint system carefully models many of modern data-intensive application scenarios with privacy preserving issues, such as publish-subscribe systems [6], distributed Data Mining tools [8], advanced sensor network analysis tools [18], OLAP [3] and so forth. Also, beyond practical evidences, theoretical aspects confirm the benefits encompassed by our framework. In fact, it should be noted that our research is fully contextualized in the *Statistical Disclosure Control* (SDC) scientific field proposed by Domingo-Ferrer in [13], which is a widely-recognized authoritative research contribution. Specifically, based on the SDC framework, *the privacy preservation of a given data domain is obtained via trade-offing the accuracy and privacy of data*. The main idea of such an approach is that of admitting the need for data provisioning while, at the same time, the need for privacy of data. In fact, *full data hiding* or *full data camouflaging* are both useless, as well as publishing completely-disclosed data sets. Therefore, trade-offing accuracy and privacy of data is a reasonable solution to the research challenge we address. In this context, two meaningful measures for evaluating the accuracy and privacy preservation capabilities of an arbitrary method/technique have been introduced [13]. The first one is referred as *Information Loss* (IL). IL allows us to estimate the loss of information (i.e., the *accuracy decrease*) due to a given privacy preserving method/technique. The second one is the *Disclosure Risk* (DR). DR allows us to estimate the risk of disclosing sensitive data due to a given privacy preserving method/technique. The main difference between our proposed framework and SDC relies in the fact that SDC is focused on the privacy preservation of *statistical databases* [27], whose privacy preserving (e.g., [1]) and *inference control* (e.g., [11]) issues have been deeply investigated, whereas our framework specifically considers the privacy preservation of OLAP data cubes, but rather it is inspired to the SDC philosophy. Although several similarities between statistical databases and data cubes have been recognized in past research efforts [27], many differences still occur [27] so that we believe that a *specific* SDC privacy preservation method for data cubes must be considered, with *specific* features targeted to the OLAP context. The latter one is the main contribution of our research.

Starting from these considerations, in this paper we propose a robust sampling-based framework for privacy preserving OLAP, and provide a comprehensive experimental evaluation of this framework on synthetic data cubes. These data cubes allow us to easily control all the functional characteristics of a cube (e.g., dimension number, size, *sparseness coefficient* etc), thus leading to the achievement of a reliable and multi-perspective experimental assessment. Performance of the proposed framework is also compared with the one of the method *Zero-Sum* [29], which can be reasonably considered as the state-of-the-art for perturbation-based privacy preserving OLAP techniques, *under several perspectives of analysis* that encompass (*i*) the *quality* of the final synopsis data cube in accomplishing the privacy and accuracy constraints, (*ii*) the *effectiveness* of the final synopsis data cube in providing retrieved approximate answers having the desired degree of privacy, and, finally, (*iii*) the *sensitivity* of the final synopsis data cube under the ranging of the space bound constraint, which is a critical parameter of our framework. Specifically, as we demonstrate in our experimental assessment, the study of the latter property confirms the *robustness* of our framework, i.e. its *low dependency on configuration parameters*. Our experimental results underline the benefits due to the privacy preserving OLAP framework we propose, and state that performance of this framework outperforms the one of the comparison method Zero-Sum.

2 Related Work

Privacy Preserving OLAP (PPOLAP) [3] is a specialized case of *Privacy Preserving Data Mining* (PPDM) [2]. While PPDM concerns with the privacy of data during DM activities (e.g., clustering, classification, frequent item set mining, pattern discovery, association rule discovery etc), PPOLAP deals with the problem of preserving the privacy of data cells of a given data cube during typical OLAP activities such as performing classical operators (e.g., roll-up and drill-down) or evaluating complex

OLAP queries (e.g., *range-* [19], *top-k* [33], and *iceberg* [15] queries). With respect to PPDM, PPOLAP introduces more *semantics* into the privacy preservation due to its well-known knowledge-intensive tools such as multidimensionality and multi-resolution of data, and hierarchies. Preliminary studies on privacy preserving issues in Data Warehousing and OLAP can be found in [5,24]. In the following, we review the two kinds of privacy preserving OLAP techniques mentioned in Sect. 1, i.e. restriction-based and perturbation-based techniques.

Restriction-based techniques limit the queries that can be posed to the OLAP server in order to preserve the privacy of data cells. The underlying problem is related to the issue of *auditing queries in statistical databases*, which consists in analyzing the past (answered) queries in order to determine whether these answers can be composed by a malicious user to infer sensitive knowledge in the form of answers to forbidden (i.e., unauthorized) queries. Therefore, in order to understand which kind of queries must be forbidden, a restriction-based technique needs to audit queries posed to the target data (e.g., OLAP) server during an adequately-wide interval of time. Auditing queries in statistical databases is the conceptual and theoretical basis of auditing queries in OLAP systems. Interesting auditing techniques for queries against statistical databases have been proposed in [12], which introduces a model for auditing average and median queries, and [7], which proposes a technique for handling the past history of SUM queries in order to reduce the sequence of answered queries, to privacy preservation purposes. Also, [7] describes how to check the *compromisability* of the underlying statistical database when using the reduced sequence. The proposed auditing technique is called *Audit Expert*. More recently, few approaches focusing on the problem of auditing techniques for OLAP data cubes and queries appeared. Among all, we recall: (*i*) the work of Zhang *et al.* [34], which propose an interesting *information theoretic approach* that simply counts the number of cells already covered to answer previous queries in order to establish if a *new* query should be answered or not; (*ii*) the work of Malvestuto *et al.* [22], which introduce a novel notation for auditing range-SUM queries (i.e., an OLAP-like class of queries) against statistical databases making use of *Integer Linear Programming* (ILP) tools for detecting if a *new* range-SUM query can be answered safely or not.

Restriction-based techniques are also directly related to inference control methods, which are an alternative to avoid privacy breaches. [31,32] represent significant research contribution in this direction, and have heavily influenced the privacy preserving OLAP research community. In [31,32], authors propose a novel technique for limiting inference breaches in OLAP systems via detecting *cardinality-based* sufficient conditions over cuboids, in order to make data cubes safe with respect to malicious users. Specifically, the proposed technique combines access control and inference control techniques, being (*i*) first ones based on the hierarchical nature of data cubes in terms of *cuboid lattice* [16] and multi-resolution of data, and (*ii*) second ones based on directly applying restriction to *coarser aggregations* of data cubes, and then removing remaining inferences that can be still derived.

Perturbation-based techniques add random noise at various levels of the target database. Agrawal *et al.* [3] first propose the notion of PPOLAP. They define a PPOLAP model over data partitioned across multiple clients using a *randomization approach* on the basis of which (*i*) clients perturb tuples which with they participate to the distributed partition in order to gain *row-level privacy*, and (*ii*) server is capable of evaluating OLAP queries against perturbed tables via reconstructing original distributions of attributes involved by such queries. In [3], authors demonstrate that the proposed approach is safe against privacy breaches. Hua *et al.* [20] propose a different approach to preserve the privacy of OLAP data cubes. They argue that hiding parts of data that could cause inference of sensitive cuboids is enough in order to achieve the notion of "secure" data cubes. While a strong point of the proposed approach is represented by its simplicity, authors do not provide sufficient experimental analysis to prove in which measure the data hiding phase affects the target data cube. Sung *et al.* [29] propose a *random data distortion* technique, called Zero-Sum method, for preserving secret information of *individual* data cells while providing accurate answers to range queries over original aggregates. Roughly speaking, data distortion consists in iteratively altering the values of individual data cells of the target data cube in such a way as to maintain the *marginal sums* of data cells along rows and columns of the cube equal to zero. This ensures the privacy of individual data cells, and the correctness of answers to range queries.

Due to different, specific motivations, both restriction-based and perturbation-based techniques are ineffective for OLAP. Specifically, restriction-based techniques cannot be applied to OLAP systems since the nature of such systems is intrinsically *interactive*, and based on a wide set of operators and query classes. On the other hand, perturbation-based techniques cannot be applied in OLAP systems since they introduce excessive computational overheads when executed on massive data cubes, as they focus on the privacy of singleton data cube cells rather than the one of aggregations on the basis of a proper privacy OLAP notion.

3 Theoretical Model

A *data cube* $\mathcal{A}$ is a tuple $\mathcal{A} = \langle \mathcal{D}, \mathcal{L}, \mathcal{H}, \mathcal{M} \rangle$, such that: (*i*) $\mathcal{D}$ is the data domain of $\mathcal{A}$ containing (OLAP) data cells, which are the elementary aggregations of $\mathcal{A}$ computed against the relational data source S; (*ii*) $\mathcal{L}$ is the set of *dimensions* of $\mathcal{A}$, i.e. the *functional attributes* with respect to which the underlying OLAP analysis is defined (in other words, $\mathcal{L}$ is the set of attributes with respect to which relational tuples in S are aggregated); (*iii*) $\mathcal{H}$ is the set of *hierarchies* related to the dimensions of $\mathcal{A}$, i.e. hierarchical representations of the functional attributes shaped in the form of general trees; (*iv*) $\mathcal{M}$ is the set of *measures* of $\mathcal{A}$, i.e. the *attributes of interest* for the underlying OLAP analysis (in other words, $\mathcal{M}$ is the set of attributes with respect to which SQL aggregations stored in data cells of $\mathcal{A}$ are computed). Given these definitions, (*i*) $|\mathcal{L}|$ denotes the number of dimensions of $\mathcal{A}$, (*ii*) d denotes a generic dimension of $\mathcal{A}$, (*iii*) $|d|$ the cardinality of d, and (*iv*) $\mathcal{H}(d)$ the hierarchy of d. Finally, for the sake of simplicity, we assume data cubes having a unique measure (i.e., $|\mathcal{M}| = 1$). However, extending schemes, models and algorithms proposed in this paper to deal with data cubes having *multiple measures* (i.e., $|\mathcal{M}| > 1$) is straightforward.

Given an $|\mathcal{L}|$-dimensional data cube $\mathcal{A}$, an *m-dimensional range-query* Q against $\mathcal{A}$, with $m \leq |\mathcal{L}|$, is a tuple $Q = \langle R_{k_0}, R_{k_1}, ..., R_{k_{m-1}}, A \rangle$, such that: (*i*) R_{k_i} denotes a *contiguous* range defined on the dimension d_{k_i} of $\mathcal{A}$, with k_i belonging to the range

$[0, |\mathcal{L}| - 1]$, and (ii) A is a SQL aggregation operator. Applied to $\mathcal{A}$, Q returns the A-based aggregation computed over the set of data cells in $\mathcal{A}$ contained within the multidimensional sub-domain of $\mathcal{A}$ bounded by ranges $R_{k_0}, R_{k_1}, ..., R_{k_{m-1}}$. Range-SUM queries, which return the SUM of the involved data cells, are trendy examples of range queries. In our framework, we take into consideration range-SUM queries, as SUM aggregations are very popular in OLAP, and efficiently support other SQL aggregations (e.g., COUNT, AVG etc) as well as summarized knowledge extraction from massive amounts of data.

Given a query Q against a data cube $\mathcal{A}$, the *query region* of Q, denoted by $R(Q)$, is defined as the sub-domain of $\mathcal{A}$ bounded by ranges $R_{k_0}, R_{k_1}, ..., R_{k_{m-1}}$ of Q.

Given an n-dimensional data domain $\mathcal{D}$, the *volume* of $\mathcal{D}$, denoted by $\|\mathcal{D}\|$, is defined as follows: $\|\mathcal{D}\| = |d_0| \times |d_1| \times ... \times |d_{n-1}|$, such that $|d_i|$ is the cardinality of the dimension d_i of $\mathcal{D}$. This definition can also be extended to a multidimensional data cube $\mathcal{A}$, thus introducing the volume of $\mathcal{A}$, $\|\mathcal{A}\|$, and to a multidimensional range query Q, thus introducing the volume of Q, $\|Q\|$. The latter parameter is also recognized-in-literature as the selectivity of Q.

Given an $|\mathcal{L}|$-dimensional data cube $\mathcal{A}$, the privacy grid $\mathcal{P}(\mathcal{A})$ of $\mathcal{A}$ is a tuple $\mathcal{P}(\mathcal{A}) = \langle \Delta\ell_0, \Delta\ell_1, ..., \Delta\ell_{|\mathcal{L}|-1} \rangle$, such that $\Delta\ell_k$ is a range partitioning the dimension d_k of $\mathcal{A}$, with k belonging to $[0, |\mathcal{L}| - 1]$, in a $\Delta\ell_k$-based (one-dimensional) partition. By combining the partitions along *all* the dimensions of $\mathcal{A}$, we finally obtain $\mathcal{P}(\mathcal{A})$ as a *regular partition* of $R(\mathcal{A})$ (the multidimensional region associated to $\mathcal{A}$) composed by the so-called *grid regions* $\mathcal{R}_{\mathcal{P}(\mathcal{A}),k} = \lfloor \Delta\ell_{0,k}; \Delta\ell_{1,k}; ...; \Delta\ell_{|\mathcal{L}|-1,k} \rfloor$. Formally, $\mathcal{P}(\mathcal{A})$ can also be defined as a *collection* of (grid) regions, i.e. $\mathcal{P}(\mathcal{A}) = \{\mathcal{R}_{\mathcal{P}(\mathcal{A}),0}, \mathcal{R}_{\mathcal{P}(\mathcal{A}),1}, ..., \mathcal{R}_{\mathcal{P}(\mathcal{A}),|\mathcal{P}(\mathcal{A})|-1}\}$.

As *accuracy metrics* for answers to queries, we make use of the *relative query error* between exact and approximate answers, which is a well-recognized-in-literature measure of quality for approximate query answering techniques in OLAP (e.g., [6,13]). Formally, given a query Q, we denote as $A(Q)$ the exact answer to Q (i.e., the answer to Q evaluated against the original data cube $\mathcal{A}$), and as $\tilde{A}(Q)$ the approximate answer to Q (i.e., the answer to Q evaluated against the synopsis data cube $\mathcal{A}'$). Therefore, the relative query error $E_Q(Q)$ between $A(Q)$ and $\tilde{A}(Q)$ is defined as follows: $E_Q(Q) = \dfrac{|A(Q) - \tilde{A}(Q)|}{A(Q)}$.

Since we deal with the problem of ensuring the privacy preservation of OLAP aggregations, our privacy metrics takes into consideration how sensitive information can be discovered from aggregate data, and tries to contrast this possibility. To this end, we first study how sensitive aggregations can be discovered from the knowledge about exact answers, and metadata about data cubes and queries. Starting from the knowledge about the target data cube $\mathcal{A}$ (e.g., range sizes, OLAP hierarchies etc), and the knowledge about a given query Q (i.e., the volume of Q, $\|Q\|$, and the exact answer to Q, $A(Q)$), *it is possible to infer knowledge about sensitive ranges of data contained within $R(Q)$.* For instance, it is possible to derive the average value of the contribution throughout which each elementary data cell of $\mathcal{A}$ within $R(Q)$ contributes to $A(Q)$. We name this quantity as *singleton aggregation of Q,* denoted by $I(Q)$. $I(Q)$ is defined as follows: $I(Q) = \dfrac{A(Q)}{\|Q\|}$.

It is easy to understand that, in turn, starting from the knowledge about $I(Q)$, it is possible to *progressively* discover aggregations of larger ranges of data within $R(Q)$, rather than those stored within the elementary data cell, thus inferring further knowledge.

Secondly, we study how OLAP client applications can discover sensitive aggregations from the knowledge about approximate answers, and, similarly to the previous case, from the knowledge about data cube and query metadata. Starting from the knowledge about the synopsis data cube $\mathcal{A}'$, and the knowledge about a given query Q, it is possible to derive an *estimation* on $I(Q)$, denoted by $\tilde{I}(Q)$, as follows: $\tilde{I}(Q) = \dfrac{\tilde{A}(Q)}{S(Q)}$, such that $S(Q)$ is the *number of samples* extracted from $R(Q)$ to compute $\mathcal{A}'$ (note that $S(Q) < \|Q\|$). The relative difference between $I(Q)$ and $\tilde{I}(Q)$, named as *relative inference error* and denoted by $E_I(Q)$, gives us a metrics for the privacy of $\tilde{A}(Q)$, defined as follows: $E_I(Q) = \dfrac{|I(Q) - \tilde{I}(Q)|}{I(Q)}$.

Indeed, while OLAP client applications are aware about the definition and metadata of both the target data cube and queries, the number of samples $S(Q)$ for each query Q is not disclosed to them. As a consequence, in order to model this facet of our framework, we introduce the *user-perceived singleton aggregation,* denoted by $\tilde{I}_U(Q)$, which is the *effective* singleton aggregation *perceived* by external applications on the basis of the knowledge made available to them. $\tilde{I}_U(Q)$ is defined as follows: $\tilde{I}_U(Q) = \dfrac{\tilde{A}(Q)}{\|Q\|}$. Based on $\tilde{I}_U(Q)$, we derive the definition of the *relative user-perceived inference error* $E_I^U(Q)$, as follows: $E_I^U(Q) = \dfrac{|I(Q) - \tilde{I}_U(Q)|}{I(Q)}$.

It is trivial to demonstrate that $\tilde{I}_U(Q)$ provides a better estimation of the singleton aggregation of Q rather than the one provided by $\tilde{I}(Q)$, as $\tilde{I}_U(Q)$ is evaluated with respect to *all* the items contained within $R(Q)$ (i.e., $\|Q\|$), whereas $\tilde{I}(Q)$ is evaluated with respect to the number of samples extracted from $R(Q)$ (i.e., $S(Q)$). In other words, $\tilde{I}_U(Q)$ is an *upper bound* for $\tilde{I}(Q)$. Therefore, in our framework we consider $\tilde{I}(Q)$ to compute the synopsis data cube, whereas we consider $\tilde{I}_U(Q)$ to model inference issues on the OLAP client application side.

Here we highlight that, in our proposed framework, *the privacy OLAP notion is built upon $I(Q)$*. According to our approach, the final goal is *maximizing the relative inference error* $E_I^U(Q)$, as this condition means-in-practice that OLAP client applications retrieve from the synopsis data cube $\mathcal{A}'$ "sampled" singleton aggregations (i.e., $\tilde{I}_U(Q)$) that are *very different* from the "real" singleton aggregations (i.e., $I(Q)$) of the original data cube $\mathcal{A}$ (i.e., $\tilde{I}_U(Q) \neq I(Q)$) This way, the privacy of OLAP aggregations of $\mathcal{A}$ is preserved.

Similarly to related proposals appeared in literature recently [29], in our framework we introduce the privacy threshold Φ_I that gives us a *lower bound* for the relative user-perceived inference error $E_I^U(Q)$ due to evaluating a given query Q against the synopsis data cube $\mathcal{A}'$. Therefore, the privacy constraint can formally be modeled as follows: $E_I^U(Q) \geq \Phi_I$. Above all, Φ_I allows us to meaningfully model and treat privacy OLAP issues at a *rigorous mathematical/statistical plane*. Application-wise, Φ_I is set by OLAP client applications, and the privacy preserving OLAP engine must accomplish this requirement accordingly, while also ensuring the accuracy of approximate answers. The issue of determining how to set this (user-defined) parameter is a non-trivial engagement. Intuitively enough, we set this parameter in terms of a *percentage value*, as this way its semantics can be immediately captured by OLAP client applications, which indeed do not know the exact values of singleton aggregations. Without going into more details, we highlight that this approach is similar to the one adopted by a plethora of research experiences in the context of approximate query answering techniques, which make use of a widely-accepted *query error threshold* (belonging to the interval [15, 20] %) as reference for the accuracy of answers (e.g., see [10]).

4 Computing the Privacy Preserving Synopsis Data Cube

Algorithm computeSPPDataCube implements our technique for computing the synopsis data cube $\mathcal{A}'$, given the following input parameters: (*i*) the target data cube $\mathcal{A}$; (*ii*) the space bound $\mathcal{B}$; (*iii*) the integer parameter δ (described next); (*iv*) the privacy threshold Φ_I; (*v*) the typical query-workload QWL on $\mathcal{A}$. Basically, computeSPPDataCube is a multi-step algorithm. We next describe each step of computeSPPDataCube separately.

Computing the Privacy Grid. The first step of computeSPPDataCube consists in computing the privacy grid $\mathcal{P}(\mathcal{A})$ for $\mathcal{A}$. This task finally aims at determining the range $\Delta\ell_k$ for each dimension d_k of $\mathcal{A}$, with k belonging to $[0, |\mathcal{L}| - 1]$. In turn, this allows us to obtain the volume of grid regions $\mathcal{R}_{\mathcal{P}(\mathcal{A}),k}$, $\|\mathcal{R}_{\mathcal{P}(\mathcal{A}),k}\|$ by regularly partitioning $\mathcal{A}$. As stated in Sect. 1., in our framework, we determine $\Delta\ell_k$ as an adequately-small fraction of the selectivity of queries in QWL. Let S_T be the "typical" selectivity of queries in QWL, and $\|\mathcal{R}_{\mathcal{P}(\mathcal{A}),k}\|$ be the volume of $\mathcal{R}_{\mathcal{P}(\mathcal{A}),k}$. If $\|\mathcal{R}_{\mathcal{P}(\mathcal{A}),k}\| \ll S_T$, then $\mathcal{A}'$ can be computed *by using the grid region as the elementary reasoning unit, and* *adopting a resolution level lower than the resolution level of queries against* $\mathcal{A}$. This allows us to achieve an effective information gain during the computation of $\mathcal{A}'$. In fact, if the grid region $\mathcal{R}_{\mathcal{P}(\mathcal{A}),k}$ is sampled in such a way as to satisfy the privacy constraint, while ensuring the accuracy of approximate answers that involve $\mathcal{R}_{\mathcal{P}(\mathcal{A}),k}$, then the *same* properties can also be inherited by input queries on $\mathcal{A}$ as well, being the latter queries "defined" on top of grid regions.

On the other hand, it should be noted that adopting the alternative approach of sampling the regions defined by queries in *QWL* directly, without referring to the reasoning layer of grid regions, would cause that, due to the space bound constraint $\mathcal{B}$, a sub-set of query regions of *QWL* will be sampled by means of an adequately-wide set of samples, whereas the remaining query regions of *QWL* will be sampled by means of a lower number of samples (*under-sampling*), or, even, not sampled at all. It is a matter of fact to notice that the latter situation would lead to an "unfair" synopsis data cube $\mathcal{A}'$, i.e. a synopsis data cube such that queries involving some regions of $\mathcal{A}'$ are characterized by low privacy and low accuracy, whereas queries involving other regions of $\mathcal{A}'$ are characterized by high privacy and high accuracy. Contrary to this, in our framework we aim at obtaining a "fair" synopsis data cube $\mathcal{A}'$, i.e. a synopsis data cube able to accommodate a large number of queries while satisfying the privacy constraint, and also ensuring the accuracy of retrieved (approximate) answers.

How to determine S_T *from the given query-workload QWL?* A reasonable solution consists in selecting S_T *by composing all the most frequent query ranges in QWL.* Notice that these ranges can be easily gathered by means of popular active/monitoring components of conventional OLAP server platforms. Overall, this strategy allows us to easily obtain a very-reliable "representative" value of selectivity of queries in *QWL*. However, determining S_T is an orthogonal aspect for our proposed framework, so that other different strategies can be devised, and straightforwardly integrated within the core layer of the framework.

The Greedy Strategy. The second step of computeSPPDataCube embeds a *greedy strategy* for sampling the input data cube $\mathcal{A}$ in order to obtain the synopsis data cube $\mathcal{A}'$. The greedy choice is dictated by the space bound constraint $\mathcal{B}$ that imposes us to compute a "best-effort" synopsis data cube $\mathcal{A}'$, i.e. a synopsis data cube such that simultaneously (*i*) satisfies the privacy constraint, (*ii*) ensures the accuracy of approximate answers, and (*iii*) fits within $\mathcal{B}$. The reasoning unit of the sampling phase of computeSPPDataCube is the grid region $\mathcal{R}_{\mathcal{P}(\mathcal{A}),k}$, meaning that, at each iteration j and until $\mathcal{B}$ is not consumed, computeSPPDataCube greedily selects from $\mathcal{P}(\mathcal{A})$ a grid region, denoted by $\mathcal{R}^{j}_{\mathcal{P}(\mathcal{A}),k}$, and extracts from $\mathcal{R}^{j}_{\mathcal{P}(\mathcal{A}),k}$ a set of samples, denoted by $S(\mathcal{R}^{j}_{\mathcal{P}(\mathcal{A}),k})$. By iterating the above-illustrated task for each one of the selected grid regions, the final synopsis data cube $\mathcal{A}'$ is obtained.

computeSPPDataCube adopts a *greedy criterion* to select the grid region to be sampled. This criterion considers the properties of *data distributions* associated to grid regions in $\mathcal{P}(\mathcal{A})$, and *selects the most skewed grid region among the available ones* (i.e., the regions of $\mathcal{P}(\mathcal{A})$ *not* chosen during previous iterations of the algorithm). Without any loss of generality, given a grid region $\mathcal{R}_{\mathcal{P}(\mathcal{A}),k}$ in $\mathcal{P}(\mathcal{A})$ the associated data distribution, denoted by $\mathcal{F}(\mathcal{R}_{\mathcal{P}(\mathcal{A}),k})$, can be reasonably intended as a *multidimensional distribution*, following the nature of regions (which, in turn, are defined on top of multidimensional data cubes). The main idea that underlines our proposed greedy criterion is based on assuming that in order to "describe" a skewed grid region, i.e. a grid region $\mathcal{R}_{\mathcal{P}(\mathcal{A}),k}$ whose data distribution $\mathcal{F}(\mathcal{R}_{\mathcal{P}(\mathcal{A}),k})$ is skewed (e.g., distributed according to a *Zipf* distribution with *asymmetric peaks*), we need a number of samples greater than the number of samples that are necessary in order to "describe" a uniform grid region, i.e. a grid region $\mathcal{R}_{\mathcal{P}(\mathcal{A}),k}$ whose data distribution $\mathcal{F}(\mathcal{R}_{\mathcal{P}(\mathcal{A}),k})$ is *Uniform* (that is, values of $\mathcal{F}(\mathcal{R}_{\mathcal{P}(\mathcal{A}),k})$ are regularly distributed around the average value of $\mathcal{F}(\mathcal{R}_{\mathcal{P}(\mathcal{A}),k})$).

In order to determine if a given data distribution $\mathcal{F}$ is skewed or not, we adopt a well-established theoretical result of the literature [28]. According to [28], given a data distribution $\mathcal{F}$, $\mathcal{F}$ is considered as skewed if the *skewness value* of $\mathcal{F}$, denoted by $\gamma_1(\mathcal{F})$, is greater than its standard deviation, denoted by $\sigma(\gamma_1(\mathcal{F}))$, by a factor equal to 2.6 (i.e., $\gamma_1(\mathcal{F}) > 2.6 \cdot \sigma(\gamma_1(\mathcal{F}))$). $\gamma_1(\mathcal{F})$ can be computed as follows [23]: $\gamma_1(\mathcal{F}) = \dfrac{(\mu_3(\mathcal{F}))^2}{(\mu_2(\mathcal{F}))^3}$, such that $\mu_r(\mathcal{F})$ denotes the rth *central moment* of $\mathcal{F}$, defined as follows [23]: $\mu_r(\mathcal{F}) = \sum_{k=0}^{q-1}(k-\mu)^r \cdot \mathcal{F}(k)$, where q is the number of samples of $\mathcal{F}$ (i.e., data items of $\mathcal{F}$) and μ is the mean value of $\mathcal{F}$. [28] also provides us with a method for computing the standard deviation of the skewness. According to [28], $\sigma(\gamma_1(\mathcal{F}))$ can be computed as follows: $\sigma(\gamma_1(\mathcal{F})) = \sqrt{\dfrac{6}{q}}$.

On the basis of results of [28], we introduce the so-called *characteristic function* $\Psi(\mathcal{F})$, which allows us to determine if a given data distribution $\mathcal{F}$ is skewed ($\Psi(\mathcal{F}) = 1$) or uniform ($\Psi(\mathcal{F}) = 0$). $\Psi(\mathcal{F})$ is defined as follows:

$$\Psi(\mathcal{F}) = \begin{cases} 1 & if \quad \dfrac{\sqrt{q} \cdot \left(\sum_{k=0}^{q-1}(k-\mu)^3 \cdot \mathcal{F}(k)\right)^2}{\sqrt{6} \cdot \left(\sum_{k=0}^{q-1}(k-\mu)^2 \cdot \mathcal{F}(k)\right)^3} > 2.6 \\ 0 & otherwise \end{cases} \tag{1}$$

such that q and μ are the number of data items and the mean value of $\mathcal{F}$, respectively. As regards performance issues, it should be noted that (1) can be easily implemented within a software component having low computational cost.

Sampling the Grid Regions. Now we focus the attention on how the set of s amples $S(\mathcal{R}_{\mathcal{P}(\mathcal{A}),k}^j)$ is extracted from the grid region $\mathcal{R}_{\mathcal{P}(\mathcal{A}),k}^j$ at the iteration j of computeSPPDataCube, being $\mathcal{R}^j_{P(\mathcal{A}),k}$ selected by means of the greedy criterion described above. Recall that sampling the grid region is the baseline operation for computing the final synopsis data cube $\mathcal{A}'$. In particular, as regards the sampling strategy we adopt the classical *Uniform sampling*, i.e. based on a conventional Uniform generating distribution. Briefly, this sampling strategy works as follows. Given a one-dimensional data domain $\mathcal{D}$ whose definition interval is: $[I_{min}, I_{max}]$, with $I_{max} > I_{min}$, the ith sample is extracted according to a two-step task: (*i*) random sample an indexer i in $[I_{min}, I_{max}]$ by means of a Uniform distribution defined on the range $[I_{min}, I_{max}]$ (i.e., $i = Unif(I_{min}, I_{max})$); (*ii*) return the sample $\mathcal{D}[i]$. Given an n-dimensional data domain $\mathcal{D}$, the ith sample is extracted via iterating the above-illustrated task for each of the n dimensions of $\mathcal{D}$. Also, we make use of *sampling without duplicates*, i.e. at each random extraction we ensure that the sampled indexer has not been picked before.

Given the grid region $\mathcal{R}^j_{P(\mathcal{A}),k}$ at the iteration j of computeSPPDataCube, we first consider the corresponding range-SUM query, denoted by $Q^j_{P(\mathcal{A}),k}$, whose multidimensional range is equal to the range of $\mathcal{R}^j_{P(\mathcal{A}),k}$. Then, on the basis of a *metrics-driven approach*, given an integer parameter δ, such that $\delta > 0$, *we iteratively sample* $\mathcal{R}^j_{P(\mathcal{A}),k}$ *by extracting δ-sized sub-sets of samples from* $\mathcal{R}^j_{P(\mathcal{A}),k}$ *until one of the following two conditions becomes true: (i) the privacy constraint on* $\mathcal{R}^j_{P(\mathcal{A}),k}$ *is satisfied (i.e.,* $E_I(Q^j_{P(\mathcal{A}),k}) \geq \Phi_I$ *), or (ii)* $\mathcal{B}$ *is consumed (i.e.,* $\mathcal{B} = 0$ *)*. It is a matter of fact to note that δ represents the size of a sort of *buffer* used during sampling. This solution avoids excessive computational overheads that instead would be caused if sampling is performed on massive-in-size data cube without buffering. The nature of sampling used in our framework carefully takes into account the nature of OLAP queries considered (i.e., range-SUM queries), and the requirements of the privacy constraint. Let $\mathcal{V}^j_{P(\mathcal{A}),k}$ be the average value of $\mathcal{R}^j_{P(\mathcal{A}),k}$, we introduce the sub-set $\mathcal{U}^j_{P(\mathcal{A}),k}$ of data cells in $\mathcal{R}^j_{P(\mathcal{A}),k}$ whose values are greater than $\mathcal{V}^j_{P(\mathcal{A}),k}$, i.e. $\mathcal{U}^j_{P(\mathcal{A}),k} = \{C \in \mathcal{R}^j_{P(\mathcal{A}),k} \mid val(C) > \mathcal{V}^j_{P(\mathcal{A}),k}\}$, and we apply the Uniform sampling on $\mathcal{U}^j_{P(\mathcal{A}),k}$ rather than $\mathcal{R}^j_{P(\mathcal{A}),k}$. It is easy to understand that this particular sampling strategy *naturally* allows us to obtain an approximate answer to $Q^j_{P(\mathcal{A}),k}$ having a good degree of approximation, and, *at the same time*, a high degree of privacy (in other words, our sampling strategy is in favor of the satisfaction of the constraint $E_I(Q^j_{P(\mathcal{A}),k}) \geq \Phi_I$). As above-highlighted, these properties are in turn inherited by input queries on $\mathcal{A}$.

5 Experimental Results

In order to test the performance of our proposed privacy preserving OLAP technique, we conducted an extensive series of experiments on several classes of synthetic data cubes. Ranging the input parameters (such as dimension number, size, sparseness coefficient etc) is the major benefit coming from using synthetic data cubes instead of real-life ones. As highlighted in Sect. 1, in our comprehensive experimental assessment we considered several perspectives of analysis oriented to test the quality, the effectiveness, and the sensitivity of our proposed technique, respectively. Also, we compared the performance of our technique with the one of the method Zero-Sum [29], the state-of-the-art privacy preserving OLAP perturbation-based technique.

As regards the data layer of our experimental framework, we considered the case of two-dimensional data cubes, which well covers the goals of a reliable experimental evaluation focused on evaluating privacy preservation capabilities. Indeed, we also tested our privacy preserving OLAP technique on more probing multi-dimensional data cubes, and the observed results are very similar to those experienced on two-dimensional data cubes (showed in this Section). For this reason, here we present the results on two-dimensional data cubes. In particular, we engineered two kinds of two-dimensional (synthetic) data cubes: *CVA* and *SKEW*. In the first kind of data cubes (i.e., *CVA*), data cells are generated according to a *Uniform* distribution defined on a given range $[U_{min}, U_{max}]$, with $U_{min} < U_{max}$. In other words, for such data cubes the *Continuous Value Assumption* (CVA) [9] holds. CVA assumes that data cells are uniformly distributed over the target domain. In the second kind of data cubes (i.e., *SKEW*), data cells are generated according to a *Zipf* distribution defined on a given parameter z, with z in [0, 1]. In our experimental framework, the parameter D denotes the kind of generating data distributions. D can thus assume the following values: {Uniform, Zipf}. For what regards data cube size, since we deal with two-dimensional data cubes we introduce the cardinality of each dimension of the data cube, denoted by L_0 and L_1, respectively (i.e., $|d_0| = L_0$ and $|d_1| = L_1$). Accordingly, we denote as K_0 and K_1 the range sizes of grid regions of the privacy grid, respectively. Finally, to obtain close-to-real-life data cubes, we also introduce the sparseness coefficient s, which measures the percentage ratio of non-null data cells with respect to the total number of data cells of a given data cube.

Other parameters of our experimental framework are the following: (*i*) B, which models the space bound constraint $\mathcal{B}$; (*ii*) P, which models the privacy threshold Φ_l. Also, for each perspective of analysis captured in our experimental assessment, we introduce an ad-hoc metrics.

In the quality analysis, we inherit the factors introduced by Sung *et al.* in [29] (the method Zero-Sum), namely the *privacy factor* and the *accuracy factor*. Let (*i*) $\mathcal{A}$ be the input data cube, (*ii*) $\mathcal{A}'$ be the synopsis data cube, (*iii*) $Y\{\mathbf{k}\}$ be data cube cell having $\mathbf{k}$ as multidimensional indexer, with $Y = \{\mathcal{A}, \mathcal{A}'\}$, the privacy factor F_P measures the *average amount of distorted data cells contained in blocks of* $\mathcal{A}'$. A *block* in the method Zero-Sum is a sub-cube with respect to which marginal sums of perturbed data cells along rows and columns are maintained equal to zero. F_P is defined as follows [29]:

$$F_P(\mathcal{A}, \mathcal{A}') = \frac{1}{\|\mathcal{A}\|} \cdot \sum_{\mathbf{k}=0}^{\|\mathcal{A}\|-1} \frac{|\mathcal{A}'\{\mathbf{k}\} - \mathcal{A}\{\mathbf{k}\}|}{|\mathcal{A}\{\mathbf{k}\}|} \qquad (2)$$

In other words, F_P provides us with a measure on *how much good* the privacy preservation of $\mathcal{A}'$ is. Being Zero-Sum a method oriented to data cells, and our technique instead based on the privacy OLAP notion, when measuring (2) on our synopsis data cube, a slight change is needed. First, the concept of block underlying (2) is meaningfully changed with the concept of grid region (as, indeed, they are very similar natively). Secondly, if $\mathcal{A}'\{\mathbf{k}\}$ has not been sampled (i.e., $\mathcal{A}'\{\mathbf{k}\}$ = NULL), *then we change $\mathcal{A}'\{\mathbf{k}\}$ with the corresponding singleton aggregation computed with respect to the grid region that contains $\mathcal{A}'\{\mathbf{k}\}$.*

The accuracy factor F_A is instead defined in dependence of a given query Q on the synopsis data cube $\mathcal{A}'$, as follows:

$$F_A(Q) = 2^{\frac{|A(Q) - \tilde{A}(Q)|}{|A(Q)|}} \tag{3}$$

such that $A(Q)$ is the exact answer to Q and $\tilde{A}(Q)$ is the approximate answer to Q (note that (3) is very similar to the classical definition – see Sect. 3). In other words, F_A provides us with a measure on *how much good* the degree of approximation ensured by $\mathcal{A}'$ for a given query Q is. Since we are interested into a *global* testing of the accuracy of synopsis data cubes, we extend (3) to an input query-workload QWL, as follows:

$$F_A(QWL) = \frac{1}{|QWL|} \sum_{k=0}^{|QWL|} F_A(Q_k) \tag{4}$$

In the quality analysis, we set QWL as composed by the collection of range-SUM queries corresponding to blocks for the case of the method Zero-Sum, and to grid regions for the case of our technique.

Fig 1 shows the experimental results of the quality analysis for what regards the privacy factor (Fig 1 (*a*) on a *CVA* data cube and Fig 1 (*b*) on a *SKEW* data cube) and the accuracy factor (Fig 1 (*c*) on a *CVA* data cube and Fig 1 (*d*) on a *SKEW* data cube) with respect to the sparseness coefficient s of synthetic data cubes, respectively. Our technique is here labeled as SPPOLAP.

In the effectiveness analysis, given a query-workload QWL, we study the *average relative user-perceived inference error* $\overline{E}_I^U(QWL)$ due to evaluating queries in QWL against the synopsis data cube $\mathcal{A}'$. $\overline{E}_I^U(QWL)$ is defined as follows:

$$\overline{E}_I^U(QWL) = \frac{1}{|QWL|} \sum_{k=0}^{|QWL|} E_I^U(Q_k) \tag{5}$$

such that $E_I^U(Q_k)$ is the relative user-perceived inference error due to evaluating the query Q_k in QWL against the synopsis data cube $\mathcal{A}'$. In our experimental framework, queries in QWL are synthetically generated as those queries that completely "span" the target synthetic data cube, and having selectivity S equal to a fixed percentage value of the volume of the data cube. Fig 2 shows the experimental results of the effectiveness analysis on a *CVA* data cube (Fig 2 (*a*)) and a *SKEW* data cube (Fig 2 (*b*)) with respect to the query selectivity S, respectively.

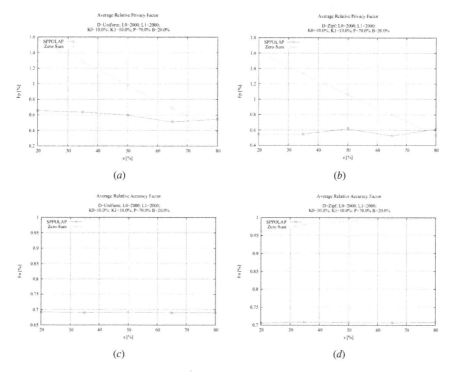

Fig. 1. Experimental results of the quality analysis for the privacy factor ((a) on a CVA data cube, (b) on a SKEW data cube) and the accuracy factor ((c) on a CVA data cube, (d) on a SKEW data cube) with respect to the sparseness coefficient of synthetic data cubes

Finally, Fig. 2 shows the same set of previous experiments (i.e., quality analysis and effectiveness analysis) when ranging the space bound constraint $\mathcal{B}$. This allows us to perform the sensitivity analysis of our technique, i.e. studying its dependency on a so-critical parameter like the space bound available to house the synopsis data cube.

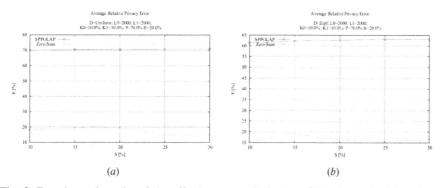

Fig. 2. Experimental results of the effectiveness analysis on a CVA data cube (a) and on a SKEW data cube (b) with respect to the selectivity of queries

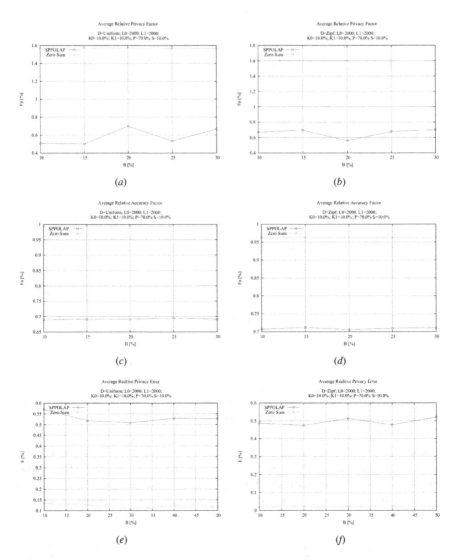

Fig. 3. Experimental results of the sensitivity analysis on a CVA data cube ((a), (c), (e)) and on a SKEW data cube ((b), (d), (f)) with respect to the space bound

From the analysis of the experimental results of our privacy preserving OLAP technique in comparison with the method Zero-Sum, it clearly follows that, for what regards the accuracy of synopsis data cubes, the performance of our technique is comparable with the one of Zero-Sum. Contrary to this, for what regards the privacy of synopsis data cubes, the performance of our technique is significantly better than the one of Zero-Sum. This confirms to us the relevance of our research contribution. Besides this, our technique introduces computational overheads that are clearly lower than those due to the method Zero-Sum, as the latter method is data-cell-oriented and neglects to consider any privacy OLAP notion (like ours). Finally, although based on the privacy OLAP notion, our comprehensive experimental evaluation states that our technique is also lowly dependent on the space bound constraint, which is a critical parameter of any data-intensive processing technique in OLAP.

6   Conclusions and Future Work

In this paper, we have presented a complete privacy preserving OLAP framework that encompasses several research innovations beyond capabilities of actual privacy preserving OLAP techniques. In particular, we have introduced the so-called privacy OLAP notion, which can be reasonably intended as a significant research contribution to the field. Also, our proposed framework is theoretically well-founded and technically sound. In addition to this, a comprehensive experimental evaluation has further confirmed the benefits deriving from adopting our proposed framework for the goal of preserving the privacy of OLAP data cubes, also in comparison with the method Zero-Sum, the state-of-the-art privacy preserving OLAP perturbation-based technique. Future work is mainly focused on the problem of extending the privacy OLAP notion towards *complex* OLAP aggregations, beyond conventional ones (e.g., SUM, COUNT, AVG).

References

1. Adam, N.R., Wortmann, J.C.: Security-Control Methods for Statistical Databases: A Comparative Study. ACM Computing Surveys 21(4), 515–556 (1989)
2. Agrawal, R., Srikant, R.: Privacy-Preserving Data Mining. In: ACM SIGMOD, pp. 439–450 (2000)
3. Agrawal, R., Srikant, R., Thomas, D.: Privacy-Preserving OLAP. In: ACM SIGMOD, pp. 251–262 (2005)
4. Beck, L.L.: A Security Mechanism for Statistical Databases. ACM Transactions on Database Systems 5(3), 316–338 (1980)
5. Bhargava, B.K.: Security in Data Warehousing. In: Kambayashi, Y., Mohania, M., Tjoa, A.M. (eds.) DaWaK 2000. LNCS, vol. 1874, pp. 287–289. Springer, Heidelberg (2000)
6. Bonifati, A., Cuzzocrea, A.: XℓPPX: a Lightweight Framework for Privacy Preserving P2P XML Databases in Very Large Publish-Subscribe Systems. In: Psaila, G., Wagner, R. (eds.) EC-Web 2007. LNCS, vol. 4655, pp. 21–34. Springer, Heidelberg (2007)
7. Chin, F.Y., Ozsoyoglu, G.: Auditing and Inference Control in Statistical Databases. IEEE Transactions on Software Engineering 8(6), 574–582 (1982)
8. Clifton, C., Kantarcioglu, M., Lin, X., Vaidya, J., Zhu, M.: Tools for Privacy Preserving Distributed Data Mining. SIGKDD Explorations 4(2), 28–34 (2003)
9. Colliat, G.: OLAP, Relational, and Multidimensional Database Systems. ACM SIGMOD Record 25(3), 64–69 (1996)
10. Cuzzocrea, A.: Overcoming Limitations of Approximate Query Answering in OLAP. In: IEEE IDEAS, pp. 200–209 (2005)
11. Denning, D.E., Schlorer, J.: Inference Controls for Statistical Databases. IEEE Computer 16(7), 69–82 (1983)

12. Dobkin, D., Jones, A.K., Lipton, R.J.: Secure Databases: Protection against User Influence. ACM Transactions on Database Systems 4(1), 97–106 (1979)
13. Domingo-Ferrer, J. (ed.): Inference Control in Statistical Databases: From Theory to Practice. Springer, Heidelberg (2002)
14. Dwork, C.: Differential Privacy: A Survey of Results. In: Agrawal, M., Du, D., Duan, Z., Li, A. (eds.) TAMC 2008. LNCS, vol. 4978, pp. 1–19. Springer, Heidelberg (2008)
15. Fang, M., Shivakumar, N., Garcia-Molina, H., Motwani, R., Ullman, J.D.: Computing Iceberg Queries Efficiently. In: VLDB, pp. 299–310 (1998)
16. Gray, J., Chaudhuri, S., Bosworth, A., Layman, A., Reichart, D., Venkatrao, M., Pellow, F., Pirahesh, H.: Data Cube: A Relational Aggregation Operator Generalizing Group-By, Cross-Tabs, and Sub-Totals. Data Mining and Knowledge Discovery 1(1), 29–54 (1997)
17. Griffiths, P., Wade, B.W.: An Authorization Mechanism for a Relational Database System. ACM Transactions on Database Systems 1(3), 242–255 (1976)
18. He, W., Liu, X., Nguyen, H., Nahrstedt, K., Abdelzaher, T.: PDA: Privacy-Preserving Data Aggregation in Wireless Sensor Networks. In: IEEE INFOCOM, pp. 2045–2053 (2007)
19. Ho, C.-T., Agrawal, R., Megiddo, N., Srikant, R.: Range Queries in OLAP Data Cubes. In: ACM SIGMOD, pp. 73–88 (1997)
20. Hua, M., Zhang, S., Wang, W., Zhou, H., Shi, B.: FMC: An Approach for Privacy Preserving OLAP. In: Tjoa, A.M., Trujillo, J. (eds.) DaWaK 2005. LNCS, vol. 3589, pp. 408–417. Springer, Heidelberg (2005)
21. Jajodia, S., Samarati, P., Sapino, M.L., Subrahmanian, V.S.: Flexible Support for Multiple Access Control Policies. ACM Transactions on Database Systems 26(4), 1–57 (2001)
22. Malvestuto, F.M., Mezzani, M., Moscarini, M.: Auditing Sum-Queries to Make a Statistical Database Secure. ACM Transactions on Information and System Security 9(1), 31–60 (2006)
23. Papoulis, A.: Probability, Random Variables, and Stochastic Processes, 2nd edn. McGraw-Hill, New York (1984)
24. Pernul, G., Priebe, T.: Towards OLAP Security Design – Survey and Research Issues. In: ACM DOLAP, pp. 114–121 (2000)
25. Sandhu, R.S., Coyne, E.J., Feinstein, H.L., Youman, C.E.: Role-based Access Control Models. IEEE Computer 29(2), 38–47 (1996)
26. Schlorer, J.: Security of Statistical Databases: Multidimensional Transformation. ACM Transactions on Database Systems 6(1), 95–112 (1981)
27. Shoshani, A.: OLAP and Statistical Databases: Similarities and Differences. In: ACM PODS, pp. 185–196 (1997)
28. Stuart, A., Ord, J.K.: Kendall Advanced Theory of Statistics, 6th edn. Distribution Theory, vol. 1. Oxford University Press, Oxford (1998)
29. Sung, S.Y., Liu, Y., Xiong, H., Ng, P.A.: Privacy Preservation for Data Cubes. Knowledge and Information Systems 9(1), 38–61 (2006)
30. Sweeney, L.: k-Anonymity: A Model for Protecting Privacy. International Journal on Uncertainty Fuzziness and Knowledge-based Systems 10(5), 557–570 (2002)
31. Wang, L., Jajodia, S., Wijesekera, D.: Securing OLAP Data Cubes against Privacy Breaches. In: IEEE SSP, pp. 161–175 (2004)
32. Wang, L., Wijesekera, D., Jajodia, S.: Cardinality-based Inference Control in Data Cubes. Journal of Computer Security 12(5), 655–692 (2004)
33. Xin, D., Han, J., Cheng, H., Li, X.: Answering Top-k Queries with Multi-Dimensional Selections: The Ranking Cube Approach. In: VLDB, pp. 463–475 (2006)
34. Zhang, N., Zhao, W., Chen, J.: Cardinality-based Inference Control in OLAP Systems: An Information Theoretic Approach. In: ACM DOLAP, pp. 59–64 (2004)

Generalization-Based Privacy-Preserving Data Collection

Lijie Zhang and Weining Zhang

Department of Computer Science, University of Texas at San Antonio
{lijez,wzhang}@cs.utsa.edu

Abstract. In privacy-preserving data mining, there is a need to consider on-line data collection applications in a client-server-to-user (CS2U) model, in which a trusted server can help clients create and disseminate anonymous data. Existing privacy-preserving data publishing (PPDP) and privacy-preserving data collection (PPDC) methods do not sufficiently address the needs of these applications. In this paper, we present a novel PPDC method that lets respondents (clients) use generalization to create anonymous data in the CS2U model. Generalization is widely used for PPDP but has not been used for PPDC. We propose a new probabilistic privacy measure to model a distribution attack and use it to define the respondent's problem (RP) for finding an optimal anonymous tuple. We show that RP is NP-hard and present a heuristic algorithm for it. Our method is compared with a number of existing PPDC and PPDP methods in experiments based on two UCI datasets and two utility measures. Preliminary results show that our method can better protect against the distribution attack and provide good balance between privacy and data utility.

1 Introduction

Consider the following scenario. A medical researcher, Steve, wants to study via data mining how cancer patients at various stages use non-prescript medicine. Due to public concerns of privacy [1], he wants to perform privacy preserving data mining (PPDM) on anonymous data. However, it is difficult to identify a source of data, since typically cancer hospitals do not document the use of non-prescript medicine and pharmacies do not document treatment of cancer patients. It is also difficult to integrate anonymous data from hospitals and pharmacies because the data do not contain personal identities. A possible solution is for Steve to use an on-line data collection service, such as a well-known survey website.

Let us consider how such a website can provide anonymous data to Steve. Perhaps the simplest solution is the privacy-preserving data publishing (PPDP) that has been extensively reported in the literature. PPDP is based on a server-to-user (S2U) model, in which one or more server (data owner such as a hospital, bank, or government agency) releases to the user (data miner) anonymous data that are obtained from the original data by perturbation [2, 3] or generalization [4, 5, 6, 7, 8, 9, 10]. Recently, generalization-based PPDP methods have gained wide acceptance and have efficient implementations [9, 11].

I.-Y. Song, J. Eder, and T.M. Nguyen (Eds.): DaWaK 2008, LNCS 5182, pp. 115–124, 2008.

To provide Steve with anonymous data via PPDP, the website has to ask respondents to submit original data that contain sensitive personal information. Unlike in a typical PPDP environment, such as a hospital where patients have to provide their private information to receive services (i.e., diagnosis or treatment), on-line data collection applications rely on voluntary submission of information. We believe that not all people are willing to voluntarily submit original data all the time, even if they know that the server will not try to breach their privacy. Thus, PPDP alone does not provide a sufficient solution for the website. We also need ways for respondents to submit anonymous data.

In the literature, privacy preserving data collection (PPDC) has been proposed for respondents to submit anonymous data. PPDC is based on a client-to-user (C2U) model in which each client (respondent) submits directly to the user (data miner) an anonymous tuple that is obtained from the original tuple using perturbation [12, 13, 14]. However, existing PPDC methods have a number of problems due to the restriction of C2U model. For example, they may fail under a PCA-based attack [15] or a distribution attack (see Section 4.1).

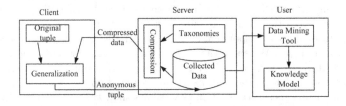

Fig. 1. PPDC in a CS2U Model

1.1   Our Contributions

To address these problems, we consider PPDC in a client-server-to-user (CS2U) model (see Figure 1) that combines S2U and C2U models. In this model, clients (respondents) create anonymous tuples with the help of a trusted server (such as the website) and submit anonymous tuples to the user (data miner) via the server. We propose a novel generalization-based PPDC method for the CS2U model. Our contributions are as follows.

1. We propose a framework for generalization-based PPDC in a CS2U model.
2. We define a new probabilistic privacy measure based on a distribution attack and use it to define the respondent's problem of finding an optimal anonymous tuple.
3. We show that this respondent's problem is NP-hard and propose a heuristic algorithm to solve it. Our generalization algorithm is able to create anonymous tuples using data already collected (rather than the entire set of original data required by existing PPDP methods) without modifying any collected tuple. To reduce the communication cost, we also present two compression methods that reduce the amount of data sent by the server to clients.

4. We compare our method with two PPDC methods and two PPDP methods in a number of experiments using two UCI datasets and two utility measures, including the information loss, which measures utility of a single anonymous tuple, and the classification accuracy of decision trees, which measures the utility of the set of collected tuples. Preliminary results show that our method balances the privacy and the utility better than existing methods do.

1.2   Related Work

Our work differs from well-known generalization-based PPDP methods such as k-anonymity [4, 6] and ℓ-diversity [7] in several ways. First, our algorithm only requires to access collected anonymous tuples, but their algorithms need all original tuples. Second, we consider a distribution attack, which is more general than attacks they consider. Third, we may generalize both QI and SA, but they generalize only QI.

Our work also differs from existing PPDC methods, including those based on random perturbation, such as Warner's method [12], which randomizes tuple with binary attributes, and MRR method [13], which randomizes tuples with multi-valued attributes, and those based on linear algebra, such as EPR method [14]. First, we use generalization rather than perturbation. Second, we utilizes a trusted server but they do not. Although EPR also uses a server to provide a perturbation matrix for respondents to create perturbed tuples, it distrusts the server. Third, our method protects against distribution attack, but theirs do not.

1.3   Road Map

The rest of this paper is organized as follows. In Section 2, we described a distribution attack, a privacy measure, a utility measure, and the Respondent's Problem. In Section 2.3, we show that the RP problem is NP-hard and present a heuristic algorithm to solve it. Section 4 presents preliminary results of our experiments, and Section 5 concludes the paper.

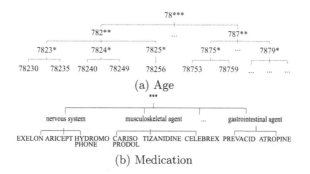

(a) Age

(b) Medication

Fig. 2. Attribute taxonomies

2   Respondent's Problem

Data tuples have a set $\mathcal{A}$ of attributes, where each attribute A has a finite set of values and a taxonomy T_A in the form of a tree (see Figure 2 for taxonomies of some attributes). Each taxonomy defines a relation $\succ$ (*more general than*) and a relation $\succeq$ (*covers*) over the values. For two values v_1 and v_2, $v_1 \succ v_2$ if v_1 is an ancestor of v_2; and $v_1 \succeq v_2$ if $v_1 \succ v_2$ or $v_1 = v_2$. We call a tuple a *base tuple* if its components are leaf values and let $\mathcal{B}$ be the set of base tuples. Both $\succ$ and $\succeq$ can be extended to tuples and components of tuples.

In our framework (see Figure 1), there is one collector (server) and a finite number of respondents (clients). Each respondent has one original tuple (which is a base tuple) and will independently create (and submit to the collector) at most one anonymous tuple. We emphasize that anonymous tuples can be submitted in any order. The collector provides respondents the set of anonymous tuples already collected. The adversary can be a respondent or the user, but not the collector.

Name	Age	Zipcode	Industry	Medication	Age	Zipcode	Industry	Medication
Mike	[26,30]	78240	Univ.	EXELON	[26,30]	7824*	Univ.	nervous
Andrew	[31,35]	78256	Electronics	Celebrex	[31,35]	78256	Retail	muscloskeletal
Denny	[26,30]	78249	Univ.	atropine	[26,30]	7824*	Education	atropine
Amy	[31,35]	78230	Furniture	EXELON	[31,35]	7823*	Furniture	nervous
Angela	[51,55]	78249	Residential	Celebrex	[51,55]	78249	Building	muscloskeletal
Leo	[31,35]	78256	Furniture	Celebrex	[31,35]	78256	Furniture	muscloskeletal
Mark	[31,35]	78235	Furniture	atropine	[31,35]	7823*	Retail	atropine

(a) Original tuples   (b) Anonymous Tuples

Fig. 3. Original vs Anonymous Data

Example 1. In Figure 3, original tuples of a set of respondents are listed in table (a) for convenience. Table (a) itself does not exist in the CS2U model. Table (b) lists anonymous tuples collected by the server at the end of PPDC process. These tuples are obtained using the algorithm described in Section 3.2. Suppose anonymous tuples are created and received in the same order as tuples are listed in table (a), the anonymous tuple of Denny (the third tuple of table (b)) will be created based on the first two tuples of table (b), that of Amy based on the first three tuples of table (b), and so on.

2.1   Linking Probability

Tuples are in the form of <QI, SA>, where QI (quasi-identifier) is a set of attributes, such as Age and Zipcode, which if joined with some publicly available database, such as the voter registry, may reveal identities of some respondents; and SA (sensitive attributes) is a set of attributes[1], such as Medication, that

---

[1] Without loss of generality, we assume that SA is a single attribute.

needs to be protected. Let O be a set of original tuples[2,3] and o be a random tuple from O. The probability for o to be $<q, a>$ is $p(q, a) = \frac{f_{q,a}}{|O|}$, where $f_{q,a}$ is the frequency of $<q, a>$ in O.

The adversary is interested in the conditional probability $Pr[SA = a|QI = q]$ (or simply $Pr[a|q]$), which measures the strength of a link between (the QI value of) a respondent and an SA value. Obviously, $Pr[a|q] = \frac{p(q,a)}{p(q)} = \frac{f_{q,a}}{f_q}$ and $f_q = \sum_{a \in \mathcal{B}[SA]} f_{q,a}$. Since respondents do not submit original tuples, the adversary does not know the probabilities of original tuples. However, with knowledge about a set G of collected anonymous tuples, taxonomies and value distributions of attributes, the adversary can estimate the linking probability $Pr[a|o[QI]]$ of any respondent o for any base SA value a. Specifically, the adversary can identify a *anonymous group*, which is a set of anonymous tuples whose QI values cover $o[QI]$, i.e., $G_o = \{t \mid t \in G, t[QI] \succeq o[QI]\}$, and then estimate $Pr[a|o[QI]]$ as follows.

$$Pr[a \mid o[QI], G] = \frac{\hat{f}_{o[QI],a}}{\hat{f}_{o[QI]}} = \frac{\sum_{t \in G_o} \phi_{QI}(o, t) \cdot \phi_{SA}(a, t[SA])}{\sum_{t \in G_o} \phi_{QI}(o, t)} \tag{1}$$

where $\hat{f}_{o[QI],a}$ and $\hat{f}_{o[QI]}$ estimate $f_{o[QI],a}$ and $f_{o[QI]}$, respectively. To emphasize that the estimates are based on the collected data, G is listed as a condition. We call this the *distribution attack*, which can succeed if the probability distribution is severely skewed.

We now explain Eq. (1). Since tuples in the anonymous group cover the $f_{o[QI]}$ occurrences of tuple $<o[QI], *>$, we can imagine spreading $f_{o[QI]}$ units of support of $<o[QI], *>$ over tuples in G_o. Each tuple in G_o contributes a fraction of a unit that is inversely proportional to the number of base tuples it covers. Thus, we have the following estimate.

$$\hat{f}_{o[QI]} = \sum_{t \in G_o} (\prod_{A \in QI} \frac{l(o[A], t[A])}{l(t[A])}) = \sum_{t \in G_o} \phi_{QI}(o, t) \tag{2}$$

Here, a weight can be assigned to each leaf value in taxonomies according to value distributions of attributes, and $l(v)$ (resp. $l(v, v')$) is the total weight of leaf nodes of node v (resp. leaf nodes shared by nodes v and v'). Intuitively, $\phi_{QI}(o, t)$ is the fraction contributed by t . Similarly, we also have

$$\hat{f}_{o[QI],a} = \sum_{t \in G_o} \frac{l(a, t[SA])}{l(t[SA])} \prod_{A \in QI} \frac{l(o[A], t[A])}{l(t[A])} = \sum_{t \in G_o} \phi_{SA}(a, t[SA])\phi_{QI}(o, t) \tag{3}$$

Definition 1. *Let τ be a privacy threshold, o be a respondent (who has not yet submitted a tuple), G be the set of tuples collected (prior to o's submission), and t be the anonymous tuple from o. We say that (the submission of) t protects o if $Pv(o, G \cup \{t\}) \geq \tau$, where $Pv(o, G') = 1 - \max_{a \in \mathcal{B}[SA]}\{Pr[a \mid o[QI], G']\}$ is the privacy of o with respect to dataset G'.*

---

[2] For convenience, we use respondent and original tuple interchangeably.
[3] Since no personal identity is available, all datasets are multi-sets with possibly duplicate tuples.

2.2 Loss of Information

To find optimal anonymous tuples for respondents, we use a generic measure of utility of a tuple: the *loss of information of a tuple* t, defined as $L(t) = \sum_{A \in \mathcal{A}} L(t[A])$, where $L(v)$ is the *loss of information* of a value v. There are many ways to measure the *loss of information of a value* v. For example, $L(v) = \frac{I(v)-1}{I(r_A)}$, where $I(v)$ is the total number of leaves of nodes v in the taxonomy and r_A is the root of the taxonomy.

2.3 Problem Statement

Let o be a respondent, G be the set of anonymous tuples collected before o submits a tuple, and τ be a privacy threshold. The Respondent's Problem (RP) is to find a generalized tuple t, such that, $t = \text{argmin}_{t' \succeq o}\{L(t')\}$, subject to $Pv(o, G \cup \{t\}) \geq \tau$.

3 Analysis and Solution

3.1 Theoretical Results

The following theorem characterizes a possible solution (i.e., an anonymous tuple that protects the respondent). Due to space limit, proofs are omitted here, but they can be found in a full paper.

Theorem 1. *An anonymous tuple* $t \succeq o$ *can protect respondent* o *if and only if* $\phi_{QI}(o,t)[\phi_{SA}(a, t[SA]) - (1 - \tau)] \leq M_a$, $\forall a \in \mathcal{B}[SA]$, *where* $M_a = (1 - \tau) \sum_{t' \in G_o} \phi_{QI}(o, t') - \sum_{t' \in G_o}[\phi_{QI}(o, t') \cdot \phi_{SA}(a, t'[SA])]$ *is the* margin of protection *wrt* o *and* a.

Given a set of collected tuples and an SA value v, both M_a and $\phi_{SA}(a, v)$ are fully determined for every base SA value a. We can partition base SA values into 8 subsets depending on whether each of M_a and $\phi_{SA}(a, v) - (1 - \tau)$ is equal to, less than, or greater than 0. Each subset (D_i) imposes a lower bound (D_i^-) and an upper bound (D_i^+) on $\phi_{QI}(o, t)$. We use these bounds to determine the existence of a solution to RP that has SA value v.

Corollary 1. *A generalized tuple with a given SA value* $v \succeq o[SA]$ *can protect respondent* o *iff* $m_1 = 0 < \phi_{QI}(o, t) \leq m_2$ *or* $0 < m_1 \leq \phi_{QI}(o, t) \leq m_2$, *where* $m_1 = \max\{D_i^- \mid i = 1, \ldots, 8\}$ *and* $m_2 = \min\{D_i^+ \mid i = 1, \ldots, 8\}$.

Thus, we can partition general tuples that cover o into subsets based on their SA values and test each partition according to Corollary 1. For each partition that contains a solution, we must solve the following RP-FixSA (i.e., RP with fixed SA value) problem to find an optimal solution tuple in the partition.

$$t = \text{argmin}_{t'}\{L(t') \mid t'[SA] = v, t' \succeq o\} \text{ subject to} \qquad (4)$$
$$\phi_{QI}(o, t)(\phi_{SA}(a, v) + \tau - 1) \leq M_a, \forall a \in \mathcal{B}[SA] \text{ such that } v \succeq a$$

However, this problem is *NP-hard* (in terms of the number of attributes and the size of attribute taxonomies). Consequently, RP is also *NP-hard*.

Theorem 2. *RP-FixSA is NP-hard.*

Corollary 2. *The Respondent's Problem is NP-hard.*

3.2   Create Anonymous Tuples

The above analysis leads to the algorithm in Figure 4, which solves RP heuristically at the client-side by first finding anonymous tuples that are possible solutions, and then choosing one of these tuples that minimizes information loss. Specifically, steps 6-7 test whether a partition may contain a solution according to Corollary 1 and findOptimalQIValue() in step 8 solves RP-FixSA.

Input: a private tuple o, a privacy threshold τ, the set G of collected tuples
Output: tuple $t \succeq o$ that satisfies $Pv(o, G \cup \{t\}) \geq \tau$ and minimizes $L(t)$
Method:
1. compute anonymous group G_o;
2. $T = \phi$;
3. for each base SA value a, compute margin of protection M_a;
4. for each SA value v in the path from $o[SA]$ to root do
5. $t[SA] = v$;
6.   compute m_1 and m_2;
7.   if $(m_1 = 0$ and $m_1 < m_2)$ or $(m_1 > 0$ and $m_1 \leq m_2)$
8. $t =$ findOptimalQIValue(t, o, G_o, m_2);
9.   if $t \neq null$
10.   if $\phi_{QI}(o, t) < m_1$
11. $t =$ specializeQI(m_1);
12.   if $m_1 \leq \phi_{QI}(o, t) \leq m_2$
13. $T = T \cup t$;
14. $t = \text{argmin}_{t' \in T}\{L(t')\}$;
15. return t;

Fig. 4. Algorithm: Generalize a Tuple for a Respondent (GTR)

3.3   Compress Collected Data

On the server-side, we reduce the communication cost by compressing the collected data with a simple incremental compression method. We represent each group of duplicate tuples as a tuple-count pair (t, c_t). By encoding attribute values, unique tuples are represented as numeric vectors with a total order. The tuple in each tuple-count pair is represented either by its index in this total order or by a compact binary form, giving two compression methods.

4   Experiments

We implemented five algorithms: the GTR algorithm, two existing PPDC methods: multi-variant random response (MRR) [13] and eigenvector perturbation

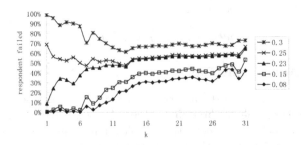

Fig. 5. Percentage of Respondents Not Protected by EPR (Adult dataset)

response (EPR) [14], and two well-known PPDP methods: Local Encoding k-anonymity (LRKA) [16] and ℓ-diversity (LD)[7]. We used two datasets from the UCI Machining Learning Repository [17]: the Adult and the Nursery, and two utility measures: information loss and classification accuracy of decision trees (learned using the ID3 decision tree mining algorithm [18]) in our experiments.

4.1 Protection against Distribution Attack

In this experiment, we investigated whether existing PPDC methods can protect against the distribution attack. Figure 5 shows the percentage of respondents (based on Adult dataset) who are not protected by EPR under a distribution attack (i.e, the privacy does not satisfy $Pv(o, G) \geq \tau$). For most values of τ, EPR fails to protect all respondents no matter what value its own privacy parameter k^* is. For example, for $\tau = 0.3$, EPR can protect at most 39% of respondents for $k^* = 31$, and less if k^* is smaller (despite that smaller k^* is thought to improve the privacy for EPR). If $k^* = 1$, EPR can protect only 0.1% of respondents. Similarly, MRR (not shown here) can only protect 40 to 60% of respondents. As a baseline of comparison, GTR always protects 100% of respondents for all values of τ, because it will not submit any data for a respondent if that person's privacy cannot be guaranteed.

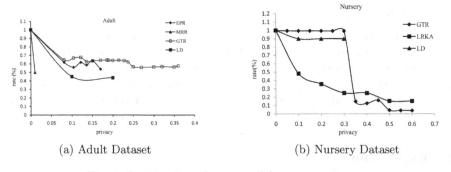

(a) Adult Dataset   (b) Nursery Dataset

Fig. 6. Decision Tree Accuracy of Anonymous Data

4.2   Utility of Collected Data

In this experiment, we compared utility of data obtained by different methods under comparable privacy guarantees. In Figure 6, privacy (x-axis) is measured by $Pv(o, G) \geq \tau$. We only show ranges of τ in which SA are not generalized to the root. LRKA does not appear in (a) because its results do not satisfy any τ no matter what the value of k is. For the same reason, EPR and MRR do not appear in (b). The utility (y-axis) is measured by the relative classification accuracy, that is the accuracy of decision trees learned from anonymous data divided by accuracy of decision trees learned from the original data. As shown in Figure 6, GTR outperforms these existing methods when $\tau \leq 0.36$. When $\tau > 0.4$, all methods produce very poor data because anonymous tuples are too general. The results for information loss are similar and omitted due to space limit.

5   Conclusions

In this paper, we present a novel PPDC method that lets respondents (clients) use generalization to create anonymous data in a CS2U model. Generalization has not been used for PPDC. We propose a new probabilistic privacy measure to model a distribution attack and use it to define the respondent's problem (RP) for finding an optimal anonymous tuple. We show that RP is NP-hard and present a heuristic algorithm for it. Our method is compared with a number of existing PPDC and PPDP methods in a number of experiments based on two UCI datasets and two utility measures. Preliminary results show that our method can better protect against the distribution attack and provide good balance between privacy and data utility.

Acknowledgement

The authors wish to thank the five anonymous reviewers for their constructive comments, which helped us to improve the quality of the paper. The work of Weining Zhang was supported in part by NSF grant IIS-0524612.

References

1. Cranor, L. (ed.): Communication of ACM. Special Issue on Internet Privacy vol. 42(2) (1999)
2. Agrawal, R., Srikant, R.: Privacy-preserving data mining. In: ACM SIGMOD International Conference on Management of Data, pp. 439–450. ACM, New York (2000)
3. Evfimievski, A., Gehrke, J., Srikant, R.: Limiting privacy breaching in privacy preserving data mining. In: ACM Symposium on Principles of Database Systems, pp. 211–222. ACM, New York (2003)

4. Samarati, P., Sweeney, L.: Protecting privacy when disclosing information:k-anonymity and its enforcement through generalization and suppression. In: Proc. of the IEEE Symposium on Research in Security and Privacy (1998)
5. Aggarwal, C.C.: On k-anonymity and the curse of dimensionality. In: International Conference on Very Large Data Bases, pp. 901–909 (2005)
6. Yang, Z., Zhong, S., Wright, R.N.: Anonymity-preserving data collection. In: International Conference on Knowledge Discovery and Data Mining, pp. 334–343 (2005)
7. Machanavajjhala, A., Gehrke, J., Kifer, D., Venkitasubramaniam, M.: ℓ-diversity: Privacy beyond k-anonymity. In: IEEE International Conference on Data Engineering (2006)
8. Li, N., Li, T.: t-closeness: Privacy beyond k-anonymity and l-diversity. In: IEEE International Conference on Data Engineering (2007)
9. LeFevre, K., DeWitt, D.J., Ramakrishnan, R.: Mondrian multidimensional k-anonymity. In: IEEE International Conference on Data Engineering (2006)
10. LeFevre, K., DeWitt, D.J., Ramakrishnan, R.: Incognito: Efficient fulldomain k-anonymity. In: ACM SIGMOD International Conference on Management of Data (2005)
11. Ghinita, G., Karras, P., Kalnis, P., Mamoulis, N.: Fast data anonymization with low information loss. In: International Conference on Very Large Data Bases, pp. 758–769 (2007)
12. Warner, S.L.: Randomized response: A survey technique for eliminating evasive answer bias. Journal of American Statistical Association 57, 622–627 (1965)
13. Du, W., Zhan, Z.: Using randomized response techniques for privacy-preserving data mining. In: International Conference on Knowledge Discovery and Data Mining (2003)
14. Zhang, N., Wang, S., Zhao, W.: A new scheme on privacy-preserving data classification. In: International Conference on Knowledge Discovery and Data Mining, pp. 374–382 (2005)
15. Huang, Z., Du, W., Chen, B.: Deriving private informaiton from randomized data. In: ACM SIGMOD International Conference on Management of Data, pp. 37–47 (2005)
16. Du, Y., Xia, T., Tao, Y., Zhang, D., Zhu, F.: On multidimensional k-anonymity with local recoding generalization. In: IEEE International Conference on Data Engineering (2007)
17. The uci machine learning repository, http://mlearn.ics.uci.edu/MLRepository.html
18. Han, J., Kamber, M.: Data Mining Concepts and Techniques. Elsevier, Amsterdam (2006)

Processing Aggregate Queries on Spatial OLAP Data

Kenneth Choi and Wo-Shun Luk

School of Computing Science, Simon Fraser University, Canada
kennethc@sfu.ca, woshun@sfu.ca

Abstract. Spatial OLAP refers to the confluence of two technologies: spatial data management and OLAP for non-spatial data. An example query on spatial OLAP data is to find the total amount of gas sold within a period of time, of certain type of gas, to a certain type of customers, by gas stations located within a query window on a map. A framework has been established in [8], which converts this query into a set of queries for a general-purpose ROLAP system. However, there has been little done at the query optimization level, once they are submitted to the OLAP system. In this paper, we develop and implement 3 query processing strategies of very different nature on our experimental MOLAP system. Detailed experimental performance data are presented, using a real-life spatial database with 1/3 million of spatial objects.

1 Introduction

Spatial data handling has been a regular feature of many mainstream database systems. Recently, researchers have begun to focus on processing of spatial data for decision making purposes. Appropriately, this area of research is often called spatial OLAP (OnLine Analytic Processing). Spatial OLAP arises from the confluence of two popular database technologies: i.e., spatial data management and OLAP technology, which is up to now mostly concerned with non-spatial data. Already there is considerable interest in exploiting the potentials of spatial OLAP in GIS (Geographic Information Systems) applications, (e.g., [1] and [6]), although GIS researchers do not always define spatial OLAP exactly as the database researchers do.

Much research about spatial data management focuses on querying spatial data. A spatial query is usually in the form of a query window, which is a 2-dimensional "rectangle". The answer to this query is the set of spatial objects located inside the rectangle. A spatial OLAP query typically requests aggregate information about the non-spatial aspects of the spatial objects inside the query window.

There are two approaches to the spatial OLAP querying. The choice appears to be application dependent. In [4] and [5], an application of traffic supervision is considered. The system is to track the positions of cars on road. A typical query to the system is to find the total number of cars on the road segment inside a query window. An aggregate R-tree solution is proposed to store the necessary information for update and query answering. This R-tree augments a regular R-tree with aggregate information, e.g., the number of cars, stored in each tree node. An alternate approach is to represent the spatial data as a regular dimension along with non-spatial dimensions. An OLAP cube is built for these dimensions. A separate R-tree is built, for the

I.-Y. Song, J. Eder, and T.M. Nguyen (Eds.): DaWaK 2008, LNCS 5182, pp. 125–134, 2008.

purpose of locating interesting spatial objects inside the query window. This approach is adopted by two papers, e.g., [7], [8]. The example application described in these two papers is about gas sales by gas station in a neighborhood. The database contains details of each gas sale: gas station ID, gas type, customer ID, transaction time, sales amount and quantity per transaction. In addition to the non-spatial data, the location of each gas station, in x-y coordinates, is also stored. A typical spatial query is: "what are the total sales of each type of gas for taxis (as customer) at gas stations within a query window?" To process queries of this type efficiently, a regular OLAP is built to store the non-spatial data, and an R-tree is built for the spatial data.

This paper considers the issues related to processing aggregate queries on spatial data. A framework is proposed in [8], for converting a spatial OLAP query into a set of regular ROLAP (relational OLAP) queries for processing on a general-purpose OLAP system. The main theme of [8] is to show that processing spatial queries may be more efficient if the aggregate data have been pre-computed. This research is to focus on query optimization inside an OLAP system. Working with experimental OLAP database system [3], which features a multidimensional interface (MOLAP), it is relatively easy, in comparison to ROLAP, to implement a specific query processing strategy and assess quantitatively the direct impact of the strategy on query perform-ance. Two query processing strategies are studied in this paper. The first one selects only a subset of spatial objects inside the query rectangle to be processed against the non-spatial predicate on a regular OLAP system, i.e., with pre-computed aggregates. The second query processing strategy does not rely on any pre-computed aggregates at all, but instead, on computation of the required aggregates on-the-fly.

The rest of this paper is organized as follows. Section 2 provides an overview of spatial data model and spatial query. In Section 3, an architecture for spatial query processing is presented, which allows us to elaborate how our approach to spatial query processing differs from the approaches described in [7] and [8], i.e., the intro-duction of query engine in our approach. In Section 4, we present three different ver-sions of query engines, one of which is a generic version. In Section 5, experimental results are presented to compare the performance of these versions of query engine, against the generic version. Section 6 is the conclusion.

2   Spatial Data Model and Spatial Query

An OLAP database is seen as a *cube* which is defined in this paper to be a k-dimensional array of cells, where k > 0. Each dimension D_i of a cube has $|D_i|$ mem-bers, $1<=i<=k$, which are organized as a *hierarchy*. The members at the leaf level are called *primary* members. All other members in a higher level of the dimension hierar-chy are called *group* members. The root of hierarchy is named after the dimension name. The hierarchy is a *tree* hierarchy, where a member is assumed to have exactly one parent, except for the root, which has no parent. In particular, there is exactly one path between a group member and any of its descendants.

A *spatial* OLAP database is in structure identical to a general-purpose MOLAP database. The main difference between the two is about the design of the cube. In a spatial OLAP database, one of the dimensions, say the kth one, is designated as the spatial-object (SO) dimension. Associated with a primary member in the SO dimension is an ID of spatial object. This is called *primary spatial object ID*, or primary SOID in short. In this paper, a spatial object is assumed to be 2-dimensional, which is identified by a minimum bounded rectangle (MBR). An MBR is represented by the x-y coordinates of its upper-right and lower-left corners. A 2D point, with x-y coordinates, is considered to be a degenerate MBR. The hierarchy associated with the SO dimension is a spatial index, which is assumed to be an R-tree in this paper. A group SOID corresponds to a non-leaf node in the R-tree. Associated with this group member, say G, is an MBR which is the MBR of all rectangles associated with the child nodes of G. The root node of this R-tree is called the *map*. To comply with the definition of OLAP cube, each node of an R-tree has exactly one parent, except the root, which has no parent.

A cell in the cube is identified uniquely by a k-tuple, which is composed of its coordinates along the k dimensions. A cell is a *group* cell if at least one coordinate of the cell is a group member of some dimension; otherwise it is a *primary* cell. A cell stores some numeric values, which are called *measures*. Measures of all primary cells are input from a data source(s). The measure of a group cell has to be derived from the measures of some of the primary cells. In this paper, we assume that all measures of all group cells are pre-computed. The process of pre-computing all group members is called aggregation. Over the last decade, there have been a large number of aggregation algorithms published in the literature. A detailed description of an aggregation algorithm that is closely aligned to this paper can be found in [3]. Clearly, only a tiny fraction of cells in the multi-dimensional array are non-empty. One may submit a query to the Spatial OLAP database to retrieve the measures stored in the cell. This query is essentially the set of k coordinates of the cell.

As an example of the spatial data model, consider a real estate database about the properties in the San Diego, area. The input data consists of data records, each of which describes a property, with the following attributes: *property type, year built, number of bedroom, lot size, region* (9-digit zipcode), *property ID, tax value, and property value*. The first 5 attributes are non-spatial dimensions, property ID is a SO dimension, and the last two are measures. For a non-spatial dimension, its dimension hierarchy is provided by the user, who groups the primary members together according to the requirement of the application. For example, the 'year built' dimension hierarchy has a group member called *modern*, which has as its child members, 1970, ..., 2008. Associated with a property ID are the longitude and latitude of its location in the map. A group member of the property ID dimension is an ID for a non-leaf node in the R-tree. Associated with this group member is the longitude-latitude pairs of the top-right and bottom-left points of its MBR.

A **spatial OLAP query** has a *select* clause and two predicates: *spatial* and *non-spatial* ones. The select clause, just like the one in standard SQL, contains a subset of measure attributes, and their associated distributive aggregation function(s). We assume here the measure is always a designated measure, e.g., tax assessment, and *sum()* is the distributive aggregation function. For brevity, we represent a spatial predicate as a query window, which is a rectangle on the map. Likewise, the non-spatial predicate is represented by a vector $<p_1, .., p_{k-1}>$, such that p_i, $1<=i<=k-1$, is a member of dimensional hierarchy D_i. The answer to this spatial query will be the measures, which are specified in the select clause, of the cells that satisfy both predicates. A cell satisfies the spatial predicate if the spatial location of associated spatial object, by virtue of its x-y coordinates, is found to be inside the query window. A cell satisfies the non-spatial predicate if each of its non-spatial coordinate, v_i, $1<=i<=k-1$, is equal to, or a descendant of, p_i. Note that the spatial object represented in the cell could be a leaf, or non-leaf node in the R-tree.

An example of a spatial query is to find the sum of tax values of all properties that lies inside a query window represented by two longitude-latitude pairs. The non- • spatial predicate is: 'property type' = *property type* AND 'year built' = *modern* AND 'bedroom count' = *3* AND 'lot size' = *48000* AND 'region' = *91901-2400*. A cell, whose property locates inside the query window, of property type 'RSFR', built in year 2003, with 3 bedrooms, on a lot with 48000 sq. ft, with 91901-2400 as the 9-digit zip code, will satisfy both predicates.

3   Architecture for Spatial Query Processing

With the architecture for spatial query processing is shown Fig. 1, we will show how a spatial query is processed, step by step.

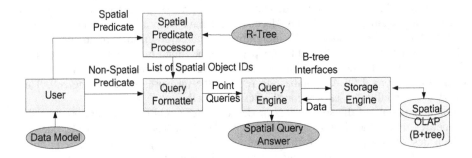

Fig. 1. Architecture for Spatial Query Processing

We start with a **user** submitting a spatial query in the format described in Section 2. It is not quite ready for processing against a general-purpose OLAP database system. The non-spatial predicate needs to be augmented with each of the SOIDs that represent the spatial objects inside the query window. This is the job of the **spatial predicate processor**, which calls the R-tree interface to retrieve a list of nodes in the R-tree whose MBR's overlap with the query window. There are two types of nodes returned: the intersected and embedded nodes. An embedded node is one which is embedded within the confine of the query window, e.g. R_1 in Fig. 2. Since P_5 and P_6 are embedded in R_1, they are not returned; R_1 is. Intersected nodes are non-embedded ones, e.g., R_2 in Fig. 2. Within R_2, primary spatial objects such P_1, P_2, and P_3, are embedded in the query window, but not P_4.

The papers [7] and [8] present different solutions on how these intersected nodes are processed by the Spatial Predicate Processor. In [8], the list of primary SOID's of all intersected nodes inside the query window, e.g. P_1, P_2, and P_3, is returned, together

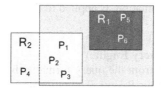

Fig. 2. Embedded & Intersectes Nodes

with the selected group SOIDs corresponding to the embedded nodes, such as R_1, to the **query formatter**. In [7] however, a heuristic is applied, so that only group SOID's are returned. The heuristic is to derive an estimate on how much percentage of the measure of an intersected node should be included in the answer.

Assuming that the user is interested in only exact query answer, we opt for the approach in [8]. However in [8], there is little about how these queries are processed by the underlying general-purpose ROLAP database system. In this paper, we introduce a module called **query engine** (see Fig. 1), which oversees how queries coming from the query formatter are processed, before they are processed by the storage engine. The answer to the spatial query will be made available by the query engine. In Section 4, we will develop three versions of query engines.

At the bottom layer of the query processing architecture is the database **storage engine** which features two interfaces: sequential and indexed ones. The main task of the storage engine is to search the **B+tree**, which stores the contents of the cube. Only non-empty cells are stored. Each such cell is assigned a unique key, called *B-key* in the following manner.

Let $D_1, ..., D_k$ be the 1st to kth dimensions, and $|D_i|$ be the total number of primary and group members for D_i. The dimensions are so named such that $|D_i| <= |D_{i+1}|$, for $1 <= i <= k-1$. All members of each dimension, say dimension i, are mapped, on 1-to-1 basis, into a range of integers, $\{1, ..., |D_i|\}$. The B-key of a (k-dimensional) cell, with the numeric coordinates $(v_1, ..., v_k)$, is given by the formula $v_1 * |D_2| * ... * |D_k| + v_2 * |D_3| * ... * |D_k| + ... + v_k$. Note that this key can be converted back uniquely to the coordinates of the cell. Associated with every non-empty cell is a cell record, which consists of its B-key and its measures. A B+tree is built with B-key as the index, where all cell records are stored on the leaf nodes. This storage structure is sometimes called integrated B+tree clustered on the B-key [2]. A query to the spatial OLAP database, which is in the form of a cell address, is first converted into the B-key of the target cell by the query engine before the call to the storage engine.

4   Query Engines

Generic

This query engine is the simplest of all, and serves as the baseline, to which the performance of all other versions of query engine is compared. It processes the input queries, one at a time. Each query is sent directly to the storage engine. The answer returned by the storage engine is the required measure associated with the query. The aggregation function, e.g., summation, is applied to the retrieved measure, to yield the query result.

SIQUE (Sequential-Index Query Engine)

Let the set of queries coming from the query formatter be called *Q-Batch*, which are sorted in ascending order of their B-keys. The rationale behind SIQUE is the recognition that a significant portion of queries in Q-batch will fail to satisfy the non-spatial predicate. More importantly, these queries will return empty cells. For example, if a property X has 4 bedrooms, the query asking for the tax value of the property with the same SOID as X, with 3 bedrooms, will return an empty cell. Since the B+tree contains only the non-empty cell, the cell corresponding to the query will not exist in the B+tree. Is there a way of screening out those queries from the Q-Batch that are destined to return empty cells, without resorting to storage engine? At present, we don't have a method that guarantees absolute success, i.e., each query the query engine sends to storage engine, will always return an non-empty cell, but the SIQUE we propose here will raise the success rate, by a significant margin, over that of the Generic query engine, which attempts every query in the Q-Batch.

Consider a simple example to show why SIQUE works. Let us suppose the queries in the Q-Batch have the following B-keys: 10, 13, 14, 16, and 20. The B+tree contains in its leaf nodes, cells with B-keys as 15, 16, and 23. The job of SIQUE is to locate the intersection of two sets with as few as queries attempted as possible. SIQUE always starts with the first query, with B-key 10 in the Q-batch. There is no match in B+tree. Suppose the storage engine is modified such that it always returns the B-key of the cell that is next to 10, i.e., 15. With this information, SIQUE skips next two queries (with B-keys 13 and 14), and tries the query with the B-key next to 15, which is 16. This time around there is a matching cell in the B+tree, and the storage engine returns the next B-key, which is 23. Since 23 is greater than any of the B-keys in Q-batch, the merge process is over. There are two queries attempted (with B-keys 10 and 16), and one successful match. That is 60% reduction in number of attempted queries in comparison with the Generic query engine.

Due to space limitation, we will provide only a rough idea how this merge process works, as depicted in Fig. 3. The *query range* is defined as the interval between the smallest and largest B-keys in the Q-batch. Let CR-batch be cell records, in sorted order, in the B-tree whose B-keys fall into the query range. The merge process begins with finding the answer for the first query, Q_0:

While (Q is inside the query range):
1. If the answer to Q is found in CR_0, CR_0 will be retrieved;
2. CR_1 = the CR with the smallest B-key among all CR's whose B-keys > Q;
3. If CR_1 is outside the query range, quit;
4. IF there exists Q_i such that $Q_i = CR_1$ THEN Q = Q_i; ELSE Q = Q_{i+1}, the query with smallest B-key among all Q's whose B-keys > CR_1;

This merge process makes certain that the number of point queries actually processed by the Storage Engine is fewer than either the queries in the Q-Batch, or cell records in the CR-Batch. The actual reduction is apparently very dependent on the contents of the Q-Batch and CR-Batch.

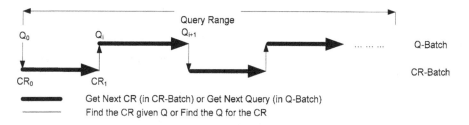

Fig. 3. Merge Process for SIQUE

PICTQUE (Primary Cell Table Query Engine)

The design of this query engine is motivated by the method of processing queries with primary SOID's described in [8]. There, an OLAP databases is constructed for processing queries with primary SOID's. We observe that the size of the set of the all primary cell records is typically only a tiny fraction of the size of the OLAP database. Instead of pre-computing the aggregates for the OLAP database as in [8], we propose to compute the aggregates on-the-fly from only the primary cells in the cube. There will be less I/O time, at the expense of increase in CPU time due to aggregate computation.

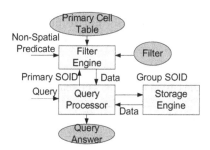

Fig. 4. Architecture of PICTQUE(a)

The architecture as shown in Fig. 1 is modified, with introduction of 3 new components, which are shown in Fig. 4. The component named **Primary Cell Table** (PICT) is a flat file containing all primary records, each of which has a B-key and measures. The **Filter Engine** receives a primary SOID from the Query Engine, and performs a search on the PICT to locate the associated primary cell record. Once retrieved, this record will be subject to a filtering process in Filter Engine to determine whether this record will satisfy the non-spatial predicate. To speed up this process, a data structure called **Filter** has been pre-computed. It contains the ancestor/descendant relationships of all pairs of members in each dimension. Below are the steps for the filtering process:

1. Let the non-spatial predicate be presented as $(p_1, \ldots p_{k-1})$ (see Section 2);
2. For a given primary cell record, decode the B-key into $(v_1, \ldots, v_{k-1}, v_k)$
3. For each i, $1<=i<=k-1$, check whether v_i is a descendant of p_i, or p_i itself.
4. If the check is true for all i's, return the measure(s) in the primary cell record.

The PICTQUE works as follows. Queries are fed into the PICTQUE from the query formatter. If the SOID contained in a query is a group one, it is sent to storage engine to retrieve the answer. Otherwise, it is sent to Filter Engine. Upon return from either engine, the measure will be included in the query answer. The running time of the query engine has three components, given n primary SOID's: the time on n table look-ups on the PICT, the time to decode n 64-bit integers, and n*(k-1) table look-ups on the Filter.

To demonstrate the efficiency of this method, we propose another version of this query engine, where only primary SOID's are used. In other words, all group SOID's are decomposed into primary SOID's, which are then fed into PICTQUE. No calls need to be initiated to storage engine; hence, no access is made to spatial OLAP database. This version of PICTQUE is called PICTQUE(b), to be distinguished from the previous version which is now called PICTQUE(a).

5   Experimental Evaluation and Analysis

In this section, we will present the performance of the four versions of Query Engine, i.e., Generic, SIQUE, PICTQUE(a), PICTQUE(b). For performance testing, the programs are compiled on Visual Studio .NET 2005 on Windows Vista. The hardware we use is Intel-based Core 2 Duo machine which is equipped with 3 GB main memory.

The sample spatial database we use for the experiment is the San Diego real estate database, the contents of which are described in Section 1. The associated dimension hierarchies are shown in Table 1. Note that the dimension hierarchy of Property ID is a 5-level R-tree.

Table 1. Dimension Hierarchies of Real Estate Database

Dimension	Property Type	Year Built	Bedroom Count	Lot Size	Region	Property ID
Levels	2	3	2	3	4	5
Primary Members	32	106	119	1,103	3,138	336,266
Total Members	33	111	120	1,108	3,231	358,078

Five spatial queries are constructed with different non-spatial predicates and query windows. The statistics about the number of spatial objects (i.e., properties or neighborhoods of properties) are shown in Table 2.

Table 2. Numbers of Primary & Group SOID's Inside Query Window

Spatial Query#	Primary SOID's	Group SOID's	Total Primary SOID's
1	2,637	990	81,597
2	4,028	1,135	93,809
3	2,394	1,102	124,110
4	5,979	1,811	166,093
5	1,847	805	259,007

Next, we will compare the performance of these query engines. The running time of each query engine required to process these 5 queries are measured. The buffer is clear of any database cache before the testing of each query engine begins. Consequently, the first query takes much longer time to execute than the rest.

We will first compare the running times of Generic and SIQUE, which are shown in Table 3. A 'real' query is one that returns non-empty cells. The reduction ratio, which is calculated to be percentage of queries that SIQUE has bypassed, is a measure of the efficiency of the SIQUE over the Generic query engine. Clearly the SIQUE outperforms Generic for every spatial query by a wide margin, due to the fact that SIQUE attains on average an 80% reduction ratio.

Table 3. Running Time (in ms): Generic vs. SIQUE

Query #	Query Statistics		Generic		SIQUE		
	Total Queries	'Real' Queries	Running Time	Attempted	Running Time	Attempted	Reduction Ratio
1	3,627	814	577	3,627	531	1,106	70%
2	5,163	1,015	203	5,163	63	1,367	74%
3	3,496	256	124	3,496	31	495	86%
4	7,790	36	219	7,790	15	130	98%
5	2,652	677	109	2,652	47	806	70%
Avg.			246		137		80%

Measurement of running time of PICTQUE(a) and PICTQUE(b) is a more complex. Unlike the Generic or SIQUE, these two query engines require two pieces of data structures: PICT and Filter (see Fig. 4), which together is much smaller in footprint than the spatial OLAP, but much more frequently used. One may read them from the disk on demand, or pre-load them entirely before the querying time. As shown in Table 4, the running time of neither of them does well in comparison with SIQUE, when pre-loading time is included. However, if the cost of pre-loading is allowed to be amortized over more spatial queries, PICTQUE(b) will have the best performance. For example, for 100 similar queries, the average running time, including pre-load, would be about 100 milliseconds, the fastest of all 4 versions.

Table 4. Running Times (in ms): PICTQUE(a) vs. PICTQUE(b)

Spatial Query#	On-Demand		Pre-load	
	PICTQUE(a)	PICTQUE(b)	PICTQUE(a)	PICTQUE(b)
Preloading	0	0	8,455	8,455
1	437	4,056	296	15
2	281	4,774	78	15
3	203	4,602	94	16
4	296	5,959	94	16
5	187	12,714	94	15
avg.	280	6,421	131	15
avg + preloading	280	6,421	1,822	1,706

6 Conclusion

Aggregate queries on spatial OLAP data can be processed accordingly the spatial query processing framework established in [7] or [8]. A query is first pre-processed against a spatial predicate, and the resultant spatial objects are then sent to a standard, non-spatial OLAP system for further processing against a non-spatial predicate. While this approach leverages the power of an off-the-shelf OLAP system for aggregation processing, it however may not be always the most efficient one. In this paper, we introduce the concept of query engine, which is to implement customized query processing strategies, before the queries are submitted to the OLAP system. One such query engine, SIQUE, turns out to be much superior in performance to the standard method, when both of them run against a real-life spatial data. Another query engine, PICTQUE, has the best performance, when pre-loading time is amortized over a large number of queries. Absent the cost of constructing and maintaining the spatial OLAP database, the PICTQUE as a query processing strategy deserves serious considerations for specific spatial OLAP applications.

References

[1] Kheops Technologies Inc., JMAP Spatial OLAP, http://www.kheops-tech.com/en/jmap/solap.jsp

[2] Kifer, M., Bernstein, A., Lewis, P.: Database Systems, 2nd edn. Addison Wesley, Reading (2006)

[3] Luk, W.: A B-tree Based Scheme for OLAP Aggregation and Query Processing (submitted for publication)

[4] Mamoulis, N., Badiras, S., Kalnis, P.: Evaluation of Top-k OLAP Queries Using Aggregate R-Trees. In: Bauzer Medeiros, C., Egenhofer, M.J., Bertino, E. (eds.) SSTD 2005. LNCS, vol. 3633. Springer, Heidelberg (2005)

[5] Papadias, D., Kalnis, P., Zhang, J., Tao, Y.: Efficient OLAP Operations in Spatial Data Warehouses. In: Jensen, C.S., Schneider, M., Seeger, B., Tsotras, V.J. (eds.) SSTD 2001. LNCS, vol. 2121. Springer, Heidelberg (2001)

[6] Pestana, G., Silval, M.: Multidimensional Modeling based on Spatial, Temporal and Spatio-Temporal Stereotypes. In: ESRI International User Conference, San Diego, Calif (2005)

[7] Rao, F., Zhang, L., Yu, X., Li, Y., Chen, Y.: Spatial Hierarchy and OLAP-Favored Search in Spatial Data Warehouse. In: DOLAP 2003, New Orleans, Louisiana (2003)

[8] Zhang, L., Li, Y., Rao, F., Yu, X., Chen, Y., Liu, D.: An Approach to Enabling Spatial OLAP by Aggregating on Spatial Hierarchy. In: Kambayashi, Y., Mohania, M., Wöß, W. (eds.) DaWaK 2003. LNCS, vol. 2737. Springer, Heidelberg (2003)

Efficient Incremental Maintenance of Derived Relations and BLAST Computations in Bioinformatics Data Warehouses

Gabriela Turcu[1], Svetlozar Nestorov[2,3], and Ian Foster[1,2,3]

[1] Department of Computer Science, University of Chicago
[2] Computation Institute, University of Chicago & Argonne National Laboratory
[3] Mathematics & Computer Science Division, Argonne National Laboratory
{gabri,evtimov}@cs.uchicago.edu, foster@mcs.anl.gov

Abstract. In the data driven field of bioinformatics, data warehouses have emerged as common solutions to facilitate data analysis. The uncertainty, complexity and change rate of biological data underscore the importance of capturing its evolution. To capture information about our database's evolution, we incorporate a temporal dimension in our data model, which we implement by means of lifespan timestamps attached to every tuple in the warehouse. This temporal information allows us to keep a full history of the warehouse and recreate any past version for purposes of auditing. Equally importantly, this information facilitates the incremental maintenance of the warehouse. We maintain the warehouse incrementally not only for relations derived by applying the standard relational operators but also for computed relations. In particular, we consider computed relations obtained through external BLAST sequence alignment computations, which are often identified as a bottleneck in the integrated warehouse maintenance process. Our experiments with subsets of protein sequences from the NCBI non-redundant database demonstrate at least 10-fold speedups for realistic target space size increases of 1% to 5%.

1 Introduction

Like many modern sciences, bioinformatics is now mainly data driven [20]. Much effort is placed into developing the means to integrate, analyze and visualize the available data so that its full scientific potential can be realized. Several challenges arise from the complexity and volume of the data and the computational needs that its integration entails. Furthermore, the changing nature and inherent uncertainty of this data prompts keeping track of its evolution.

We build a relational data warehouse to integrate biological sequence information together with derived data and results of applying the BLAST [5] sequence alignment tool to this data. We extend the warehouse schema to allow expression of valid time for each tuple in the warehouse. In this way we can maintain a full history of our data and recreate any past version on demand. We use finite differencing techniques to maintain incrementally both derived relations obtained employing the standard relational operators and relations created through external

I.-Y. Song, J. Eder, and T.M. Nguyen (Eds.): DaWaK 2008, LNCS 5182, pp. 135–145, 2008.

BLAST computations. Using incremental maintenance techniques, we demonstrate that for realistic scenarios of up to 5% percent increase in the size of the biological sequence tables involved in BLAST computations we obtain up to a 10-fold speedup.

2   Background and Related Work

2.1   Biological Data Integration

Several core approaches to biological data integration have been employed in a wide range of systems: *navigational, mediator-based* and *warehouse-based.* Navigational integration requires the user to query data sources individually and follow the links provided. This approach is intuitive and easy to learn, but leaves the integration burden to the user, is sensitive to naming clashes and does not scale well. Examples of such systems include Entrez [1], LinkDB [3] and the Sequence Retrieval System [4].

Mediator-based integration defines a common data model for the integrated sources, while the data resides in the original sources. The mediator is the central part of such a system: it receives the user's query, it translates it to queries that the individual data sources can understand and finally integrates the source's replies into an answer to the original query. The mapping between the global schema and the local schemas of the data sources can be done either by writing the mediator relations in terms of source relations or by expressing the source relations in terms of the global schema. The former solution does not scale very well, while the latter is burdensome on the individual data sources. This solution allows the user to express complex queries in a concise way and provides up-to-date information from the various sources. However, there are several difficulties with this approach: the mismatches of the integrated data are only resolved at query time, the system is sensitive to the data sources becoming unavailable and to network delays, the performance of the system being that of the slowest resource. The complexity of generating and maintaining the source wrappers is one of the main causes that hinder the success of these systems. An example of such a system is K2 [6].

A third approach and the one we adopt is warehousing integration. Analogous to the mediator-based integration, a common data model is defined for the integrated sources. In this case however, the data is brought together under one roof into the warehouse by performing extract-transform-load (ETL) operations. This solution leads to a much better system performance: query optimization can be done locally and views can be built so that user queries are answered faster. Communication latency to access the data sources is eliminated. The system is more reliable since there is less dependency on the network or source availability. The ETL processes offer a valuable opportunity to filter the data. Reconciliation is done at load time and this not only saves response time but helps ensure the correctness of the query reply. The most challenging difficulty of this solution is keeping the warehouse up to date. This is further complicated in the case of biological data warehouses where in most cases the integrated data sources do not display change notification capabilities and more often updates do not reflect minimal changes but come as periodic data dumps instead. Another difficulty is the sensitivity to schema modifications which results in the ETL software being rewritten every time. Furthermore, the creation and maintenance of a warehouse is costly and time consuming.

2.2   Warehousing and Warehouse Maintenance

A data warehouse is defined as a collection of non-volatile data that has been extracted, filtered and integrated from various original data sources with the purpose of supporting subject-oriented data analysis [12].

The maintenance of a data warehouse involves periodic refreshing to keep the data faithful to the original sources as well as purging with the purpose of, for example, eliminating a part of the accumulated historical data from the warehouse. The warehouse refresh can be performed by *full recomputation* or *incremental update* [21]. The full recomputation approach involves simply repeating the ETL operations used to create the warehouse. This process does not involve additional complexity, but is often unacceptably time consuming. An alternative incremental approach determines the changes (additions and deletions) that occur in the original data sources and propagates those increments throughout the warehouse. When increments are relatively small, this solution is advantageous in spite of the additional complexity introduced by the update expressions.

A relational data warehouse can be viewed as a set of physically stored derived relations that are defined as functions over base relations of a relational database. In other words, we can think of a relational data warehouse as a *materialized view* and by extension, relate the warehouse maintenance problem with the well studied problem of materialized view maintenance.

Incremental view maintenance has received a lot of attention from the research community([10,22,11,17,8]). The solutions proposed differ in terms of the amount of information required to perform maintenance, the type of modifications supported (e.g. insert, delete, update), the view language complexity handled (Select-Project-Join (SPJ) views, recursive views, aggregate views) [10].

The class of algorithms we employ is based on the finite differencing program optimization introduced by Paige [15]. Finite differencing tries to replace a global, time-consuming, repeated calculation by incremental modifications done locally. Based on the finite differencing idea, Qian and Wiederhold [17] and Griffin et al [9] derive incremental recomputation rules to update active relational expressions with minimal computational effort. A subset of the relational algebra operators are considered to construct the relational expressions incrementally maintained. Updates are modeled as disjunct sets of deleted/inserted tuples.

2.3   Temporal Data Model

A *temporal database* is a database that incorporates some aspects of time (excluding user-defined time), trying to compensate for the drawback of conventional databases which offer a *snapshot* view of the modeled reality. Two orthogonal dimensions of time are usually considered in temporal databases: *valid time* and *transaction time* [19]. The valid time refers to the time when a fact associated with the database is believed to be true in the real world modeled by the database. The transaction time refers to the time when an entity is current in the database. Thus, transaction time captures the history of an entity within the database. Valid or transaction times can be represented by instantaneous (event) timestamps, interval timestamps or by means of temporal elements (finite sets of intervals) [7,14] and can be associated with individual/groups of attributes/tuples.

2.4   BLAST Computations

BLAST is the de facto standard for biological sequence alignment in computational biology. It obtains increased speed by sacrificing the sensitivity of the dynamic programming algorithm traditionally used to tackle sequence alignment [18]. It matches a set of *query sequences* against a set of *target sequences*, reporting only those alignments that receive a high score according to a prechosen scoring system and are also statistically significant. For each alignment an expect value (*evalue*) is reported, representing the number of alignments expected to be obtained by chance alone with a score at least as high in a search space of the same size. The lower the evalue, the more significant the respective alignment. The evalue of an alignment is calculated based on alignment statistics introduced by Karlin and Altschul [13]:

$$E = K \times m \times n \times e^{-\lambda \times S_R} \tag{1}$$

where m and n represent the effective lengths of the query and the target spaces respectively, K and λ represent Karlin-Altschul parameters and S_R represents the raw score of the alignment obtained directly based on the chosen scoring system.

3   Warehouse Design and Maintenance

3.1   Warehouse Design

Figure 1 describes the overall design of our system and the data warehouse. We use MySQL Server v. 5.0.45 for our implementation.

Base Relations. The source data we consider is represented as a set of GFF3 files containing sequence information made available by NCBI and the non-redundant (nr) database. We adopt the GFF3 format [2] as the exchange format in our system. A GFF3 parser implemented in Python extracts the information of interest into load files that are then loaded into the warehouse and become our base relations. Some of the base relations we store are shown below with their primary key attributes highlighted.

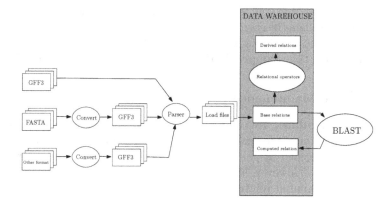

Fig. 1. System and data warehouse overview

features (seqID, featureID, source, type, start, end, score, strand, phase, translationID,ncbi_annotation, creation_time, expiration_time)

feature_dbxrefs (seqID, featureID, dbxref, creation_time, expiration_time)

nr_sequence_fastas (seqID, fasta, creation_time, expiration_time)

Derived Relations. The derived relations we consider are obtained from the base relations of the warehouse to which we apply the standard relational algebra operators (select, project, join). The bioinformatics applications we have encountered so far did not involve aggregation, so our examples and experiments focus on derived relations without aggregation. However, our incremental maintenance framework can be adapted to handle *incrementally computable* aggregate functions [16].

Computed Relations. The relation *blast_results* is created from results obtained by running the BLAST tool on the sequences in the *nr_sequence_fastas* base table. It has the following schema, where the attributes in the primary key have been highlighted:

blast_results (query_ID, subject_ID, identity, align_length, mismatches, gap_openings, query_start, query_end, subject_start, subject_end, e_value, bit_score, creation_time, expiration_time)

These results are loaded into the warehouse directly from the output file produced by the BLAST tool with the -m 8 option.

Temporal Aspects. We extend the schema of all stored relations with two additional attributes (*creation_time* and *expiration_time*) to express the transaction-time lifespan of all tuples in the warehouse. We employ the TIMESTAMP data type provided in MySQL 5.0.45 for both these fields. The *creation_time* attribute represents the time when the corresponding tuple has become current in the warehouse. The *expiration_time* has a NULL value for all the tuples that are current in the warehouse, or the timestamp of the time instant when the tuple became invalid.

3.2 Warehouse Maintenance

Base Relations Update. We consider an update to our warehouse represented as a new set of sequence information files. The update of the base relations can be then simply performed by going through the ETL steps and replacing all the base relations tuples. However, in order to support incremental maintenance for the derived and computed relations in our warehouse and to track the history of the base relations, we need to determine the set of insertions and deletions of tuples in consecutive versions of the base relations. Diff algorithms can be used on consecutive sets of GFF3 files to compute the differences. A more convenient solution is to parse the input files and load the information in the warehouse and compute the additions and deletions using the database system.

In the following, let R represent the current instance of a base relation and R^V its subset of valid tuples (having *expiration_time=NULL*). Let R_{new} be the new timeless (without the the *creation_time* and *expiration_time* attributes) instance of R. For any relation X with a schema of the form $(a_1, \ldots, a_n, creation\_time, expiration\_time)$ let $\overline{X}$ be the corresponding relation where the *creation_time* and *expiration_time* attributes have been projected out. Let δ_R^+ and δ_R^- be the set of timeless tuples that need to be inserted and, respectively, deleted from R in order to update it to R_{new}. Then we have:

$$\delta_R^+ = R_{new} - \overline{R^V} \tag{2}$$

$$\delta_R^- = \overline{R^V} - R_{new} \tag{3}$$

Once the two deltas are computed, we follow two steps to complete the update: invalidate (by setting the attribute *expiration_time = current_session_time*) the tuples of R that are current and whose timeless version are also found in δ_R^- and insert in R the tuples found in δ_R^+ with the time information *creation_time = current_session_time* and *expiration_time = NULL*.

We notice that the following is true: $\delta_R^+ \cap \delta_R^- = \emptyset$. To prove this, considering a tuple $t \in \delta_R^+$, it follows that $t \in R_{new} \wedge t \notin \overline{R^V}$. Since $\overline{R^V} - R_{new} \subseteq \overline{R^V}$ and $t \notin \overline{R^V}$, we have $t \notin \overline{R^V} - R_{new}$, that is $t \notin \delta_R^-$. Therefore $\delta_R^+ \cap \delta_R^- = \emptyset$ and it follows that the two update steps can be performed in any order.

One update step computes $\overline{R^V} \cap \delta_R^- = \overline{R^V} \cap (\overline{R^V} - R_{new}) = \overline{R^V} - R_{new}$ invalidating those tuples of $\overline{R^V}$ that can also be found in δ_R^-. This computation step is performed correctly as there is a unique tuple in $\overline{R^V}$ that corresponds to a tuple in δ_R^- conditioned on no duplicates in any instance of R_{new}.

The next update step consists of adding to stored relation R those tuples that are new, that is the tuples which are found in δ_R^+ expanded with time information accordingly: *creation_time=current_session_start* and *expiration_time=NULL*.

This step will yield our intended result: the tuples expanded with time information can be successfully inserted into relation R as the *creation_time* attribute is part of the primary key for relation R. For example, a tuple identical in its timeless form to a previously invalidated tuple can be added again into the warehouse. The *creation_time* key attribute distinguishes the new tuple from the expired tuple that was created at a different time and is no longer current in the warehouse.

Derived Relations. We express the derived relations that are the target of our maintenance as expressions involving the base relations as operands and the basic relational algebra operators (select, project, join). We construct our update expressions according to the solution proposed by Qian and Wiederhold and Griffin et al [17,9]. For example in the case of a derived relation S that is a selection of a base relation R: $S = \sigma_C(R)$. When we have an update to the relation R then S becomes $S + \delta_S^+ - \delta_S^-$, where $\delta_S^+ = \sigma_C(\delta_R^+)$ and $\delta_S^- = \sigma_C(\delta_R^-)$. Similar update expressions can be written for arbitrary derived relations that are constructed employing the standard relational operators.

BLAST Computations. We consider BLAST computations on subsets of the nr protein data from our warehouse. For a query sequence space Q and a target sequence space T, we want to apply incremental maintenance techniques to alignment results by deleting the alignments that are due to alignment of δ_Q^- to T and $Q - \delta_Q^-$ to δ_T^- and adding in alignments from comparing $Q - \delta_Q^-$ to δ_T^+ and δ_Q^+ to $T - \delta_T^- + \delta_T^+$. This however would not be correct. Recall that equation 1 shows a dependency of the calculated evalue on the size of the search space and that the BLAST tool reports only the high scoring alignments that have an *evalue* below a user specified threshold. So if the query or target search space vary the evalue of an alignment would be different and if the evalue threshold is increased between updates then we cannot reuse results for the previously known sequences as they would not be complete. The interplay between the user specified evalue threshold and computed evalue dependency on search space size leads us to ask the restrictions that the evalue threshold stays constant or decreases, the search space size increases between updates and all other BLAST computation parameters are constant so that the scores of the alignments are preserved. Under these assumptions, we consider the adjustment of evalues. Considering that BLAST reports bit scores ($S_B = (\lambda S_R - \ln K)/\ln 2$) instead of raw scores, we note that by a simple substitution we can rewrite equation 1 as a function of the bit score:

$$E = \frac{m \times n}{2^{S_B}} \tag{4}$$

The $m \times n$ product represents the effective search space which incorporates a length adjustment for both the query and target spaces to factor in the lower likelihood that an alignment starts near the edge of a sequence. We compute this product separately for every (query sequence, target space) pair. Under the stated restrictions, we can now perform incremental update by aligning $Q - \delta_Q^-$ to δ_T^+ and δ_Q^+ to $T + \delta_T^+ - \delta_T^-$ and adjusting the evalues for alignments of $Q - \delta_Q^-$ to $T - \delta_T^-$ according to equation 4.

4   Experimental Results

4.1   Derived Relations

We consider updating incrementally all the derived relations in our warehouse. For example, for our *nr_sequence_fastas* relation (call it R) we could have a derived relation (call it S) of the form *SELECT * FROM nr_sequence_fastas WHERE seqId* $<=$ *'z'*. Below we show the results for increments of various sizes for relation R. $INC(o)$ represents the delta computation for the base relation R. $INC(select)$ represents the delta computation and update of the derived relation S.

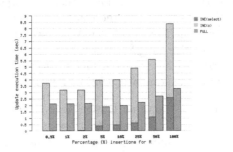

Fig. 2. Comparison of INC and FULL approaches for select relation

The incremental solution is in this case more expensive due to the delta computation overhead. The increments computation is however performed only once as the increments can be propagated over any expression involving the base relation for which they are computed. In a common scenario where several users define materialized views on top of the base relation, this update method becomes advantageous.

4.2   BLAST Computations

We compare the update options for BLAST computation results: *FULL* (full recomputation of BLAST results) and *INC* (incremental BLAST results recomputation as described above).

Q = ct, Positive Increment for T. Figure 3 presents the results for the situation when the query space is constant and there is only a δ_T^+ increment. The query space contains 100 sequences throughout the update, while the target space is initially of size 100000 and then increases by up to 100%.

We note that even for the doubling of the target space size to 200000 sequences, we still have a significant speedup: the incremental update computation would take only roughly 65% of the time with a small overhead.

Q = ct, Increments and Decrements for T. Figure 4 shows the situation in which Q is constant, $|Q| = 10$ and initially we have $|T| = 100000$ and there are both decrements and increments to T such that the overall change to T stays at 10%.

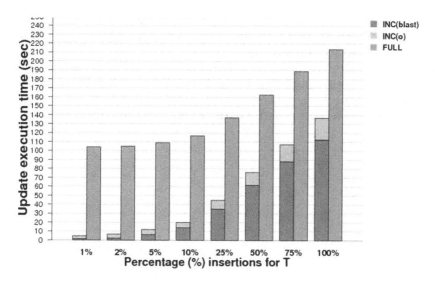

Fig. 3. Comparison of INC and FULL approaches for $|Q|=100$, initial $|T|=100000$

Here we observe that in most cases we do not gain a significant advantage by using the incremental update approach. This is in most cases expected. In the extreme, for a decrement of 100% of T, we would expect that we incur a large overhead by computing the large increments and spend the same time in both approaches to actually yield the alignments for the target space that has been completely renewed. For a realistic scenario in which the increments represent a small portion of the total size of the spaces involved, we can still see a significant gain.

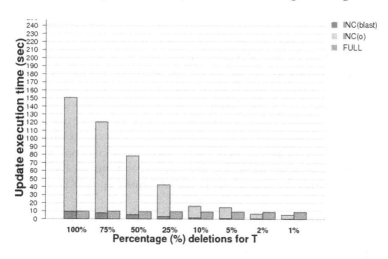

Fig. 4. Comparison of INC and FULL approaches for $|Q|=10$, initial $|T|=100000$, increments and decrements to T

Q, T Vary, Positive Increments. Figure 5 considers the situation in which both Q and T are of equal size and evolve only by positive increments to as much as 100% of their original size.

As figure 5 shows, even when both the query and target spaces double, we still have a significant gain and a relatively small overhead. In reality the most common scenario would place the increments in the 1% to 5% range, for which the incremental update approach takes a quarter the time the traditional approach does.

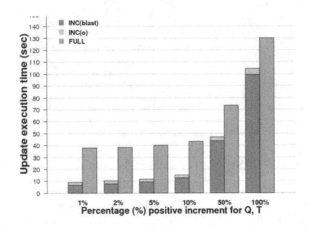

Fig. 5. Comparison of INC and FULL approaches for $|Q| = |T| = 1000$, increments to T,Q

5   Summary

We have presented the incremental maintenance of the derived and computed relations of our integrated bioinformatics data warehouse and the intersection of these techniques with the temporal extension to our warehouse schema. Incorporating temporal information in the warehouse allows us to store its history and recreate any past version on demand. We aim to extend the incremental maintenance techniques to warehouse relations that result from external computations. In particular, we consider BLAST computation results, whose maintenance has been identified as a bottleneck in the update process of integrated bioinformatics warehouses. We show that under a set of restrictions, we can apply incremental maintenance techniques to BLAST results and obtain up to a 10-fold update speedup when the target sequence space size increases up to 5%.

References

1. NCBI Entrez, http://www.ncbi.nlm.nih.gov/Entrez
2. GFF3 specification, http://www.sequenceontology.org/gff3.shtml
3. Kyoto University Bioinformatics Center, LinkDB system,
   http://www.genome.ad.jp/dbget/linkdb.html

4. EBI SRS, http://srs.ebi.ac.uk
5. Altschul, S.F., Gish, W., Miller, W., Myers, E.W., Lipman, D.J.: Basic local alignment search tool. J. Mol. Biol. 215, 403–410 (1990)
6. Davidson, S.B., Crabtree, J., Brunk, B.P., Schug, J., Tannen, V., Overton, G.C., Stoeckert Jr., C.G.: K2/Kleisli and GUS: Experiments in integrated access to genomic data sources. IBM Systems Journal 40 (2001)
7. Dyreson, C., Grandi, F., Käfer, W., Kline, N., Lorentzos, N., Mitsopoulos, Y., Montanari, A., Nonen, D., Peressi, E., Pernici, B., Roddick, J.F., Sarda, N.L., Scalas, M.R., Segev, A., Snodgrass, R.T., Soo, M.D., Tansel, A., Tiberio, P., Wiederhold, G.: A consensus glossary of temporal database concepts. SIGMOD Rec. 23, 52–64 (1994)
8. Griffin, T., Libkin, L.: Incremental maintenance of views with duplicates. In: SIGMOD 1995: Proceedings of the 1995 ACM SIGMOD international conference on Management of data, pp. 328–339 (1995)
9. Griffin, T., Libkin, L., Trickey, H.: An Improved Algorithm for the Incremental Recomputation of Active Relational Expressions. IEEE Transactions on Knowledge and Data Engineering 9, 508–511 (1997)
10. Gupta, A., Mumick, I.S.: Maintenance of Materialized Views: Problems, Techniques and Applications. IEEE Quarterly Bulletin on Data Engineering; Special Issue on Materialized Views and Data Warehousing 18, 3–18 (1995)
11. Gupta, A., Mumick, I.S., Subrahmanian, V.S.: Maintaining views incrementally. In: SIGMOD 1993: Proceedings of the 1993 ACM SIGMOD international conference on Management of data, pp. 157–166 (1993)
12. Inmon, W.H.: Building the Data Warehouse. John Wiley & Sons, Inc., Chichester (1992)
13. Karlin, S., Altschul, S.F.: Methods for assessing the statistical significance of molecular sequence features by using general scoring schemes. Proc. Natl. Acad. Sci. U S A 87, 2264–2268 (1990)
14. Özsoyoğlu, G., Snodgrass, R.T.: Temporal and Real-Time Databases: A Survey. IEEE Transactions on Knowledge and Data Engineering 7, 513–532 (1995)
15. Paige, R., Koenig, S.: Finite Differencing of Computable Expressions. ACM Trans. Program. Lang. Syst. 4, 402–454 (1982)
16. Palpanas, T., Sidle, R., Cochrane, R., Pirahesh, H.: Incremental maintenance for non-distributive aggregate functions. In: VLDB (2002)
17. Qian, X., Wiederhold, G.: Incremental Recomputation of Active Relational Expressions. IEEE Transactions on Knowledge and Data Engineering 3, 337–341 (1991)
18. Smith, T.F., Waterman, M.S.: Identification of common molecular subsequences. Journal of Molecular Biology 147, 195–197 (1981)
19. Snodgrass, R.T., Ahn, I.: A taxonomy of time databases. In: SIGMOD 1985: Proceedings of the 1985 ACM SIGMOD international conference on Management of data, pp. 236–246 (1985)
20. Szalay, A., Gray, J.: 2020 Computing: Science in an exponential world. Nature 440, 413–414 (2006)
21. Ullman, J.D., Garcia-Molina, H., Widom, J.: Database Systems: The Complete Book. Prentice Hall PTR (2001)
22. Vista, D.: Optimizing incremental view maintenance expressions in relational databases. University of Toronto (1997)

Mining Conditional Cardinality Patterns for Data Warehouse Query Optimization

Mikołaj Morzy[1] and Marcin Krystek[2]

[1] Institute of Computing Science
Poznan University of Technology
Piotrowo 2, 60-965 Poznan, Poland
Mikolaj.Morzy@put.poznan.pl
[2] Poznan Supercomputing and Networking Center
Noskowskiego 10, 61-704 Poznan, Poland
mkrystek@man.poznan.pl

Abstract. Data mining algorithms are often embedded in more complex systems, serving as the provider of data for internal decision making within these systems. In this paper we address an interesting problem of using data mining techniques for database query optimization. We introduce the concept of conditional cardinality patterns and design an algorithm to compute the required values for a given database schema. However applicable to any database system, our solution is best suited for data warehouse environments due to the special characteristics of both database schemata being used and queries being asked. We verify our proposal experimentally by running our algorithm against the state-of-the-art database query optimizer. The results of conducted experiments show that our algorithm outperforms traditional cost-based query optimizer with respect to the accuracy of cardinality estimation for a wide range of queries.

1 Introduction

Knowledge discovery is traditionally defined as a non-trivial process of of finding valid, novel, useful, and ultimately understandable patterns and regularities in very large data volumes, whereas data mining is considered a crucial step in the knowledge discovery process, consisting of the application of a given algorithm to a given dataset in order to obtain the initial set of patterns [6]. In this paper we show how data mining can be exploited to enhance the query optimization process. For this purpose we embed a data mining algorithm into the query optimizer of the relational database management system. Somehow contrary to traditional data mining, where domain experts are usually required to assess the quality of discovered patterns or to fine-tune algorithm parameters, embedded data mining solutions do not allow external intervention into the process. Therefore, we make the following assumptions regarding presented solution. Firstly, the data being fed into the algorithm must be cleaned and of high quality as no iteration between data mining phase and data pre-processing phase are possible.

I.-Y. Song, J. Eder, and T.M. Nguyen (Eds.): DaWaK 2008, LNCS 5182, pp. 146–155, 2008.

The results of embedded data mining algorithm must be represented in the form that allows automatic validation and verification, because all results are being instantly consumed by subsequent process steps and no human validation or verification of patterns is possible. Finally, data mining algorithms being embedded outside of knowledge discovery process must not require sophisticated configuration and parametrization, ideally, these should be either zero-conf algorithms or auto-conf algorithms.

In this paper we present a way to embed a data mining algorithm into the query optimization process in order to enhance the quality of estimates made by the optimizer. We analyze the schema and discover potential join conditions between database tables based on referential constraints. For all tables that can be meaningfully joined, we compute conditional cardinalities. Our method works best in the data warehouse environment. In the case of queries asked against a snowflake schema or star schema, our ability to accurately predict cardinalities of attributes for a fixed set of other attribute values (i.e., a set of dimension attribute values) in the presence of many joins and selection criteria helps to discover better query execution plans. The original contribution of the paper is the following. We develop a simple knowledge pattern, called the conditional cardinality. We design an algorithm that identifies suitable pairs of attributes that should be included in conditional cardinality computation. The identification of these pairs of attributes is performed by mining the database dictionary. Finally, we show how the discovered conditional cardinality counts can be used to better estimate the query result size. We verify our proposal experimentally using the state-of-the-art database query optimizer from Oracle 10g RDBMS to prove the validity and efficacy of the approach.

The paper is organized as follows. In Section 2 we briefly discuss related work. Section 3 introduces the concept of conditional cardinality count. We present our algorithm for computing conditional cardinalities in Section 4 and we report on the results of the experimental evaluation of our proposal in Section 5. We conclude this paper in Section 6 with a brief summary and a future work agenda.

2 Related Work

There are numerous works on both query optimization and data mining, but, surprisingly, few works have been published on using data mining techniques to enhance query optimization process [1,9]. Most research focused on enhancing existing statistics, usually by the means of detailed histograms. An interesting idea appeared in [8], that consisted in using approximate histograms that could be incrementally refreshed to reflect the updates to the underlying data. Specialized histograms for different application domains, have been proposed, including data warehouses. For instance, in [3] the concept of using query results for multidimensional histogram maintenance is raised. The dynamic aspect of histograms is addressed in [5] where the authors develop an algorithm for incremental maintenance of the histogram.

Another research domain that influenced our work concerned using complex statistics in query optimization. The use of Bayesian Networks in query optimization is postulated in [7]. The need to reject the attribute value independence assumption is advocated in [10]. Finally, the idea of using query expression statistics for query optimization has been proposed in [2] and a framework for automatic statistics management was presented in [11].

Concepts presented in this paper are similar to the concepts introduced in [4], where the estimation of the aggregated view cardinality is performed using k-dependencies. The main difference is that k-dependencies represent a-priori information derived from the application domain, whereas conditional cardinality patterns are computed automatically from the data.

3   Conditional Cardinality

In this section we formally introduce the concept of conditional cardinality patterns. Let R,S denote database relations, and let $A \in R$, $B \in S$ be attributes A, B of relations R,S, respectively. Let $val(R.A)$ be the number of distinct values of the attribute A in relation R. Traditionally, the selectivity factor for an attribute A is defined as $sel(R.A) = \frac{1}{val(R.A)}$. Let n denote the number of tuples resulting from joining relations R and S on some equality join condition, presumably, using a foreign key constraint. We are interested in finding the number of distinct values of the attribute B in $R \bowtie S$ for a fixed value of the attribute A. Let $\{a_1, a_2, \ldots, a_m\}$ be the values of the attribute A, and let $card(B|a_i) = |\{t \in R \bowtie S : R.A = a_i\}|$ denote the number of distinct values of the attribute B in $R \bowtie S$ where $A = a_i$.

The *conditional cardinality* of the attribute $B \in S$ conditioned on the attribute $A \in R$ is the averaged number of distinct values of the attribute B appearing in the result of the join $R \bowtie S$ for a fixed value of A and is given by

$$card(B|A) = \frac{1}{m} \sum_{i=1}^{m} card(B|a_i)$$

Using conditional cardinality allows for more accurate estimation of the cardinality of a query. Having computed $card(B|A)$ we can estimate the size of the result of a query Q of the form SELECT * FROM R JOIN S WHERE R.A = 'a' AND S.B = 'b' to be

$$card(Q) = \frac{sel(R.A) * n}{card(B|A)} \quad or, \ equally, \quad card(Q) = \frac{sel(S.B) * n}{card(A|B)}$$

Note that we do not consider the quality of the estimation of n, the cardinality of $R \bowtie S$ and we do not require a specific type of join (e.g., a natural join or an outer join). Computing all conditional cardinalities between any pair of attributes from the schema is obviously unfeasible and computationally prohibitively expensive. We compute compute conditional cardinalities only for pairs of joinable attributes, i.e., pairs of attributes from tables that can be joined by one-to-one, one-to-many, or many-to-many relationship. We refer to such conditional cardinalities as *conditional cardinality patterns*. We are well aware that using a grandiose term *pattern* to describe such a simple measure may spur criticism and may seem unmerited. We use this term purposefully to stress the fact that these varying counts are fundamental in the entire cardinality estimation procedure.

4   Algorithm for Estimating Query Result Size

In this section we present an algorithm for estimating the number of tuples returned by an SQL query using conditional cardinality patterns. Our algorithm works under the following three assumptions. Firstly, the database is in either star or snowflake schema. Secondly, only equality conditions are allowed to appear in the query, both for joining tables and for issuing selection criteria. Lastly, queries consist of JOIN and WHERE clauses only, with no subqueries, in-line views, set containment operators, range operators, and such. Below we present the outline of the algorithm. These assumptions reflect the current state of our research, but we expect to relax them as more research is conducted and formulas for arbitrary selection conditions are determined.

4.1   Algorithm Steps

1. Split attributes from the WHERE clause into two element ordered sets $S_i = (A, B)$. In each pair (A, B) the attributes must come from different tables, which are dimensions of the same fact table. If, during set creation, this condition can not be ensured, then the left-over attributes should be placed in singleton sets.
2. Let n be the number of all fact tables used in join clause of query. All n fact tables should be ordered in the way that ensures that the i-th fact table will be in one-to-many relationship with (i+1)-th fact table, for $i = 1, \ldots, n$.
3. For each pair of joinable tables, calculate the join cardinality N. In order to do so, select pairs of attributes from the i-th fact table and one of its dimensions. If it is the first execution of step 3. and step 6. was not executed yet, then N should be equal to the join cardinality of tables from which the attributes come from. Otherwise, if step 6. was executed, N should be equal to the value returned in step 6.
4. Let $A \in R$. C_i is an attribute such that: $C_i \in S_j, S_j \neq R$, where R and S_j are dimensions of the same fact table and C_i belongs to the pair of attributes, which were analyzed in the previous iteration of the algorithm. Let k be the number of such attributes. Then, the selectivity of the attribute A is given by

$$\max \left\{ sel(A), \frac{1}{card(A|C_1)}, \frac{1}{card(A|C_2)}, \ldots, \frac{1}{card(A|C_k)} \right\}$$

5. calculate the cardinality of the result of the query as

$$L = \frac{N * sel(A)}{card(B|A)}$$

If the set of attributes has only one element, then L should be computed as $L = N * sel(A)$

6. If there are other pairs of attributes that have not been evaluated for the i-th fact table then go to step 3. Perform computation for the next pair of attributes, with the exception that N should be set to the value returned in step 6.

7. If there exist fact tables that have not been analyzed, then let S be a dimensional table joining current i-th fact table with the (i+1)-th fact table. Let us use the following notation. Let n denote the number of tuples in the table S, let m denote the number of tuples in the (i+1)-th fact table. For current value of L, computed in step 6. do: $L = \frac{L*m}{n}$ and go to step 2.

Computed value L is the estimation of the number of tuples returned by the query.

4.2 Preprocessing

Analyzing database schema is the first step in the preprocessing procedure. The discovery of relationships between user tables is based on data dictionary view. The view contains information about all user table names, constrain names and their types. Because we are looking for tables that are joined by one-to-many or many-to-many relationships, we only consider tables with primary key and foreign key constraints. Data dictionary is looked up in search of the list of tables that remain in a one-to-many or a many-to-many relationship. Cardinality of the result of join operation on two tables is one of the start parameters in estimation algorithm. To avoid computing this value each time the algorithm analyzes a pair of attributes drawn from these tables, we count the cardinality of the join operation and store this value together with table relationship information.

4.3 Gathering Statistics

Estimation algorithm is based on statistical information about data stored in the database. Conditional cardinality patterns are an extension of traditional statistics gathered in database dictionary. The estimation process makes an implicit assumption that values of all attributes have constant distribution. This assumption is seldom true. Disjunctive attribute values distort estimation process, so we have decided to identify such values and process them in a special way. Let $c_i = |\{r \in R : r.A = a_i\}|$ denote the number of tuples in the relation R, for which the attribute A has value a_i, and let p denote the threshold above which an attribute value is considered disjunctive. Let $\bar{c} = \frac{1}{m} \sum_i c_i$ be the average number of tuples for one distinct value of the attribute A, and let σ_c be the standard deviation of c computed over all possible values of the attribute A. For each $A \in R$ and for each value a_i of A we compute the z-score of c. If z-score falls outside of the range $\langle -p, p \rangle$, then we consider the value a_i of the attribute A as disjunctive. The choice of p is arbitrary and should be guided by the Chebyshev's inequality which states that "no more than $\frac{1}{k^2}$ of the values are more than k standard deviations from the mean". The user should set the value of p accordingly to the desired sensitivity to atypical attribute values. For all attributes we collect information about the number of its distinct values. We also gather information about minimum, maximum and average value of each numerical attribute.

4.4   Conditional Cardinality

Condition cardinality is computed for each ordered pair of attributes (A, B). Pair generation process must ensure that both attributes belong to different relations which remain in one-to-many or many-to-many relationship. Relation pairs and correct attributes can be chosen based on information gathered during preprocessing procedure. By creating a cartesian product of attributes, one can easily find all possible pairs of attributes of the two relations. Because pairs are ordered and $(A, B) \neq (B, A)$, so for each pair we must also create its mirror pair, by inverting attribute positions in the pair. Finally, for each pair we check if the first attribute in the pair has any disjunctive value. If so, then this pair is cloned and saved with annotation about which disjunctive value it applies to. Otherwise the pair is saved without any annotation. Condition cardinality is computed accordingly to its definition from Section 3 for each pair of attributes. If a pair has an annotation about disjunctive value of its first attribute, then condition cardinality is computed as the number of distinct values of second attribute in a joined result relation, where the first attribute is equal to the annotated value.

4.5   Query Generator

Query generator is a simple tool, prepared specially for the experiment. Queries created by the generator consist of three clauses: a SELECT, JOIN and WHERE clauses. The complexity of the last two clauses is controlled by input parameters. From all available tables one table is randomly chosen and it is inserted into the query's JOIN clause. If this table is a fact table, then all its dimensions are also inserted in the JOIN clause. Next, we choose some other fact table that can be joined with current one (fact tables are in a many-to-many relationship) and repeat the insertion procedure. If the next fact table does not exist or the JOIN clause is long enough, then we assume that generation of the JOIN clause is finished. For each table in the JOIN clause we select all attributes that meet our requirements. To satisfy each condition type cardinality, some attributes are taken randomly from this set, and put in the query's WHERE clause in an appropriate form. Finally, the SELECT clause consists of one random attribute form each table in the query's JOIN clause.

5   Experiments

The first goal of the experiment was to prove the existence of correlation between the estimated and the real number of tuples. The second goal of the experiment

Table 1. Statistics of the SH schema

table	size before reducing	size after reducing
COUNTRIES	23	18
CUSTOMERS	55500	1808
PRODUCTS	72	72
CHANNELS	5	4
TIMES	1826	1431
PROMOTIONS	503	4
SALES	918843	75922
COSTS	82112	81425

was the comparison of the accuracy gained by using conditional cardinality patterns with the accuracy of the leading commercial solution. As our testbed we have chosen Oracle 10g database management system. All experiments were conducted on a PC with openSUSE 10.2 GNU/Linux and Oracle 10g Enterprise Edition 10.2 database. All queries were generated on top of the default Sales History (SH) data warehouse schema pre-installed in the database. Because of time and technical constraints, the size of the default SH schema was reduced. All tuples from SALES table were grouped by customer id, and only groups counting between 4 and 70 tuples were retained. Next, we have reduced sizes of all dimension tables by deleting tuples missing from SALES table. Statistics of the original and reduced SH schema are presented in Table 1.

During the experiment 130 different queries were generated, 44 among them returned at least one tuple and 86 returned no tuples. Query generator allowed only for equality conditions in generated queries. Equality conditions are very selective, therefore, the number of conditions allowed in the WHERE clause was set to 10% of all attributes that could have been used in the clause. On average, this setting resulted in WHERE clauses of 6 conditions (not including table join conditions). To ensure optimal conditions for Oracle optimizer all possible statistics were gathered for schema SH. For the estimation algorithm, detail information about table attributes was gathered, as described above, and conditional cardinality for all possible pairs of attributes was computed.

For each query we compute three values: the true number of tuples returned by the query (denoted real), the estimated number of tuples using condition cardinality patterns (denoted cond.card), and the number of tuples estimated by the Oracle optimizer (denoted oracle). Estimations vary from 0 to over 15 000 000. To present results in a more readable form, queries were divided into bins, depending on the their true cardinality. The first bin contains queries that return no rows, the second bin contains queries returning up to 100 rows, the third bin contains queries returning up to 1000 rows, and so on. For each bin data aggregation was performed in the following way: evaluated values were scaled relatively to the biggest value from the bin. This dominating value was assumed to represent 100%, and the remaining two values were calculated as the percentage of the biggest value. The results depicted in Figure 1 are averaged over all queries.

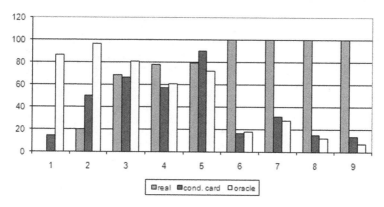

Fig. 1. Precision of cardinality estimates

Based on data received from the test, the Pearson correlation coefficient has been computed between true (random variable X) and estimated (random variable Y) number of returned tuples. The correlation coefficient is $r_{XY} = 0.978$. To prove that the correlation is statistically significant, a t-test has been performed with the null hypothesis H_0 of r_{XY} being insignificant. For the confidence range of 99% the critical value is $t_0 = 2.6148$, whereas the t statistics yields

$$t = \frac{r_{XY}}{1 - r_{XY}^2} * \sqrt{n - 2} = 52.75$$

Because $t > t_0$, we reject the null hypothesis and we embrace the opposite hypothesis of the correlation coefficient being significant.

In our experiment many queries return no rows. To assure that a large fraction of random variable X values being 0 does not bias the test, we have repeated it for only non-zero values of the variable X. This time the Pearson correlation coefficient was $r'_{XY} = 0.982$ and the t-statistics as $t' = 33.91$, so the null hypothesis could have been rejected with confidence level of 99%.

Figure 2 presents mistakes committed by estimations. Based on results from Figure 1, each mistake was calculated as the absolute difference between true query cardinality and the respective estimation. To compare the quality of estimates generated by condition cardinality patterns and the Oracle optimizer let us use the following notation. Let x denote the true number of tuples returned by the query, let y denote the number of tuples estimated by using condition cardinality patterns, and let z denote the number of tuples estimated by the Oracle optimizer. In addition, let Δ be the measurement of estimation quality defined as follows: $\Delta_{xy} = |x - y|$ and $\Delta_{xz} = |x - z|$. All estimates tend to minimize the difference Δ, therefore, for each comparison we find the winner as follows: if $\Delta_{xy} < \Delta_{xz}$ then condition cardinality estimation wins, otherwise if $\Delta_{xz} < \Delta_{xy}$, then the Oracle optimizer wins, otherwise if $\Delta_{xy} = \Delta_{xz}$, then we announce the tie. Table 2 summarizes the comparison of the condition cardinality estimation with the Oracle optimizer estimation as the number of wins, losses, and ties.

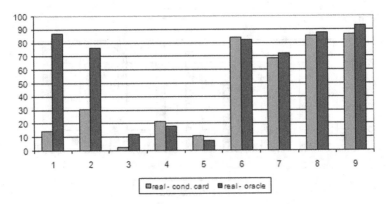

Fig. 2. Comparison of estimate differences

Table 2. Comparison of condition cardinality with the Oracle optimizer

returned rows	$\Delta_{xy} < \Delta_{xz}$	$\Delta_{xz} < \Delta_{xy}$	total
$x = 0$	72	14	86
$x \neq 0$	29	15	44
total	101	29	130

We can conclude that for 78% of queries condition cardinality estimates are better than the Oracle optimizer. On the downside, we have noticed that when conditional cardinality method miscalculates the cardinality of the query result, usually, the committed error is much larger than the error made by the Oracle query optimizer. We attribute this to the way conditional cardinality is propagated through conditions in the query. When a mistake is made early in the estimation process, this mistake is amplified by subsequent estimations for the remaining selectors, which results in rather a formidable error.

6   Conclusions

In this paper we have used data mining for the query optimization process. We have developed a simple knowledge model, conditional cardinality patterns, and we have designed an algorithm for identifying promising pairs of attributes. We have used discovered patterns to improve the accuracy of cardinality estimation for typical data warehouse queries. Our experiments were conducted against the state-of-the-art query optimizer from Oracle 10g database management system. The results of conducted experiments show clear advantage of using conditional cardinality patterns in the data warehouse query optimization process. This paper reports on the results of the preliminary research conducted in the field. In the future we intend to extend the framework to handle aggregation queries. We also plan to further utilize database dictionary to mine patterns that might be useful for other database related tasks, such as database maintenance, storage optimization or user management.

References

1. Bennett, K., Ferris, M.C., Ioannidis, Y.E.: A genetic algorithm for database query optimization. In: Belew, R., Booker, L. (eds.) Proceedings of the Fourth International Conference on Genetic Algorithms, San Mateo, CA, pp. 400–407. Morgan Kaufmann, San Francisco (1991)
2. Bruno, N., Chaudhuri, S.: Exploiting statistics on query expressions for optimization. In: SIGMOD 2002: Proc. of the 2002 ACM SIGMOD international conference on Management of data, pp. 263–274. ACM, New York (2002)
3. Bruno, N., Chaudhuri, S., Gravano, L.: Stholes: a multidimensional workload-aware histogram. In: SIGMOD 2001: Proc. of the 2001 ACM SIGMOD int'l conference on Management of data, pp. 211–222. ACM, New York (2001)
4. Ciaccia, P., Golfarelli, M., Rizzi, S.: On estimating the cardinality of aggregate views. In: Design and Management of Data Warehouses, p. 12 (2001)
5. Donjerkovic, D., Ioannidis, Y., Ramakrishnan, R.: Dynamic histograms: Capturing evolving data sets. In: ICDE 2000: Proc. of the 16th Int. Conference on Data Engineering, Washington, DC, USA, p. 86. IEEE Computer Society, Los Alamitos (2000)
6. Fayyad, U.M., Piatetsky-Shapiro, G., Smyth, P., Uthurusamy, R.: Advances in Knowledge Discovery and Data Mining. AAAI/MIT Press (1996)
7. Getoor, L., Taskar, B., Koller, D.: Selectivity estimation using probabilistic models. SIGMOD Rec. 30(2), 461–472 (2001)
8. Gibbons, P.B., Matias, Y., Poosala, V.: Fast incremental maintenance of approximate histograms. In: VLDB 1997: Proceedings of the 23rd International Conference on Very Large Data Bases, San Francisco, CA, USA, pp. 466–475. Morgan Kaufmann Publishers Inc., San Francisco (1997)
9. Hsu, C.-N., Knoblock, C.A.: Rule induction for semantic query optimization. In: 11th Int. Conf. on Machine Learning, pp. 112–120. Morgan Kaufmann, San Francisco (1994)
10. Poosala, V., Ioannidis, Y.E.: Selectivity estimation without the attribute value independence assumption. In: VLDB 1997: Proceedings of the 23rd International Conference on Very Large Data Bases, San Francisco, CA, USA, pp. 486–495. Morgan Kaufmann Publishers Inc., San Francisco (1997)
11. Chaudhuri, S., Narasayya, V.: Automating statistics management for query optimizers. In: ICDE 2000: Proceedings of the 16th International Conference on Data Engineering, Washington, DC, USA, p. 339. IEEE Computer Society, Los Alamitos (2000)

Up and Down: Mining Multidimensional Sequential Patterns Using Hierarchies

Marc Plantevit, Anne Laurent, and Maguelonne Teisseire

LIRMM, Univ. Montpellier 2, CNRS, 161 rue Ada, 34392 Montpellier, France
{plantevi,laurent,teisseire}@lirmm.fr

Abstract. Data warehouses contain large volumes of time-variant data stored to help analysis. Despite the evolution of OLAP analysis tools and methods, it is still impossible for decision makers to find data mining tools taking the specificity of the data (e.g. multidimensionality, hierarchies, time-variant) into account. In this paper, we propose an original method to automatically extract sequential patterns with respect to hierarchies. This method extracts patterns that describe the inner trends by displaying patterns that either go from precise knowledge to general knowledge or go from general knowledge to precise knowledge. For instance, one rule exhibited could be *data contain first many sales of coke in Paris and lemonade in London for the same date, followed by a large number of sales of soft drinks in Europe*, which is said to be *divergent* (as precise results like coke precede general ones like soft drinks). On the opposite, rules like *data contain first many sales of soft drinks in Europe and chips in London for the same date, followed by a large number of sales of coke in Paris* are said to be *convergent*. In this paper, we define the concepts related to this original method as well as the associated algorithms. The experiments which we carried out show the interest of our proposal.

1 Introduction

Data warehouses collect large volume of data through time for decision making purpose. As soon as data are described through time, sequential pattern mining is well adapted [1]. Indeed, sequential patterns aim at describing the main trends from a database based on correlations between events through time. However, sequential patterns are mined among only one dimension whereas databases can contain several dimensions. Therefore, they have recently been extended to multidimensional sequential patterns in order to handle this multidimensionality [7,8,10]. Even if multidimensional sequential pattern mining provides a better view of source data for decision support, these methods cannot totally take advantage of the framework of multidimensional databases. In particular, they do not consider hierarchies. Note that mining rules at very high levels of granularity leads to trivial rules whereas mining rules at very low levels of granularity is not always possible because the support value would be too low. The algorithm HYPE [9] defined to take hierarchies into account in the extraction of

I.-Y. Song, J. Eder, and T.M. Nguyen (Eds.): DaWaK 2008, LNCS 5182, pp. 156–165, 2008.

multidimensional sequential patterns has some drawbacks. This approach does not allow to discover patterns such as *"When coke sales increase in U.K, soft drink sales increase one month later in E.U"*. Indeed, the two multidimensional items of the sequence $(U.K, Coke)$ and $(Europe, Soft\ drink)$ are comparable (*i.e.* $(U.K, Coke)$ is more specific than $(Europe, Soft\ drink)$). Note that mining all possible combinations of items is impracticable because of the size of the search space. To the best of our knowledge, there is no sequence mining approach that proposes to take hierarchy into account in a multidimensional framework such that *comparable* items can appear in the discovered sequences.

In this paper, we propose the concepts of *convergent* and *divergent* multidimensional sequences. They provide a more complete knowledge extraction that is better adapted to the main specificity of multidimensional frameworks. Thus, the generation of patterns is either from general items to specific items or from specific items to general ones in order to limit the number of candidate patterns. These new kinds of multidimensional sequences allow to mine longer sequences by modulating the degree of precision/generalization among them. A convergent sequence goes from general knowledge to precise knowledge. As an example, *"when soft drink sales increase in USA, coke sales increase on the west coast whereas lemonade sales increase on the east cost"* is a convergent sequence. A divergent sequence goes from precise knowledge to general knowledge. For instance, *"many sales of beer in London and of wine in France are followed by many sales of alcoholic drinks in Europe"* is a divergent sequence.

The rest of this paper is organized as follows. Preliminary concepts and related work are described in Section 2. We define the convergent and divergent multidimensional sequences and algorithms that allow their discovery in Section 3. Some experiments carried out on synthetic and real data are reported in Section 4. In the last Section, we give some conclusions and perspectives for future researches.

2   Related Work: Multidimensional Sequential Patterns and Hierarchies

Combining several analysis dimensions allows to extract knowledge that well describe the data. [7] was the first work dealing with multidimensional sequential pattern mining. The purchased products are not only described by *date_id* and *customer_id* as in classic sequential pattern mining, but according to a set of dimensions such as *Cust-Grp, City, Cust-Age, etc.*. This approach mines sequences that are defined among only one dimension (*product*). These sequences are described by a multidimensional pattern. Thus, it is impossible to mine combinations of multidimensional pattern through time.

[8] proposes to mine such *inter pattern* multidimensional sequences. Discovered patterns do not only combine several analysis dimensions. These dimensions are combined through temporal dimensions (e.g. time). As an example, in the pattern *"lemonade sales increase in N.Y. then coke sales increase in L.A."*, *NY* appears before *LA* and *lemonade* before *coke*.

In [10], the authors mine for sequential patterns in the specific framework of Web Usage Mining. Even though they consider three dimensions (pages, sessions, days), these dimensions are very particular since they belong to a single hierarchized dimension.

This approach provides a better time management but does not fit to multidimensional framework.

Few approaches handle both hierarchy and multidimensionality in sequential pattern mining. In [10] , dimensions are just used to represent time, so multidimensionality is not really handled. HYPE ([9]) allows the mining of sequences that are defined among different levels of hierarchy. HYPE provides the discovery of rules as *"when drink sales increase in Europe, carbonated water exports increase in France whereas soft drink exports increase in USA"* where different levels of hierarchy are present in the multidimensional sequence. However, this proposal cannot extract sequences with items that are defined on the same dimensions but with different granularities such as $(London, Coke)$ and $(Europe, Soft\ drink)$. Indeed, this approach mines multidimensional sequential patterns from the most specific items in order to preserve its scalability.

3   CD_M2S Convergent or Divergent Multidimensional Sequential Patterns

In this section, we introduce an original concept. Indeed, human mind often thinks in two different and symmetrical ways. Thinking runs from general to specific or from specific to general. We try to replicate these types of reasoning in the knowledge that we want to extract. We introduce the concept of convergent and divergent sequences. First, we present the preliminary definitions associated to multidimensional sequential patterns and hierarchies. We then describe the convergent and divergent patterns and the associated algorithms.

3.1   Preliminary Definitions

Let SDB be a set of *multidimensional data sequences*. Each element of data sequences is defined on a set of m *analysis dimensions* denoted by D_A. Each dimension $D_i \in D_A$ is associated with a domain of values, denoted by $Dom(D_i)$. For every dimension D_i, we assume that $Dom(D_i)$ contains a specific value denoted by ALL_i.

We assume that each dimension $D_i \in D_A$ is associated with a *hierarchy* H_i. Every hierarchy H_i is a direct acyclic graph (DAG) whose nodes are elements of $Dom(D_i)$ and whose *root* is ALL_i. As usual, the edges of such a DAG can be seen as *is-a* relationships. The *specialization* relation corresponds to a top-down path in T_i, *i.e.* a path connecting two nodes when scanning T_i from the root to the leaves. We note that when no hierarchy is defined for a dimension D_i, we consider H_i as being a tree whose root is ALL_i and whose leaves are all the elements of $Dom(D_i) \setminus \{ALL_i\}$.

Every element (item) e_i of a multidimensional data sequence is a tuple $t = (d_1, \ldots, d_m)$ such that for every $i = 1, \ldots, m$, $d_i \in Dom(D_i)$ and d_i is a leaf in H_i. In other words, data sequences are defined at the finest levels of the hierarchies associated to D_A.

The multidimensional sequence database in Table 1 is used to illustrate the different concepts and definitions. It describes the purchases of products carried out in various cities of the world for three different companies identified by an S_{ID}. Items of data sequences are defined on two dimensions: *Place* and *Product*. Dimension *Place* is associated to hierarchy H_{Place} whose root is ALL_{Place}. Element of dimension *Place* are defined through several levels of hierarchy: $ALL_{Place} > Continent > Country > City$. Dimension *Product* is associated to hierarchy $H_{Product}$ whose root is $ALL_{Product}$ and *Coke* and *Wine* are leaves. Part of the hierarchies is illustrated Fig. 3.1.

Table 1. Set of multidimensional data sequences SDB

S_{ID}	Multidimensional data sequences
S_1	$\langle\{(Paris, Coke)\}\{(Paris, Coke)\}\{(London, Coke)\}\{(Tokyo, Coke)\}\rangle$
S_2	$\langle\{(Paris, Coke)\}\{(Lyon, Wine)(Paris, Coke)\}\{(Turin, Coke)\}\{(N.Y, Coke)\}\rangle$
S_3	$\langle\{(N.Y, Wine)\}\{(L.A, Coke)\}\{(Paris, Wine), (London, Wine)\}\rangle$

A *multidimensional item* $a = (d_1, \ldots, d_m)$ is a tuple defined on D_A such that $\exists d_i \neq ALL_i$. It is important to note that a multidimensional item can be defined with any value at any level of the hierarchies associated to the analysis dimensions. For instance, $(Europe, Coke)$ and $(N.Y, Wine)$ are two multidimensional items.

Since multidimensional items are defined at different levels of hierarchies, it is possible to compare them using a specificity relation. Let $a = (d_1, \ldots, d_m)$ and $a' = (d'_1, \ldots, d'_m)$ be two multidimensional: (i) e is said to be *more general* than a' ($a' \subseteq a$) if $\forall d_i$, d_i is an ancestor of $d_{i'}$ in H_i or $d_i = d'_i$; (ii) a is said to be *more specific* than a' ($a \subseteq a'$) if $\forall d_i$, d_i is a descendant of d'_i in H_i or $d_i = d'_i$; (iii) a and a' are said to be *incomparable* if there is no relation between them ($a \not\subseteq a'$ and $a' \not\subseteq a$).

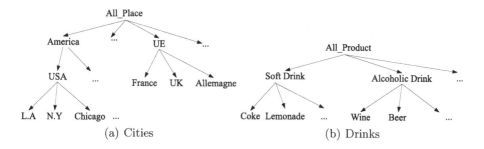

(a) Cities   (b) Drinks

Fig. 1. Part of Hierarchies ALL_{Place} and $H_{Product}$

A *multidimensional itemset* $i = \{a_1, \ldots, a_k\}$ is a non-empty set of multidimensional items such that for all distinct i, j in $\{1 \ldots k\}$, a_i and a_j are incomparable. For instance, $\{(France, Wine), (U.K, Wine)\}$ is a multidimensional itemset. $\{(Europe, ALL_{Product}), (London, Wine)\}$ is not a multidimensional itemset because $(London, Wine) \subseteq (Europe, ALL_{Product})$.

A *multidimensional sequence* $s = \langle i_1, \ldots, i_l \rangle$ is an ordered list of multidimensional itemsets. $\langle \{(USA, Wine)\}\{(France, Wine)(U.K, Wine)\}\rangle$ is a multidimensional sequence associated to the database SDB Table 1.

A multidimensional data sequence S *supports* a multidimensional sequence $s = \langle i_1, \ldots, i_l \rangle$ if for every item a_i of every itemset i_j, there exists an item a_i' in S such that $a_i' \subseteq a_i$ with respect to the ordered relation (itemset i_1 must be discovered before itemset i_2, etc.). According to Table 1, data sequence S_2 supports the sequence $s = \langle \{(France, Coke)\}\{(Europe, Wine)\}\{(USA, Coke)\}\rangle$.

The support of a sequence s is the number of data sequences of SDB that support s. Given a user-defined minimum support threshold *minsup*, a sequence is said to be *frequent* if its support is greater than or equal to *minsup*.

Given a set of multidimensional data sequences SDB that are defined on a set of dimension D_A, the problem of mining *multidimensional sequential patterns* is to discover all multidimensional sequences that have a support greater than or equal to the user specified minimum support threshold *minsup*.

3.2   Convergent and Divergent Multidimensional Sequences

So far, we have defined the problem of mining multidimensional sequential patterns. We have also noticed that taking hierarchies into account provides relations between multidimensional items. Now, we can introduce the concept of convergent and divergent sequences.

Definition 1 (Divergent Sequence). *A sequence $s = \langle i_1, \ldots, i_k \rangle$ is said to be a divergent sequences if for every item $e_j \in i_k$, $\nexists e_{j'}' \in i_{k'}$ such that $k' < k$ and $e_j \subset e_{j'}'$.*

In other words, for each item e of the sequence, there does not exist more general item contained before e in the sequence. The sequence $\langle \{(Paris, Coke)\}, \{(France, Coke)(U.K, Coke)\}, \{(Europe, Coke)\}\{(ALL_{Place}, Coke)\}\rangle$ is a divergent sequence.

Definition 2 (Convergent Sequence). *A sequence $s = \langle i_1, \ldots, i_k \rangle$ is said to be a convergent sequence if for every item $e_j \in i_k$, $\nexists e_{j'}' \in i_{k'}$ such that $k' < k$ and $e_{j'}' \subset e_j$.*

For each item e of the sequence, there does not exist more specific item contained before e in the sequence. The sequence $\langle \{(ALL_{Place}, Wine)\}, \{(Europe, Wine)\}, \{(Italy, Wine)(France, Wine)\}\rangle$ is a convergent sequence.

3.3   Algorithm

Ordering the items in the itemsets of the sequences is a fundamental step to avoid the already examined cases. Existing methods that are based on different paradigms (*pattern growth* ([6]), *Apriori*([1,5,11,2])), are not directly applicable in a multidimensional framework. Indeed, items, that are not defined with the finest level of hierarchy, are not explicitly present in the database. Such items are retrieved by inference since there is no associated tuple in the database.

We then introduce functions to locally handle all items and not only items that are present in data sequences. An itemset is said to be *extended* if it is equal to its closure according to the relation of specialization ($\subseteq$). This notion allows to take all items into account. In order to enhance the management of items, we introduce a *lexicographicospecific order* (lgs) that is an alpha-numeric order according to the precision degree of the item. Thus, the most specific items are the first to be handled. We have to define a function LGS-Closure that transforms an itemset of a data sequence into its extended itemset that contain all the items that can be inferred. As an example, $LGS\text{-}Closure(\{(Paris, Coke)\}) =$ $\{(Paris, Coke), (Paris, ALL_{Product}, (France, Coke), (France, ALL_{Product}), (Europe, Coke), (Europe, ALL_{Product}), (ALL_{Place}, Coke)\}$. We note that the tuple $(ALL_{Place}, ALL_{Product})$ is not considered by definition of multidimensional item.

The extraction of frequent items can be done on each extended itemset. In pattern growth approaches, sequences are extracted by greedily adding a frequent item to a frequent sequence. It is thus necessary to define an efficient way for extending sequences from the last itemset of the sequences. For this purpose, we define the function $LGS\text{-}Closure_X$ where X is an itemset that contains "forbidden items" (*i.e.* every items e_i of the last itemset of the prefix sequence and all items comparable to e_i). As an example, $\{(ALL_{Place}, Wine)\} =$ $LGS\text{-}Closure_{\{(Europe, ALL_{Product})\}}(\{(Paris, Wine), (London, Wine)\})$.

Divergent sequences are discovered thanks to algorithm CD_M2S. To mine all divergent sequences on SDB, routine $CD\_M2S(\langle\rangle, SDB, \emptyset, minsup)$ is called.

This algorithm is pattern growth based [6]. Instead of scanning the whole database, level by level as Apriori based methods, the database is projected according to the *prefix sequence*. This data projection projection is quite different from [6]. Indeed, we have to handle all possible items, so the projection has to take the itemsets of the data sequences that contain the discovered item, and not only the item as in [6].

Two kinds of items can be extracted from the projected database:
1. Items that are added in a new itemset of the prefix sequence α. These items are mined thanks to LGS-Closure.
2. Items (denoted by _e) that are added in the last itemset of the prefix sequence α. In this case, we use the function LGS-Closure$_X$ where X is the last itemset of the sequence α.

This algorithm allows the extraction of divergent sequences. To mine convergent sequences, it is necessary to use the same algorithms but on an **inverted database**. Indeed, beginning by the end (invert the ordered relation) of the data sequences and re-invert the discovered patterns allows a discovery from general to particular case.

Algorithm 1. CD_M2S

Data: Prefix sequence α, projected database $SDB|_\alpha$, set of current frequent sequences FS , minimum support threshold $minsup$

Result: Set of divergent frequent sequences with prefix α

begin

    if $\alpha \neq \langle\rangle$ **then**

        $FS \leftarrow FS \cup \{\alpha\}$;

    $LF \qquad \leftarrow \qquad \{e \quad s.t. \quad support(e, SDB|_\alpha) \quad \geq \quad minsup \quad$ and $\nexists e' \ s.t. \ support(e', SDB|_\alpha) \geq minsup$ and $e' \subset e\}$;

    foreach *items* $e \in LF$ **do**

        $\alpha' \leftarrow \alpha.e$;

        Call $CD\_M2S(\alpha', SDB|_{\alpha'}, FS, minsup)$;

end

---

4   Experiments

In this Section, we report experiments on both synthetic and real data.

Synthetic Data:

Experiments were carried out on a synthetic database. This database contains $10,000$ data sequences (with an average of 47 itemsets) over 5 analysis dimensions. Some hierarchical relations are defined between elements of each analysis dimension. In this paper, we report the behavior of our approach (number of patterns, runtime) according to several parameters (support threshold, $|D_A|$, degree and depth of hierarchies).

Figures Fig. 2(a) and Fig. 2(b) respectively report the runtime and the number of frequent sequences according to the minimum support threshold. The number of sequences tends to increase when the support decreases. The runtime follows the same behavior. However, it is possible that the number of frequent sequences decreases when the support decreases. Indeed, some more specific items can appear. Furthermore, a more general item is faster inferred in a data sequence than a more specific one. Therefore, we can obtain a smaller number of frequent sequences.

Figures Fig. 2(c) and Fig. 2(d) report the runtime and the number of frequent sequences according to the depth of the hierarchies on the analysis dimension. Adding one level in the hierarchies provides more precise data (*soda* becomes *pepsi* or *coca*). There is thus more different values in the database. CD_M2S is robust front of the specialization phenomena. Indeed, even if the data become very specific (5 different levels in the hierarchy), our approach allows to extract some sequences that are described among several level of hierarchy. We can notice that the runtime increases when the number of levels of the hierarchies increases. This is due to the number of potentially frequent items that increases.

Figures Fig. 2(e) and Fig. 2(f) report the behavior (runtime and number of sequences) of our approach according to the degree of the hierarchies. Increasing the degree of a hierarchy provides more specific data (adding a son or an

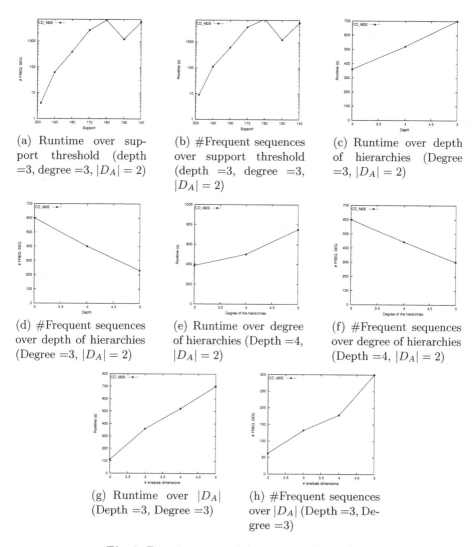

(a) Runtime over support threshold (depth =3, degree =3, $|D_A| = 2$)

(b) #Frequent sequences over support threshold (depth =3, degree =3, $|D_A| = 2$)

(c) Runtime over depth of hierarchies (Degree =3, $|D_A| = 2$)

(d) #Frequent sequences over depth of hierarchies (Degree =3, $|D_A| = 2$)

(e) Runtime over degree of hierarchies (Depth =4, $|D_A| = 2$)

(f) #Frequent sequences over degree of hierarchies (Depth =4, $|D_A| = 2$)

(g) Runtime over $|D_A|$ (Depth =3, Degree =3)

(h) #Frequent sequences over $|D_A|$ (Depth =3, Degree =3)

Fig. 2. Experiments carried out on synthetic data

instance). Our approach allows to discovery knowledge while hierarchies become more specific. However, the cost of the extraction (runtime) is more expensive.

Figures Fig. 2(g) and Fig. 2(h) report the runtime and the number of sequences according to the number of analysis dimensions ($|D_A|$). Adding some analysis dimensions generates an increase of the number of frequent sequences and the runtime.

These experiments on synthetic data show the robustness of our approach according to the diversity of the data (D_A, degree and depth of the hierarchies, etc.). Considering a more diverse source data leads to a more important extraction cost that stays acceptable.

Real Data:
We report experiments on logs of toy game [3] which is an *Eleusis* based card game. This game was created in order to simulate the activity of the scientific discovering (publications, refutations, experiments). The problem in Robert Abbot's Eleusis card game [4] is to find a secret law hidden from the players and determining the valid sequences of cards that can be played during the game. [3] proposes a new version in which humans are helped by machines to produce a theory. Players have to discover a rule (for instance a rule is *"two successive cards must have two different colors"* and a positive example according to this rule is *"ace of hearts followed by king of spades"*). A player wins points if he produces a positive example or he publishes his theory. He also earns points if he refutes a theory published by another player. He loses points if his theory is disproved by another player.

The hidden rules are card sequences. Each sequence contains a left part and a right part. Each part can contain several cards. We have described this problem according to several analysis dimensions: one dimension for the card values (king, queen, ..., ace); one dimension for the card colors (heart, diamond, spade and clubs); one dimension for the position of the card in the sequence(right or left) and one dimension for the oracle answer (true, wrong).

We can obtain convergent and divergent rules. A divergent sequence is: *"For the secret rule water lily, players frequently propose the following cards: three of spade, ace of spade, a odd card of spade and finally a black odd card."*. A convergent sequence is: *"For the secret rule lily, players frequently propose a red card, a card of heart, a numbered card of heart."*. We notice that these rules are relevant for the expert and they cannot be extracted with classical algorithm.

5   Conclusion

We proposed an original method to extract multidimensional sequences that are defined on several levels of hierarchy according to different points of view: from general to particular (convergent) or from particular to general (divergent). We thus defined the concepts of convergent and divergent multidimensional sequences. We also introduced the algorithm CD_M2S that is pattern growth based. Some experiments on synthetic and real data show the interest of our approach. Note that this proposal is totally different from [9]. Indeed, in this paper, we focus on mining for special sequences: divergent or convergent sequences. Such sequences mean that comparable items can appear together within a convergent or divergent sequence whereas they cannot in $HYPE$. Furthermore, $HYPE$ algorithm is APriori based whereas algorithm CD_M2S is pattern-growth based.

This work offers several perspectives. First, divergent sequence can model special behaviours like buzz or the appearance of a seminal paper that leads to lot of publications and applications. Convergent sequences can model behaviors that become specialized through time like the appearance of new scientific topics or marketing products. Therefore, It would be very interesting to focus on the discovery and the prediction of such behaviors. Second, the efficiency of the extraction can be enhanced thanks to condensed representations (closed patterns, etc.) that provide some properties to efficiently prune the search space. Furthermore, other propositions can be done on the management of the hierarchies. We can imagine a modular management of hierarchies where some levels of the hierarchies would be more important (minimal and maximal levels on some hierarchies in order to be not too general or too specific) to fit user needs and to preserve the scalability of the extraction.

References

1. Agrawal, R., Srikant, R.: Mining sequential patterns. In: Yu, P.S., Chen, A.L.P. (eds.) ICDE 1995, pp. 3–14. IEEE Computer Society, Los Alamitos (1995)
2. Ayres, J., Flannick, J., Gehrke, J., Yiu, T.: Sequential pattern mining using a bitmap representation. In: KDD, pp. 429–435 (2002)
3. Dartnell, C., Sallantin, J.: Assisting scientific discovery with an adaptive problem solver. In: Discovery Science, pp. 99–112 (2005)
4. Gardner, M.: Mathematical games. Scientific American (1959)
5. Masseglia, F., Cathala, F., Poncelet, P.: The psp approach for mining sequential patterns. In: Zytkow, J.M., Quafafou, M. (eds.) PKDD 1998. LNCS, vol. 1510, pp. 176–184. Springer, Heidelberg (1998)
6. Pei, J., Han, J., Mortazavi-Asl, B., Wang, J., Pinto, H., Chen, Q., Dayal, U., Hsu, M.-C.: Mining sequential patterns by pattern-growth: The prefixspan approach. IEEE Transactions on Knowledge and Data Engineering 16(10) (2004)
7. Pinto, H., Han, J., Pei, J., Wang, K., Chen, Q., Dayal, U.: Multi-dimensional sequential pattern mining. In: CIKM 2001, pp. 81–88. ACM, New York (2001)
8. Plantevit, M., Choong, Y.W., Laurent, A., Laurent, D., Teisseire, M.: M^2SP: Mining Sequential Patterns Among Several Dimensions. In: Jorge, A.M., Torgo, L., Brazdil, P.B., Camacho, R., Gama, J. (eds.) PKDD 2005. LNCS (LNAI), vol. 3721, pp. 205–216. Springer, Heidelberg (2005)
9. Plantevit, M., Laurent, A., Teisseire, M.: Hype: mining hierarchical sequential patterns. In: DOLAP, pp. 19–26 (2006)
10. Yu, C.-C., Chen, Y.-L.: Mining sequential patterns from multidimensional sequence data. IEEE Transactions on Knowledge and Data Engineering 17(1), 136–140 (2005)
11. Zaki, M.J.: Spade: An efficient algorithm for mining frequent sequences. Machine Learning 42(1/2), 31–60 (2001)

Efficient *K*-Means Clustering Using Accelerated Graphics Processors

S.A. Arul Shalom [1], Manoranjan Dash[1], and Minh Tue[2]

[1] School of Computer Engineering, Nanyang Technological University,
50 Nanyang Avenue, Singapore
[2] NUS High School of Mathematics and Science, Clementi Avenue 3, Singapore
{sal10001,asmdash}@ntu.edu.sg, h0630082@nus.edu.sg

Abstract. We exploit the parallel architecture of the Graphics Processing Unit (GPU) used in desktops to efficiently implement the traditional *K*-means algorithm. Our approach in clustering avoids the need for data and cluster information transfer between the GPU and CPU in between the iterations. In this paper we present the novelties in our approach and techniques employed to represent data, compute distances, centroids and identify the cluster elements using the GPU. We measure performance using the metric: computational time per iteration. Our implementation of *k*-means clustering on an Nvidia 5900 graphics processor is 4 to 12 times faster than the CPU and 7 to 22 times faster on the Nvidia 8500 graphics processor for various data sizes. We also achieved 12 to 64 times speed gain on the 5900 and 20 to 140 times speed gains on the 8500 graphics processor in computational time per iteration for evaluations with various cluster sizes.

Keywords: *K*-means clustering, GPGPU, Computational efficiency.

1 Introduction

Commodity Graphics Processors in today's PC world are highly parallel with extreme computational powers. Modern GPU is capable of processing tens of millions of vertices per second and rasterize hundreds of millions of fragments per second. The schematic in Figure 1 shows the vertex transformation and fragment texturing and coloring stages in a typical graphics pipeline. Due to the fact that these processors are economically affordable and programmable for implementing iterative algorithms there is high possibility that desktop computers with such GPUs will soon be capable of performing fast and efficient computing.

1.1 Graphics Processors for General Purpose Computations

The factors that enable the processing power of GPUs are the inherent parallel architecture, peak memory bandwidth, high floating-point operations and the various hardware stages with programmable processors. High-end graphics processors such as Nvidia GeForce7900 GTX and GeForce8800 Ultra have peak memory bandwidth of about 51.2GB/sec and 103.68GB/sec respectively. The floating-point operations

I.-Y. Song, J. Eder, and T.M. Nguyen (Eds.): DaWaK 2008, LNCS 5182, pp. 166–175, 2008.

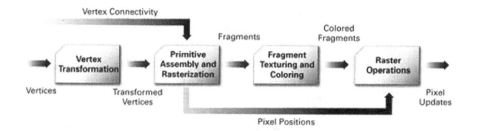

Fig. 1. The Graphics Hardware Pipeline

possible are over 200 GFlops in Nvidia GeForce7900 GTX and 345 GFlops in Nvidia GeForce8800 Ultra. The performance of the GPU in computing general-purpose algorithms depends heavily on how the algorithms are arranged so as to exploit the parallel data processing power of the GPU. In our study, we have used Nvidia's GeForce FX 5900 XT processor and a GeForce 8500 GT processor. In graphics processing, the GPU receives commands for display in the form of vertices and connectivity details from the CPU. Today's GPUs have very high memory bandwidth and parallel internal processors, which are capable to process streams of incoming data. These processing is performed in either the vertex or the fragment processor in the GPU using specific shader programs. Computations in the GPU processors are data independent, which means that the processing that occurs in each processor, is independent of the data used by other processors. Currently, there is lot of research focus in the arena of implementing general-purpose computations in the GPU (GPGPU) to leverage on speed w.r.t unit cost function [9]. Within the GPU, the fragment processors support parallel texture operations and are able to process floating-point vectors. Implementations of GPGPU are challenging and mostly utilize the texture processing abilities of the fragment processor of the GPU.

1.2 An Approach to Implement *K*-Means Algorithm in GPU

In this paper we present an efficient implementation of the *k*-means clustering algorithm completely in the GPU. We realize this by using the multi-pass rendering and multi-shader capabilities of the GPU. This is done by maximizing the use of textures and minimizing the use of shader program constants [3]. In this implementation we have minimized the use of GPU shader constants thus improving the performance as well as reducing the data transactions between the CPU and the GPU. Handling data transfers between the necessary textures within the GPU is much more efficient than using shader constants. This is mainly due to the high memory bandwidth available in the GPU pipeline. Since all the steps of *k*-means clustering could be implemented in the GPU, the transferring of data back to the CPU during the iterations is avoided [13]. The programmable capabilities of the GPU have been thus exploited to efficiently implement *k*-means clustering in the GPU.

2 Existing *K*-Means Clustering Methods Using Graphics Processors

Two existing *k*-means implementations are analyzed in this section [2, 3]. We find out how the current GPU hardware architecture could be further exploited to overcome some of the limitations faced in these implementations [2, 3, 13].

2.1 The *K*-Means Implementation with Textures and Shader Constants

The *k*-means iterative clustering method and few of its variants have been implemented in the GPU [3]. The result is a speed-up in clustering between 1.5 to 3 times compared to the CPU implementation. In this implementation each input point is stored in a single texel of a texture as a float and mapped over several textures. The cluster data is stored in fragment shader constants. The fragment processor accesses the textures to obtain the data points and the fragment constant to obtain the cluster information. The fragment processors and depth tests are used in this implementation. The functioning of the depth buffer is programmed in such a way that the resultant distance is written into the depth buffer and the cluster ID (label) of the cluster that is closest to the data point is written into the color value. The cluster data in the fragment shader constants are updated after the iteration. After all the iterations are complete the cluster IDs are read back to the CPU. In the above implementation it is notable that it is faster to read and to write the cluster data in the fragment shader constants than accessing the data in the textures. But this limits the performance when there are many clusters. Moreover, in this implementation only one clusters' data is stored in the fragment shader constants at any one time. This leads to the fact that the fragment shader constants are to be accessed very frequently for distance computations, comparisons and centroid updating during the iterations, which restrains the possible parallelism in updating clusters. The use of the depth buffer is limited to fixed-point values; which requires the distance metric to be scaled in a range of 0 to 1.

2.2 The *K*-Means Implementation Using Stencil and Depth Buffers

The implementation of *k*-means clustering on commodity GPUs has been presented in [2]. The motivation is to reduce the number of operations in the fragment processor. Data points to be clustered are stored in textures. The distance of each data point to every centroid is computed in parallel. The cluster label of each data point is identified after the computation of distances of each data point to all centroids is done. The computed distances are assigned to the depth buffer by the fragment program. The current distance in the depth buffer is compared with the distance that arrives from the fragment program. Writing the distances into the depth buffer continues until the distance computations to all the centroids are completed. Stencil buffers are defined and maintained for each cluster to keep track of the nearest label. The fragment program is written in such a way that the stencil buffer of each centroid contains the label of the nearest data points and the depth buffer has the corresponding distance values. In this implementation multiple depth buffers and stencil buffers are required to match the number of initial cluster centroids. This becomes a bottleneck when the number of clusters is high. Moreover the use of the depth buffer is limited to fixed-point values; thus the distance metric needs to be scaled and rescaled back for the iterations.

3 Harnessing the Power of GPU in Clustering

Our implementation is done using OpenGL as the Application Programming Interface (API) [4], and the operational kernels are invoked via shader programs, using the Graphics Library Shading Language (GLSL) [11]. The Single Instruction Multiple Data (SMID) technique is employed to achieve data or vector level parallelism in the fragment processor.

Few technical concepts that harness the GPU into a usable parallel processor are:

- ❖ Independence of texture elements (texels) that can be accessed by fragment shaders
- ❖ Mapping of the texture into a required geometric shape
- ❖ Shader as a set of instructions (kernels) that modify mapped texels
- ❖ Shader execution with draw command to output to the memory buffer.

3.1 Efficient Use of GPU Hardware for Parallel Computation

The input data sets are stored in 2 dimensional (2D) textures of size $\sqrt{N}$ x $\sqrt{N}$, where N is the total number of observations. In this approach we use the Luminance format of the texture elements. Luminance texture format allows the texture to store a single floating-point value per texel. The initial centroids, the distances and the other relevant information are stored in the textures. The new cluster centroids are transferred back to the CPU and the stop condition is checked after the iteration.

An important concept in GPGPU is stream programming or parallel programming model. In this model, all data is represented as a stream or "an ordered set of data of the same data type" [10]. The key to maximize parallel processing is to allow multiple computation units to operate on the data simultaneously and to ensure

Table 1. GLSL codes for distance computations

```
char* shader3 = \
"#extension GL_ARB_texture_rectangle : enable\n" \
        "uniform sampler2DRect textureX;" \
        "uniform sampler2DRect textureY;" \
        "uniform float cx;" \
        "uniform float cy;" \
        "void main(void) { " \
"   float x = texture2DRect(textureX,
gl_TexCoord[0].st).x;" \
"   float y = texture2DRect(textureY,
gl_TexCoord[0].st).x;" \
"   gl_FragColor.x = sqrt((x-cx)*(x-cx)+(y-cy)*(y-
cy));"\
        "}";
```

that each unit operates efficiently. This model of parallelism can be achieved easily in the GPU hardware data path [10]. A kernel is defined as an instruction or computation that operates on the entire stream. The kernel is applied to each element of data in the hardware path. The result is dependent only on that particular element and independent of the rest of the stream. We implement the distance metric calculations in parallel and the GLSL fragment codes are listed in Table 1.

The code listing in Table 1 works as follows: The first line is an OpenGL extension that allows the 2 dimensional (2D) texture targets to support data sizes, which are "non-power-of-twos" (NPOT). The next two lines are used in GLSL to access the NPOT 2D rectangular texture sources in parallel. The codes within the main are used to read the data from each texel in the textures and the code line with gl_FragColor performs the actual computation simultaneously on data that is read from each of the texel. The resultant parameter is stored in the target textures.

4   Efficient Implementation of K-Means Clustering in the GPU

In our implementation of k-means clustering we have kept the use of fragment shader constants to a very minimum. Identifying the data point that belongs to each cluster and maintaining the identity of that cluster is achieved by using individual textures. These textures are labeled so as to know the data points that belong to the cluster. In this way, we are able to keep track of the cluster elements, execute cluster operations in parallel and minimize use of shader constants. No practical issues were noticed by defining sufficient textures to handle a large number of textures, say up to 32 clusters. No programmable limitation is foreseen to extend the implementation to handle large number of clusters.

4.1   Distance Computation in GPU

The data that needs to be clustered are stored in 2D textures of size $\sqrt{N}$ x $\sqrt{N}$. The kernels are applied to all the elements in the textures using fragment programs. Figure 2 shows that the distances of each data point to the clusters are stored in individual textures. The Euclidean distances of the k-clusters to each of the data point is calculated and is stored in k distance textures. Every corresponding texel coordinate in each of these distance textures has the distance from the same point to the kth cluster. To compute the minimum distance, all the distance textures are simultaneously loaded as read textures. The *Dmin* texture (minimum distance texture) is defined and initiated as the write texture. The minimum of each corresponding texel is executed via a fragment shader program and the resultant is written into the *Dmin* texture. The *Dmin* texture has the distance of each data point to its nearest centroid.

4.2   Clusters Identification, Labeling, Centroid Computations, Updating in GPU

To identify and label the data point, which is closest to the cluster centroid, each kth cluster distance texture is compared with the *Dmin* texture. The kernel, which is executed as a fragment shader program, compares the value in the cluster distance texture

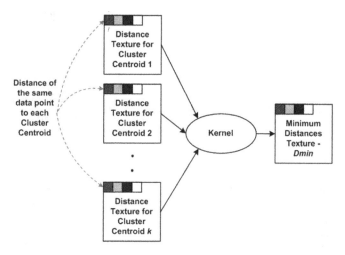

Fig. 2. Cluster Distance Textures and Minimum Distance Texture

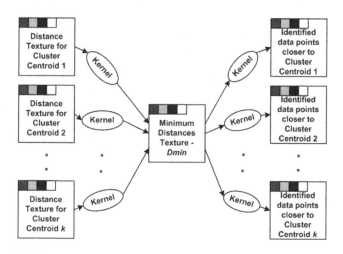

Fig. 3. Identification of Cluster Groups from the Distance Textures

to the corresponding texel value in the *Dmin* texture. Figure 3 shows kernel operations on the textures. If the distance values in both the textures are same, the output buffer is written with a value of "1" if not, "0". This kernel is repeated for all the *k*-cluster distance textures and that results in *k*-group textures. The new cluster centroids are computed based on the elements in each of the cluster group textures. In order to compute the new centroid the number of elements in the group is counted followed by the sum of the data values of the coordinates and subsequently the average is computed. Figure 4 shows the kernel operation on the cluster group textures and the texture with the input data points.

These operations are accomplished by executing a kernel, which effectively performs reduction operation on the textures. As a result each cluster has its own output

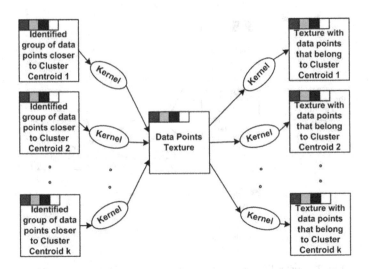

Fig. 4. Labeling of Data Points using a Texture for each Cluster

texture and the data points in that texture belong to the corresponding cluster. The number of passes required to compute each new centroid is given as the $Log_2\sqrt{N}$, where N is the number of data points. To make these computations efficient the read and write textures are swapped after every execution of the kernel.

The new centroids are computed in three steps using the texture reduction technique [5, 6]. The centroid for each cluster is then calculated. At the end of the kernel operation, each new centroid will be stored in a texture. After all the iterations are complete the centroids and the final data points of each cluster are transferred to the CPU.

5   Experimentations and Evaluations

5.1   The Experimental Setup for GPU Computations

A set of clustering data from the "Intelligent data storage for data-intensive analytical systems" [12] was used to test the efficiency of the implementation. The timings taken by the CPU and GPU were measured for every run. A 1.5 GHz Pentium IV CPU with a nVIDIA GeForce FX 5900 XT processor and a 3 GHz Pentium IV CPU with a nVIDIA GeForce 8500 GT processor were used in the experiment. The 8500 GT GPU has 16 textures processing texels to pixels at a memory clock rate of 800 MHz and 512MB of video memory. The peak memory bandwidth is 12.8 GB/sec. The 5900 XT is the older version of the 7900 GTX GPU. The 5900 engine has only 8 textures processing texels to pixels at a memory clock rate of 700 MHz and 256MB of video memory. The peak memory bandwidth is 22.4 GB/sec. The 8500 processor has a PCI E -16x interface with the CPU for data transfer whereas the 5900 have an AGP 8X communication slot. The clustering algorithms were implemented using the OpenGL API with embedded shader programs. The shader programs were developed using GLSL.

5.2   The Analysis of Results: CPU w.r.t GPU Computational Execution Time

The CPU and the GPU computational execution time were measured to compare the speed. The GPU execution time included the steps such as loading the data into the GPU textures, complete computations, assigning data observations to the cluster textures and transferring back the cluster information and centroids to the CPU. Using the "covtype" [12] dataset, we varied the number of clusters and the data size. We discuss the computational time in seconds / iteration as a performance metric for comparison, computed based on measured computational time and the number of iterations. Figure 5 shows the performance w.r.t the number of data points for the various processors used.

Tremendous gain in performance (time per iteration) is noticed while using the GPU over the CPU. It is quite evident that the CPU performance is affected by the data size whereas GPU shows little or no drop in performance with increased data size. The implementation in the 5900 GPU gains about 4 to 12 times in speed than its CPU counterpart. This is at least 3 times faster than the previous implementation [3]. The implementation of the same algorithm in the 8500 GPU gains speed by about 30 times than its CPU counterpart.

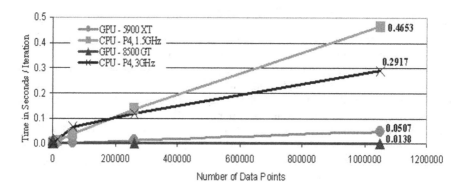

Fig. 5. Performance of *K*-means Clustering based on Data Size

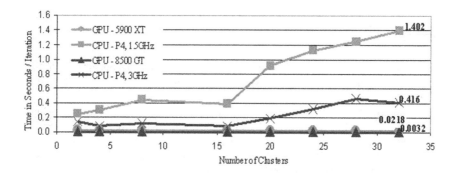

Fig. 6. Performance of *K*-means Clustering based on number of Clusters

Figure 6 shows the performance w.r.t the number of clusters. It is obvious that when the number of clusters increases the performance of the implementation in the GPU shows no or very little drop. For number of clusters less than 20, the implementation in the 5900 GPU gains about 10 to 20 times in computational speed than its CPU counterpart. The gain in computational time is more than 50 times when there are more than 20 clusters. When there are 32 clusters in the data ser, the 8500 GPU is faster about 130 times than its CPU counterpart.

6 Conclusions and Future Work

In this paper, we have presented an effective implementation of k-means algorithm in the GPU. The performance of this implementation has surpassed the CPU implementations by few tens to about a hundred. It has also succeeded the previous implementation of the k-means clustering in the GPU with significant gains in computational resources. By efficient use of textures in the fragment processor, the use of constants to update cluster data has been eliminated. Moreover all cluster centroids can be updated in parallel using textures after the iterations. Thus the necessity to transfer data and results between the GPU and CPU during the computations has been avoided. The results are encouraging and have made the k-means clustering algorithm much more efficient.

The parallel processing capabilities of the GPU will be further exploited to implement clustering of Gene expressions. This will involve the scaling of the k-means implementation to accommodate more dimensions. Similar approach in identification of data points in the clusters and computing cluster centroids using fragment shader and textures will be applied to Hierarchical clustering methods. Clustering techniques such as Fuzzy c-Means and variants of Hierarchical agglomerative clustering algorithm will be implemented to show higher computational efficiencies. There is a physical limitation in the size of the texture and the maximum number of textures available for simultaneous computation. These hardware limitations will have to be considered in future implementations of computational algorithms in GPU.

Recently, NVIDIA has released CUDA (Compute Unified Device Architecture), which is a technology that allows programmers to code algorithm into the 8000 series of GeForce graphics processors directly; an advantage for non-graphics programmers.

References

1. Fluck, O., Aharon, S., Cremers, D., Rousson, M.: GPU histogram computation. In: International Conference on Computer Graphics and Interactive Techniques, ACM SIGGRAPH (2006)
2. Cao, F., Tung, A.K.H., Zhou, A.: Scalable Clustering Using Graphics Processors. In: Yu, J.X., Kitsuregawa, M., Leong, H.-V. (eds.) WAIM 2006. LNCS, vol. 4016. Springer, Heidelberg (2006)
3. Hall, J.D., Hart, J.C.: GPU Acceleration of Iterative Clustering. In: The ACM Workshop on GPC on GPU & SIGRAPH (2004)
4. Richard, W.S., Lipchak, B.: OpenGL SuperBible. Sams Publishing (2005)

5. Göddeke, D.: Basic Math Tutorial. Retrieved from GPGPU (2007),
   `http://www.mathematik.uni-dortmund.de/~goeddeke/`
   `gpgpu/tutorial.html`
6. Göddeke, D.: Reduction Tutorial. Retrieved from GPGPU (2007),
   `http://www.mathematik.uni-dortmund.de/~goeddeke/`
   `gpgpu/tutorial2.html`
7. NVIDIA. GPU Gems 2. ADDISON-WESLEY (2006)
8. Zhang, Q., Zhang, Y.: Hierarchical clustering of gene expression profiles with graphics hardware acceleration. Pattern Recognition Letters 27, 676–681 (2006)
9. Owens, J.D., Luebke, D., Govindaraju, N., Harris, M., Krüger, J., Lefohn, A.E., Purcell, T.J.: A Survey of General-Purpose Computation on Graphics Hardware. In: Eurographics (2006)
10. Owens, J.D.: Streaming Architectures and Technology Trends. In: GPU Gems2, ch. 29, pp. 457–470 (2004)
11. Rost, R.J.: OpenGL® Shading Language, 2nd edn. Addison Wesley Professional, Reading (2006)
12. Intelligent Data Storage for Data-Intensive Analytical Systems. Real Data Set covtype Downloaded from D·Star (2007),
    `http://uisacad2.uis.edu/dstar/data/clusteringdata.html`
13. Takizawa, H., Kobayashi, H.: Hierarchical parallel processing of large scale data clustering on a PC cluster with GPU co-processing. J. Supercomputing 36, 219–234 (2006)

Extracting Knowledge from Life Courses: Clustering and Visualization

Nicolas S. Müller, Alexis Gabadinho, Gilbert Ritschard, and Matthias Studer

Department of Econometrics and Laboratory of Demography
University of Geneva
{nicolas.muller,gilbert.ritschard,matthias.studer,
alexis.gabadinho}@unige.ch

Abstract. This article presents some of the facilities offered by our TraMineR R-package for clustering and visualizing sequence data. Firstly, we discuss our implementation of the optimal matching algorithm for evaluating the distance between two sequences and its use for generating a distance matrix for the whole sequence data set. Once such a matrix is obtained, we may use it as input for a cluster analysis, which can be done straightforwardly with any method available in the R statistical environment. Then we present three kinds of plots for visualizing the characteristics of the obtained clusters: an aggregated plot depicting the average sequential behavior of cluster members; an sequence index plot that shows the diversity inside clusters and an original frequency plot that highlights the frequencies of the n most frequent sequences. TraMineR was designed for analysing sequences representing life courses and our presentation is illustrated on such a real world data set. The material presented should also be of interest for other kind of sequential data such as DNA analysis or web logs.

1 Introduction

This paper[1] discusses some of the tools made available in TraMineR[2], a package that we developed for the R statistical environment. TraMineR, which stands for life trajectory miner, is a toolbox for the analysis and visualization of sequential data. Though TraMineR is mainly intended for analyzing life courses, most of its features may be of interest for other kind of sequential data, such as DNA sequences or web logs for instance. The features discussed in this paper include the computation of the optimal matching (OM) distance between any two sequences, also known as the edit distance, and three kinds of graphical representation of sets of sequences. The OM distances can be used for clustering

---

[1] This study has been realized within the Swiss National Science Foundation project SNSF 100012-113998/1. The empirical results are based on data collected within the "Living in Switzerland: 1999-2020" project steered by the Swiss Household Panel (www.swisspanel.ch) of the University of Neuchâtel and the Swiss Statistical Office.
[2] TraMineR is downloadable from http://mephisto.unige.ch/biomining/.

I.-Y. Song, J. Eder, and T.M. Nguyen (Eds.): DaWaK 2008, LNCS 5182, pp. 176–185, 2008.

the sequences, and the graphical representations for characterizing the obtained clusters.

The methods and graphics are presented through a real world example. More specifically, we consider life courses of Swiss people born during the first half of the 20th century, using data from a retrospective survey carried out by the Swiss Household Panel. The life courses are made up of constituent events in familial life such as the leaving from the family home, the birth of the first child, the first marriage or the first divorce. By using these events as our basis, it is possible to look at individual life courses in the form of a sequence of states, where each event that occurs in a person's life course corresponds to a change of state.

This article is divided in the following way. The first part presents the source data and the necessary transformations to build state sequences from events. The second part presents the optimal matching method and the operations cost problematic. The third part concerns the visualization tools and their functioning. The fourth part reviews quickly the existing software for optimal matching analysis and presents our TraMineR R-package. We finally conclude on the possibilities that such methods bring to the social sciences.

2   Source Data

The data we extract from the answers to a survey is shown in a table where each line represents an individual and each column a variable (Table 1).

Table 1. An example of data showing familial life course events

ind.	birth	leaving home	marriage	child	divorce
1	1974	1992	1994	1996	n/a

The move to a sequential representation is not without interest. The difficulty lies in representing a combination of events which either took place or did not, at a certain age, by a unique state. In a more formal manner, we define the state of a person at a given age as information on events that have already happened. From any given state, one can say which events have already taken place. One or more events which occur during the year t will cause the individual to go from the state he was in at $t-1$ to a new state. The definition of the states from events remains, however, strongly dependent from the type of data and the aim of the study. A simple way to proceed would be to create a state for each combination of events. By so doing, the number of states would rise to 2^n for n events, which renders the interpretation difficult whenever many events have to be taken into consideration. We have therefore chosen to group certain events in accordance with the research objectives.

For the purpose of this research, we retained four events which constitute familial life: the departure from the family home, the first marriage, the first divorce and the birth of the first child. Table 2 shows the encoding of the states which we have drawn up in relation to the four selected events. The number of events was reduced from 16 to 8, notably by eliminating impossible states (all those which contain a divorce without a previous marriage), or by combining two states that would be too rare (for example state 2 concerns married individuals who have not left the family home regardless of whether or not they have any children). By referring to this list of states and to the example shown in Table 1, the result of the creation of a sequence of familial life is found in Table 3.

Table 2. List of states

	leaving home	marriage	children	divorce
0	no	no	no	no
1	yes	no	no	no
2	no	yes	yes/no	no
3	yes	yes	no	no
4	no	no	yes	no
5	yes	no	yes	no
6	yes	yes	yes	no
7	yes/no	yes	yes/no	yes

Table 3. An example of data as a sequence of states

individual	1974	...	1991	1992	1993	1994	1995	1996	1997	1998	...
1	0	...	0	1	1	3	3	6	6	6	...

The data used in this article comes from the retrospective biographical survey carried out by the Swiss Household Panel (www.swisspanel.ch) in 2002. We have retained the individuals who were at least 30 years old at the time of the survey in order to have only complete sequences between the ages of 15 and 30. In this way, our sample is made up of 4318 individuals born between 1909 and 1972.

3   Optimal Matching Method

The method we used in this work to analyze sequences is the optimal matching (OM) method. The algorithm used is inspired by the methods of sequence alignment and dynamic programming used in molecular biology, especially for the comparison of proteins, or sequences of DNA thought to be homologous [1; 2]. This method of working was devised in order to enable the rapid calculation of numerous sequences in order to find the correspondence between them. The first algorithms of OM based on the Levenshtein distance [3] appeared at the beginning of the 1970s and their first use in social science goes back to the article by Abbott and Forrest on their application to historical data [4]. We owe numerous methodological articles on the use of these methods in social sciences, especially in sociology, to Abbott [5; 6]. The interest in applying this method to a life course lies in the possibility to use the OM distances to do a cluster analysis.

3.1   Method Description

The OM method uses the Needleman-Wunsch algorithm to compute a distance between two sequences based on the Levenshtein distance [3; 2]. Take Ω, the set of possible operations, and $a[\omega]$ the result of the application of the operations $\omega \in \Omega$ on the sequence a. We take into account 3 types of operations: the insertion of an element, the suppression of an element, or the substitution of one element by another. We attribute a cost $c(\omega)$ to each ω which corresponds to the cost of implementing the operation $\omega \in \Omega$. The OM distance between a sequence a and a sequence b can be formulated as follows: $d(a, b) = \min\{c[\omega_1, ..., \omega_k] \mid b = a[\omega_1, ..., \omega_k], \omega \in \Omega, k \geq 0\}$, with $c[\omega_1, ...\omega_k] = \sum_{i=1}^{k} c[\omega_i]$. In other words, for each pair of sequences, we look for the combination of operations with the lowest total cost that renders both sequences identical.

The Needleman-Wunsch algorithm for finding the minimal distance $d(a, b)$ can be summarized in 4 points:

1. A matrix D of size m = (length of $a + 1$) and n = (length of $b + 1$) is created.
2. The value of cell $d_{0,0}$ is set to 0 and the values of the first column and line are computed using the following recursive equation:

$$d_{i,0} = d_{i-1,0} + c[\omega(a_i, \phi)] \qquad (1)$$

$$d_{0,j} = d_{0,j-1} + c[\omega(\phi, b_j)] \qquad (2)$$

   where $c[\omega(a_i, \phi)] = c[\omega(\phi, b_j)]$ is the cost of an insertion/deletion operation.
3. The remaining cells are recursively filled with the following equation:

$$d_{i,j} = \min \begin{cases} d_{i,j-1} + c[\omega(\phi, b_j)] \\ d_{i-1,j-1} + c[\omega(a_i, b_j)] \\ d_{i-1,j} + c[\omega(a_i, \phi)] \end{cases} \qquad (3)$$

   where $c[\omega(a_i, b_j)]$ is the cost of replacing the ith state of sequence a by the jth state of sequence b.
4. The minimal distance is then found in the cell $d_{m,n}$.

This algorithm is detailed in several other publications ([2; 7; 1]).

Example. We will now present a visual example of how the distance between a sequence a (ECD) and sequence b ($ABCD$) is computed by the algorithm. Firstly, the matrix D of size 3×4 is initialized with its cell $d_{0,0} = 0$. To simplify the example, we attribute a cost of 1 to all operations $\omega \in \Omega$. The first line and the first column are recursively filled according to the equations (2) and (1). The resulting matrix is on the left panel of Table 4. Each cell is then recursively defined by equation (3). For example, cell $d_{1,1}$ takes its value from the minimum of these three possibilities: $d_{0,0} + c[\omega(E, A)]$, $d_{0,1} + c[\omega(E, \phi)]$ or $d_{1,0} + c[\omega(\phi, A)]$. The $d_{1,1}$ value is then 1, corresponding to a substitution of A by E ($\omega(E, A)$). After filling the entire matrix, we find that the minimal sum of costs to transform sequence a into sequence b is 2, corresponding to the substitution of A by E and then the deletion of B ($\omega(B, \phi)$).

Table 4. On the left: initial matrix, on the right: completed matrix

		A	B	C	D
	0	1	2	3	4
E	1				
C	2				
D	3				

		A	B	C	D
	0	1	2	3	4
E	1	1	2	3	4
C	2	2	2	2	3
D	3	3	3	3	2

3.2   Cost of the Edit Operations

As previously seen, a cost c can be ascribed to the operations $w \in \Omega$. The costs of substitution, in which we were particularly interested, can be represented in the form of a symmetrical matrix which defines a value for each pair of states. In the context of a use in social science, it is extremely difficult to base the attribution of these values on a theoretical model. Such practice has given rise to a debate [8]. It is indeed difficult to determine the cost of transforming one state into another, yet it is interesting and sometimes of fundamental importance to be able to differentiate between these costs. In order to do that, we used the following method: the cost of transforming a state i into a state j is calculated in terms of the rate of longitudinal transition: $c[w(i,j)] = c[w(j,i)] = 2 - p(i_t|j_{t-1}) - p(j_t|i_{t-1})$. The basic cost is fixed at 2 (so that the substitution cost for a transition that is never observed is the highest) and the greater the transition probability $p(i_t|j_{t-1})$ of going from state i to state j, and vice versa, the more the cost decreases. Thus, the substitutions corresponding to the most frequently observed transitions will be less costly than those which never occur. Another method offered by the software T-COFFEE/SALTT [9], involves iteratively calculating an optimal substitution costs matrix (Gauthier et al. [10]), but did not give satisfying results on our dataset.

In this paper, a value of 1 was attributed to the cost of insertion and deletion in the solution based on the rate of transition. The reason for this choice was to favour the operations of insertion or deletion in the case where a sequence has the same states than another one but one or two years later or sooner. This allows to keep a small distance between this kind of sequences.

3.3   Clustering Using the Optimal Matching Distances

We are now able to produce a matrix for distances measuring the differences between individuals' life courses. This can be used in a agglomerative hierarchical clustering using the Ward criterion. We chose a 4 clusters solution for two reasons: in the first place, the dendrogram resulting from the hierarchical clustering shows a clear split at the 4 clusters level; in the second place, the 4 clusters receive good interpretation both visually and through logistic regression models.

4   Visualization of the Clustering Results

Life state sequences are difficult to visualize for several reasons. Firstly, states are categorical values that cannot be easily plotted over time like time series.

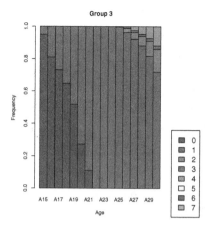

Fig. 1. Relative frequencies of individuals in each state at each age (cluster 3)

Furthermore, it is not possible to calculate an "average" sequence for representing each cluster. Plotting individual life sequences like time series, i.e. with time on the X axis and the arbitrarily sorted states on the Y axis, would become unreadable when a lot of sequences are drawn. That is why we need specific plots for visualizing life state sequences.

In this paper, we consider three kinds of plots. The first one can be used to get a quick overview of the clusters, such as shown for cluster 3 on Fig. 1. For each possible age, from 15 to 30, we find the proportion of individuals being in each different states (from 0 to 7). These plots give only a very general overview of the clusters as it shows only aggregated results. That is why we need additional types of graphics.

The *frequencies plots* represent each state of the life sequences of the n most frequent sequences. Fig. 2 shows the 10 most frequent sequences in each of the 4 clusters, as they were defined by the hierarchical clustering. Each color corresponds to a state, as they are described in Table 2. The height of each sequence is proportional to its relative frequency.

Alternatively, we can visualize each sequence by a stacked horizontal line. We obtain this way a so-called indexplot [11]. This is similar to the previous representation, except that all the sequences are drawn. In order to get a better view of what kind of sequences are in a cluster, it is useful to sort them before plotting them. In this case, we have computed an optimal matching distance between a random sequence in the cluster and all the other ones, following an idea from Brzinsky-Fay et al. [12]. The example from Fig. 3 highlights clearly the benefit of sorting the sequences for getting a usable representation of the cluster.

As we can see, the first kind of plot (Fig. 1) is not an accurate overview of the content of the clusters. That is why using graphs based on individual data is important to better understand the clustering results and to observe the diversity inside each cluster.

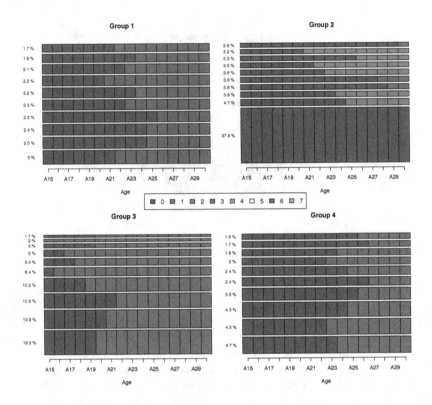

Fig. 2. 10 most frequent sequences in each cluster

5   The TraMineR R Package

We have implemented the methods for computing distances and visualizing the sequences in a package named "TraMineR" for the R statistical software environment. This section reviews some of the other softwares available for sequence analysis. It then introduces the TraMineR package and shortly comments about the complexity of the different methods.

5.1   Review of Existing Softwares

The major problem for the social scientist who wants to do sequence analysis is the lack of implementation of these methods in standard statistical software. So far, among the most widely used statistical packages, only Stata and SAS provide optimal matching analysis through third-party modules [12; 13]. The software TDA [7] is also able to compute optimal matching distances but is no longer under development. The same is true for OPTIMIZE developed by Abbott. The packages T-COFFEE/SALTT [9] and CHESA [14] compute, among other things, optimal matching distances, but provide no plotting tools.

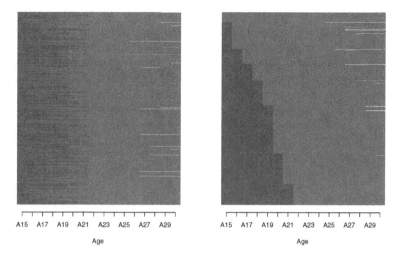

Fig. 3. Unsorted (left) and sorted (right) indexplots of group 3 (blue is state 0, red is state 1)

Within the academic field, R is becoming an important tool for several reasons: it is open-source software, its programming language is high-level and derived from S (a widespread statistical programming language) and most of the statistical methods are accessible through modules. That is why we decided to create a module in R for sequence analysis. Indeed, the plotting sub-system is efficient and flexible enough to allow us to create custom graphics like the ones presented in this paper. It also permits to directly access and interact with many other already implemented methods, such as hierarchical clustering, logistic regression or multi-dimensional scaling. The purpose of TraMineR is to allow the user to do everything related to sequence analysis within the R software environment, starting from sequence data creation to visualizing the results of a clustering. This paper is centered on the optimal matching method, however TraMineR offers other distances or measures such as those proposed by Elzinga ([15; 16]).

5.2   Complexity of the Algorithm

The main algorithm, as we have said before, is the optimal matching method to compute distances between sequences. As the distances are symmetrical (the distance between A and B equals the one between B and A), the number of distances computed is $\frac{n \times (n-1)}{2}$. To compute the distance between two sequences, the algorithm builds a matrix whose size is defined by the length of the sequences. The solution is found by filling all the cells of this matrix with the recursive equation (3). The complexity of this part of the algorithm is $O(\ell_1 \times \ell_2)$, where ℓ_1 is the length of the first sequence and ℓ_2 the length of the second one. Assuming that all the sequences are of equal length, as this is the case in this paper, then the total complexity is $O(n^2 \times \ell^2)$. The plotting methods require only the drawing of the sequences and thus are of complexity $O(n)$.

In order to speed up the computation of distances, the Needleman-Wunsch algorithm is written in C instead of the R programming language; as a result, a R shared library is created and the method is called from R. The C version is on our test machine about 15 times faster than the one written in R. In order to reduce the number of distances to be computed, we implemented a process that considers only one instance of multiple instances of a same sequence. Our life sequences data set contains for instance only 841 different sequences among 4318. The number of distances computed was thus reduced from $\frac{4318 \times 4317}{2} = 9320403$ to $\frac{841 \times 840}{2} = 353220$. The computation took less than 15 seconds on our test machine (Intel Core2Duo@2GHZ).

6   Conclusion

We presented in this paper some of the salient tools offered by our TraMineR R-package for extracting useful knowledge from sequential data. More specifically, we discussed OM, one among the methods implemented for evaluating the closeness between sequences describing life courses. We presented also three kinds of complementary graphics for visualizing sequence data. The first of them gives some kind of average view of a set of sequences, the second one depicts the frequencies of the more frequent sequences and the third one reveals the within group discrepancy. If the first and third kind of plots have already been proposed in the literature, the frequency plots of sequences is an original contribution. Such visualization tools prove to be of great help for the experts. They highly facilitate the drawing of significant interpretations regarding the structuring of events and their timing. TraMineR is still under development and should in the near future include many other features, among which descriptive indicators of individual sequences as well as of set of sequences and frequent episodes mining tools. Since TraMineR runs in R it is also quite straightforward to use for instance the computed distances with procedures such as multidimensional scaling offered by other R packages. By offering the social scientist, and indeed any interested end user, the possibility to run all her/his analyses of sequences and produce all her/his graphics within a same free and open-source software, TraMineR should help popularize sequential analysis in the social sciences.

Bibliography

[1] Deonier, R., Tavaré, S., Waterman, M.: Computational Genome Analysis: an Introduction. Springer, Heidelberg (2005)
[2] Needleman, S.B., Wunsch, C.: General method applicable to the search for similarities in the animo acid sequence of two proteins. Journal of Molecular Biology 48, 443–453 (1970)
[3] Kruskal, J.: An overview of sequence comparison. In: Time warps, string edits, and macromolecules. The theory and practice of sequence comparison, pp. 1–44. Adison-Wesley, Don Mills (1983)
[4] Abbott, A., Forrest, J.: Optimal matching methods for historical sequences. Journal of Interdisciplinary History 16, 471–494 (1986)

[5] Abbott, A., Hrycak, A.: Measuring resemblance in sequence data: An optimal matching analaysis of musician's careers. American Journal of Sociolgy 96(1), 144–185 (1990)

[6] Abbott, A., Tsay, A.: Sequence analysis and optimal matching methods in sociology, Review and prospect. Sociological Methods and Research 29(1), 3–33 (2000) (With discussion, pp 34-76)

[7] Rohwer, G., Pötter, U.: TDA user's manual. Software, Ruhr-Universität Bochum, Fakultät für Sozialwissenschaften, Bochum (2002)

[8] Wu, L.: Some comments on sequence analysis and optimal matching methods in sociology: Review and prospect. Sociological Methods and Research 29, 41–64 (2000)

[9] Notredame, C., Bucher, P., Gauthier, J.A., Widmer, E.: T-COFFEE/SALTT: User guide and reference manual (2005), Available at, http://www.tcoffee.org/saltt

[10] Gauthier, J.A., Widmer, E.D., Bucher, P., Notredame, C.: How much does it cost? Optimization of costs in sequence analysis of social science data. Sociological Methods and Research (forthcoming, 2008)

[11] Scherer, S.: Early career patterns: A comparison of Great Britain and West Germany. European Sociological Review 17(2), 119–144 (2001)

[12] Brzinsky-Fay, C., Kohler, U., Luniak, M.: Sequence analysis with Stata. The Stata Journal 6(4), 435–460 (2006)

[13] Lesnard, L.: Describing social rhythms with optimal matching (2007)

[14] Elzinga, C.H.: CHESA 2.1 User manual. User guide, Dept of Social Science Research methods, Vrije Universiteit, Amsterdam (2007)

[15] Elzinga, C.H.: Sequence similarity: A nonaligning technique. Sociological Methods & Research 32, 3–29 (2003)

[16] Elzinga, C.H.: Combinatorial representations of token sequences. Journal of Classification 22(22), 87–118 (2005)

A Hybrid Clustering Algorithm Based on Multi-swarm Constriction PSO and GRASP

Yannis Marinakis[1], Magdalene Marinaki[2], and Nikolaos Matsatsinis[1]

[1] Decision Support Systems Laboratory, Department of Production Engineering and Management, Technical University of Crete, 73100 Chania, Greece
marinakis@ergasya.tuc.gr, nikos@ergasya.tuc.gr
[2] Industrial Systems Control Laboratory, Department of Production Engineering and Management, Technical University of Crete, 73100 Chania, Greece
magda@dssl.tuc.gr

Abstract. This paper presents a new hybrid algorithm, which is based on the concepts of Particle Swarm Optimization (PSO) and Greedy Randomized Adaptive Search Procedure (GRASP), for optimally clustering N objects into K clusters. The proposed algorithm is a two phase algorithm which combines a Multi-Swarm Constriction Particle Swarm Optimization algorithm for the solution of the feature selection problem and a GRASP algorithm for the solution of the clustering problem. In this paper in PSO, multiple swarms are used in order to give to the algorithm more exploration and exploitation abilities as the different swarms have the possibility to explore different parts of the solution space and, also, a constriction factor is used for controlling the behaviour of particles in each swarm. The performance of the algorithm is compared with other popular metaheuristic methods like classic genetic algorithms, tabu search, GRASP, ant colony optimization and particle swarm optimization. In order to assess the efficacy of the proposed algorithm, this methodology is evaluated on datasets from the UCI Machine Learning Repository. The high performance of the proposed algorithm is achieved as the algorithm gives very good results and in some instances the percentage of the corrected clustered samples is very high and is larger than 98%.

Keywords: Particle Swarm Optimization, Greedy Randomized Adaptive Search Procedure, Clustering Analysis.

1 Introduction

Particle Swarm Optimization (PSO) is a population-based swarm intelligence algorithm that was originally proposed by Kennedy and Eberhart [11] that simulates the social behavior of social organisms by using the physical movements of the individuals in the swarm. Its mechanism enhances and adapts to the global and local exploration. Most applications of PSO have concentrated on the optimization in continuous space while some work has been done to the discrete optimization [12,22]. Recent complete surveys for the Particle Swarm Optimization can be found in [1,20]. The Particle Swarm Optimization (PSO)

I.-Y. Song, J. Eder, and T.M. Nguyen (Eds.): DaWaK 2008, LNCS 5182, pp. 186–195, 2008.

is a very popular optimization method and its wide use, mainly during the last years, is due to the number of advantages that this method has compared to other optimization methods. Some of the key advantages are that this method does not need the calculation of derivatives, that the knowledge of good solutions is retained by all particles and that particles in the swarm share information between them. PSO is less sensitive to the nature of the objective function, can be used for stochastic objective functions and can easily escape from local minima. Concerning its implementation, PSO can easily be programmed, has few parameters to regulate and the assessment of the optimum is independent of the initial solution. Clerc and Kennedy [3] proposed a constriction factor in order to prevent explosion, ensure convergence and to eliminate the parameter that restricts the velocities of the particles. Usually, in PSO-based algorithms only one swarm is used. Recently, a number of works have been conducted that use more than one swarms either using the classic PSO [2,19,23] or using some variations of the classic PSO like the method called TRIBES [4]. These methods have more exploration and exploitation abilities due to the fact that the different swarms have the possibility to explore different parts of the solution space.

In this paper, a new hybrid nature inspired intelligent technique, that uses a Constriction Particle Swarm Optimization algorithm with Multiple Swarms and a metaheuristic algorithm, the Greedy Randomized Adaptive Search Procedure (GRASP) [6] is presented and analyzed in detail for the solution of the clustering problem. More precisely, the proposed Multi-Swarm Constriction Particle Swarm Optimization - Greedy Randomized Adaptive Search Procedure (MSCPSO-GRASP) algorithm uses the MSCPSO for the feature selection phase of the clustering algorithm while for the clustering phase the GRASP algorithm is applied. The main difference of this algorithm from the other algorithms that use more than one swarms is a feedback procedure that is used in order, initially, to distribute the information (i.e., the good solutions of each swarm) in all other swarms and, afterwards, to help all the particles of all swarms to follow this information in order to finally find a new better solution. In order to assess the efficacy of the proposed algorithm, this methodology is evaluated on datasets from the UCI Machine Learning Repository. Also, the method is compared with the results of a number of other metaheuristic algorithms for clustering analysis that use, mainly, hybridization techniques that incorporate a Tabu Search based algorithm [7], a Genetic based algorithm [8] and an Ant Colony Optimization (ACO) algorithm [5]. Also, comparisons are performed using a Hybrid Particle Swarm Optimization - GRASP algorithm [13] in order to give the efficiency of the proposed algorithm compared to an algorithm that is based on the classic Particle Swarm Optimization and to show the fact that with the use of multiple swarms it is more probable to find a better local minimum as the algorithm has the possibility to explore a wider solution space. The rest of this paper is organized as follows: In the next section the proposed Hybrid MSCPSO-GRASP algorithm is presented and analyzed in detail. In section 3, the analytical computational results for the datasets used in this study are presented while in the last section conclusions and future research are given.

2   The Proposed Hybrid MSCPSO-GRASP for Clustering

2.1   Clustering Problem

Clustering analysis identifies clusters (groups) embedded in the data, where each cluster consists of objects that are similar to one another and dissimilar to objects in other clusters ([9,21,24]). The typical cluster analysis consists of four steps (with a feedback pathway) which are the *feature selection*, that is an optimization problem, where the problem is to search through the space of feature subsets to identify the optimal or near-optimal one with respect to a performance measure (see [10] for feature selection algorithms) or *feature extraction* where some transformations are used in order to generate useful and novel features from the original ones, the *clustering algorithm design* or *selection* that is usually combined with the selection of a corresponding proximity measure ([9,21]) and with the construction of a clustering criterion function which makes the partition of clusters a well defined optimization problem (see ([9,21,24]) for an analytical survey of the clustering algorithms), the *cluster validation* where external indices, internal indices, and relative indices are used for cluster validity analysis ([9,24]) and the *results interpretation* where experts in the relevant fields interpret the data partition in order to guarantee the reliability of the extracted knowledge ([24]).

More precisely, the problem of clustering N objects (patterns) into K clusters is considered. In particular the problem is stated as follows: Given N objects in R^n, allocate each object to one of K clusters such that the sum of squared Euclidean distances between each object and the center of its belonging cluster (which is also to be found) for every such allocated object is minimized. For the mathematical description of the clustering problem see ([13,14,15,16,17,18].)

The proposed algorithm (Hybrid MSCPSO-GRASP) for the solution of the clustering problem is a two phase algorithm which combines a Multi-Swarm Constriction Particle Swarm Optimization (MSCPSO) algorithm for the solution of the feature selection problem and a Greedy Randomized Adaptive Search Procedure (GRASP) for the solution of the clustering problem. In this algorithm, the activated features are calculated by the MSCPSO (see 2.2) and the fitness (quality) of each particle is calculated by the clustering algorithm (see 2.3). The clustering algorithm has the possibility to solve the clustering problem with known or unknown number of clusters (see 2.3 and [13,14,15,16,17,18]).

2.2   MSCPSO for the Feature Selection Problem

For the feature selection problem, a MSCPSO algorithm is used. In each of the swarms, the main procedures of the Constriction Particle Swarm Optimization [3] algorithm are used. Thus, each swarm finds a local optimum applying the Constriction Particle Swarm Optimization Procedure. This local optimum is located in a different point of the solution space. Then, the best particles (solutions) of each swarm create a new swarm in which again a Constriction Particle Swarm Optimization (CPSO) procedure is applied. The creation of the new swarm helps the distribution of the information, i.e. the movement direction of each swarm. The application of the CPSO algorithm in the new swarm has the possibility to create even better solutions that are restored backwards to the other swarms, i.e. the particles that constitute the new swarm return their new positions and velocities to their original swarms. Then, the whole procedure is repeated until a prespecified number of iterations. When this number has been reached the best particle from all swarms is the optimum.

In each swarm the algorithm first randomly initializes a swarm of particles. The position of each particle is represented by a d-dimensional vector in problem space $s_i = (s_{i1}, s_{i2}, ..., s_{id})$, $i = 1, 2, ..., M$ (M is the population size), and its performance is evaluated on the predefined fitness function. Thus, each particle is randomly placed in the d-dimensional space as a candidate solution (in the feature selection problem d corresponds to the number of activated features). The velocity of the i-th particle $v_i = (v_{i1}, v_{i2}, ..., v_{id})$ is defined as the change of its position. The flying direction of each particle is the dynamical interaction of individual and social flying experience. The algorithm completes the optimization through following the personal best solution of each particle and the global best value of the whole swarm. Each particle adjusts its trajectory toward its own previous best position and the previous best position attained by any particle of the swarm, namely p_{id} and p_{gd}. In the discrete space, a particle moves in a state space restricted to zero and one on each dimension where each v_i represents the probability of bit s_i taking the value 1. Thus, the particles' trajectories are defined as the changes in the probability and v_i is a measure of individual's current probability of taking 1. If the velocity is higher it is more likely to choose 1, and lower values favor choosing 0. A sigmoid function is applied to transform the velocity from real number space to probability space:

$$sig(v_{id}) = \frac{1}{1 + exp(-v_{id})} \tag{1}$$

In the binary version of CPSO, the velocities and positions of particles are updated using the following formulas [3]:

$$v_{id}(t + 1) = \chi(v_{id}(t) + c_1 rand1 (p_{id} - s_{id}(t)) + c_2 rand2 (p_{gd} - s_{id}(t))) \tag{2}$$

where

$$\chi = \frac{2}{|2 - c - \sqrt{c^2 - 4c}|} \text{ and } c = c_1 + c_2, c > 4 \tag{3}$$

$$s_{id}(t + 1) = \begin{cases} 1, & \text{if } rand3 < sig(v_{id}) \\ 0, & \text{if } rand3 >= sig(v_{id}) \end{cases} \tag{4}$$

where $p_{id} = (p_{i1d}, ..., p_{ind})$ is the best position encountered by i-th particle so far; p_{gd} represents the best position found by any member in the whole swarm population; t is iteration counter; s_{id} is the valued of the d-th dimension of particle s_i, and $s_{id} \in \{0, 1\}$; v_{id} is the corresponding velocity; $sig(v_{id})$ is calculated according to the Equation (1), c_1 and c_2 are acceleration coefficients; $rand1$, $rand2$ and $rand3$ are three random numbers in $[0, 1]$. With this formulation, the velocity limit, V_{max}, is no longer necessary.

2.3   GRASP for the Clustering Problem

As it was mentioned earlier in the clustering phase of the proposed algorithm a **Greedy Randomized Adaptive Search Procedure (GRASP)** ([6]) is used which is an iterative two phase search algorithm (a **construction phase** and a **local search phase**). An initial solution (i.e. an initial clustering of the samples in the clusters) is constructed step by step and, then, this solution is exposed for improvement in the local search phase of the algorithm. The first problem that we had to face was the selection of the number of the clusters. Thus, the algorithm works with two different ways.

   If the number of clusters is known a priori, then a number of samples equal to the number of clusters are selected randomly as the initial clusters. In this case, as the iterations of GRASP increased the number of clusters do not change. In each iteration, different samples (equal to the number of clusters) are selected as initial clusters. Afterwards, the RCL is created (the Restricted Candidate List - RCL is the list that is used for the selection of the next element that will be chosen to be inserted to the current solution). The probabilistic component of a **GRASP** is characterized by randomly choosing one of the best candidates in the list but not necessarily the top candidate. In our implementation, the best promising candidate samples are selected to create the RCL. The samples in the list are ordered taking into account the distance of each sample from all centers of the clusters and the ordering is from the smallest to the largest distance. From this list, the first D samples (D is a parameter of the problem) are selected in order to form the final RCL. The candidate sample for inclusion in the solution is selected randomly from the RCL using a random number generator. Finally, the RCL is readjusted in every iteration by recalculating all the distances based on the new centers and replacing the sample which has been included in the solution by another sample that does not belong to the RCL, namely the $(D + t_1)$th sample where t_1 is the number of the current iteration. When all the samples have been assigned to clusters a local search strategy is applied in order to improve the solution. The local search works as follows: For each sample the probability of its reassignment in a different cluster is examined by calculating the distance of the sample from the centers. If a sample is reassigned to a different cluster the new centers are calculated. The local search phase stops when in an iteration no sample is reassigned. If the number of clusters is unknown then, initially a number of samples are selected randomly as the initial clusters. Now, as the iterations of GRASP increased the number of clusters changes but cannot become less than two. In each iteration a different number of clusters can be found. The creation of the initial solutions and the local search phase work as in the previous case. The only difference compared to the previous case concerns the use of the validity measure in order to choose the best solution ([13,14,15,16,17,18]).

3   Computational Results

The performance of the proposed methodology is tested on 9 benchmark instances taken from the UCI Machine Learning Repository. The datasets from the UCI Machine Learning Repository were chosen to include a wide range of domains and their characteristics are given in Table 1. The data varies in terms of the number of observations from very small samples (Iris with 150 observations) up to larger data sets (Spambase with 4601 observations). Also, there are data sets with two and three clusters. In one case (Breast Cancer Wisconsin) the data set is appeared with different size of observations because in this data set there is a number of missing values. The problem of missing values was faced with two different ways. In the first way where all the observations are used we took the mean values of all the observations in the corresponding feature while in the second way where we have less values in the observations we did not take into account the observations that they had missing values. Some data sets involve only numerical features, and the remaining include both numerical and categorical features (in parentheses in Table 1). For each data set, Table 1 reports the total number of features and the number of categorical features in parentheses. The parameter settings for MSCPSO-GRASP algorithm were selected after thorough empirical testing and they are: The number of swarms is set equal to 5, the number of particles in each swarm is set equal to 20, the number of generations is set equal to 100, the size of RCL is set equal to 50, the number of GRASP's iterations is equal to 100 and the coefficients are $c_1 = 2.05, c_2 = 2.05$. The algorithm was implemented in Fortran 90 and was compiled using the Lahey f95 compiler on a Centrino Mobile Intel Pentium M 750 at 1.86 GHz, running Suse Linux 9.1.

Table 1. Data Sets Characteristics

Data Sets	Observations	Features	Clusters
Australian Credit (AC)	690	14(8)	2
Breast Cancer Wisconsin 1 (BCW1)	699	9	2
Breast Cancer Wisconsin 2 (BCW2)	683	9	2
Heart Disease (HD)	270	13(7)	2
Hepatitis 1 (Hep1)	155	19 (13)	2
Ionosphere (Ion)	351	34	2
Spambase(spam)	4601	57	2
Iris	150	4	3
Wine	178	13	3

The objective of the computational experiments is to show the performance of the proposed algorithm in searching for a reduced set of features with high clustering of the data. The purpose of feature variable selection is to find the smallest set of features that can result in satisfactory predictive performance. Because of the curse of dimensionality, it is often necessary and beneficial to limit the number of input features in order to have a good predictive and less computationally intensive model. In general there are $2^{NF} - 1$, where NF denotes the number of features, possible feature combinations and, thus, in our cases the problem with the fewest number of feature combinations is the Iris (namely $2^4 - 1$), while the most difficult problem is the Spambase where the number of feature combinations is $2^{57} - 1$.

A comparison with the classic k-means and other metaheuristic approaches for the solution of the clustering problem is presented in Table 2. In this Table, besides the proposed algorithm, ten other algorithms are used for the solution of the feature subset selection problem and for the clustering problem. In the first group of algorithms, besides the proposed algorithm, the results of three other algorithms are presented where in two of them (Memetic-GRASP and HBMO-GRASP) in the feature selection phase a Memetic algorithm and a Honey Bees Mating Optimization (HBMO) are used, respectively, while in the clustering phase a GRASP algorithm is used. The other algorithm of the first group uses a PSO algorithm in the feature selection phase and an ACO algorithm in the clustering phase. In the second group of algorithms, all algorithms use the GRASP algorithm in the clustering phase while in the feature selection phase a PSO algorithm, an ACO algorithm, a genetic algorithm and a tabu search algorithm are used, respectively. In the third group of algorithms, initially the classic k-means algorithm is used for the clustering problem using all features, while in the rest columns an ACO algorithm is used in both phases and, finally, a PSO algorithm is used in both phases. The parameters and the implementation details of all of the algorithms presented in the comparisons are analyzed in papers [13,14,15,16,17,18].

From this table, it can be observed that the Hybrid MSCPSO-GRASP algorithm performs better (has the largest number of correct clustered samples) than the other ten algorithms in all instances. It should be mentioned that in some instances the differences in the results between the Hybrid MSCPSO-GRASP algorithm and the other ten algorithms are very significant. Mainly, for the two data sets that have the largest number of features compared to the other data sets, i.e. in the Ionosphere data set the percentage of corrected clustered samples for the Hybrid MSCPSO-GRASP algorithm is 88.88% while for all the other methods the percentage varies between 70.65% to 88.03% and in the Spambase data set the percentage of corrected clustered samples for the Hybrid MSCPSO-GRASP algorithm is 90.17% while for all the other methods the percentage varies between 82.80% to 87.54%. It should, also, be noted that a hybridization algorithm performs always better than a no hybridized algorithm. More precisely, the only five algorithms that are competitive in almost all instances with the proposed Hybrid MSCPSO-GRASP algorithm are the Hybrid HBMO-GRASP, the Hybrid Memetic-GRASP, the Hybrid PSO - ACO, the Hybrid PSO - GRASP and the Hybrid ACO - GRASP algorithms. These results prove the significance of the solution of the feature selection problem in the clustering algorithm as when more sophisticated methods for the solution of this problem (Particle Swarm Optimization, Honey Bees Mating Optimization, and Ant Colony Optimization) were used the performance of the clustering algorithm was improved. The significance of the solution of the feature selection problem using the Multi Swarm Constriction Particle Swarm Optimization Algorithm is, also,

Table 2. Results of the algorithms

Instance	MSCPSO-GRASP		Memetic-GRASP		PSO-ACO		HBMO-GRASP	
	Selected Feat.	Correct Clustered	Selected Features	Correct Clustered	Selected Feat.	Correct Clustered	Selected Feat.	Correct Clustered
BCW2	5	667(97.65%)	5	664(97.21%)	5	664(97.21%)	5	664(97.21%)
Hep1	5	142(91.65%)	9	139(89.67%)	6	139(89.67%)	5	140(90.32%)
AC	7	610(88.40%)	8	604(87.53%)	8	604(87.53%)	8	604(87.53%)
BCW1	5	681(97.42%)	8	677(96.85%)	5	677(96.85%)	5	677(96.85%)
Ion	5	312(88.88%)	5	305(86.89%)	7	302(86.03%)	8	309 (88.03%)
spam	28	4149(90.17%)	32	4019(87.35%)	39	4012(87.19%)	31	4028 (87.54%)
HD	6	243(90.00%)	9	236(87.41%)	9	235(87.03%)	8	237(87.77%)
Iris	3	147(98.00%)	3	146(97.33%)	3	146(97.33%)	3	146(97.33%)
Wine	6	177(99.43%)	7	176(98.87%)	7	176(98.87%)	7	176(98.87%)

Instance	PSO-GRASP		ACO-GRASP		Genetic-GRASP		Tabu-GRASP	
	Selected Features	Correct Clustered	Selected Feat.	Correct Clustered	Sel. Feat.	Correct Clustered	Sel. Feat.	Correct Clustered
BCW2	5	662(96.92%)	5	662(96.92%)	5	662(96.92%)	6	661(96.77%)
Hep1	7	135(87.09%)	9	134(86.45%)	9	134(86.45%)	10	132(85.16%)
AC	8	604(87.53%)	8	603(87.39%)	8	602(87.24%)	9	599(86.81%)
BCW1	5	676(96.70%)	5	676(96.70%)	5	676(96.70%)	8	674(96.42%)
Ion	11	300(85.47%)	2	291(82.90%)	17	266(75.78%)	4	263(74.92%)
spam	51	4009(87.13%)	56	3993(86.78%)	56	3938(85.59%)	34	3810(82.80%)
HD	9	232(85.92%)	9	232(85.92%)	7	231(85.55%)	9	227(84.07%)
Iris	3	145(96.67%)	3	145(96.67%)	4	145(96.67%)	3	145(96.67%)
Wine	7	176(98.87%)	8	176(98.87%)	7	175(98.31%)	7	174(97.75%)

Instance	k-Means		ACO		PSO	
	Sel Feat.	Correct Clustered	Selected Features	Correct Clustered	Sel. Feat.	Correct Clustered
BCW2	9	654(95.74%)	5	662(96.92%)	5	662(96.92%)
Hep1	19	121(78.06%)	9	133(85.80%)	10	132(85.16%)
AC	14	580(84.05%)	8	601(87.10%)	8	602(87.24%)
BCW1	9	672(96.13%)	8	674(96.42%)	8	674(96.42%)
Ion	34	248(70.65%)	16	258(73.50%)	12	261(74.35%)
spam	57	3958(86.02%)	41	3967(86.22%)	37	3960(86.06%)
HD	13	220(81.48%)	9	227(84.07%)	9	227(84.07%)
Iris	4	144(96%)	3	145(96.67%)	3	145(96.67%)
Wine	13	172(96.92%)	7	174(97.75%)	7	174(97.75%)

demonstrated by the fact that with this algorithm the best solution was found by using less features than the other algorithms used in the comparisons. More precisely, in the most difficult instance, the Spambase instance, the proposed algorithm needed 28 features in order to find the optimal solution, while the other nine algorithms (in the k-means the feature selection problem was not solved) the algorithms needed between 31 - 56 features to find their best solution.

But the most important observation is the improvement of the results of the Multi Swarm Constriction PSO - GRASP algorithm from the results of the hybrid PSO - GRASP algorithm [13]. The improvement in all instances is between 0.56 and 4.56 percentage units. There is a significant improvement even for the instances where the PSO - GRASP algorithm has found a solution with more than 98% in the percentage of corrected clustered data. Although the hybrid PSO-GRASP algorithm gives inferior results compared to the results of almost all other nature inspired algorithms used in the comparisons as the algorithm could not find a better local optimum, the use of Multi Swarm Constriction PSO - GRASP algorithm not only improves the results of the hybrid PSO - GRASP algorithm but it, also, gives better results from all hybridized algorithms that we have used for the solution of the same problems. Thus, it is demonstrated that the use of multiple swarms in the Particle Swarm Optimization can lead the search in unexplored regions of the solution space and, thus, to find a better local optimum. The basic reason for this issue is that in the case of multiple swarms all the swarm are moved in different regions and the exploration and exploitation abilities of the algorithm are increased. It should, also, be mentioned that the algorithm was tested with two options: with known and unknown number of clusters. When the number of clusters was unknown and, thus, in each iteration of the algorithm different initial values of clusters were selected, the algorithm always converged to the optimal number of clusters and with the same results as in the case where the number of clusters was known.

4   Conclusions and Future Research

In this paper, a new metaheuristic algorithm, the Hybrid MSCPSO-GRASP, is proposed for solving the Clustering Problem. This algorithm is a two phase algorithm which combines a Multi-Swarm Constriction Particle Swarm Optimization algorithm for the solution of the feature selection problem and a Greedy Randomized Adaptive Search Procedure for the solution of the clustering problem. A number of other metaheuristic algorithms for the solution of the problem were also used for comparison purposes. The performance of the proposed algorithm was tested using various benchmark datasets from UCI Machine Learning Repository. The significance of the solution of the clustering problem by the proposed algorithm is demonstrated by the fact that the percentage of the correct clustered samples is very high and in some instances is larger than 98%. Also, the use of the multiple swarms and of the constriction factor is needed in order the algorithm to find more easily the local optimum compared to the classic PSO. Future research is intended to be focused in using different algorithms both to the feature selection phase and to the clustering algorithm phase.

References

1. Banks, A., Vincent, J., Anyakoha, C.: A review of particle swarm optimization. Part I: background and development. Natural Computing 6(4), 467–484 (2007)
2. Brits, R., Engelbrecht, A.P., van den Bergh, F.: Locating multiple optima using particle swarm optimization. Applied Mathematics and Computation 189, 1859–1883 (2007)
3. Clerc, M., Kennedy, J.: The particle swarm: explosion, stability and convergence in a multi-dimensional complex space. IEEE Transactions on Evolutionary Computation 6, 58–73 (2002)
4. Clerc: Particle Swarm Optimization. ISTE, London (2006)
5. Dorigo, M., Stutzle, T.: Ant Colony Optimization. A Bradford Book, MIT Press, Cambridge (2004)
6. Feo, T.A., Resende, M.G.C.: Greedy randomized adaptive search procedure. Journal of Global Optimization 6, 109–133 (1995)
7. Glover, F.: Tabu Search I. ORSA Journal on Computing 1(3), 190–206 (1989)
8. Goldberg, D.E.: Genetic Algorithms in Search, Optimization, and Machine Learning. Addison-Wesley Publishing Company, Inc., Massachussets (1989)

9. Jain, A.K., Murty, M.N., Flynn, P.J.: Data Clustering: A Review. ACM Computing Surveys 31(3), 264–323 (1999)
10. Jain, A., Zongker, D.: Feature Selection: Evaluation, application, and Small Sample Performance. IEEE Transactions on Pattern Analysis and Machine Intelligence 19, 153–158 (1997)
11. Kennedy, J., Eberhart, R.: Particle swarm optimization. In: Proceedings of 1995 IEEE International Conference on Neural Networks, vol. 4, pp. 1942–1948 (1995)
12. Kennedy, J., Eberhart, R.: A discrete binary version of the particle swarm algorithm. In: Proceedings of 1997 IEEE International Conference on Systems, Man, and Cybernetics vol. 5, pp. 4104–4108 (1997)
13. Marinakis, Y., Marinaki, M., Matsatsinis, N.: A Hybrid Particle Swarm Optimization Algorithm for Cluster Analysis. In: Song, I.-Y., Eder, J., Nguyen, T.M. (eds.) DaWaK 2007. LNCS, vol. 4654, pp. 241–250. Springer, Heidelberg (2007)
14. Marinakis, Y., Marinaki, M., Matsatsinis, N.: A Stochastic Nature Inspired Metaheuristic for Clustering Analysis. International Journal of Business Intelligence and Clustering Analysis 3(1), 30–44 (2008)
15. Marinakis, Y., Marinaki, M., Matsatsinis, N.: A Hybrid Clustering Algorithm based on Honey Bees Mating Optimization and Greedy Randomized Adaptive Search Procedure. In: Learning and Intelligence Optimization - LION 2007. LNCS. Springer, Heidelberg (in print, 2007)
16. Marinakis, Y., Marinaki, M., Matsatsinis, N., Zopounidis, C.: A Memetic-GRASP Algorithm for Clustering. In: 10th International Conference on Enterprise Informations Systems, Barcelona, Spain, 13 -16 June (2008)
17. Marinakis, Y., Marinaki, M., Doumpos, M., Matsatsinis, N., Zopounidis, C.: A Hybrid ACO-GRASP Algorithm for Clustering Analysis. Annals of Operations Research (submitted, 2007)
18. Marinakis, Y., Marinaki, M., Doumpos, M., Matsatsinis, N., Zopounidis, C.: A Hybrid Stochastic Genetic - GRASP Algorithm for Clustering Analysis. Operational Research: An International Journal 8(1), 33–46 (2008)
19. Niu, B., Zhu, Y., He, X., Wu, H.: MCPSO: A multi-swarm cooperative particle swarm optimizer. Applied Mathematics and Computation 185, 1050–1062 (2007)
20. Poli, R., Kennedy, J., Blackwell, T.: Particle Swarm Optimization. An Overview. Swarm Intelligence 1, 33–57 (2007)
21. Rokach, L., Maimon, O.: Clustering Methods. In: Maimon, O., Rokach, L. (eds.) Data Mining and Knowledge Discovery Handbook, pp. 321–352. Springer, New York (2005)
22. Shi, Y., Eberhart, R.: A modified particle swarm optimizer. In: Proceedings of 1998 IEEE World Congress on Computational Intelligence, pp. 69–73 (1998)
23. Tillett, T., Rao, T.M., Sahin, F., Rao, R.: Darwinian Particle Swarm Optimization. In: Proceedings of the 2nd Indian International Conference on Artificial Intelligence, Pune, India, pp. 1474–1487 (2005)
24. Xu, R., Wunsch II, D.: Survey of Clustering Algorithms. IEEE Transactions on Neural Networks 16(3), 645–678 (2005)

Personalizing Navigation in Folksonomies Using Hierarchical Tag Clustering

Jonathan Gemmell, Andriy Shepitsen, Bamshad Mobasher, and Robin Burke

Center for Web Intelligence
School of Computing, DePaul University
Chicago, Illinois, USA
{jgemmell,ashepits,mobasher,rburke}@cti.depaul.edu

Abstract. The popularity of collaborative tagging, otherwise known as "folk-sonomies", emanate from the flexibility they afford users in navigating large information spaces for resources, tags, or other users, unencumbered by a pre-defined navigational or conceptual hierarchy. Despite its advantages, social tagging also increases user overhead in search and navigation: users are free to apply any tag they wish to a resource, often resulting in a large number of tags that are redundant, ambiguous, or idiosyncratic. Data mining techniques such as clustering provide a means to overcome this problem by learning aggregate user models, and thus reducing noise. In this paper we propose a method to personalize search and navigation based on unsupervised hierarchical agglomerative tag clustering. Given a user profile, represented as a vector of tags, the learned tag clusters provide the nexus between the user and those resources that correspond more closely to the user's intent. We validate this assertion through extensive evaluation of the proposed algorithm using data from a real collaborative tagging Web site.

Keywords: collaborative tagging, hierarchical clustering, personalization.

1 Introduction

Collaborative tagging is an emerging trend allowing Internet users to manage, share and annotate online resources. Two of the most prominent examples are del.icio.us[1] in which users bookmark URLS and Flickr[2] which allows users to upload, share and manage pictures. The foundation of collaborative tagging is the annotation; a user describes a resource with a tag. A collection of annotations results in a complex network of interrelated users, resources and tags, commonly referred to as a folksonomy [8]. The freedom to explore this large information space of resources, tags, or even other users is central to the utility and popularity of collaborative tagging. Tags make it easy and intuitive to retrieve previously viewed resources [5]. Tagging allows users to categorized resources by several terms, rather than one directory or a single branch of an ontology [10]. Users may enjoy the social aspects of collaborative tagging [3]. Because collaborative tagging applications reap the insights of many users rather than a few "experts",

---

[1] del.icio.us

[2] www.flickr.com

I.-Y. Song, J. Eder, and T.M. Nguyen (Eds.): DaWaK 2008, LNCS 5182, pp. 196–205, 2008.

they are more dynamic and able to incorporate a changing vocabulary or absorb new trends quickly [16].

Even though collaborative tagging offers many benefits, it also presents unique challenges for search and navigation. To reduce the entry cost, most collaborative tagging applications permit unsupervised tagging. As a result folksonomies often contain a great deal of noise such as tag redundancy in which several tags have the same meaning or tag ambiguity in which a single tag has many different meanings. This noise can confound users as they navigate the folksonomy.

Data mining techniques such as clustering provide a means to overcome these problems. Through clustering, redundant tags can be aggregated; the combined trend of a cluster can be more easily detected than the effect of a single tag. The effect of ambiguity can also be diminished, since the uncertainty of a single tag in a cluster can be overwhelmed by the additive effects of the rest of the tags. Furthermore, an advantage of collaborative tagging applications is the richness of the user profiles [18]. These profiles are central for personalization algorithms. By associating a user's profile with a particular cluster, we may surmise the user's interest in the topics represented by the clusters.

Previous work has shown that clustering can be useful in collaborative tagging. Tags can be aggregated into clusters with coherent topic areas. For example, in [2] and [6], they suggested tag clustering for search and taxonomy generation, respectively. In [17], clustering techniques were used for tag recommendation. In [16], tag clusters are presumed to be representative of the resource content, moving the Internet closer to the Semantic Web. Resource recommendation has benefited from tag clustering [11] in which an affinity between a user and a set of tag clusters was calculated. In [1], the user's annotation history is incorporated in web search.

In this paper, we propose an algorithm to personalize search and navigation in folksonomies based on unsupervised hierarchical agglomerative tag clustering. By measuring the importance of a tag cluster to a user, the user's interests can be better understood. Likewise, by associating resources with tag clusters, resources relevant to the topics can be identified. Using the tag clusters as intermediaries between a user and a resource, we infer the relevance of the resource to the user and re-rank the results of a basic search thereby personalizing the user experience. Alternative efforts for search and ranking were made in [7], in which the structure of the folksonomy is utilized.

We outline basic approaches used for search in folksonomies in Section 2 and motivate the need for personalization. A modified version of hierarchical agglomerative clustering method as well as a personalization algorithm are then presented in Section 3. In Section 4, we evaluate our personalization algorithm and compare the effectiveness of the hierarchical clustering algorithm in that context.

2 Search and Navigation in Folksonomies

A folksonomy can be described as a four-tuple $D = \langle U, R, T, A \rangle$ where, U is a set of users, R is a set of resources, T is a set of tags, and A is a set of annotations represented as user-tag-resource triples:

$$A \subseteq \{\langle u, r, t \rangle : u \in U, r \in R, t \in T\} \tag{1}$$

A folksonomy can, therefore, be viewed as a tripartite hyper-graph [9] with users, tags, and resources represented as nodes and the annotations represented as hyper-edges connecting a user, a tag and a resource.

2.1 Standard Search in Folksonomies

Contrary to traditional Internet applications, a search in a collaborative tagging application is performed with a tag rather than a keyword. Most often the tag is selected through the user interface. Applications vary in the way they handle ranking of results when a user selects a tag: recency, authority, linkage, or vector space models. In this work we focus on the vector space model [13] adapted from the information retrieval discipline to work with folksonomies. Each user is modeled as a vector, u, over the set of tags, where the weight, $w_u(t_i)$, in each dimension corresponds to the importance of a particular tag, t_i. We define u as: $u = \langle w_u(t_1), w_u(t_2)...w_u(t_{|T|}) \rangle$. Resources can also be modeled as a vector, r, over the set of tags with weights, $w_r(t_i)$.

In calculating the vector weights, a variety of measures can be used. The *tag frequency*, *tf*, for a tag, t, and a resource, r is the number of times the resource has been annotated with the query tag. We define *tf* as: $tf(t, r) = |\{a = \langle u, r, t \rangle \in A : u \in U\}|$. Likewise, the well known *term frequency * inverse document frequency* [12] can be modified for folksonomies as used as weights. The *tf*idf* multiplies the aforementioned frequency by the relative distinctiveness of the tag, measured as the log of the total number of resources, N, divided by the number of resources to which the query tag was applied, n_t. We define *tf*idf* as: $tf\text{*}idf(t, r) = tf(t, r) * \log(N/n_t)$.

With either term weighting approach, a similarity measure between a query, q, represented as a vector over the set tags, and a resource, r, also modeled as a vector over the set tags, can be calculated. In this work, since search or navigation is often initiated by selecting a single tag from the user interface, we assume the query is a vector with only one tag. A standard search may calculate the cosine similarity [14] of the query, q, and a resource, r, where the weight is either the *tf* or *tf*idf* and then return a list of resources ordered by their similarity to the query.

$$cos(q, r) = \frac{\sum_{t \in T} w_q(t) * w_r(t)}{\sqrt{\sum_{t \in T} w_q(t)^2} * \sqrt{\sum_{t \in T} w_r(t)^2}} \tag{2}$$

Since the query is modeled as a vector containing only one tag, this equation may be further simplified. However, we have provided the full equation so that it can be applicable in a more general setting.

2.2 Need for Personalization

While personalization has been shown to increase the utility of other Web applications, the need for personalization in collaborative tagging is even more critical. Noise in the folksonomy, such as tag redundancy and tag ambiguity hinder the user's navigation. Redundant tags can hinder algorithms that depend on calculating similarity between resources. Ambiguous tags can result in the overestimation of the similarity of resources that are in fact unrelated. Tag clustering helps to facilitate personalization. By aggregating tags into clusters with similar meaning, tag redundancy can be assuaged since the

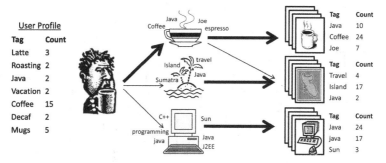

Fig. 1. Clusters represent coherent topic areas and serve as intermediaries between a user and the resources

trend for a cluster can be more easily identified than the effect of a single tag. Ambiguity will also be remedied to some degree, since a cluster of tags will assume the aggregate meaning and overshadow any ambiguous meaning a single tag may have. Further, by using clusters to represent topic areas, the user's interest in that topic can be more easily quantified. The user's intended meaning for ambiguous tags can be inferred through the analysis of the user profile.

3   Personalized Search Based on Tag Clustering

For tag clustering we use a modified version of the traditional hierarchical agglomerative clustering algorithm   [4]. Tag clusters will serve as intermediaries between a user and the resources. The user's connection to each cluster is calculated. Likewise, a measure is calculated from each cluster to all resources. By using clusters as the nexus between a user and a resource, the user's interest in the resource can be inferred. This information allows us to reorder the results of a basic search, personalizing each user's view of the information space.

For example, in Figure 1, a hypothetical user searching based on the tag "Java" has a strong connection to the cluster of coffee related tags, and weaker connections to the other two clusters. Likewise, resources dealing with coffee has a strong relation to the coffee cluster. The user's interest in those resources can, therefore, be inferred. However, the coffee cluster has a weak relation to resources dealing with travel to Java and Sumatra. The relevance of those resources to the user is consequently minimal.

3.1   Modified Hierarchical Agglomerative Tag Clustering

Hierarchical tag clustering requires a similarity measure between tags. The cosine similarity between two tags, t and s, may be calculated by treating each tag as a vector over the set of resources and using *tag frequency* or *tag frequency * inverse document frequency* as the weights in the vectors.

$$cos(t, s) = \frac{\sum_{r \in R} w_r(t) * w_r(s)}{\sqrt{\sum_{r \in R} w_r(t)^2} * \sqrt{\sum_{r \in R} w_r(s)^2}} \tag{3}$$

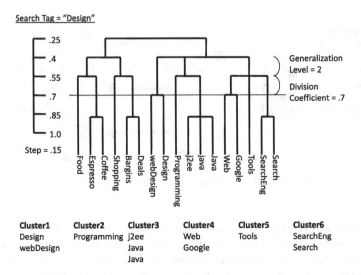

Fig. 2. An example of hierarchical tag clustering

Once the cosine similarities are calculated, it is possible to construct the clusters. As the hierarchical clustering algorithm begins each tag forms a singleton cluster. Then, during each stage of the procedure, clusters of tags are joined together depending on the level of similarity between the clusters. Several clusters may be aggregated at once. This is done for many iterations until all tags have been aggregated into one cluster. To compute the similarity between clusters, a centroid for each cluster is calculated and treated as though it were a single tag. The result is a hierarchical clustering of the tags such as the example depicted in Figure 2.

Hierarchical clustering has several parameters that require tuning in order to achieve optimum results in the personalization routine. The parameter *step* is the decrement by which the similarity threshold is lowered. At each iteration, clusters of tags are aggregated if the similarity between them meets a minimum threshold. This threshold is lowered by *step* until it reaches 0. By modifying this parameter the granularity of the hierarchy can be controlled. An ideal value for *step* would aggregate tags slowly enough to capture the conceptual hierarchy between individual clusters.

In order to break the hierarchy into distinct clusters, a *division coefficient* is chosen as a cutoff point. Any cluster below this similarity threshold is considered an independent cluster in the personalization routine. Selecting a value near one will result in many small clusters with high internal similarity. Alternatively, selecting a small value will result in few larger clusters, with lower internal similarity.

We extend the standard hierarchical agglomerative algorithm by introducing the *generalization level*. Normally, all clusters below the *division coefficient* would be used, but in this modification only those clusters descendent of the selected tag are used. The *generalization level* allows the algorithm to return more general tag clusters for the hierarchy. Instead of using only the descendants of the selected tag, a larger branch of the hierarchy is used by first traveling up the tree the specified number of levels. If the *generalization level* is set too low, it is possible to overlook a cluster representing relevant resources. However, if the algorithm includes too many clusters not related to the user's selected tag, irrelevant topic areas can be introduced.

For example, in the hypothetical hierarchy of clusters depicted in Figure 2, if the user selects the tag "Design," the algorithm will first identify the level at which this tag was added to the hierarchy. In this case, it was added when the similarity threshold was lowered to .7. With a *generalization coefficient* of 2, the algorithm proceeds up two levels in the hierarchy. Finally, the *division coefficient* is used to break the branch of the hierarchy into distinct clusters.

3.2 Personalization Algorithm Based on Tag Clustering

There are three inputs to a personalized search: the selected tag, the user profile and the discovered hierarchical clusters. The output of the algorithm is an ordered set of resources. For each cluster, c, the user's interest is calculated as the ratio of times the user, u, annotated a resource with a tag from that cluster over the total number of annotations by that user. We denote this weight as $uc\_w(u,c)$ and defined it as:

$$uc\_w(u,c) = \frac{|\{a = \langle u, r, t \rangle \in A : r \in R, t \in c\}|}{|\{a = \langle u, r, t \rangle \in A : r \in R, t \in T\}|} \tag{4}$$

Also, the relation of a resource, r, to a cluster is calculated as the ratio of times the resource was annotated with a tag from the cluster over the total number of times the resource was annotated. We call this weight $rc\_w(r,c)$ and defined it as:

$$rc\_w(r,c) = \frac{|\{a = \langle u, r, t \rangle \in A : u \in U, t \in c\}|}{|\{a = \langle u, r, t \rangle \in A : u \in U, t \in T\}|} \tag{5}$$

The relevance of the resource to the user, *relevance(u,r)*, is calculated from the sum of the product of these weights over the set of all clusters, C. This measure is defined as:

$$relevance(u,r) = \sum_{c \in C} uc\_w(u,c) * rc\_w(r,c) \tag{6}$$

Intuitively, each cluster can be viewed as the representation of a topic area. If a user's interests parallels closely the subject matter of a resource, the value for *relevance(u,r)* will be correspondingly high.

A basic search is performed on the query, q, using the vector space model and either *tag frequency* or *term frequency * inverse document frequency*. A similarity, *rankscore(q,r)*, is calculated for every resource in the dataset using the cosine similarity measure. A personalized similarity is calculated for each resource by multiplying the cosine similarity by the relevance of the resource to the user. We denote this similarity as *p_rankscore(u,q,r)* and define it as:

$$p\_rankscore(u,q,r) = rankscore(q,r) * relevance(u,r) \tag{7}$$

Once the *p_rankscore(u,q,r)*, has been calculated for each resource, the resources are returned to the user in descending order of the score. While the weights from clusters to resources will be constant regardless of the user, the weights connecting the users to the clusters will differ based on the user profile. Consequently, the results will be personalized.

4   Experimental Evaluation

We validate our approach through extensive evaluation of the proposed algorithm using data from a real collaborative tagging Web site. A Web crawler was used to extract data from del.icio.us from 5/26/2007 to 06/15/2007, iterating over users. In this collaborative tagging application, the resources are Web pages. The dataset contains 29,918 users, 6,403,442 resources and 1,035,177 tags. There are 47,184,492 annotations with one user, resource and tag. Two random samples of 5,000 users were taken from the dataset. Five-fold cross validation was performed on each sample. For each fold, 20% of the users were partitioned from the rest as test users. Clustering was completed using the data from the remaining 80% of the users. From each user in the test set, 10% of the user's annotations were randomly selected as test cases. Each case consisted of a user, tag and resource.

The basic search requires only a tag as an input. The cosine similarity was calculated for all resources to the test tag, and the resources were then ordered. The rank of the resource in the basic search, r_b, was recorded. The personalized search requires the test tag, the test user profile, and a set of discovered clusters. The relevance of the resources to the test user was calculated and was used to re-rank the resources. The new rank of the test resource in the personalized search, r_p, was recorded. Since the user has annotated this resource, we assume that it is relevant to the user. In order to judge the improvement provided by the personalized search, the difference in the inverse of the two ranks can be used, imp [15]. It is defined as:

$$imp = \frac{1}{r_p} - \frac{1}{r_b} \tag{8}$$

If the personalized approach re-ranks the resource nearer the top of the list, the improvement will be positive, or if the personalized approach re-ranks the resource further down the list, the improvement will be negative. The maximum improvement is 1. However, it is nearly impossible to achieve such a dramatic improvement unless the resource is re-ranked from complete obscurity to the top position.

The choice of tf or $tf * idf$ played an important role. In all cases $tf * idf$ is superior. This is likely the result of the normalization that occurs. The two weighting techniques appear to have nearly identical trends in the tuning of the parameters.

The parameter *step* is the decrement by which the similarity threshold is lowered. In Figure 3, increasing its value results in diminished performance, as tags are aggregated too quickly. Yet, if the value for step is too low, the granularity of the derived clusters can become too fine grained and over-specialization can occur. In these experiments, best results were achieved with a value of 0.004. The same value was used when testing the other parameters.

The *division coefficient* defines the level where the hierarchy is dissected into individual clusters. Since, the personalization algorithm relies on tag clusters to serve as the intermediary between users and resources and presupposes the clusters represent distinct well defined topics, the selection of the *division coefficient* is integral to the success of the personalization algorithm. Intuitively, the goal of tuning the *division coefficient* is to discover the optimum level of specificity. As shown in Figure 4, the optimum value for this dataset is approximately 0.1. This is also the value used when testing other parameters.

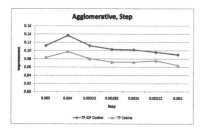

Fig. 3. The effect of *step* on the personalization algorithm

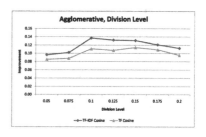

Fig. 4. The effect of *division level* on the personalization algorithm

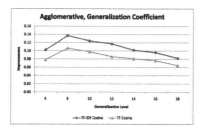

Fig. 5. The effect of *generalization level* on the personalization algorithm

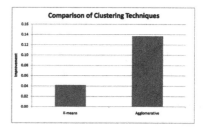

Fig. 6. A comparison of agglomerative and *k*-means clustering

The importance of the *generalization level* is demonstrated in Figure 5. For these experiments the optimum value for the *generalization level* is 8. This is also the default value when testing other parameters. The *generalization level* has important ramifications related to the user's motivation. If the user is searching for something specific, a low *generalization level* may be chosen, focusing on clusters more directly related to the selected tag. However, if the user is browsing through the folksonomy, a higher *generalization level* may be appropriate. It may increase serendipity and introduce topics unknown but nevertheless interesting to the user.

If the *generalization level* is set very high, the algorithm will behave like a standard hierarchical agglomerative clustering algorithm, returning all clusters below the *division coefficient*. In this case the performance of the personalization drops precipitously, underscoring the importance of the modifications to the algorithm.

In order to ascertain the relative value of the modified hierarchical agglomerative clustering technique, k-means clustering was also tested. The k-means clustering technique requires the input parameter, k, the predetermined number of clusters; we tried several values for k and found a value of 350 to yield best results. Best results for hierarchical clustering reached .137, whereas k-means achieved an improvement of only .042 as shown in Figure 6. Moreover, the standard deviation of the K-means clustering was .37. It was .28 for hierarchical agglomerative clustering. Not only did hierarchical clustering prove to offer the most benefit to the personalization algorithm, but it was also the most reliable. The poor results of k-means clustering may be attributed to its inability to identify innocuous tags. Every tag is placed in a cluster and given equal weight. Hierarchical clustering, on the other hand, uses a similarity threshold to identify strongly related tags and can ignore weak similarities. These innocuous tags may be adding noise to the k-means clusters.

In sum, the evidence demonstrates that tag clusters can serve as effective intermediaries between users and resources thereby facilitating personalization. The modified hierarchical agglomerative clustering offers a level of customization not offered by other techniques. The parameter *step* effects the granularity of the hierarchy, while the *division coefficient* offers a means to select the cluster specificity. Moreover, the modifications to the algorithm coupled with the *generalization level* allow the algorithm to incorporate clusters strongly related to the user's selected tag or take a broader view of the hierarchy thereby promoting serendipitous discovery.

5   Conclusions

In this work, we have proposed a personalization algorithm for folksonomies based on modified hierarchical tag clustering. The clusters bridge the gap between users and resources, offering a means to infer the user's interest in the resource. Standard search results based on cosine similarity are reordered using the relevance of the resource to the user. The clustering technique is independent of the personalization algorithm. Parameters for the hierarchical clustering method were tuned, demonstrating that the quality of the tag clusters greatly affects the performance of the personalization technique. By using clusters directly related to the selected tag, the modified hierarchical agglomerative clustering proved to offer better improvement in our experimental results as well as greater flexibility.

Acknowledgments

This work was supported in part by the National Science Foundation Cyber Trust program under Grant IIS-0430303 and a grant from the Department of Education, Graduate Assistance in the Area of National Need, P200A070536.

References

1. Bao, S., Xue, G., Wu, X., Yu, Y., Fei, B., Su, Z.: Optimizing web search using social annotations. In: Proceedings of the 16th international conference on World Wide Web, pp. 501–510 (2007)
2. Begelman, G., Keller, P., Smadja, F.: Automated Tag Clustering: Improving search and exploration in the tag space. In: Proceedings of the Collaborative Web Tagging Workshop at WWW, vol. 6 (2006)
3. Choy, S., Lui, A.: Web Information Retrieval in Collaborative Tagging Systems. In: Proceedings of the 2006 IEEE/WIC/ACM International Conference on Web Intelligence, pp. 352–355 (2006)
4. Gower, J., Ross, G.: Minimum Spanning Trees and Single Linkage Cluster Analysis. Applied Statistics 18(1), 54–64 (1969)

5. Hammond, T., Hannay, T., Lund, B., Scott, J.: Social Bookmarking Tools (I). D-Lib Magazine 11(4), 1082–9873 (2005)
6. Heymann, P., Garcia-Molina, H.: Collaborative Creation of Communal Hierarchical Taxonomies in Social Tagging Systems. Technical report, Technical Report 2006-10, Computer Science Department (April 2006)
7. Hotho, A., Jaschke, R., Schmitz, C., Stumme, G.: Information retrieval in folksonomies: Search and ranking. The Semantic Web: Research and Applications 4011, 411–426 (2006)
8. Mathes, A.: Folksonomies-Cooperative Classification and Communication Through Shared Metadata. Computer Mediated Communication (Doctoral Seminar), Graduate School of Library and Information Science, University of Illinois Urbana-Champaign (December 2004)
9. Mika, P.: Ontologies are us: A unified model of social networks and semantics. Web Semantics: Science, Services and Agents on the World Wide Web 5(1), 5–15 (2007)
10. Millen, D., Feinberg, J., Kerr, B.: Dogear: Social bookmarking in the enterprise. In: Proceedings of the Special Interest Group on Computer-Human Interaction conference on Human Factors in computing systems, pp. 111–120 (2006)
11. Niwa, S., Doi, T., Honiden, S.: Web Page Recommender System based on Folksonomy Mining for ITNG 2006 Submissions. In: Proceedings of the Third International Conference on Information Technology: New Generations, pp. 388–393 (2006)
12. Salton, G., Buckley, C.: Term-weighting approaches in automatic text retrieval. Information Processing and Management: an International Journal 24(5), 513–523 (1988)
13. Salton, G., Wong, A., Yang, C.: A vector space model for automatic indexing. Communications of the ACM 18(11), 613–620 (1975)
14. Van Rijsbergen, C.: Information Retrieval, Butterworth-Heinemann Newton, MA, USA (1979)
15. Voorhees, E.: The Text Retrieval Conference-8 Question Answering Track Report. Proceedings of TREC 8, 77–82 (1999)
16. Wu, X., Zhang, L., Yu, Y.: Exploring social annotations for the semantic web. In: Proceedings of the 15th international conference on World Wide Web, pp. 417–426 (May 2006)
17. Xu, Z., Fu, Y., Mao, J., Su, D.: Towards the semantic web: Collaborative tag suggestions. In: Collaborative Web Tagging Workshop at WWW 2006, Edinburgh, Scotland (May 2006)
18. Yan, R., Natsev, A., Campbell, M.: An efficient manual image annotation approach based on tagging and browsing. In: Workshop on multimedia information retrieval on The many faces of multimedia semantics, pp. 13–20 (2007)

Clustered Dynamic Conditional Correlation Multivariate GARCH Model

Tu Zhou and Laiwan Chan

The Chinese University of Hong Kong
Shatin, N.T., Hong Kong
{tzhou,lwchan}@cse.cuhk.edu.hk

Abstract. The time-varying correlations between multivariate financial time series have been intensively studied. For example DCC and Block-DCC models have been proposed. In this paper, we present a novel Clustered DCC model which extends the previous models by incorporating clustering techniques. Instead of using the same parameters for all time series, a cluster structure is produced based on the autocorrelations of standardized residuals, in which clustered entries sharing the same dynamics. We compare and investigate different clustering methods using synthetic data. To verify the effectiveness of the whole proposed model, we conduct experiments on a set of Hong Kong stock daily returns, and the results outperform the original DCC GARCH model as well as Block-DCC model.

Keywords: multivariate time series analysis, GARCH, DCC.

1 Introduction

Correlation analysis is important to identify interacting pairs of time series across multiple time series data sets. In financial time series analysis, the correlations are always critical inputs for the common tasks of financial management, such as asset allocation and risk management. In the last two decades, the reliable estimates of correlations between financial time series have been intensively studied and many variant models have been proposed to describe the correlations between time series.

Ever since 1990, Constant Conditional Correlation model (CCC) [7] was proposed to solve the high number of parameters problem in multivariate time series models. In [10], Engle presented a new class of models Dynamic Conditional Correlation (DCC) that allows the correlations to change over time. However the dynamics are constrained to be identical for all the correlations. This constraint is released by Block-DCC model [3,4] by introducing a block-diagonal structure to capture the important dependence. [12,14,16] can be referred to for detailed survey.

In Block-DCC model, each block structure, indicating similar time series group, is simply and manually determined according to the business nature. It assumes that stocks in the same business section perform similarly and thus

I.-Y. Song, J. Eder, and T.M. Nguyen (Eds.): DaWaK 2008, LNCS 5182, pp. 206–216, 2008.

constrains the dynamics of correlations within the same section to be equal. On one hand, this constraint is too tight to model real world data accurately. On the other hand, correlation is a concept involving a pair of stocks, but the block structure only group single stocks.

Our model eliminates these constraints by clustering similar stock pairs rather than single stocks, which significantly improves the flexibility and fitness of DCC models. Both DCC GARCH [9] and Generalized DCC [10] can be regarded as special cases of our model.

The reminder of the paper is organized as follows: Section 2 gives a brief review of DCC GARCH model and extended Block-DCC GARCH model. In section 3, Clustered DCC model is proposed. In section 4, synthetic data are generated for clustering algorithms selection and accuracy evaluation. In section 5, we adopt Maximum Likelihood Estimation to estimate model parameters and Box-Pierce Test [15] is introduced for model evaluation. In section 6, experimental result and a portfolio application on Hong Kong stock market are presented compared with previous models. A succinct conclusion will be given in the end.

2   Related Work

First of all, we give a background introduction of famous Generalized Autoregressive Conditional Heteroscedasticity (GARCH) model.

The GARCH model [6] has been widely used to describe financial time series. Take the most used univariate GARCH(1,1) model for example. The process $\{y_t\}$ is modeled by:

$$y_t = \sigma_t \varepsilon_t \tag{1}$$

The standardized residuals ε_t is i.i.d.$(0,1)$ and σ_t can be expressed in terms of previous y_t and σ_t.

$$h_t = \sigma_t^2 = \omega + \alpha y_{t-1}^2 + \beta \sigma_{t-1}^2 \tag{2}$$

where $\omega \geq 0, \alpha > 0, \beta > 0, \alpha + \beta < 1$ and (y_t, σ_t) is a strictly stationary solution of (1) and (2). h_t is so called the volatility of the time series. This model captures the time-varying volatility clustering effect of financial time series. The GARCH(1,1) model, simple though, is the most widely used and proven to be successful in modeling conditional variance. Through out this paper, GARCH(1,1) is used.

2.1   Multivariate DCC GARCH Model

Given N-dimension assets, the multivariate DCC GARCH(1,1) Model [9] is defined as

$$R_t = (Q_t^*)^{-1} Q_t (Q_t^*)^{-1} \tag{3}$$

where the $N \times N$ conditional covariance matrix $Q_t = \{q_{t,ij}\}$ is given by:

$$Q_t = (1 - \alpha - \beta)\bar{Q} + \alpha \varepsilon_{t-1} \varepsilon_{t-1}^T + \beta Q_{t-1} \tag{4}$$

ε_t is a vector of univariate residuals defined in (1). $\bar{Q}$ is the $N \times N$ unconditional variance matrix of ε_t, in line with standard univariate GARCH result.
$Q_t^* = diag(\sqrt{q_{t,11}}, \sqrt{q_{t,22}}, \ldots \sqrt{q_{t,NN}})$ is introduced to ensure that R_t is correlation matrix. Positive definitiveness of the DCC-GARCH is controlled by the
correlation function and depends on parameter restrictions, namely α and β are
non-negative scalar numbers satisfying $\alpha + \beta < 1$.

The drawback of DCC model is that parameters α and β are scalars, so
that all the conditional correlations obey the same dynamics. This problem was
addressed and the DCC model was generalized by Engle [10], who suggested the
following Generalized DCC trying to solve the constraints of equal dynamics for
all correlations

$$Q_t = (\iota\iota' - A - B) \circ \bar{Q} + A \circ \varepsilon_{t-1}\varepsilon_{t-1}^T + B \circ Q_{t-1} \tag{5}$$

where ι is a vector of ones and $\circ$ is the Hadamard product of two identically sized
matrices, which is computed simply via element by element multiplication. A, B
are $N \times N$ symmetric matrices each composed of $N \times (N+1)/2$ different parameters. However the full matrix generalization incurs large increase in the number
of parameters, especially when N is high. It is hard to maintain tractability,
which make the model unattractive. In this respect, Block DCC model [3,4,5]
was proposed.

2.2   Block-DCC GARCH Model

The idea of Block-DCC GARCH model is to group the assets based on their
business nature and constrain the dynamics to be the same within each section.
The model has the same formulation as (5). But N assets are manually grouped
into w sets of dimension $m_1, m_2, \ldots m_w$ respectively. ι_j indicates a column vector of ones of j dimension. Therefore the parameter matrix A is given below.
Similarly can matrix B be expressed. A specific example is given in the right
with $N = 3, w = 2$ and $m_1 = 2, m_2 = 1$. For a DCC GARCH(1,1) model, only
$w + w(w - 1)/2$ parameters is needed for each parameter matrix.

$$A = \begin{bmatrix} \alpha_{11}\iota(m_1)\iota(m_1)' & \alpha_{12}\iota(m_1)\iota(m_2)' & \ldots & \alpha_{w1}\iota(m_1)\iota(m_w)' \\ \vdots & \vdots & \ddots & \vdots \\ \alpha_{w1}\iota(m_w)\iota(m_1)' & \alpha_{w2}\iota(m_w)\iota(m_2)' & \ldots & \alpha_{ww}\iota(m_w)\iota(m_w)' \end{bmatrix} \begin{bmatrix} \alpha_{11} & \alpha_{11} & \alpha_{12} \\ \alpha_{11} & \alpha_{11} & \alpha_{12} \\ \alpha_{12} & \alpha_{12} & \alpha_{22} \end{bmatrix} \tag{6}$$

The problem of Block-DCC model lies in the manual sectorial allocation approach. For example, in [4] the Italian Mibtel general index is grouped into
three major sectors: Industrials, Services and Finance. This grouping method
is not reasonable across all assets. Stocks in the same sector can perform dis-
tinctively along the time period. In addition, correlation is a concept involving
a pair of stocks, grouping similar single stocks does not make sense. In next
section we propose a Clustered DCC GARCH Model which cluster similar stock
pairs.

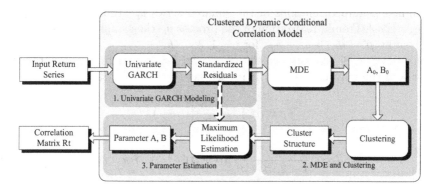

Fig. 1. CDCC Model Framework

3   Clustered DCC GARCH Model

To clarify our model, a framework is depicted in Fig. 1. The rectangles represent data while rounded rectangles indicate operation. In phase 1, through univariate GARCH modeling, standardized residuals ε_t are obtained. In phase 2, Minimum Distance Estimation (MDE) is applied on ε_t to obtain the parameters matrix $\boldsymbol{A}_0$ and $\boldsymbol{B}_0$, which are then clustered to form a cluster structure ζ. These are detailed discussed in this section. ε_t and ζ are then plugged in Phase 3 to estimate parameter matrix $\mathbf{A}$ and $\mathbf{B}$. The main difference of our model from DCC models is the extra part Phase 2, which employ the cluster structure. Both univariate GARCH model and multivariate correlation model are employing Maximum Likelihood Estimation, which will be given in section 5.1.

The CDCC model derives the dynamic correlation of stock pair (i, j) as

$$R_{t,ij} = Q_{t,ij}/\sqrt{Q_{t,ii}Q_{t,jj}} \tag{7}$$

$$Q_{t,ij} = (1 - A_{ij} - B_{ij})\bar{Q}_{ij} + A_{ij}\varepsilon_{t-1,i}\varepsilon_{t-1,j} + B_{ij}Q_{t-1,ij} \tag{8}$$

For k-cluster CDCC, $A_{ij} = \alpha_s, B_{ij} = \beta_s$ when stock pair (i, j) is in cluster s, $s = 1, \ldots, k$.

3.1   Minimum Distance Estimation (MDE)

To cluster similar dynamics of stock pairs, we aim to coarsely estimate parameters for each pair. However, directly applying Maximum Likelihood Estimation on all possible pairs of stocks will cause extremely high computation complexity. In contrast, MDE [1], by minimizing the Mahalanobis distance of a vector of sample autocorrelations from the corresponding population autocorrelations, provides more efficient estimation, and requires no strong distributional assumptions. For completeness we briefly review MDE here. When $\alpha + \beta < 1$, the GARCH(1,1) process can be represented as:

$$\sigma_t^2 = \sigma^2(1 - \alpha - \beta) + \alpha y_{t-1}^2 + \beta \sigma_{t-1}^2 \tag{9}$$

The first g autocorrelations of the squared series y_t^2 are in the vector $\rho' = [\rho_1, \rho_2, \ldots, \rho_g]$. From a realization of the process y_t, the sample autocorrelations are given by $\hat{\rho}' = [\hat{\rho}_1, \hat{\rho}_2, \ldots, \hat{\rho}_g]$, for $t = 1, 2, \ldots, T$.

According to [8] the autocorrelation ρ' can be derived from the parameter vector $\boldsymbol{\lambda}$ of the univariate GARCH model, here specifically α, β.

$$\rho_1 = (\alpha + \frac{\alpha^2 \beta}{1 - 2\alpha\beta - \beta^2}) \tag{10}$$

$$\rho_k = (\alpha + \frac{\alpha^2 \beta}{1 - 2\alpha\beta - \beta^2})(\alpha + \beta)^{k-1} \quad for \quad k \geq 2 \tag{11}$$

with constrain $3\alpha^2 + 2\alpha\beta + \beta^2 < 1$.

There is convergence in distribution: $\sqrt{T} \cdot (\hat{\rho} - \rho) \Rightarrow N(0, \mathbf{C})$ where $\mathbf{C}$ is the $g \times g$ matrix with (i,j)th element given by

$$c_{ij} = \sum_{k=1}^{\infty} (\rho_{k+i} + \rho_{k-i} - 2\rho_i\rho_k)(\rho_{k+j} + \rho_{k-j} - 2\rho_j\rho_k) \tag{12}$$

With the theoretical support above, the parameters of a stable GARCH(1,1) model can be estimated from the autocorrelations of the squared process.

$$\hat{\boldsymbol{\lambda}} = \arg \min\{S\} = \arg \min\{(\hat{\rho} - \rho(\boldsymbol{\lambda}))^T \boldsymbol{W} (\hat{\rho} - \rho(\boldsymbol{\lambda}))\} \tag{13}$$

The optimal weighting matrix is $\mathbf{W} = \mathbf{C}^{-1}$. A consistent estimator of $\mathbf{C}$ is $\hat{\mathbf{C}}$, the $g \times g$ sample counterpart of $\mathbf{C}$. Thus, in practice, the desired parameter vector $\boldsymbol{\lambda}$ is obtained by minimizing $S = (\hat{\rho} - \rho(\boldsymbol{\lambda}))^T \hat{C}^{-1}(\hat{\rho} - \rho(\boldsymbol{\lambda}))$

3.2   Clustered DCC (CDCC) Based on MDE

Since the variance matrix $\boldsymbol{Q}_t$ in (8) has similar derivation as (9), we apply MDE on the autocorrelations of the cross product of two standardized residual series to estimate parameters α, β for each series pair (i, j). (For $i, j = 1, 2, \ldots, N, i \neq j$, there are totally $\mathcal{N} = N(N-1)/2$ pairs.) The α, β parameter samples obtained form two $N \times N$ symmetric matrices $\boldsymbol{A}_0$ and $\boldsymbol{B}_0$ excluding diagonal entries. We obtain a cluster structure ζ by clustering the $\mathcal{N} \times 2$ parameters into k clusters. Different from (6), the clusters does not result in neatly partitioned blocks. Entries in the same group usually scatter in the matrix. Here we use another specific example to illustrate the structure.

$$\zeta = \begin{bmatrix} - & \diamond & \diamond & \heartsuit \\ \diamond & - & \triangle & \heartsuit \\ \diamond & \triangle & - & \diamond \\ \heartsuit & \heartsuit & \diamond & - \end{bmatrix}$$

Cluster	Mark	Entries
1	$\diamond$	(1,2) (1,3) (3,4)
2	$\heartsuit$	(1,4) (2,4)
3	$\triangle$	(2,3)

Fig. 2. Clustering Result Example

Assume we have $N = 4$ stocks, thus there will be $4 \times (4-1)/2 = 6$ pairs of stocks. Suppose the 6 pairs are clustered into 3 clusters, see Fig. 2. Then the clusters structure denoted as ζ is:

For k clusters, k distinctive parameters are filling into $N \times N$ matrix according to the structure ζ to form $\mathbf{A}/\mathbf{B}$. The (i,i) pair are not taken into clustering, because the diagonal entry of $\mathbf{Q}_t$ in (4) represents the variance of a single stock, while the off-diagonal entries indicate the covariance between two stocks, and they obey different dynamics. To maintain $\mathbf{Q}_t$ a covariance matrix, the diagonal of $\mathbf{A}, \mathbf{B}$ will take the same parameters in line with the subscripts. For example, when computing $Q_{t,ij}$, parameters in A_{ii} and A_{jj} will be the same as A_{ij}. This helps maintain the positive definitiveness of $\mathbf{R}_t$.

The DCC GARCH model is single (the least) cluster case of our model, and the Generalized DCC GARCH model can be regarded as $N \times (N-1)/2$ (the most) clusters case. Our novel model highly generalizes multivariate correlation models. It is more flexible than DCC model by differentiating various dynamics among stock pairs. At the meantime, it saves considerable amount of parameters and raises efficiency compared to Generalized DCC GARCH.

4   Clustering Method Selection

In order to choose a suitable clustering algorithm, we generate synthetic data to evaluate the accuracy of different clustering methods.

The generation process is the reverse course of the model. First of all we initialize $N \times N$ parameters matrix $\mathbf{A}$ and $\mathbf{B}$ with randomly generated cluster structure ζ'. The unconditional covariance matrix $\bar{\mathbf{Q}}$ is obtained by randomly select $N \times N$ principle submatrix from a covariance matrix of real data. Then for each day t, $\mathbf{Q}_t, \mathbf{R}_t$ can be derived from (7),(8)iteratively. The standardized residuals are composed by $\varepsilon_t = \mathbf{R}_t^{1/2} \epsilon_t$, where ϵ_t is randomly generating N dimension i.i.d.$(0,1)$ data. By the linear transformation, $\mathbf{R}_t$ becomes the correlation matrix of standardized residuals ε_t.

Once ε_t are generated, MDE and Clustering are applied to obtain the estimated cluster structure $\hat{\zeta}$. To evaluate the correctness of $\hat{\zeta}$ from original ζ', we define the clustering accuracy as the ratio of the correctly clustered sample number to total amount of samples. With the help of matching matrix, it is computed by the trace of matching matrix over sum of the matrix. Given an example in Table 1, totally 27 samples, $accuracy = (5+3+11)/27 = 0.7037$. We regard one single process of synthetic data generation and accuracy estimation as one iteration.

Table 1. Matching Matrix Example

	Cluster 1	Cluster 2	Cluster 3
Cluster 1	5	3	0
Cluster 2	2	3	1
Cluster 3	0	2	11

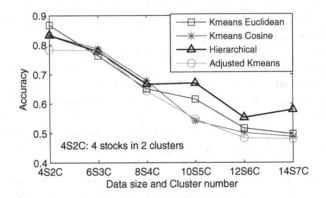

Fig. 3. Clustering Accuracy of Different Methods and Data Sets

Fig. 3 shows the accuracy of different methods with different number of stocks and clusters. The clustering methods are: Kmeans using Euclidean/Cosine distance; Hierarchical Clustering; and Adjusted Kmeans. Considering the remarkable scale difference of α and β, the Adjusted Kmeans refers to Kmeans applying on sample data that are adjusted to the same scale. For statistic purpose, we run 50 iterations for each specific stock and cluster case (corresponding to the horizontal axis tick in Fig. 3) to compute the average accuracy.

There is no significant performance difference between the algorithms when the stock and cluster numbers are small. However Hierarchical clustering algorithm outperforms other algorithms when stock and cluster numbers become larger. Based on this result, the Hierarchical method is adopted in the real world data application shown in section 6.

5   Model Estimation and Testing Method

In this section, Maximum Likelihood Estimation is introduced to estimate parameters of the proposed model. In addition, we introduce Box Pierce Statistic Test for model fitness evaluation.

5.1   Maximum Likelihood Estimation

Following the DCC model [9], we adopt Quasi-Maximum Likelihood estimation. Corresponding to Fig. 1, univariate GARCH models are estimated for each return series, in Phase 1. The standardized residuals obtained from Phase 1 and the cluster structure obtained from Phase 2 are used to estimate the parameters of the dynamic correlation. Denote the parameters of the univariate GARCH models and the parameters of the dynamic correlation by θ_1 and θ_2 respectively. For N-dimensional process $\boldsymbol{y}_t$, the likelihood of the model can be written as:

$$LogL(\theta_1, \theta_2 | \boldsymbol{y}_t) = -\frac{1}{2} \sum_{t=1}^{T} [Nlog(2\pi) + log(|\boldsymbol{Q}_t|) + \boldsymbol{y}_t \boldsymbol{Q}_t^{-1} \boldsymbol{y}_t^T] \qquad (14)$$

In Phase 1, R_t is replaced with I_N, an identity matrix of size N.

$$LogL(\theta_1|\boldsymbol{y}_t) = -\frac{1}{2}\sum_{t=1}^{T}[Nlog(2\pi) + log(\boldsymbol{I}_N) + 2log(|\boldsymbol{Q}_t^*|) + \boldsymbol{y}_t\boldsymbol{Q}_t^{*-1}\boldsymbol{I}_N^{-1}\boldsymbol{Q}_t^{*-1}\boldsymbol{y}_t^T]$$

In Phase 3, log likelihood is conditional on the parameters estimated in Phase 1

$$LogL(\theta_2|\hat{\theta}_1, \boldsymbol{y}_t) = -\frac{1}{2}\sum_{t=1}^{T}[Nlog(2\pi) + log(\boldsymbol{R}_t) + 2log(|\boldsymbol{Q}_t^*|) + \boldsymbol{\varepsilon}_t\boldsymbol{R}_t^{-1}\boldsymbol{\varepsilon}_t^T] \quad (15)$$

$\boldsymbol{\varepsilon}_t = \boldsymbol{Q}_t^{*-1}\boldsymbol{y}_t$ are standardized residuals. It is easier to exclude the constant terms and simply maximize:

$$LogL(\theta_2|\boldsymbol{\varepsilon}_t) = -\frac{1}{2}\sum_{t=1}^{T}[log(|\boldsymbol{R}_t|) + \boldsymbol{\varepsilon}_t^T\boldsymbol{R}_t^{-1}\boldsymbol{\varepsilon}_t] \quad (16)$$

Parameters obtained in Phase 3 are plugged into $\mathbf{A}, \mathbf{B}$ according to the cluster structure to form the parameter matrix of the model.

5.2   Box Pierce Statistic Test

To assess the fitness of the models, we use the Box-Pierce statistic test [11,15] to check the cross-product of the standardized residuals. Let $\hat{\varepsilon}_{ti}$ be the standardized residual for the i-th series, put

$$c_{t,ij} = \begin{cases} \hat{\varepsilon}_{ti}^2 - 1 & i = j \\ \hat{\varepsilon}_{ti}\hat{\varepsilon}_{tj} - \hat{R}_{t,ij} & i \neq j \end{cases} \quad (17)$$

where conditional correlation $\hat{R}_{t,ij} = \hat{Q}_{t,ij}/\sqrt{\hat{Q}_{t,ii}\hat{Q}_{t,jj}}$ is the estimated $R_{t,ij}$. If the multivariate conditional model fits the data, there should be no autocorrelation in $\{c_{t,ij}, t \geq 1\}$ for any fixed i, j $(i, j = 1, \ldots, N)$. Define

$$B(i, j; M) = N\sum_{k=1}^{M}\varrho_{ij,k}^2 \quad (18)$$

$\varrho_{ij,k}$ is the sample autocorrelation of $c_{t,ij}$ at lag k. It is intuitively clear that the large value of $B(i, j; M)$ suggests model inadequacy. M is set as 5 for experiment in next section. For each stock pair (i, j), $B(i, j; 5)$ is computed.

6   Experimental Result on Real World Data

6.1   Model Comparison and Analysis

The data set used for simulation in this paper contains the daily divident/split adjusted closing return series of 14 stocks selected from Hang Seng Index constitutes. These stocks are: { 0001.HK, 0002.HK, 0003.HK, 0004.HK, 0005.HK,

0006.HK, 0010.HK, 0011.HK, 0012.HK, 0013.HK, 0016.HK, 0019.HK, 0023.HK, 0293.HK}. The data period ranges from Jan 1990 to Sep 2007, result in 4135 observations for each stock. In order to investigate various changing dynamics in this long term period, we chop the data into five overlapping segments: day $1 + 500(i - 1)$ to $1500 + 500(i - 1)$ as training data, and day $1501 + 500(i - 1)$ to $2000 + 500(i - 1)$ as testing data, $i = 0, \ldots, 4$. For Block-DCC model, the 14 stocks are manually classed into two major sectors: Service and Finance (including banks and real estate company), as shown in Table 2.

Table 2. 14 Hong Kong stocks section

Service		0002.HK 0003.HK 0006.HK	0013.HK 0293.HK
Finance	Real estate	0001.HK 0012.HK 0004.HK 0016.HK 0010.HK 0019.HK	
	Banks	0005.HK 0011.HK	0023.HK

To compare Clustered DCC (CDCC) model with original DCC and Block-DCC models, Quasi Maximum Likelihood (QML) and Box-Pierce (BP) Test are conducted on five segments, and the average results are presented in Table 3 grouping by different number of clusters.

It is apparent that the CDCC model outperforms DCC and Block-DCC models in both training data and testing data in terms of both QML and Box-Pierce Test. Interestingly the result of all CDCC models is nearly in direct proportion with the number of clusters. When the clusters increase, QML increases and BP value decreases, indicating higher model fitness. Therefore how to determine the number of clusters will be a compromise between model fitness and computation efficiency.

Table 3. Summary of different models

	DCC	Block	CDCC				
Cluster	1	2	2	3	4	5	6
QML training data	-24008.2	-24007.7	-24004.8	-24003.6	-24003	-24000.6	-23996.6
QML testing data	-7820.72	-7819.13	-7816.96	-7816.92	-7815.54	-7814.9	-7814.04
BP training data	5.2257	5.2246	5.2224	5.2140	5.2146	5.2109	5.1993
BP testing data	5.0428	5.0335	4.9441	4.9437	4.9437	4.9448	4.9422

6.2 Portfolio Risk Application

To further verify the effectiveness of the proposed model we conduct a portfolio selection application on the same data set. The optimal portfolio derived from a better model should have better performance in terms of risk-return trade-off. We apply the classical optimal portfolio theory [13], which is to minimize the risk and maximize the portfolio return.

Let r_t be the vector of asset returns with conditional mean μ_t and conditional covariance matrix Σ_t at time t. ω_t is a vector of non-negative weights sum up as 1. The risk of the portfolio is measured by the variance of the r_t^p, which can be expressed as $\omega_t^T \Sigma_t \omega_t$. Given the covariance matrix is also the correlation matrix of the standardized residuals ε_t, the risk can be derived as $\omega_t^T R_t \omega_t$.

The application is conducted on five segments of training data each size 500, based on a 1500 length moving window. For given time t, the conditional mean μ_t is estimated by moving average of historical returns and R_t is predicted by different models using data from $t - 1500$ up to $t - 1$. The optimal portfolio weight $\hat{\omega}_t$ is obtained by solving $min\{\omega_t^T R_t \omega_t / \omega_t^T \mu_t\}$. The realized portfolio return is $r_t^p = \hat{\omega}_t^T r_t$. The above steps repeat and the portfolio rebalances every trading day.

Table 4. Average Daily Portfolio Return

Model	Seg 1	Seg 2	Seg 3	Seg 4	Seg 5	Mean	Annual Return
			(10^{-3})				
AvgPort	-0.860	0.125	-0.314	1.040	0.526	0.1034	2.62%
DCC	-0.286	-0.098	-0.259	0.978	0.685	0.2042	5.23%
CDCC	-0.173	-0.025	-0.131	0.989	0.701	0.2722	7.04%

Table 4 shows the average of realized daily portfolio return for each segment, and the mean value of 5 segments, all . Annual Return is computeds using the mean daily return for approximately 250 trading days per year. As a reference, AvgPort is portfolio with even weight. Six-cluster CDCC model is used in this experiment, and it achieves considerable higher return than DCC model. We also notice that CDCC model has better performance compared to average market when the whole market is down.

7   Conclusion

In this paper, we propose CDCC model to improve the fitness of dynamic conditional correlation estimation. The proposed cluster structure raises the flexibility of DCC model yet still maintains the parameter parsimony of the model. Original DCC [9] and Generalized DCC model [10] can both be regarded as special cases of our model. The CDCC model can be widely used in financial applications such as portfolio selection and risk management. A simple portfolio application is demonstrated and our model achieves considerable improvement over DCC model in terms of portfolio return.

References

1. Baillie, R.T., Chung, H.: Estimation of GARCH Models from the Autocorrelations of the Squares of a Process. Journal of Time Series Analysis 22, 631–650 (2001)
2. Bauwens, L., Rombouts, J.: Bayesian clustering of many GARCH models. Econometric Reviews Special Issue Bayesian Dynamic Econometrics (2003)
3. Billio, M., Caporin, M., Gobbo, M.: Block Dynamic Conditional Correlation Multivariate GARCH models. Greta Working Paper (2003)
4. Billio, M., Caporin, M., Gobbo, M.: Flexible Dynamic Conditional Correlation multivariate GARCH models for asset allocation. Applied Financial Economics Letters 2, 123–130 (2006)
5. Billio, M., Caporin, M.: A Generalized Dynamic Conditional Correlation Model for Portfolio Risk Evaluation. Working Paper of the Department of Economics of the Ca' Foscari University of Venice (2006)
6. Bollerslev, T.: Generalized Autoregressive Conditional Heteroskedasticity. Journal of Econometrics 31, 307–327 (1986)
7. Bollerslev, T.: Modelling the coherence in short-run nominal exchange rates: a multivareiate generalized ARCH approach. Review of Economic and Statistics 72, 498–505 (1999)
8. Ding, Z., Granger, C.W.J.: Modeling volatility persistence of speculative returns: A new approach. Journal of Econometrics 73, 185–215 (1996)
9. Engle, R.F., Sheppard, K.: Theoretical and Empirical properties of Dynamic Conditional Correlation Multivariate GARCH. Working paper, University of California, San Diego, CA (2001)
10. Engle, R.F.: Dynamic Conditional Correlation - a simple class of multivariate GARCH. Journal of Business and Economics Statistics 17, 425–446 (2002)
11. Fa, J., Wang, M., Yao, Q.: Modeling Multivariate Volatilities via Conditionally Uncorrelated Components. Working Paper,Department of Statistics,London School of Economics and Political Science (2005)
12. Laurent, S., Bauwens, L., Jeroen, V.K.: Rombouts: Multivariate GARCH models: a survey. Journal of Applied Econometrics 21, 79–109 (2006)
13. Markowitz, H.: Portfolio Selection. Journal of Finance 7, 77–91 (1952)
14. Serban, M., Brockwell, A., Lehoczky, J., Srivastava, S.: Modeling the Dynamic Dependence Structure in Multivariate Financial Time Series. Journal of Time Series Analysis (2006)
15. Tse, Y.K., Tsui, A.K.C.: A note on diagnosing multivariate conditional heteroscedasticity models. Journal of Time Series Analysis 20, 679–691 (1999)
16. Vargas, G.A.: An Asymmetric Block Dynamic Conditional Correlation Multivariate GARCH Model. Philippines Statistician (2006)

Document Clustering by Semantic Smoothing and Dynamic Growing Cell Structure (DynGCS) for Biomedical Literature

Min Song[1], Xiaohua Hu[2], Illhoi Yoo[3], and Eric Koppel[4]

[1] Information Systems, New Jersey Institute of Technology, USA
Min.song@njit.edu
[2] College of Information Science and Technology, Drexel University, USA
Tony.hu@cis.drexel.edu
[3] Department of Health Management and Informatics, School of Medicine,
University of Missouri-Columbia, USA
yooil@health.missouri.edu
[4] Computer Science, New Jersey Institute of Technology
erk7@njit.edu

Abstract. The general goal of clustering is to group data elements such that the intra-group similarities are high and the inter-group similarities are low. In this paper, we propose a novel hybrid clustering technique that incorporates semantic smoothing of document models into a neural network framework. Recently it has been reported that the semantic smoothing model enhances the retrieval quality in Information Retrieval (IR). Inspired by that, we apply the context-sensitive semantic smoothing model to boost accuracy of clustering that is generated by a dynamic growing cell structure algorithm, a variation of the neural network technique. We evaluated the proposed technique on article sets from MEDLINE, the largest biomedical digital library in Biomedicine. Our experimental evaluations show that the proposed algorithm significantly improves the clustering quality over the traditional clustering techniques.

Keywords: document clustering, feature selection, neural network.

1 Introduction

Document clustering is a research area that concerns organizing textual documents into meaningful groups that represent topics in document collections in an unsupervised fashion. There have been rigorous attempts to integrate document clustering into various different areas of Text Mining and Information Retrieval. In today's world, where documents are stored and retrieved electronically, document clustering has been given focal attention: it assists the users in discovering hidden similarity and key concepts. In addition, the size of text collections is increasing rapidly. To handle the increasing size of document collections, a clustering algorithm has to not only solve the incremental problem but it must also have high efficiency in a large dataset. In addition, a document is typically represented as a long vector in feature space. A feature space may have more than thousands of dimensions, even in small-sized of text collections and a feature vector is often very sparse. Documents with similar

I.-Y. Song, J. Eder, and T.M. Nguyen (Eds.): DaWaK 2008, LNCS 5182, pp. 217–226, 2008.
© Springer-Verlag Berlin Heidelberg 2008

concepts are grouped into the same cluster and clusters with the similar concepts are located nearby on a map. This distinguishes neural clustering from traditional statistical cluster analysis, which only assigns objects to clusters but ignores the relationship between clusters. Most document clustering algorithms require a form of preprocessing of data. Consequentially, unimportant features are eliminated and the original dimension is reduced to a more manageable size. Due to the sparseness problem of data space, end clustering results turn out to be of low quality. In addition, reduction of dimensionality may disturb preserving the original topological structure of the input data.

To solve these problems, we propose a context-sensitive semantic smoothing of a document model and incorporate it into Dynamic Growing Cell Structure (DynGCS). The effect of model smoothing which was originated and actively applied in Information Retrieval has not been extensively studied in the context of document clustering [11]. Most model-based clustering approaches simply use Laplacian smoothing to prevent zero probability [7, 12], while most similarity-based clustering approaches employ the heuristic TF*IDF scheme to discount the effect of general words [10]. As showed in [12], model-based clustering has several advantages over discriminative based approaches. One of the advantages of model-based approaches is that it learns generative models from the documents, with each model representing one particular document set. Due to promising results reported in model-based clustering approaches, we proposed a novel semantic smoothing technique to improve clustering quality.

DynGCS is an adaptive variant of an artificial neural network model, Self-Organizing Map (SOM), which is well suited for mapping high-dimensional data into a 2-dimensional representation space. The training process is based on weight vector adaptation with respect to the input vectors. SOM has shown to be a highly effective tool for document clustering [5]. One of the disadvantages of SOM in document clustering is its fixed size in terms of the number of units and their particular arrangement, which must be defined prior to the start of the training process. Without knowledge of the type and the organization of documents, it is difficult to get satisfying results without multiple training runs using different parameter settings. This is extremely time consuming, given the high-dimensional data representation. DynGCS solves the problem of fixed-sized structure by dynamically generating multiple layers of SOM.

The organization of this paper is as follows: Section 2 describes semantic smoothing of document model. Section 3 denotes the DynGCS algorithm. Section 4 presents results. We conclude our paper in Section 5.

2 Semantic Smoothing of Document Models

In document clustering, a TF*IDF score is often used as the dimension values of document vectors. In the context of language model, a TF*IDF scheme is roughly equivalent to the background model smoothing. Since TF*IDF is a pure probabilistic scheme, it does not convey semantics of content represented by terms and phrases. As an alternative, [6] proposes semantic smoothing where context and sense information are incorporated into the model. Latent Semantic Indexing (LSI) is an early attempt at semantic smoothing which projects documents in a corpus into a reduced space where document semantics becomes clear. LSI explores the structure of term co-occurrence with Singular Value Decomposition (SVD). However, the problem of LSI is that it increases noise while reducing the dimensionality because it is unable to recognize polysemy. In practice, it is also criticized for the lack of scalability and ability to interpret.

Our semantic smoothing technique is similar to the one proposed in [11]. Their approach utilizes multi-word phrases (e.g. "star war", "movie star") as topic signatures. Using multi-word phrases has several advantages: 1) a multi-word phrase is often unambiguous: 2) multi-word phrases can be extracted from a corpus by existing statistical approaches without human knowledge: and 3) documents are often full of multi-word phrases; thus, it is robust to smooth a document model through statistical translation of multi-word phrases in a document to individual terms. Unlike [11], we employ an information gain-based keyphrase extraction technique [9] to generate a set of phrases in documents that achieves competitive performance in biomedical data collections

Our keyphrase extraction procedure consists of two stages: building an extraction model and extracting keyphrases. The extraction model is trained for keyphrases before the keyphrase extraction technique is applied. Keyphrases are extracted by referencing the keyphrase model. Both training and test data are processed by the following three components: 1) Data Cleaning, 2) Data Tokenizing, and 3) Data Discretizing. After that, there are three features to calculate information gain: 1) TF-IDF, 2) Part-Of-Speech (POS), and 3) First Occurrence of Phrases. Based on these features, we build a discretization table and rank the candidate keyphrases based on its information gain measure. We compared information gain with other techniques such as Naïve Bayesian used in KEA [13] and found that information gain gave us the best performance in our previous experiments. In addition, we adopted the Mahalanobis Distance to measure distance among models. Mahalanobis distance takes into account the covariance among the variables in calculating distances [8]. Suppose we have indexed all documents in a given collection C with terms and phrases as illustrated in Figure 1. Note that Vp denotes phrase vector, Vw denotes word vector, and Vd denotes document vector. The translation probabilities from a keyphrase t_k to any individual term w, denoted as $p(w \mid t_k)$, are also given. Then we can easily obtain a document model below:

$$p_t(w \mid d) = \sum_k p(w \mid t_k) p_{ml}(t_k \mid d) \tag{1}$$

The likelihood of a given document generating the keyphrase t_k can be estimated with

$$p_{ml}(t_k \mid d) = \frac{c(t_k, d)}{\sum_i c(t_i, d)} \tag{2}$$

where $c(t_i, d)$ is the frequency of the keyphrase t_i in a given document d.

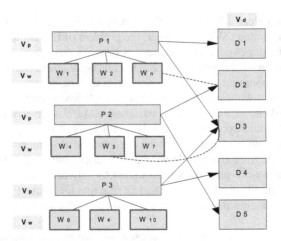

Fig. 1. Phrase/Term and Document Relation

We refer to the above model as translation model indicated in [6]. As we discussed in the introduction, the translation from multi-word phrases to individual terms would be very specific. Thus, the translation model not only weakens the effect of "general" words, but also relieves the sparseness of class-specific "core" words. However, not all topics in a document can be expressed by key phrases. If only the translation model is used, there will be serious information loss. A natural extension is to interpolate the translation model with a unigram language model below:

$$p_b(w \mid d) = (1 - \alpha) p_{ml}(w \mid d) + \alpha p(w \mid C) \qquad (3)$$

$P_{ml}(w \mid d)$ is a maximum likelihood estimator. We refer to this unigram model as simple language model or baseline language model. We use Jelinek-Mercer smoothing on the purpose of further discounting "general" words.

The final document model for clustering use is described in equation (4). It is a mixture model with two components: a simple language model and a translation model.

$$p_{bt}(w \mid d) = (1 - \lambda) p_b(w \mid d) + \lambda p_t(w \mid d) \qquad (4)$$

The translation coefficient (λ) is used to control the influence of two components in the mixture model. In our experiment, we set the background translation coefficient to 0.5. With training data, the translation coefficient can be trained by optimizing the clustering quality. After estimating a language model for each document in the corpus with context-sensitive semantic smoothing, we use the Mahalanobis Distance of two language models as the distance measure of the corresponding two documents. Given two probabilistic document models $p(w \mid d_x)$ and $p(w \mid d_y)$, the Mahalanobis Distance of $p(w \mid d_x)$ to $p(w \mid d_y)$ is defined as:

$$d(\overrightarrow{p_x}, \overrightarrow{p_y}) = \sqrt{\sum_{i=1}^{p} \frac{(p_y - p_x)^2}{\sigma_i^2}} \tag{5}$$

where σ_i is the standard deviation of the p_x over the sample set.

3 DynGCS

DynGCS is a self-organizing neural network model that incrementally builds a Dirichlet Voronoi tessellation of input space while automatically finding its structure and size. DynGCCs are used in artificial neural network (ANN) architectures. ANN can allow clusters to search and reduce search time. Unsupervised ANN learning for document clustering offers a number of advantages. Most commonly unsupervised ANN learning methods are Hard Competitive Learning (HCL), Self-Organizing Map (SOM) algorithm, and Adaptive Resonance Theory (ART). All these methods have one thing in common: the winner-takes-all concept. HCL and SOM are dependent on input data density, therefore it would not be wise to use these for purpose of improved information clustering and retrieval. The main drawback of the existing algorithms is that they require either pre-specification of the number of clusters (K-means clustering and SOM), or have the user decide the number of clusters to the user (hierarchical clustering). Particularly for a large dataset, SOM suffers from its fixed network architecture, which has motivated the development of a number of adaptive variants [12].

In this paper, we introduce an adaptive variant of SOM to resolve the aforementioned issue. DynGCC is an unsupervised clustering method that adaptively decides on the best architecture for the self-organizing map. This stems from Growing Cell Structure (GCS), introduced by Fritzke[3,4]. Fritzke introduced an incremental self-organizing network with variable topology, known as GCS based on SOM and Hebbian learning. The GCS has three main advantages over the SOM: first, the network structure is determined automatically by the input pattern. Second, the network size needs not to be predefined. Third, all parameters of the model are constant. Therefore, a "cooling schedule" is not required which is a contrast to the conventional SOM. The problem with GCS, however, is that it tends to overspill as the map grows larger.

To tackle the problem of overspill, we combine GCS with the Growing Hierarchical Self-Organizing Map (GH-SOM). GH-SOM adopts a hierarchical structure with multiple layers, where each layer consists of a set of independent self-organizing maps [2]. With the probability of adding every cell in a SOM from one layer to the next layer of the hierarchy, it shares the sample adaptation steps with GCS. There is, however, one exception that uses a decreasing learning rate and a decreasing neighborhood radius. The mean quantization error of the map is used to decide whether a new level of the hierarchy needs to be created. For instance, at level 0, the single SOM unit is assigned a weight vector m_0, such that $m_0 = [\mu 0_1, \mu 0_2, ..., \mu 0_n]^T$ is computed as the average of all input data. The mean quantization error of this single unit is computed as the following with d representing the number of input data x:

$$mqe_0 = \frac{1}{d} \| m_0 - x \| \tag{6}$$

DynGCS produces distribution-preserving mappings as other SOM related algorithms. DynGCS operates on the following principle of GCS [1]:

1) During the training stage, the number of clusters and the connections among them are dynamically assigned.
2) Adaptation strength is constant over time.
3) Adaptation occurs only in the best-matching cell (BMU) and its neighborhood.
4) Adaptation increments the signal counter for BMU and decrements the remaining cells.
5) For the adaptation of the output map to the distribution of the input vectors, insertion of new cells and deletion of existing cells occur.

The DynGCS hierarchy is constructed from and superimposed onto the standard GCS algorithm. DynGCS starts with the small architecture. The DynGCS is a hierarchical self-organizing neural network designed to preserve topological structure of input data based on GCS and Growing Hierarchical Self-Organizing Map. DynGCS grows dynamically. In each growth, the DynGCS adds two children to the leaf whose heterogeneity is less than a threshold and turns it to a node. This process goes on until the heterogeneity of all cells is less than the threshold. A learning process similar to GCS is adopted.

The DynGCS algorithm is shown in Figure 2. Initially there is only one root node. All the input data is linked to the root. The reference vector of the root node is initialized with the centroid of the data. In growth mode, two child nodes are appended to the root node. All input data linked to the root node is distributed between these child nodes by employing a learning process. Once the learning process is finished, the heterogeneities of the leaf nodes are scrutinized to decide whether expansion to another level is necessary. If another expansion is needed, then a new growth step is invoked. Two child nodes are appended to the leaf nodes if the level of heterogeneity is greater than the threshold. All input data is distributed again with the learning process and a new growth begins. This process continues until the heterogeneity of all the leaves is less than the threshold. In DynGCS, each leaf represents a cluster that includes all data linked to it. The reference vector of a leaf is the centroid of all data linked to it. Therefore, all reference vectors of the leaves form a Voronoi set of the original dataset. Each internal node represents a cluster that includes all data linked to its leaf descendants. The reference vector of an internal node is the centroid of all data linked to its leaf descendants.

The height of DynGCC is $\log_d M$, where d is the branch factor and M is the number of nodes in the hierarchy. M is $O(N)$ where N is the number of data. Let J be the average number of learning iterations for each learning process. Thus, the time complexity factor for DynGCC will be $O[\log_d N * (J * N + d * J * N)]$. Since J and d are constants, the complexity will be. $O(cN * \log_d) \sim O(N * \log_d N)$

SM/DynGCS Algorithm
1. /** Parse input data **/
2. /** Apply information gain based keyphrase extraction techniques to input data **/
3. /** Apply semantic smoothing technique to documents **/
4. /** Apply Mahalanobis Distance to feature vectors **/
5. /** Initialization **/
6. **Do**
7. **For** any leaf whose heterogeneity is greater than the threshold
8.   Changes the leaf to a node and create two descendent leaves.
9.   Initialize the reference vector of the new leaves with the node's reference vector
10.   Set the cell growing flag of the new leaves to true
11. **Do**
12. **For** each input data
13.   Find BMU winner
14.   Update reference vectors of winner and its neighborhood
15.   Increase time parameter, $t = t + 1$.
16. **While** the cell growing flag of all lowest level node are false
17. **While** the heterogeneity of all leaf nodes are less than the threshold

Fig. 2. SM/DynGCS Algorithm

4 Experiments

In this section, we report our evaluation method, data collection, and experiment results.

Evaluation Method: We evaluated the proposed algorithm by comparing clustering output with known classes as answer keys. There have been a number of comparison metrics, such as mutual information metric, misclassification index (MI), purity, confusion matrix, F-measure, and Entropy; refer to [14] for details. In our experiment we used misclassification index, purity, and Entropy as clustering evaluation metrics. Note that the smaller MI and Entropy imply the better clustering quality while the larger purity indicates the better clustering quality.

Data Collections: We used public MEDLINE data for the experiments by collecting document sets related to various diseases. We use the "MajorTopic" tag provided by MeSH to identify main concepts along with the MeSH disease terms as queries to MEDLINE. Once we retrieve the datasets, we generate various document combinations by randomly mixing the document sets whose numbers of classes are 2 to 4. The document sets used for generating the combinations are later used as answer keys on the performance measure. Refer to [14] for details on data collections and test data sets.

Experiment Results: In this section, we will report experiment results with average (μ) and standard deviation (σ). In our experiments, we compared our method, SM/DynGCS with three other algorithms such as TF*IDF/DynGCS, K-means, and SOM.

TF*IDF/DynGCS: TF*IDF, which is widely used in information retrieval, was implemented for feature vectors. It is a measure of importance for a term in a document or class. As indicated by formula 7, TF*IDF is a term frequency in a document or class, relative to overall frequency. The TF*IDF feature is a well-known weighting scheme in information retrieval.

$$W_{ij} = tf_{ij} * \log_2 \frac{N}{n} \tag{7}$$

W_{ij} weight of term t_i in document D_j and tf_{ij} is frequency of term t_j in document D_j. N is the number of documents in a collection and n is the number of documents where term t_j occurs at least once.

SOM: Self-Organizing Map (SOM) is a well accepted neural network technique in document clustering. In this experiment, we used 10,000 iterations and set the initial learning rate to 0.1. The 10 x 10 SOM is trained to cluster and the final number of cells is 250.

K-Means: K-Means is a simple but powerful unsupervised learning algorithm that solves the well known clustering problem. Because K-Means may produce different clustering results every time due to its random initializations, we ran it five times and averaged the values of clustering evaluation metrics.

Table 1. Comparison of Evaluation Metrics

	MI	Entropy	Purity
Tf*idf/DynGCS	μ:0.23 σ: 0.13	μ :0.26 σ:0.21	μ :0.92 σ:0.16
SM/DynGCS	μ :0.19 σ: 0.17	μ :0.17 σ:0.23	μ :0.93 σ:0.21
K-means	μ :0.35 σ:0.12	μ :0.31 σ:0.14	μ :0.79 σ:0.12
SOM	μ :0.28 σ:0.13	μ :0.27 σ:0.15	μ :0.88 σ:0.15

Table 1 shows the comparison of the overall clustering quality of SM/DynGCS, TF*IDF/DynGCS, K-means, and SOM. The clustering results are from 5 datasets from which the averages and standard deviations are calculated. We notice that SM/DynGCS is superior to the hierarchical algorithms.

Comparing miscalculation indexes, SM/DynGCS outperforms the other three. Compared to K-means, it is 0.16% better in terms of μ. In terms of entropy, the best performance was made by SM/DynGCS. SM/DynGCS improves accuracy by about 0.14% when compared to K-means. TF*IDF/DynGCS and SOM perform almost at the same level. With purity, the results show that SM/DynGCS performs 0.14% better than K-means. As indicated by the results, integrating semantic smoothing into clustering algorithm improves accuracy more than DynGCS with TF*IDF does.

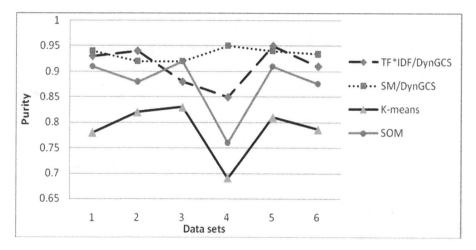

Fig. 3. Performance Comparison on Different Datasets

Figure 3 shows how four different clustering techniques perform on five different datasets. Unlike the other three techniques, TF*IDF/DynGCS, K-means, and SOM, the performance of semantic smoothing-based DynGCS is stable across five datasets.

5 Conclusion

In this paper, we proposed a hybrid clustering algorithm that combines semantic smoothing of document model and dynamic growing cell structure. We developed a context-sensitive smoothing method for document models and used Mahalanobis distance for smoothed probabilistic models as a document similarity measure for clustering. With these feature vectors, we applied a dynamic growing cell structure technique to cluster the given collections.

We performed a comparison study of SM/DynGCS with the other three techniques: TF*IDF/DynGCS, K-means, and SOM on 5 MEDLINE corpora. The experimental results indicated that our SM/DynGCS is superior to the other three approaches in terms of μ and σ. For future work, we will conduct more comprehensive experiments including other model-based partitional approaches. We also plan to apply our technique to dynamic document clustering, i.e. clustering search results.

References

1. Bruske, J., Sommer, G.: Dynamic cell structures. In: Tesauro, G., Touretzky, D., Leen, T. (eds.) Advances in Neural Information Processing Systems, vol. 7, pp. 497–504. MIT Press, Cambridge (1995)
2. Dittenbach, M., Merkl, D., Rauber, A.: The Growing Hierarchical Self-Organizing Map. In: Proc. Intl. Joint Conf. on Neural Networks (IJCNN 2000) (2000)
3. Fritzke, B.: A growing neural gas network learns topologies. In: Lean, T.K. (ed.) Advances in Neural Information Processing Systems 7, pp. 625–632. MIT Press, Cambridge (1995)

4. Fritzke, B.: Growing cell structures - a self-organizing network for unsupervised and supervised learning. Neural Networks 7(9), 1441–1460 (1994)
5. Kohonen, T., Kaski, S., Lagus, K., Salojärvi, J., Honkela, J., Paatero, V., Saarela, A.: Self organization of a massive text document collection. In: Oja, E., Kaski, S. (eds.) Kohonen Maps, pp. 171–182. Elsevier, Amsterdam (1999)
6. Lafferty, J., Zhai, C.: Document Language Models, Query Models, and Risk Minimization for Information Retrieval. In: Proceedings of the 24th ACM SIGIR Conference on Research and Development in IR, pp. 111–119 (2001)
7. Nigam, K., McCallum, A.: Text Classification from Labeled and Unlabeled Document Using EM. Machine Learning 39(2-3), 103–134
8. Pearson, R.K.: Mining imperfect data; dealing with contamination and incomplete records. In: SIAM 2005 (2005)
9. Song, M., Song, I.-Y., Hu, X.: KPSpotter: a flexible information gain-based keyphrase extraction system. In: WIDM 2003, pp. 50–53 (2003)
10. Steinbach, M., Karypis, G., Kumar, V.: A Comparison of Document Clustering Techniques. In: Workshop on Text Mining, SIGKDD (2000)
11. Zhang, X., Zhou, X., Hu, X.: Semantic Smoothing for Model-based Document Clustering. In: The 2006 IEEE International Conference on Data Mining (IEEE ICDM 2006), HongKong, December 18-22 (2006)
12. Zhong, S., Ghosh, J.: Generative Model-based Document Clustering: a Comparative Study. Knowledge and Information Systems 8(3), 374–384 (2005)
13. Witten, I.H., Paynter, G.W., Frank, E., Gutwin, C., Nevill-Manning, C.G.: KEA: Practical automatic keyphrase extraction. In: Proc. DL 1999, pp. 254–256 (1999)
14. Yoo, I., Hu, X., Song, I.-Y.: A Coherent Graph-based Semantic Clustering and Summarization Approach for Biomedical Literature and a New Summarization Evaluation Methods. BMC Bioinformatics 8(Suppl 9), S4 (2007)

Mining Serial Episode Rules with Time Lags over Multiple Data Streams

Tung-Ying Lee[1], En Tzu Wang[1], and Arbee L.P. Chen[2]

[1] Department of Computer Science, National Tsing Hua University, Hsinchu, Taiwan, R.O.C.
`u902521@alumni.nthu.edu.tw`, `m9221009@em92.ndhu.edu.tw`
[2] Department of Computer Science, National Chengchi University, Taipei, Taiwan, R.O.C.
`alpchen@cs.nccu.edu.tw`

Abstract. The problem of discovering *episode rules* from static databases has been studied for years due to its wide applications in prediction. In this paper, we make the first attempt to study a special episode rule, named *serial episode rule with a time lag* in an environment of multiple data streams. This rule can be widely used in different applications, such as traffic monitoring over multiple car passing streams in highways. Mining serial episode rules over the data stream environment is a challenge due to the high data arrival rates and the infinite length of the data streams. In this paper, we propose two methods considering different criteria on space utilization and precision to solve the problem by using a prefix tree to summarize the data streams and then traversing the prefix tree to generate the rules. A series of experiments on real data is performed to evaluate the two methods.

Keyword: Multiple data streams, Data mining, Serial episode rule, Time lag.

1 Introduction

The progress of technologies including communications and computations has led to a more convenient way of life and also brings huge amounts of commercial benefits. However, the high speed of communications and powerful capability of computations generate data as a form of *continuous data streams* rather than static persistent datasets, raising the complexity of data management. A data stream is an unbounded sequence of data continuously generated at a high speed. In such applications as network traffic management, sensor network systems and traffic management systems, we may need to handle different categories of the data streams.

Consider a scenario as follows. Roads are connected to each other in real road networks. Consequently, certain roads with heavy traffic may cause the other roads to be obstructed. Users may be interested in the following rule: *when road A and road B have heavy traffic, five minutes later, road C will most likely be congested*. This kind of rules can be applied to navigation systems, helping the users to avoid being blocked up in traffic. In the traffic monitoring systems, *flow* and *occupancy* of a road detected by sensors are employed in qualifying the traffic conditions [2]. The flow of a road is the

I.-Y. Song, J. Eder, and T.M. Nguyen (Eds.): DaWaK 2008, LNCS 5182, pp. 227–240, 2008.

number of cars passing by a sensor per minute. The occupancy of a zone is a ratio of a time interval in which the detection zone is occupied by a car. The collected data of the flow and occupancy are divided into several discrete classes [6]. Therefore, different traffic conditions can be represented by the different classes of the flow and occupancy. For example, if a road is under the condition of *low flow* and *high occupancy*, the traffic of this road can be qualified as "congested." Since many roads are simultaneously monitored by the sensors in a traffic monitoring system, a continuous multi-streams environment is formed. In order to find the rules as discussed above, a new problem of *finding serial episode rules with time lags over multiple streams* is addressed and solved in this paper. Different from our approach to design the new online mining algorithms, a framework using existing offline mining techniques to find frequent episodes from the historic data to monitor the current traffic status and to predict the coming traffic in time is proposed in [7].

We first introduce the problem of discovering episode rules from the static time series data [3] [4] [5] [10] [11] [12] in the following. Formally, an episode can be described using a *directed acyclic graph* (DAG). Each node in the graph represents an *event*. Suppose that the nodes A and B are kept in an episode E. If there is an edge from A to B, it means that B occurs after A. If there is no edge between A and B, it means that the order of the appearances of A and B is not important. The *time interval* of an occurrence of an episode E in the time series data is the interval from the occurring of the first event in E to that of the last event in E appearing in the time series. In addition, a parameter named *time bound* is set to limit the duration of the occurrence. *This means that we only concern about the occurrences of the episode, whose time interval are within the time bound.* We call these occurrences *valid occurrences* of the episode for the following discussion. The number of valid occurrences of an episode in the time series data is counted to determine whether it is *frequent*, and then from the frequent episodes to derive an episode rule. There are two ways of counting the number of valid occurrences of an episode, the *window-based strategy* and the *minimal occurrence strategy*.

The window-based strategy is to slide a window with a length equaling the time bound T over the time series data to compute the total number of the windows containing an episode E. In this strategy, a certain valid occurrence of E may be counted more than once due to being contained in the distinct windows. On the other hand, the minimal occurrence strategy is to count the number of valid occurrences satisfying the following constraint: *within its time interval, the other valid occurrences do not exist.* Given a user-defined threshold, named *minimum support*, if the number of the windows in the window-based strategy or the number of the counted valid occurrences in the minimal occurrence strategy of an episode exceeds or equals the minimum support, it is defined as *frequent*. The algorithms proposed in [10][11][12] for finding frequent episodes based on the window-based strategy and the minimal occurrence strategy are named WINEPI and MINEPI, respectively. In these approaches, only *serial episodes* and *parallel episodes* are considered. A serial episode is a sequence of events while a parallel episode is a set of events. The above methods based on the Apriori algorithm [1] are designed only for these two types of episodes because all episodes can be decomposed into their combinations. Finding frequent episodes is extended to finding *episode rules* in [11]. An episode rule describes that after an episode occurs another episode may occur. Harm et al. define a new episode rule by specifying a time lag between these two episodes and propose the MOWCATL approach to find the episode rules in multiple sequences [3][5]. The s*erial rule* discussed in [3] is a special episode rule restricted to have only two *serial episodes*. We call it *serial episode rules with time lags* (SERs) in this paper. An SER is represented in a form of $X \to_L Y$ where X and Y are *serial episodes* and L is a fixed time lag. The first serial episode X is defined as the *precursor* of the rule and the second serial episode Y is defined as the *successor*. The rule means that after X occurs, Y will probably occur in L time units later. As mining association rules, *support* and *confidence* are also used to be the measures of rule interestingness for mining SERs. The support of an SER is the number of the *tightest* interval in which the rule occurs, which is used to express the strength of the rule. The confidence of an SER is a conditional probability that the successor of the rule occurs within the time lag, given that the precursor occurs. The users can define their own interestingness of the rules by giving two thresholds, *minimum support* and *minimum confidence*. If both the support and confidence of a rule are no less than the minimum support and the minimum confidence respectively, the rule is defined as *significant*. Therefore, mining SERs is to return all the significant SERs to the users.

Refer to the scenario of the traffic prediction which motives us to *address this new problem of mining SERs over multi-streams*. We deal with the multi-streams by assuming that the events are generated from n streams with the same fixed sampling rate, and all the events generated at the same time form an *n-tuple event*. Let an *itemset* be a subset of an *n*-tuple event. The *serial episode* to be discussed in our problem is then defined as a sequence of itemsets. The challenges of solving this problem are described as follows. First, enormous amounts of episodes enumerated from the multiple streams may overload the memory utilization and incur enormous processing time. Second, in order to check whether the delay between the precursor and successor of a rule satisfies the time lag of a significant SER, the time intervals within which the episodes occur must be recorded. Obviously, as time goes by, the records for storing the time intervals of the episodes may use a huge amount of memory space.

In this paper, we propose a framework for mining significant SERs over multiple data streams, which counts the valid occurrences of the serial episodes using the minimal occurrence strategy. Different from the previous approaches which only focus on finding frequent episodes over event streams [9][13], in this paper, we also combine the serial episodes to generate the serial episode rules with time lags. Therefore, not only the support counts but also the time information of the occurrences of a serial episode need to be efficiently processed. In this framework, a prefix tree is employed to store the serial episodes enumerated from the *n*-tuple event stream. According to the considerations of different criteria on space utilization and precision, two methods storing different information in the prefix tree are proposed in this paper. Moreover, Lossy Counting [8] is applied to the two methods to save the memory required.

The remainder of the paper is organized as follows. Section 2 introduces the preliminaries and formulates the problem to be solved. The detailed algorithms are described in Section 3. The experiment results for evaluating the methods are presented in Section 4, and finally, Section 5 concludes this work.

2 Preliminaries

The definitions of the supports and confidences of SERs, the problem formulation, the data structure used in the methods, and several observations of SERs are described in this section.

2.1 Problem Formulation

Consider a centralized system which collects n *synchronized data streams* denoted as DS_1, DS_2,..., DS_n. A data stream in the system is an unbounded sequence of items (events). Moreover, the domain of the data in each data stream may be distinct from the others. The synchronized data streams mean that the data arrival rate for each data stream is consistent, that is, each data stream generates an item at the same time unit. Let $DS_j(i)$ represent the item arriving at time i and coming from the jth stream. An *n-tuple event* $R(i)$ is defined as a set of items coming from all data streams at time i, i.e. $R(i) = \{DS_1(i), DS_2(i), ..., DS_n(i)\}$. Moreover, an *itemset* is defined as a subset of an n-tuple event. Therefore, a *serial episode* discussed in this paper is described as an ordered list of itemsets, for example $S = (I_1)(I_2), ..., (I_k)$ is a serial episode, where I_1, I_2, ..., I_k are itemsets. The following first two definitions follow [11].

Definition 1: (Minimal Occurrence) *Given a serial episode S, a minimal occurrence of S can be identified by its time interval. A time interval* $[a, b]$ *is a minimal occurrence of S if it satisfies the following two constraints: 1) S occurs in the time interval* $[a, b]$ *and 2) S does not occur in any proper subinterval of* $[a, b]$, *that is, S does not occur in* $[c, d]$, *where* $a \leq c$, $d \leq b$ *and the duration of* $[a, b]$ > *the duration of* $[c, d]$. *The duration of* $[a, b]$ *is defined to equal* $b - a + 1$. *The set of all minimal occurrences of S is denoted as* $MO(S) = \{[a, b]| [a, b]$ *is a minimal occurrence of S*\}. ∎

For a minimal occurrence of a serial episode E, if it satisfies the time bound T, it is called a *valid minimal occurrence* of E in the following discussion.

Definition 2: (Support of Serial Episode) *Given a time bound T and a serial episode S, the support of S, supp(S), is the number of valid minimal occurrences of S i.e.,* $supp(S) = |\{[a, b] | [a, b] \in MO(S) \wedge (b - a + 1) \leq T\}|$. ∎

Definition 3: (Support of SER) *Given serial episode rule* $R: S_1 \rightarrow_{lag = L} S_2$ *with a time bound T, where the time lag is equal to L. The support of R representing the strength of R and helping to recognize the degree of significance of R is defined as* $supp(R) = |\{[a, b] | [a, b] \in MO(S_1) \wedge (b - a + 1) \leq T \wedge \exists [c, d] \in MO(S_2)$ *s.t.* $(d - c + 1) \leq T \wedge (c - a) = L\}|$. ∎

Definition 4: (Confidence of SER) *The confidence of the serial episode rule* $R: S_1 \rightarrow_{lag = L} S_2$ *is a conditional probability that* S_2 *occurs and satisfies the fixed time lag L, given that* S_1 *occurs, defined as* $conf(R) = supp(R)/supp(S_1)$. ∎

Given four parameters including the *maximum time lag*, *Lmax*, the *minimum support*, *minsup*, the *minimum confidence*, *minconf*, and the *time bound*, *T*. The problem of mining SERs is to find all the SERs, e.g. $R: S_1 \rightarrow_{lag = L} S_2$, satisfying the following constraints: 1) $L \leq Lmax$, 2) $supp(R) \geq N \times minsup$, where N is the number of the received n-tuple events generated from the multiple streams, and 3) $conf(R) \geq minconf$.

| 1 2 3 4 5 6 7 8 9 10 11 12 13 14 15 16 17 18 19 |
| a b b c g a b f h g a b a b g f g a b |
| A B B S G A B C H G A B A B G F G A B |

Fig. 1. An example of the 2-tuple events

Table 1. All parameters for Figure 1

Lmax	Minsup	Minconf	T
5	0.2	0.8	3

Table 2. Examples of serial episodes and serial episode rules with time lags

Serial episodes	Minimal Occurrences	Support
$(a\ A)(b\ B)$	[1, 2], [6, 7], [11, 12], [13, 14], [18, 19]	$5 > 19 \times 0.2 = 3.8$
$(g\ G)$	[5, 5], [10, 10], [15, 15], [17, 17]	$4 > 3.8$
Serial episode rules	**Minimal Occurrences**	**Support, Confidence**
$(a\ A)(b\ B) \to_4 (g\ G)$	[1, 2]→[5, 5], [6, 7]→[10, 10], [11, 12]→[15, 15], [13, 14]→[17, 17]	Supp: 4, Conf: 4/5 = 0.8

Moreover, the calculating of the supports for the serial episodes and the SERs has to take the time bound T into account.

Consider an example as in Figure 1. 19 *2-tuple* events have been received. All parameters are given as in Table 1 and some serial episodes, serial episode rules with time lags, and their corresponding minimal occurrences are listed in Table 2. The interval [1, 3] is not the minimal occurrence of $(a\ A)(b\ B)$ due to the interval [1, 2] which is a proper subset [1, 3]. The support of the serial episode $(a\ A)(b\ B)$ is 5 because each minimal occurrence of $(a\ A)(b\ B)$ are valid. By comparing all the time intervals of the valid minimal occurrences of the two serial episodes $(a\ A)(b\ B)$ and $(g\ G)$, the support of the serial episode rule R: $(a\ A)(b\ B) \to_4 (g\ G)$ is calculated.

2.2 Data Structure of the Methods: Prefix Tree

Since the serial episodes may have common prefixes, using the *prefix tree structure* to store the serial episodes is more space-efficient. The prefix tree structure consists of the nodes with labels and a root without a label. Moreover, each node in the prefix tree belongs to a *level*. The root is at the level 0 and if a node is at the level k, its children are at the level $k + 1$. The node in the prefix tree is used to represent a serial episode. By orderly combining all labels of those nodes in the path from the root to the node, the serial episode represented by the node can be derived. The label of a node has two forms including "X" and "$\_X$," where X is an item. The node with *a label of "X"* represents a serial episode in which X *follows the label of the parent of the node*. Alternatively, the node with *a label of "$\_X$"* represents a serial episode in which X *simultaneously occurs with the label of the parent of the node*. An example of using the prefix tree to represent the serial episodes is shown in Figure 2.

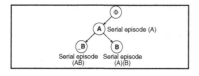

Fig. 2. An example of the prefix tree

Table 3. An example for Observation 1

Streams\Time unit	1	2	3	4
Stream 1	A	A	D	A
Stream 2	d	a	a	b

2.3 Observations of Serial Episode Rules with Time Lags

Several observations of the characteristics of SERs are described as follows, which are used in the methods detailed in Section 3. In order to simplify the display of the following discussion, the terms including S^+X and S^+_X are first introduced. Given a serial episode S, S^+X represents a serial episode S *followed by X*. The last itemset in S^+X only contains an item, X. Alternatively, S^+_X represents another serial episode similar to S. The difference between S and S^+_X is that the last itemset in S^+_X consists of the last itemset in S and X.

Observation 1: *Given a serial episode S and a newly arrived item X which is one of the items in the current n-tuple event. If we want to know whether S^+X has a new minimal occurrence, the last two minimal occurrences of S must be checked.*

Consider the data streams shown in Table 3. After the third time unit passed, the minimal occurrences of A are [1, 1] and [2, 2], that is, MO((A)) = {[1, 1], [2, 2]}. At the fourth time unit, since A and b are generated, MO((A)) becomes {[1, 1], [2, 2], [4, 4]}. We focus on the serial episode $(A)(b)$. As can be seen, the new minimal occurrence [2, 4] of the serial episode $(A)(b)$ is associated with the minimal occurrence [2, 2] of (A). If only the last minimal occurrence of A, [4, 4], is considered to join to the minimal occurrence [4, 4] of b, the minimal occurrence [2, 4] of $(A)(b)$ will be ignored. Therefore, the last two minimal occurrences of (A) should be checked.

Observation 2: *Given two serial episode rules $X\rightarrow_L(AB)$ and $X\rightarrow_LA$, according to the Apriori property [1], $supp(X\rightarrow_LA) \geq supp(X\rightarrow_L(AB))$. Therefore, the rule $X\rightarrow_L(AB)$ is not significant if the rule $X\rightarrow_LA$ does not satisfy one of the minsup and the minconf.*

Observation 3: *Given two serial episode rules $(AB)\rightarrow_L(CD)$ and $A\rightarrow_LC$, obviously, $supp(A\rightarrow_LC) \geq supp((AB)\rightarrow_L(CD))$. Therefore, the rule $(AB)\rightarrow_L(CD)$ is not significant if $supp(A\rightarrow_LC) < supp(AB) \times minconf$.*

Observation 4: *Given a serial episode rule $(A)(B)\rightarrow_L(CD)$ with a time bound T, the precursor of the rule has at most $T - 1$ types, i.e. $(A)\rightarrow_p(B)$, where $0 < p < T$. While taking the time lag into account, the types of the rules are denoted as $(A)\rightarrow_p(B)\rightarrow_{L-p}(CD)$, where $0 < p < T$ and $L - p > 0$. Obviously, the support of $(A)\rightarrow_p(B)\rightarrow_{L-p}(CD)$ must be smaller than or equal to $supp(A\rightarrow_pB)$ and $supp(B\rightarrow_{L-p}C)$. Therefore, $supp((A)(B)\rightarrow_LC) \leq \sum_p min(supp(A\rightarrow_pB), supp(B\rightarrow_{L-p}C))$. Since $supp((A)(B)\rightarrow_LC) \geq supp((A)(B)\rightarrow_L(CD))$, we infer that the rule $(A)(B)\rightarrow_L(CD)$ is not significant if $\sum_p min(supp(A\rightarrow_pB), supp(B\rightarrow_{L-p}C)) < supp(AB) \times minconf$.*

3   Methods

Two methods, LossyDL and TLT for finding significant SERs in the environment of multi- streams are proposed in this paper. Both of them use the prefix tree structure to store the information of the serial episodes. LossyDL keeps all valid minimal occurrences for each serial episode in the prefix tree. On the other hand, TLT only keeps 1) the last two valid minimal occurrences and the support of each serial episode in the prefix tree and 2) *the supports of the reduced rules in the additional tables*, to be detailed in Subsection 3.2. When the users want the SERs, to LossyDL, the valid minimal occurrences of any two serial episodes with enough supports are joined to check whether they satisfy the time lag. On the other hand, to TLT, any two serial episodes with enough supports will be combined into a *candidate rule* and then, the non-significant rules will be pruned by using the pruning strategies derived from Observations 2, 3, and 4. Since the prefix tree may increase as the time goes by, Lossy Counting [8] is used in these methods to avoid keeping the serial episodes with low supports. The detailed operations in LossyDL and TLT are respectively described in Subsections 3.1 and 3.2.

3.1 The LossyDL Method

The principle of LossyDL is to keep all the valid minimal occurrences of a serial episode under the *prefix tree* structure. Therefore, in addition to a label, each node (except the root node) of the prefix tree in LossyDL also keeps a *duration list* which is a set used to contain all the valid minimal occurrences in the MO set of its corresponding serial episode. The number of the valid minimal occurrences kept in the duration list is regarded as the support of the corresponding serial episode. Moreover, in order to avoid keeping too many serial episodes with low supports, the principle of Lossy Counting [8] is used in LossyDL. Given an error parameter ε, if the number of received n-tuple events is divisible by $\lceil 1/\varepsilon \rceil$, for each node, the oldest minimal occurrence is removed from its duration list. Moreover, the nodes with empty duration lists are removed from the prefix tree. When the users want the mining results, the SERs are generated from the prefix tree by comparing the duration lists of any two nodes with enough supports. The algorithms of LossyDL are shown in Figures 3 and 4 and explained as follows.

Input: multi-streams *DSs*, *minsup*, *minconf*, *T*, *Lmax*, and ε
Output: the prefix tree *PT*
Variable: a node *b*, the corresponding serial episode *S*, and its last two valid minimal occurrences $[t_{11}, t_{12}]$ and $[t_{21}, t_{22}]$
        in the duration list
1. Create a prefix tree *PT* with a root node containing φ
2. When an n-tuple event R_i received from *DSs* at current time i
3.   **for each** item X in R_i
4.   **for each** node b in *PT*   // bottom-up traversing
5.   **if** ($i = t_{22}$)
6.   **if** ($[t_{11}, i]$ is a valid minimal occurrence of S^+X)
7.   Append it to the duration list of S^+X
8.   **if** ($[t_{21}, t_{22}]$ is a valid minimal occurrence of S^+_X)
9.   Append it to the duration list of S^+_X
10.   **else**
11.   **if** ($[t_{21}, i]$ is a valid minimal occurrence of S^+X)
12.   Append it into the duration list of S^+X
13. **If** (the number of received records is divisible by $\lceil 1/\varepsilon \rceil$)
14.   Remove the oldest valid minimal occurrence for each node and remove the nodes with empty duration lists

---

Fig. 3. Maintaining the prefix tree in LossyDL

Initially, the root of the prefix tree is created. When a new n-tuple event arrives, all items in the n-tuple event are sequentially used to traverse the tree. The traversing order is *bottom-up*. The bottom-up traversing order means that *the nodes in the high levels are traversed before those in the low levels*. During the traversing process, the new serial episodes may be generated, that is, the new nodes may be created, and the new valid minimal occurrences for some serial episodes may be added. Suppose that $R(i)$ is an n-tuple event, where i is the current time unit and the item X is contained in $R(i)$. When X is processed, two cases need to be considered. Let d be a node corresponding to a serial episode S, the last two valid minimal occurrences of S stored in d are $[t_{11}, t_{12}]$ and $[t_{21}, t_{22}]$. When d is visited, (case 1) if t_{22} is equal to i, $[t_{21}, t_{22}]$ and $[t_{11}, i]$ become *the candidate occurrences* for S^+_X and S^+X, respectively; otherwise (case 2), $[t_{21}, i]$ is the candidate occurrence for S^+X. A candidate occurrence of S^+X (S^+_X) is recognized as valid and inserted into the duration list of the node corresponding to S^+X (S^+_X) if both of the last two valid minimal occurrences of S^+X (S^+_X) are not its proper subinterval and moreover, it satisfies the time bound T. Notice that if the node corresponding to S^+X (S^+_X) is not kept in the prefix tree, a new node with a label equal to X is created as a child of b.

Since LossyDL is rooted in Lossy Counting to reduce the memory usage, the supports of the serial episodes kept in the prefix tree and the significant SERs obtained by LossyDL must be equal to or less than their real supports. Therefore, when the users request the mining results, the duration lists of any two serial episodes with their supports equaling or exceeding $(minsup - \varepsilon) \times N$ are checked to see whether any valid minimal occurrences of the two serial episodes can be combined to contribute to the supports of a SER. Then, for each SER, $R: S_1 \rightarrow_L S_2$, satisfying the following two constraints are returned to avoid the *false dismissals*: 1) $\mathrm{supp}(R) \geq (minsup - \varepsilon) \times N$ and 2) $(\mathrm{supp}(R) + \varepsilon N)/\mathrm{supp}(S_1) \geq minconf$.

An example for maintaining the prefix tree in the LossyDL method is shown in Figures 5 and 6. Suppose that the time bound is 3 and after some items were processed, the prefix tree is as the left-hand side in Figure 5. At time unit equal to 3, an item B contained in the n-tuple event is processed and the maintenance of the prefix tree is shown in Figure 6. The nodes in the prefix tree are traversed to be compared with B. The bottom-up traversing order is shown as the right-hand side of Figure 5. In Figure 6, notice that, $[1, 3]$ is a minimal occurrence of $(A)(B)(B)$, but $[1, 3]$ is not a minimal occurrence of $(A)(B)$ because $[1, 2]$ is a proper subinterval of $[1, 3]$.

Input: stream size N, *minsup*, *minconf*, T, *Lmax*, and PT
Output: significant SERs
Variable: a time lag, *lag*, nodes N_1, N_2 and their corresponding serial episodes S_1, S_2
1. **for each** two nodes N_1, N_2 in PT
2.   **if** $(\mathrm{supp}(S_1) \geq (minsup - \varepsilon) \times N$ and $\mathrm{supp}(S_2) \geq (minsup - \varepsilon) \times N)$
3.   Check the duration lists of N_1 and N_2
4.   **for all** *lag* from 1 to *Lmax*
5.   Calculate $\mathrm{supp}(S_1 \rightarrow_{lag} S_2)$
6.   **if** $((\mathrm{supp}(S_1 \rightarrow_{lag} S_2) + \varepsilon N)/\mathrm{supp}(S_1) \geq minconf$ and $\mathrm{supp}(S_1 \rightarrow_{lag} S_2) \geq (minsup - \varepsilon) \times N)$
7.   Return $S_1 \rightarrow_{lag} S_2$

Fig. 4. Rule generation of LossyDL

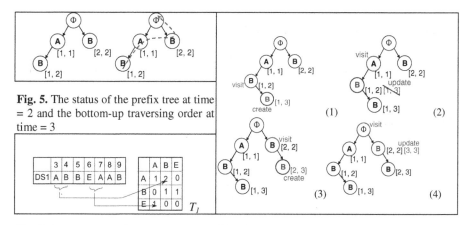

Fig. 5. The status of the prefix tree at time = 2 and the bottom-up traversing order at time = 3

Fig. 7. An example of the time lag table T_1 **Fig. 6.** An example of maintaining the prefix tree after receiving an n-tuple event at time = 3

3.2   The TLT Method

Although Lossy Counting is used to reduce the required memory space for LossyDL, the duration list in each node of the prefix tree still needs much memory space. In order to avoid keeping almost all the valid minimal occurrences for a serial episode, another method named *TLT* is proposed. A more space-efficient structure, named *time lag table*, is used in TLT to summarize the time positions of the SERs. A time lag table is a two-dimension table used to store the supports of *the reduced SERs* which are SERs with their precursors and successors both containing a single item. For example, R: $A \rightarrow_L B$ is a reduced SER, if A and B are items. The row and column of a time lag table represent the items and moreover, a time lag table with a label of L keeps the supports of the reduced SERs with a time lag of L, e.g. the entry $T_L(A, B)$ in the table T_L keeps the support of R: $A \rightarrow_L B$. An example of the time lag table T_1 is shown in Figure 7.

Instead of keeping almost the whole duration list, each node of the prefix tree in TLT keeps only the last two valid minimal occurrences and the support for its corresponding serial episode. Moreover, since the time lag of the significant SERs is at most $Lmax$, such as $S_1 \rightarrow_{Lmax} S_2$, some reduced SERs with a time lag at most $Lmax + T - 1$ (from the first itemset of S_1 to the last itemset of S_2) may exist. Therefore, in addition to the prefix tree, TLT will also keep $Lmax + T - 1$ additional time lag tables, named T_1, T_2, …, $T_{Lmax + T - 1}$, which respectively store the supports of the reduced SERs with a time lag equal to $1, 2, …, Lmax + T - 1$.

The maintenance of the prefix tree in TLT is similar to that in LossyDL, and moreover, Lossy Counting is also employed in removing the nodes with low supports in TLT. The last two valid minimal occurrences kept in the node are used as in LossyDL to check whether a new valid minimal occurrence for its child node occurs or not. If it occurs, the support of its corresponding serial episode kept in the node of the prefix tree is increased by one and moreover, the original last two minimal occurrences of its corresponding serial episode will also be updated. When the number of the received n-tuple events of the data streams is divisible by $\lceil 1/\varepsilon \rceil$, TLT decreases the support of

---

Input: N, *minsup*, *minconf*, ε, *Lmax*, *PT*, and a set of time lag tables T_i, $i = 1$ to *Lmax* + $T - 1$
Output: significant SERs
Variable: a time lag, *lag*, two nodes N_1, N_2, their corresponding serial episodes S_1, S_2, the first itemset of S_1,
      I_{11}, the first itemset of S_2, I_{21}, the second itemset of S_1, I_{12}, and a temporary variable **TempCount**
      for estimating the support boundary

1.	**for each** two nodes N_1, N_2 in PT
2.	**if** (supp(S_1) ≥ (*minsup* − ε) × N and supp(S_2) ≥ (*minsup* − ε) × N)
3.	**for all** *lag* from 1 to *Lmax*
4.	**for each** item I in I_{11} and J in I_{21}
5.	Obtain supp($I \rightarrow_{lag} J$) from T_{lag}
6.	**if** (supp($I \rightarrow_{lag} J$) < (*minsup* − ε) × N)
7.	Prune $S_1 \rightarrow_{lag} S_2$ **// from Observation 2**
8.	**if** (supp($I \rightarrow_{lag} J$) < supp(S_1) × *minconf*)
9.	Prune $S_1 \rightarrow_{lag} S_2$ **// from Observation 3**
10.	**for each** item X in I_{12}
11.	*TempCount* = 0
12.	**for each** p s.t. $0 < p < T$ and *lag* − $p > 0$
13.	*TempCount* = *TempCount* + min(supp($I \rightarrow_p X$), supp($X \rightarrow_{lag-p} J$))
14.	**if** (*TempCount* < supp(S_1) × *minconf*)
15.	Prune $S_1 \rightarrow_{lag} S_2$ **// from Observation 4**
16.	**if** ($S_1 \rightarrow_{lag} S_2$ cannot be pruned)
17.	Return $S_1 \rightarrow_{lag} S_2$

Fig. 8. Rule generation of TLT

each serial episode by one and remove the nodes with supports = 0 from the prefix tree. In addition to the prefix tree, by keeping the last *Lmax* + $T - 1$ n-tuple events of the stream, the time lag table T_1, T_2, ..., $T_{Lmax + T - 1}$ can be updated when a new n-tuple event is generated.

When the users want the significant SERs, any two serial episodes with enough supports, that is, greater than or equal to (*minsup* − ε) × N, will form a *candidate SER*. Then, the candidate SER is checked to see whether it satisfies the pruning rules derived from Observations 2, 3, and 4. That is, given a candidate SER R: $S_1 \rightarrow_{lag} S_2$, where the time lag = 1 to *Lmax*, and supp(S_1) and supp(S_2) both equal or exceed (*minsup* − ε) × N, R is returned to the users if it pass all of the pruning rules shown in Figure 8. The rule generation algorithm of TLT is described in Figure 8.

4 Performance Evaluations

This paper first addresses the problem of finding significant serial episode rules with time lags over multi-streams. Therefore, to evaluate the effectiveness of LossyDL and TLT, a series of experiments on real data are performed and the experiment results are provided in this section. All experiments are performed on a PC with the Intel Celeron(R) 2.0GHz CPU, 1GB of memory, and under the Windows XP operating system. The *memory* space utilization, the *updating time* for the summary, the *mining time* while requesting the significant SERs, and the *precision* are used to evaluate the two proposed methods. The error parameter ε is set to 0.1 × *minsup*, as usually done in the other papers [8]. The maximum time lag, *Lmax*, is set to 10.

Two datasets tested in the experiments are described as follows: 1) Dryness and Climate Indices, denoted as *PDOMEI*: the data set contains several dryness and climate indices which are derived by experts and usually used to predict droughts. For example, several serial episode rules related to these indices are discovered in [14]. We select four indices from these indices, including Standardized Precipitation Index (SPI1, SPI3) [17], Pacific Decadal Oscillation Index (PDO) [16], and Multivariate ENSO Index (MEI) [15]. *Therefore, there are four data streams with a length of 124, including SPI1, SPI3, PDO, and MEI in this data set.* Moreover, the values of these indices are divided into seven discrete categories [14], making *the total types of items are 28.* 2) Another data set is "Twin Cities' Traffic Data Archive," denoted as *Traffic*: the data set obtained from TDRL [18] is Twin Cities' traffic data near the 50th St. during the first week of February, 2006. *It contains three data streams with a length of 1440.* Each stream is generated by a sensor periodically reporting the occupancy and flow. A pair of the occupancy and the flow is divided into discrete classes and regarded as items. *There are 55 distinct items in this dataset.*

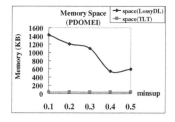
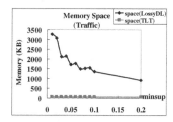

Fig. 9. Comparison of memory space between LossyDL and TLT

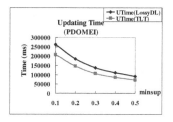
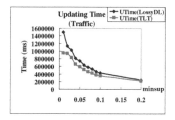

Fig. 10. Comparison of updating time between LossyDL and TLT

Since Lossy Counting is applied to both of the two methods, the required memory space of the two methods will increase as time goes by and drastically decrease when the sizes of the streams are divisible by $\lceil 1/\varepsilon \rceil$. Therefore, we concentrate on the maximal memory space used for the information of the occurrences of the serial episodes that is, the duration lists in LossyDL and the time lag tables in TLT, during the procedures, which is shown in Figure 9. As can be seen, the lower the *minsup* is, the more memory space the LossyDL method requires. This is because the error parameter ε is set to 0.1 × *minsup* and as ε decreases, the memory space required in LossyDL will increase. Alternatively, the memory space used in TLT is almost fixed under any values of the *minsup* because the sizes of the time lag tables are decided according to the number of items. Moreover, from Figure 9, it can find that the memory space used in TLT is substantially less than that in LossyDL. For example, in the Traffic dataset, the memory space used in TLT is around 64 KB, which is about 2% ~ 7% times that in LossyDL.

The Running time in the experiments can be divided into two types. One for *updating* such structures as the prefix tree, the duration lists, and the time lag tables and another one is for *mining* all the significant SERs from the structures. As shown in Figure 10, the updating time of TLT is less than that of LossyDL because increasing the counters in the time lag tables of TLT takes less time than frequently inserting/deleting the valid minimal occurrences into/from the duration lists of LossyDL. Moreover, the updating time of LossyDL and TLT roughly decreases as the *minsup* increases. Again, this is because ε is set to $0.1 \times minsup$. As ε (*minsup*) decreases, the size of the prefix trees will increase, thus making the updating time of two methods to be increased. In the Traffic dataset, the average processing time of LossyDL for each transaction is about 1.04 second. In other words, our methods can process the streaming data on time if the average arriving rate is less than 1 records/sec. This is acceptable in the traffic monitoring system because the sampling rates of the sensors cannot be set too high due to power consumption.

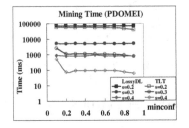 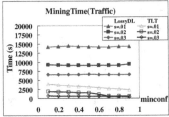

Fig. 11. Comparison of mining time between LossyDL and TLT

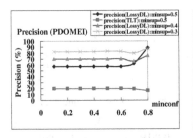 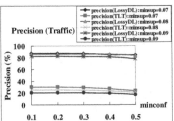

Fig. 12. Comparison of precision between LossyDL and TLT

The mining time of the two methods are shown in Figure 11. Since the lower the *minsup* (s) is, the more the serial episodes need to be checked, the mining time of the two methods will increase as the *minsup* decreases. Moreover, as can be seen, the mining time of the two methods will lightly decrease as *minconf* increases but the effects of *minconf* on the mining time seems to be marginal. This is because the mining time is almost decided according to the number of the serial episodes with enough supports. From Figure 11, it can also find that the mining time of TLT is lower than that of LossyDL because in LossyDL, the duration lists of two serial episodes need to be joined but to TLT, we only need to check the values in the time lag tables.

Since both of LossyDL and TLT are false-positive oriented approaches, their recall rates are equal to 100%. Therefore, we concentrate on the precision rates of the two methods, which are shown in Figure 12. The LossyDL method has the higher precision rates than the TLT method on both of the two datasets. This is because almost all the valid minimal occurrences are kept in the duration lists of LossyDL, making the support and confidence of a SER be precisely calculated by comparing the duration lists of the serial episodes.

5 Conclusions

In this paper, we address the problem of finding significant serial episode rules with time lags over multiple data streams and propose two methods, LossyDL and TLT, to solve the problem. The prefix tree structure is used to store the information of the support for each serial episode in the two methods. In order to limit the memory space for the prefix tree, the principle of Lossy Counting is integrated into the maintenance of the prefix tree. Since the number of the events (items) is much less than the number of the serial episodes, TLT is more space-efficient. Alternatively, since LossyDL keeps almost all the valid minimal occurrences for a serial episode, the precision of LossyDL is higher than that of TLT. In the near future, we will combine these two methods into a hybrid method to investigate the balance between the memory space used and the precision. On the other hand, we will also manage to extend the approach to mine general episode rules with time lags over multiple data streams.

References

[1] Agrawal, R., Srikant, R.: Fast Algorithms for Mining Association Rules in Large Databases. In: VLDB 1994, pp. 487–499 (1994)
[2] Hall, F.L.: Traffic stream characteristics. In: Traffic Flow Theory, U.S. Federal Highway Administration (1996)
[3] Harms, S.K., Deogun, J.S.: Sequential Association Rule Mining with Time Lags. Journal of Intelligent Information Systems 22(1), 7–22 (2004)
[4] Harms, S.K., Deogun, J., Saquer, J., Tadesse, T.: Discovering representative episodal association rules from event sequences using frequent closed episode sets and event constraints. In: ICDM 2001, pp. 603–606 (2001)
[5] Harms, S.K., Deogun, J., Tadesse, T.: Discovering Sequential Association Rules with Constraints and Time Lags in Multiple Sequences. In: ISMIS 2002, pp. 432–441 (2002)
[6] Han, J., Kamber, M.: Data Mining: Concepts and Techniques. Morgan Kaufmann, San Francisco (2001)
[7] Liu, Y., Choudhary, A., Zhou, J., Khokhar, A.: A Scalable Distributed Stream Mining System for Highway Traffic Data. In: Fürnkranz, J., Scheffer, T., Spiliopoulou, M. (eds.) PKDD 2006. LNCS (LNAI), vol. 4213, pp. 309–321. Springer, Heidelberg (2006)
[8] Manku, G.S., Motwani, R.: Approximate Frequency Counts Over Data Streams. In: VLDB 2002, pp. 346–357 (2002)

[9] Laxman, S., Sastry, P.S., Unnikrishnan, K.P.: A Fast algorithm for Finding Frequent Episodes in Event Streams. In: KDD 2007, pp. 410–419 (2007)

[10] Mannila, H., Toivonen, H.: Discovering Generalized Episodes using Minimal Occurrence. In: KDD 1996, pp. 146–151 (1996)

[11] Mannila, H., Toivonen, H., Verkamo, A.I.: Discovery of Frequent Episodes in Event Sequences. Data Mining and Knowledge Discovery 1(3), 259–289 (1997)

[12] Mannila, H., Verkamo, A.I., Toivonen, H.: Discovering Frequent Episodes in Sequences. In: KDD 1995, pp. 210–215 (1995)

[13] Mielikäinen, T.: Discovery of Serial Episodes from Streams of Events. In: SSDBM 2004, p. 447 (2004)

[14] Tadesse, T., Wilhite, D.A., Hayes, M.J.: Discovering Associations between Climatic and Oceanic Parameters to Monitor Drought in Nebraska Using Data-Mining Techniques. Journal of Climate 18(10), 1541–1550 (2005)

[15] Multivariate ENSO Index (MEI),
http://www.cdc.noaa.gov/people/klaus.wolter/MEI/

[16] Pacific Decadal Oscillation (PDO) Index,
http://jisao.washington.edu/data_sets/pdo/

[17] Standardized Precipitation Index,
http://www.drought.unl.edu/monitor/archivedspi.htm

[18] TDRL, http://tdrl1.d.umn.edu/services.htm

Efficient Approximate Mining of Frequent Patterns over Transactional Data Streams

Willie Ng and Manoranjan Dash

Centre for Advanced Information Systems,
Nanyang Technological University,
Singapore 639798
{ps7514253f,AsmDash}@ntu.edu.sg

Abstract. We investigate the problem of finding frequent patterns in a continuous stream of transactions. It is recognized that the approximate solutions are usually sufficient and many existing literature explicitly trade off accuracy for speed where the quality of the final approximate counts are governed by an error parameter, ϵ. However, the quantification of ϵ is never simple. By setting a small ϵ, we achieve good accuracy but suffer in terms of efficiency. A bigger ϵ improves the efficiency but seriously degrades the mining accuracy. To alleviate this problem, we offer an alternative which allows user to customize a set of error bounds based on his requirement. Our experimental studies show that the proposed algorithm has high precision, requires less memory and consumes less CPU time.

1 Introduction

Frequent pattern mining (FPM) has been well recognized to be fundamental to several prominent data mining tasks. In this paper, we investigate the problem of online mining for frequent patterns. Here, the goal of FPM is to uncover a set of itemsets (or any objects), whose occurrence count is at least greater than a pre-defined support threshold based on a fraction of the stream processed so far. In reality, with limited space and the need for real-time analysis, it is practically impossible to enumerate and count all the itemsets for each of the incoming transactions in the stream. With a domain of n unique items, we can generate up to 2^n distinct collections (or itemsets) in the data stream! The sheer volume of a stream over its lifetime does not allow it to be processed in the main memory or even in a secondary storage. Therefore, it is imperative to design algorithms that process the data in an online fashion with only a single scan of the data stream while operating under resource constraints.

For online mining, one typically cannot obtain the exact frequencies of all itemsets, but has to make an estimation. In general, approximate solutions in most cases may already be satisfactory to the need of users. Indeed, when faced with an infinite data set to analyze, many existing works explicitly trade off accuracy for speed where the quality of the final approximate counts are governed by an error parameter, ϵ [1,2,3]. Thus, the work related to the online mining for frequent patterns then boils down to the problem of finding the right form of data structure and related construction algorithms so that the

I.-Y. Song, J. Eder, and T.M. Nguyen (Eds.): DaWaK 2008, LNCS 5182, pp. 241–250, 2008.
© Springer-Verlag Berlin Heidelberg 2008

required frequency counts can be obtained with a bounded error for unbounded input data and limited memory.

However, before such online algorithm can be successfully implemented, we note that defining a proper value of ϵ is non-trivial [4]. Usually, we end up facing a dilemma. That is, by setting a small error bound, we achieve good accuracy but suffer in terms of efficiency. On the contrary, a bigger error bound improves the efficiency but seriously degrades the mining accuracy. In this paper, we will empirically show that even by setting ϵ to one-tenth (so called "rule of thumb") of the support threshold as suggested by [2], it does not necessarily yield an appealing result. In this paper, we propose a new solution to alleviate this drawback. We observe that in several existing papers, ϵ is uniformly fixed for all size of itemsets. To distinguish our work from them, we assign different values of ϵ to handle different itemsets based on their length. More specifically, the algorithm ensures that (1) all true frequent patterns are output; (2) for any frequent patterns having length equal or smaller than x, the error in the estimate is no more than ϵ_1; and (3) for any frequent patterns having length greater than x, the error in the estimate is no more than ϵ_2, where $\epsilon_1 \leq \epsilon_2$. The remainder of this paper is organized as follows: In Section 2, we introduce the preliminaries of discovering frequent patterns in the context of data streams. Section 3 discusses the issues pertaining to the definition of ϵ. Section 4 provides a formal treatment of this problem. Section 5 demonstrates its effectiveness through experiments on synthetic data sets. We provide an overview of the previous research in the Section 6. Finally, conclusions are presented in Section 7.

2   Preliminaries

The problem of mining frequent patterns online can be formally stated as follows: Let DS denote a transactional data stream, which is a sequence of continuously in-coming transactions, e.g., $t_1, t_2, ..., t_N$. We denote N as the number of transactions processed so far. Let $\mathcal{I} = \{i_1, i_2, ..., i_n\}$ be a set of distinct literals, called items.. Each transaction has a unique identifier (*tid*) and contains a set of items. An itemset X is a set of items such that $X \in (2^{|\mathcal{I}|} - \{\emptyset\})$ where $2^{|\mathcal{I}|}$ is the power set of $\mathcal{I}$. Let k define the length of an itemset. Thus, an itemset with k items is called a k-itemset. Alternatively, we may use the notation $|X|$ to denote the number of items in X. We write "*ABC*" for the itemset $\{A, B, C\}$ when no ambiguity arises. We assume that items in each transaction are kept sorted in their lexicographic order. Next, the frequency of an itemset X, denoted by $freq(X)$, is the number of transactions in DS that contain X. The support of an itemset X, denoted by $\sigma(X)$, is the ratio of the frequency of X to the number of transactions processed so far, i.e., $\sigma(X) = freq(X)/N$.

Definition 2.1 (*Frequent Pattern*[1]). Given a pre-defined support threshold σ_{min}, an itemset X is considered a frequent itemset (FI) if its frequency , $freq(X)$, is more than or equal to $\sigma_{min} \times N$. Let *Freqk* denote the set of frequent k-itemsets, and *AFI* denote the set of all frequent itemsets.

---

[1] In some literature, the terms *large* or *covering* have been used for *frequent*, and the term *itemset* for *pattern*.

Definition 2.2 (*Sub-frequent Pattern*). Consider an itemset X as a sub-frequent itemset (FI_{sub}) if its frequency is less than $\sigma_{min} \times N$ but not less than $\epsilon \times N$, where $\epsilon < \sigma_{min}$. Let AFI_{sub} be the set of all FI_{sub}.

Definition 2.3 (*Significant Pattern, SP*). We are only interested in frequent patterns. However, the AFI_{sub} are the potential itemsets that may become frequent as time passes by. Therefore, we need to maintain the AFI_{sub} and AFI at all times. We consider both AFI_{sub} and AFI as significant pattern (SP) and let ASP denote the set of all significant patterns.

Definition 2.4 (*Error bound*). Let ϵ be a given error parameter such that $\epsilon \in (0, \sigma_{min}]$. ϵ approximates the frequency of an itemsets such that the error is less than the true frequency by at most $\epsilon \times N$.

3   A Closer Look at ϵ

In order to provide a comprehensive understanding of our research problem we are addressing, we devote some space here for describing and analyzing the *Lossy Counting* algorithm (LCA). This algorithm is chosen for discussion due to its popularity. In fact, several recent algorithms also adopted the same error bound approach [1]. We refer the reader to [2] for more detail on the technical aspect of this algorithm. In LCA, the main features are:

1. All itemsets, whose true frequency is above $\sigma_{min} \times N$, are output (100% recall),
2. no itemset whose true frequency is less than $(\sigma_{min} - \epsilon) \times N$ is output and
3. the estimated frequencies are less than the true frequencies by at most ϵN.

The weakness of *LCA* is that its performance is greatly influenced by ϵ which is the crux of this algorithm. Note that ϵ is actually a minimum support threshold used to control the quality of the approximation of the mining result. For small ϵ, *LCA* achieves high precision, but consumes more memory and time, while for large ϵ, LCA uses small memory and runs faster at the expense of poor precision. If $\epsilon \ll \sigma_{min}$, a large number of SP needs to be generated and maintained. We refer the reader to the super exponential growth mentioned in [5].

3.1   Important Observations

To verify our analysis, we conducted a simple experiment to study the impact of changing the size of ϵ on frequent itemset mining. IBM synthetic data set [6] is used. A data stream of 2 millions transactions is generated. It has an average size of 15 and 10k unique items. We implemented the LCA. Without loss of generality, we equate the buffer size to the number of transactions processed per batch. The algorithm is run with a fixed batch size of 200k transactions. The Apriori alogrithm [6] is used to generate the true frequent patterns AFI_{true} from the data set. We fixed σ_{min} to be 0.1% and the resulted size of $|AFI_{true}| = 26136$. Two metrics, the recall and the precision are used. Given a set AFI_{true} and a set AFI_{appro} obtained by LCA, the recall is $\frac{|AFI_{true} \cap AFI_{appro}|}{|AFI_{true}|}$ and the precision is $\frac{|AFI_{true} \cap AFI_{appro}|}{|AFI_{appro}|}$. If the recall equals 1, the

Table 1. Impact of varying the value of epsilon

ϵ (%)	$|ASP|$	$|AFI_{appro}|$	$|AFI_{true}|$	Recall	Precision	Processing Time(s)
0.05	363614	258312	26136	1.00	0.10	178
0.01	1821961	54499	26136	1.00	0.48	1560
0.005	2397711	36538	26136	1.00	0.71	8649
0.003	2934985	31031	26136	1.00	0.84	28650

results returned by the algorithm contains all true results. This means no false negative. If the precision equals 1, all the results returned by the algorithm are some or all of the true results. This means no false positives are generated.

In Table 1, the first column is the error parameter (ϵ), and the second is the number of itemsets (ASP) that need to be maintained in the data structure D_{struct} while running the LCA. That means, any itemset having a frequency of ϵN or above will be in ASP. The higher the value of $|ASP|$ the more memory is consumed. The third column is the output size of the frequent itemsets $|AFI_{appro}|$. The fourth column is the true size of the frequent itemsets $|AFI_{true}|$. The next two columns are the accuracy of the output. The last column is the time spent in running the algorithm. The best point of LCA is its ability to uncover all frequent patterns. From the table, it is apparent that the LCA indeed ensures all itemsets whose true frequency is above $\sigma_{min} \times N$ are output. This implies $AFI_{true} \subseteq AFI_{appro}$. When ϵ is at 0.05% (half of σ_{min}), we see that its memory consumption and processing time is low. However, the precision is rather poor. This can be explained from the fact that any itemset in ASP is output as long as its computed frequency is at least $(\sigma_{min} - \epsilon) \times N$. If ϵ approaches σ_{min}, more false positives will be included in the final result.

Remark 1. *The experiment reflects the dilemma of false-positive oriented approach. The memory consumption increases reciprocally in terms of ϵ where ϵ controls the error bound.*

In contrast, when ϵ decreases, the cutoff value for ASP starts to reduce. This results in more and more entries being inserted into D_{struct} leading to higher memory consumption and processing time. However, when ϵ is low, $(\sigma_{min} - \epsilon) \times N$ increases, thus making it tougher for any itemset in ASP to get into AFI_{appro}. Clearly, the tighter the error bound, the higher the precision. Unfortunately, there is no clear clue as to how small ϵ must be defined. From the table, even the rule of thumb of setting $\epsilon = 0.01\%$ (one-tenth of σ_{min}) does not seem to yield appealing result. The precision is still below satisfactory. For example, out of the 54499 itemsets, 28363 of them are false positives!

3.2   Is It Worth the Overhead?

Table 1 provides only one side of the story. Additionally, we further examine the behavior of ASP when it is partitioned according to the length of the itemset. Table 2 illustrates the breakdown sizes of $|ASP|$ when ϵ is fixed at 0.01%. The first column represents the corresponding itemset length. The second column represents the number of itemsets for each k-itemset that the LCA maintained ($|ASP| = \sum |ASP_k|$). The third column denotes the break down size of all the true frequent itemsets

Table 2. Breakdown values of $|ASP|$ with $\epsilon = 0.01\%$

k-itemset	$|ASP_k|$	$|Freqk_{true}|$	Recall	Precision
1	7817	4763	1.00	0.609313
2	115214	3668	1.00	0.031836
3	217080	4692	1.00	0.021614
4	305438	4669	1.00	0.015286
5	338558	4060	1.00	0.011992
6	308944	2420	1.00	0.007833
7	237057	1196	1.00	0.005045
8	153389	499	1.00	0.003253
9	82957	142	1.00	0.001712
10	36996	25	1.00	0.000676
11	13458	2	1.00	0.000149
12	5053	0	NA	NA
Overall	1821961	26136	1.00	0.014345

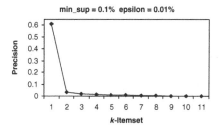

min_sup = 0.1% epsilon = 0.01%

Fig. 1. Individual precision for breakdown of $|ASP|$ with $\epsilon = 0.01\%$

($|AFI_{true}| = \sum |Freqk_{true}|$ and $Freqk_{true} \subseteq ASP_k$). The fourth and fifth column represent the recall and precision respectively. The last row in the table shows the overall performance when all the k-itemset are merged. Note that unlike Table 1, here, the precision[2] is used to assess the efficiency of LCA when the length of the itemset increases. Figure 1 gives a broader picture of the fast diminution of precision when the length of the itemset increases. We reason that this is related to the anti-monotone property. According to the anti-monotone property, an itemset X must have frequency either smaller than or equal to its subset Y, where $|X| > |Y|$. With increasing length of itemsets, the probability of meeting σ_{min} for the itemsets that meet ϵ decreases. Consider this example: itemset A, AB and ABC have support equal to 1.5%, 0.1% and 0.05% respectively. If σ_{min} is 0.5% and ϵ is 0.05%, all the three itemsets will be inserted into D_{struct}. However, only itemset A will meet σ_{min}. As shown in Figure 1, generation of excess itemsets particularly at the larger k leads to a smaller precision. This compels one to wonder why would one need to maintain, for instance when $k = 11$, 13458 entries in order to uncover only 2 true frequent itemsets!

---

[2] In other words, we are studying how many false positive are kept in the main memory. Since the recall is always 1, the precision is simply equivalent to the ratio of $|Freqk_{true}|$ to $|ASP_k|$.

4   Solutions

The example problem in the previous section has highlighted the dilemma pertaining to the definition of ϵ. In this section, we propose the customized lossy counting algorithm, denoted as CLCA, to address the problem faced when we attempt to quantify ϵ.

4.1   Customized Lossy Counting Algorithm, CLCA

In many existing literature, ϵ is uniformly fixed for all size of itemsets [2,3,7]. In principle, this may be important if one is interested in uncovering the entire set of frequent patterns regardless the length of the itemsets. However, as shown in the previous section, the longer the pattern we wish to explore, the less efficient LCA becomes. In situations where user is only keen on the itemsets having a size range from 1 to x and less interested on itemsets having length greater than x, using LCA to uncover all the frequent patterns might not be wise. For practicality, short frequent patterns are usually preferred in many applications. The rationales behind this preference are simple: (1) short patterns are more understandable and actionable; (2) short patterns are more likely to reflect the regularity of data while long patterns are more prone to reflect casual distribution of data [8,9]. In this respect, by focusing on the shorter patterns, one will be able to obtain an overall picture of the domain earlier without being overwhelmed by a large number of detailed patterns. With this understanding, we introduce CLCA to alleviate the drawback of LCA.

The basic intuition of CLCA is as follows: let smaller error bounds to find shorter frequent patterns and larger ones to find longer frequent patterns. We assign different error bounds to different itemsets based on their length. The strategy is simple. Here, a tight error bound can be used at the beginning to explore the itemsets within length from 1-x and after which, the bound is progressively relaxed as the length of the itemset increases. By assigning a tight error bound, the precisions for mining short frequent patterns will be improved but the efficiency for uncovering them will be reduced. On the contrary, a relaxed error bound will make the search for long patterns fast, but lose in accuracy. In other words, we migrate the workload for finding long frequent patterns to the front. CLCA is designed to focus on mining short frequent patterns.

Example 1. *Imagine a user wants to mine frequent patterns whose frequency is at least 0.1% of the entire stream seen so far. Then $\sigma_{min} = 0.1\%$. Due to resource constraint, he focuses his search on itemsets having a size range from 1 to 4 ($x = 4$) and willing to accept an error less than $0.1\sigma_{min}$. However, he does not wish to give up all the remaining long frequent patterns and willing to accept them as long as their error are less than $0.5\sigma_{min}$.*

The treatment for this example problem is outlined in Algorithm 1. The algorithm is similar to LCA except for the initialization. Instead of assigning a uniform ϵ for all k-itemsets, it offers an alternative, which allows user to customize the range of error bounds based on his requirement. In CLCA, the error bound of every pattern length is stored in an array *Error* of n elements. In this paper, we assume CLCA will mine up to 20-itemsets ($n = 20$). To generate the content of kth element , we require a function

Algorithm 1. Customized Lossy Counting Algorithm

```
1: N ← 0, D_struct ← ∅, NS ← ∅;
2: for k = 1 to n do
3:     Error[k] = Sig(k);    // the range is scaled to [ε₁, ε₂]
4: end for
5: for every B_size incoming transactions do
6:     N ← B_size + N;
7:     for all k such that 1 ≤ k ≤ n do
8:         Uncover all k-itemsets with frequency ≥ cutoff[k] and store them in NS;
9:         Insert NS to D_struct;
10:        Update the frequency of each k-itemsets in D_struct;
11:        if f_est(k-itemset)+Δ ≤ Error[k] × N then
12:            Eliminate this entry from D_struct;
13:        end if
14:        NS ← ∅;
15:    end for
16:    if mining results are requested then
17:        return all entries with f_est(X) ≥ (σ_min − Error[|X|]) × N;
18:    end if
19: end for
```

that could map any k to a new value between $[\epsilon_1, \epsilon_2]$. Here, this is computed using the sigmoid function $Sig(k) = \frac{1}{1+e^{s(k-h)}}$, where h is the threshold and s determines the slope of the sigmoid unit. This function is defined as a strictly increasing smooth bounded function satisfying certain concavity and asymptotic properties. The reason of using a sigmoid function is due to its unique feature of "squashing" the input values into the appropriate range. All small k values are squashed to ϵ_1 and higher ones are squashed to ϵ_2 (line 3). Figure 2 depicts the result of using the sigmoid function to compute $Error[k]$. Note that a standard sigmoid function will only output a range of 0-1, therefore we need to scale it down to $[\epsilon_1, \epsilon_2]$.

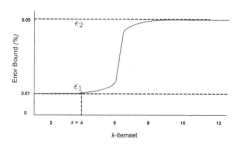

Fig. 2. Using a sigmoid function to generate the range of error bounds

Like LCA, CLCA processes the data in batches. However, to facilitate exposition of CLCA, we fixed the batch size instead of the buffer size. B_{size} denotes the number of transactions in a batch and N is updated after every new batch of transactions arrive (line 6). For each itemset of length k, we identify a corresponding cutoff frequency such that $cutoff[k] = Error[k] \times B_{size}$. The cutoff is used to determine whether a k-itemset should be inserted into D_{struct}. When processing a batch, a new set of k-itemsets having frequency $\geq cutoff[k]$ will be temporary kept in the pool NS (line 8). These are the newly uncovered set of significant patterns that need to be inserted into D_{struct}. D_{struct}

contains a set of entries $(X, f_{est}(X), \Delta)$. For insertion, a new entry is created for every itemset in NS (line 9). The frequency of the itemsets will be directly transfer to $f_{est}(X)$. $f_{est}(X)$ is the estimated frequency of X The value of Δ, which is the maximum possible error before insertion, is defined as $Error[k] \times N - cutoff[k]$. For k-itemsets which are already present in D_{struct}, we update by adding their frequency in the current batch to $f_{est}(X)$ (line 10). Deletion will be carried out after D_{struct} has been updated. For any k-itemset in D_{struct}, it will be deleted if $f_{est}(k\text{-itemset}) + \Delta \leq Error[k] \times N$. In order to avoid the combinatorial explosion of the itemsets, we apply the anti-monotone property so that if an itemset X does not exist in D_{struct}, then all the supersets of X need not be considered. When a user demands for mining result, CLCA outputs all entries, $(X, f_{est}(X), \Delta) \in D_{struct}$ if $f_{est}(X) \geq (\sigma_{min} - Error[\|X\|]) \times N$.

5 Performance Study

This section describes the experiments conducted in order to determine the effectiveness of CLCA. All experiments were performed on a 1.7GHz CPU Dell PC with 1GB of main memory and running on the Windows XP platform. The algorithm was written in C++. For efficient implementation, the algorithm uses the Trie data structure [2]. We used the synthetic data sets in our experiments and they were generated using the code from the IBM QUEST project [6]. The nomenclature of these data sets is of the form $TxxIyyDzzK$, where xx refers to the average number of items present per transaction, yy refers to the average size of the maximal potentially frequent itemsets and zz refers to the total number of transactions in K(1000's). Items were drawn from a universe of $\mathcal{I} = 10k$ unique items. The two data sets that we used are $T9I3D2000K$ and $T15I7D2000K$. Note that $T15I7D2000K$ is a much denser data set than $T9I3D2000K$ and therefore requires more time to process. We fixed the minimum support threshold $\sigma_{min} = 0.1\%$ and the batch size $B_{size} = 200K$. For comparison with LCA, we set $\epsilon = 0.1\sigma_{min}$ which is a popular choice for LCA.

Two set of error bounds, R1 and R2, were evaluated. For R1, we let $x = 3$, $\epsilon_1 = 0.03\sigma_{min}$ and $\epsilon_2 = 0.5\sigma_{min}$. For R2, we let $x = 4$, $\epsilon_1 = 0.05\sigma_{min}$ and $\epsilon_2 = 0.4\sigma_{min}$. Table 3 and 4 show the experiment results for the two data sets after processing 2 million of transactions. We observe that the time spent by CLCA on both data sets are much lesser than LCA. This can be reasoned by considering two factors: (1) ϵ_2 is much bigger than ϵ. A majority of error bounds in R1 and R2 are close or equal to ϵ_2. Higher error bounds will lead to lesser itemsets from getting into ASP. (2) x is small when compared to the maximum length of frequent patterns mined. Thus the high computation due to ϵ_1 can be compromised by setting a bigger ϵ_2.

As expected, the recall for CLCA and LCA is 1. This shows that CLCA indeed guarantee that all true frequent patterns are output. However, the precisions generated by R1 and R2 are weaker than LCA. Note that these represent only the entire set of frequent patterns. The main task of CLCA is to focus on mining short frequent patterns. Table 5 shows the strength of CLCA. Here, the precisions for mining all the frequent patterns with length ranging from 1 to x are recorded. Clearly, CLCA performs better

Table 3. Experiment results for $T9I3D2000K$

	$|ASP|$	$|AFI_{appro}|$	$|AFI_{true}|$	Recall	Precision	Time(s)
LCA	233913	12377	9418	1.00	0.760927	149
CLCA R1	149577	17092	9418	1.00	0.551018	96
CLCA R2	156378	13647	9418	1.00	0.751037	112

Table 4. Experiment results for $T15I7D2000K$

	$|ASP|$	$|AFI_{appro}|$	$|AFI_{true}|$	Recall	Precision	Time(s)
LCA	1821961	54499	26136	1.00	0.479568	1560
CLCA R1	1123565	207297	26136	1.00	0.126080	1096
CLCA R2	1400217	134221	26136	1.00	0.194724	1230

Table 5. Precision values for all frequent patterns with length ranging from 1 to x

	$1-x$	$T10I3D2000K$	$T15I7D2000K$
LCA	1-3	0.791008079	0.695626822
CLCA R1	1-3	0.933149171	0.876151689
LCA	1-4	0.77383377	0.600999865
CLCA R2	1-4	0.857219596	0.7645896

here. In addition, Figure 3 shows the execution time on $T15I7D2000K$ where CLCA performs faster than LCA at every test points . As we can see, the cumulative execution time of CLCA and LCA grows linearly with the number of transactions processed in the streams. Figure 4 further demonstrates the efficiency of CLCA. It shows the comparison in terms number of itemsets to be maintained in D_{struct}. Interestingly, for CLCA, the size of itemsets remains stable throughout the lifetime of processing the stream.

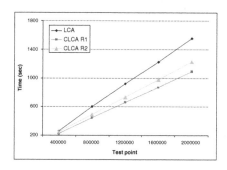

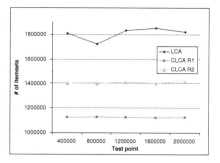

Fig. 3. Execution time on $T15I7D2000K$

Fig. 4. Number of itemsets to be maintained in D_{struct} for $T15I7D2000K$

6 Conclusions

In this paper, we propose CLCA to address the problem of quantifying ϵ. We place our emphasis on mining short frequent patterns in data stream. Instead of using fixed ϵ, we use a set of error bounds. The described algorithm is simple. Our experiments show that CLCA works better than LCA if one is more concern with mining short frequent patterns. This research raises 2 open questions. The first one concern the design of $Error[n]$: is there an optimum set of error bounds? The other one is, can we vary the error bounds for long patterns as well? We leave a more comprehensive investigation on these two questions for future research.

References

1. Cheng, J., Ke, Y., Ng, W.: A survey on algorithms for mining frequent itemsets over data streams. An International Journal of Knowledge and Information Systems (2007)
2. Manku, G.S., Motwani, R.: Approximate frequency counts over data streams. In: VLDB, pp. 346–357 (2002)
3. Giannella, C., Han, J., Pei, J., Yan, X., Yu, P.: Mining frequent patterns in data streams at multiple time granularities. In: Kargupta, H., Joshi, A., Sivakumar, K., Yesha, Y. (eds.) Next Generation Data Mining, pp. 191–212. AAAI/MIT (2003)
4. Cheng, J., Ke, Y., Ng, W.: Maintaining Frequent Itemsets over High-Speed Data Streams. In: Ng, W.K., Kitsuregawa, M., Li, J., Chang, K. (eds.) PAKDD 2006. LNCS (LNAI), vol. 3918, pp. 462–467. Springer, Heidelberg (2006)
5. Kohavi, Z.Z.R.: Real world performance of association rule algorithms. In: ACM SIGKDD (2001)
6. Agrawal, R., Srikant, R.: Fast algorithms for mining association rules. In: Proc.of the 20th VLDB conf. (1994)
7. Chen, L., Lee, W.: Finding recent frequent itemsets adaptively over online data streams. In: Proc. of ACM SIGKDD Cof., pp. 487–492 (2003)
8. Yang, L., Sanver, M.: Mining short association rules with one database scan. In: Arabnia, H.R. (ed.) Proceedings of the International Conference on Information and Knowledge Engineering, pp. 392–398. CSREA Press (2004)
9. Liu, B., Hsu, W., Ma, Y.: Pruning and summarizing the discovered associations. In: Proceedings of ACM SIGKDD International Conference in Knowledge Discovery and Data Mining, pp. 125–134 (1999)

Continuous Trend-Based Clustering in Data Streams[*]

Maria Kontaki, Apostolos N. Papadopoulos, and Yannis Manolopoulos

Department of Informatics, Aristotle University
54124 Thessaloniki, Greece
{kontaki,apostol,manolopo}@delab.csd.auth.gr

Abstract. Trend analysis of time series is an important problem since trend identification enables the prediction of the near future. In streaming time series the problem is more challenging due to the dynamic nature of the data. In this paper, we propose a method to continuously clustering a number of streaming time series based on their trend characteristics. Each streaming time series is transformed to a vector by means of the Piecewise Linear Approximation (PLA) technique. The PLA vector comprises pairs of values (timestamp, trend) denoting the starting time of the trend and the type of the trend (either UP or DOWN) respectively. A distance metric for PLA vectors is introduced. We propose split and merge criteria to continuously update the clustering information. Moreover, the proposed method handles outliers. Performance evaluation results, based on real-life and synthetic data sets, show the efficiency and scalability of the proposed scheme.

1 Introduction

The study of query processing and data mining techniques for data stream processing has recently attracted the interest of the research community [3,6], due to the fact that many applications deal with data that change frequently with respect to time. Examples of such application domains are network monitoring, financial data analysis, sensor networks, to name a few.

A class of algorithms for stream processing focuses on the recent past of data streams by applying a *sliding window*. In this way, only the last W values of each streaming time series is considered for query processing, whereas older values are considered obsolete and they are not taken into account. As it is illustrated in Figure 1, streams that are non-similar for a window of length W (left), may be similar if the window is shifted in the time axis (right). Note that, in a streaming time-series data values are ordered with respect to the arrival time. New values are appended at the end of the series.

Trend analysis has been used in the past in static and streaming time series [9,12,8]. We use trends as a base to cluster streaming time series for two reasons. First, the trend is an important characteristic of a streaming time series.

---

[*] Research supported by the PENED 2003 program, funded by the GSRT, Ministry of Development, Greece.

I.-Y. Song, J. Eder, and T.M. Nguyen (Eds.): DaWaK 2008, LNCS 5182, pp. 251–262, 2008.

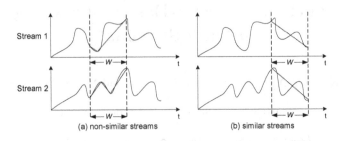

Fig. 1. Similarity using a SLIDING window of length W

In several applications the way that stream values are modified is considered important, since useful conclusions can be drawn. For example, in a stock data monitoring system it is important to know which stocks have an increasing trend and which ones have a decreasing one. Second, trend-based representation of time series is closer to human intuition. In the literature, many papers [5] use the values of the data streams and a distance function, like Euclidean distance, to cluster streams. Although the distance between a pair of streams may be large, the streams may be considered similar, if their plots are examined. Thus, distance functions on raw values are not always appropriate to cluster or to classify objects.

In this paper, we focus on the problem of continuous clustering of streaming time series based on the trends of the series as time progresses. The sliding window model is used for the clustering, i.e., the last W values of each stream are considered. Each streaming time series is represented by a Piecewise Linear Approximation (PLA). The PLA of a stream is calculated based on incremental trend identification. An appropriate distance function is introduced to quantify the dissimilarity between PLAs.

Recently, a number of methods have been proposed to attack the problem of data stream clustering [1,5]. The fundamental characteristic of the proposed methods is that they attack the problem of incremental clustering the data values of a single streaming time series, whereas our work focuses on incremental streaming time series clustering using multiple streams.

The majority of the aforementioned contributions apply variations of k-median clustering technique and therefore, the desired number of clusters must be specified. Our clustering algorithm automatically detects the number of clusters, by using the proposed *split* and *merge* criteria. As time progresses, the values, and probably the trends of streaming time series, are modified. It is possible, a split of a cluster in two different new clusters to be necessary to capture the clustering information. The proposed split criterion identifies such situations. In addition, the proposed merge criterion identifies a possible merge between two different clusters. Moreover, the proposed method handles outliers. The contributions of our work are summarized as follows:

- The PLA technique is used based on incremental trend identification, which enables the continuous representation of the time series trends.

- A distance function between PLAs is introduced.
- Continuous trend-based clustering is supported. Split and merge criteria are proposed in order to automatically detect the number of clusters.

The rest of the article is organized as follows. In Section 2, we discuss the incremental trend identification process. Section 3 presents the proposed method for continuous clustering, whereas Section 4 reports the experimental results based on real-life and synthetic data sets. Finally, Section 5 concludes the work.

2   Trend Detection

In this section, we study the problem of the incremental determination of each stream synopsis, to enable stream clustering based on trends. The basic symbols, used throughout the study, are summarized in Table 1.

Trend detection has been extensively studied in statistics and related disciplines [4,7]. In fact, there are several indicators that can be used to determine trend in a time series. Among the various approaches we choose to use the TRIX indicator [7] which is computed by means of a triple moving average on the raw stream data. We note that before trend analysis is performed, a smoothing process should be applied towards removing noise and producing a smoother curve, revealing the time series trend for a specific time interval. This smoothing is facilitated by means of the TRIX indicator, which is based on a triple exponential moving average (EMA) calculation of the logarithm of the time series values.

The EMA of period p over a streaming time series S is calculated by means of the following formula:

$$EMA_p(t) = EMA_p(t-1) + \frac{2}{1+p} \cdot (S(t) - EMA_p(t-1)) \tag{1}$$

Table 1. Basic notations used throughout the study

Symbol	Description
S, S_i	a streaming time series
PLA_x	PLA of streaming time series S_x
$PLA(i)$, $PLA_x(i)$	the i-th segment of a PLA
$PLA(i).t_{start}$, $PLA(i).t_{end}$	the starting and ending time of segment $PLA(i)$
$PLA(i).v_{start}$, $PLA(i).v_{end}$	the values of the starting and ending time of $PLA(i)$
$PLA(i).slope$	the slope of segment $PLA(i)$
cs, cs_i	a common segment between two PLAs
C, C_i	a cluster
$C.n$, $C_i.n$	the number of streaming time series of a cluster
$centroid_i$	the centroid of cluster C_i
$C.avg$, $C_i.avg$	the average DPLA distance of the streaming time series of the cluster and its centroid
nC_i	the nearest cluster of cluster C_i
W	sliding window length

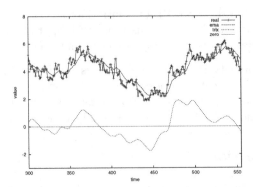

Fig. 2. Example of a time series and the corresponding $TRIX(t)$ signal

The TRIX indicator of period p over a streaming time series S is calculated by means of the following formula:

$$TRIX(t) = 100 \cdot \frac{EMA3_p(t) - EMA3_p(t-1)}{EMA3_p(t-1)} \qquad (2)$$

where $EMA3_p$ is a signal generated by the application of a triple exponential moving average of the input time series.

The signal $TRIX(t)$ oscillates around the zero line. Whenever $TRIX(t)$ crosses the zero line, it is an indication of trend change. This is exactly what we need in order to perform a trend representation of an input time series. Figure 2 illustrates an example. Note that the zero line is crossed by the $TRIX(t)$ signal, whenever there is a trend change in the input signal. Figure 2 also depicts the smoothing achieved by the application of the exponential moving average.

Definition 1
The PLA representation of a streaming time series S for a time interval of W values is a sequence of at most W-1 pairs of the form $(t, trend)$, where t denotes the starting time of the segment and *trend* denotes the trend of the stream in the specified segment (UP or DOWN).

Each time a new value arrives, the PLA is updated. Three operations (ADD, UPDATE and EXPIRE) are implemented to support incremental computation of PLAs. The ADD operation is applied when a trend change is detected and adds a new PLA point. The UPDATE operation is applied when the trend is stable and updates the timestamp of the last PLA point. The EXPIRE operation deletes the first PLA point when the first segment of the PLA expires.

3   Continuous Clustering

3.1   Distance Function

The literature is rich in distance metrics for time series. The most popular family of distance functions is the L_p norm, which is known as city-block or Manhattan norm when $p=1$ and Euclidean norm when $p=2$. A significant limitation of L_p norms is that they require the time series to have equal length. In our proposal, we compute distances between the PLAs of streaming time series, which may have different lengths. Therefore, L_p norm distance metrics cannot be used. In order to express similarity between time series of different lengths, other more sophisticated distance measures have been proposed. One such distance measure is Time Warping (TW) that allows time series to be stretched along the time axis. The disadvantage of TW is that it is computationally expensive and therefore, its use is impractical in a streaming scenario. In [10], an incremental computation of TW has been proposed, but it is limited in the computation of the distance between a static time series and a streaming time series, thus it is not suitable for our problem.

We propose the $DPLA$ distance function to overcome the above shortcomings. $DPLA$ splits the PLAs in common segments and computes the distance between each segment. The sum of distances of all segments gives the overall distance between two PLAs. A distance function for PLAs should take into account specific characteristics of the time series: 1) the trend of the segment: segments with different trends should have higher distance than segments with similar trends, and 2) the length of the segment: long segments should influence more the distance than short ones.

Before we proceed with the definition of $DPLA$, let us define the *slope* of a segment. Assume the i-th segment of a PLA starting at time $PLA(i).t_{start}$ and ending at time $PLA(i).t_{end}$. The values of $PLA(i)$ at the start and the end point are denoted as $PLA(i).v_{start}$ and $PLA(i).v_{end}$ respectively.

Definition 2
The *slope* of a segment $PLA(i)$ is the fraction of the difference of the values of the segment to the length of the segment:

$$PLA(i).slope = \frac{PLA(i).v_{end} - PLA(i).v_{start}}{PLA(i).t_{end} - PLA(i).t_{start}} \qquad (3)$$

Generally, PLAs have a different number of segments, each one of different length. Thus, in order to compare two PLAs, we use the notion of *common segment*. A common segment of PLA_x and PLA_y is defined between $max(PLA_x(i).t_{start}, PLA_y(j).t_{start})$ and $min(PLA_x(i).t_{end}, PLA_y(j).t_{end})$, where i and j are initialized to 1 and assume values up to the number of segments of PLA_x and PLA_y respectively.

For example, assume the two PLAs of Figure 3. We start with $i = j = 1$. The first common segment is defined by the maximum starting timestamp (t_1) and the minimum ending timestamp (t_2). Since we have reached the ending point of the segment belonging to PLA_2, we increase j by one. We inspect now the first segment of PLA_1 ($i = 1$) and the second segment of PLA_2 ($j = 2$). By observing Figure 3 we realize that the next common segment of PLA_1 and PLA_2 is defined by the timestamps t_2 and t_3. This process continues until we reach the end of the PLAs.

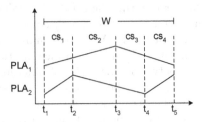

Fig. 3. Common segments of PLAs

The distance in a common segment cs defined by the i-th segment of the first PLA and the j-th segment of the second PLA is given by the following formula:

$$Dcs = |PLA_x(i).slope - PLA_y(j).slope| \cdot (cs.t_{end} - cs.t_{start}) \qquad (4)$$

Definition 3
The distance between two PLAs PLA_x and PLA_y with n common segments is given by the sum of distances of common segments:

$$DPLA(PLA_x, PLA_y) = \sum_{i=0}^{n} Dcs_i \qquad (5)$$

Notice that $DPLA$ function takes into account both the trend and the length of the segment and it can be computed incrementally.

3.2   Clustering Algorithm

Each cluster C_i has an id and a centroid which is the PLA of a streaming time series belonging to this cluster. Moreover, a cluster C stores a $C.n \times C.n$ matrix with the distances of streaming time series of the cluster. A streaming time series S_x belongs to cluster C_i, if: $\forall j \neq i$, $DPLA(PLA_x, centroid_i) \leq DPLA(PLA_x, centroid_j)$. Additionally, we keep a two-dimensional matrix with the distances of the centroids of all clusters.

First, we present the merge criterion. The average distance $C_i.avgD$ of a cluster C_i is the average $DPLA$ distance of the streaming time series of the cluster and its centroid.

Merge criterion
Two clusters C_i and C_j are merged if the sum of their average distances is higher than the half of the distance between their centroids:

$$c_i.avgD + c_j.avgD > DPLA(centroid_i, centroid_j)/2 \qquad (6)$$

To merge two clusters, we have to decide the centroid of the new cluster. We compute the distances of all PLAs of the two clusters with their centroids. The PLA which has the minimum sum of these two distances is chosen as the centroid of the new cluster.

The split criterion is more complicated. For each PLA_x of cluster C_i, we partition all other PLAs in two subsets A and B. Subset A comprises the PLAs that are close to PLA_x and subset B comprises the remaining PLAs. We separate close from distant streaming time series by using a threshold. Therefore, PLAs that their distance to PLA_x is below the threshold are assumed to be close to PLA_x. In our experiments, this threshold is set to half of the maximum distance of PLA_x to all other PLAs of C_i. The average distance between PLA_x and PLAs belonging to subset A is denoted as $PLA_x.close$, whereas the average distance between PLA_x and PLAs belonging to subset B is denoted as $PLA_x.distant$.

Split criterion
A cluster C_i splits in two different clusters if:

$$\frac{1}{C_i.n} \sum_{x=1}^{C_i.n} \frac{PLA_x.distant - PLA_x.close}{max(PLA_x.distant, PLA_x.close)} > \delta \qquad (7)$$

The above definition tries to approximate the silhouette coefficient of the new clusters. The silhouette coefficient [11] is a well-known metric for clustering evaluation, and ranges between [-1,1]. Values close to 1 indicate the existence of a good clustering whereas values below 0 indicate the absence of a clustering. The intuition for the above definition is that, if two clusters exist, then for each PLA the fraction $\frac{PLA_x.distant - PLA_x.close}{max(PLA_x.distant, PLA_x.close)}$ should be high. The parameter δ can affect the clustering significantly. Values below 0.5 may reduce the number of produced clusters, whereas large values (above 0.7) can cause consecutive splits resulting in a large number of clusters. Essentially, parameter δ controls the clustering quality and therefore, it is not necessary to change at runtime. In our experiments, we have used $\delta = 0.6$.

The centroids of the new clusters are chosen to be the PLAs that complies to the following rules: 1) $\frac{PLA.distant - PLA.close}{max(PLA.distant, PLA.close)} > \delta$ for both PLAs and 2) the $DPLA$ distance between them is the highest distance between the PLAs survived the first rule.

The outline of CTCS algorithm (Continuous Trend-based Clustering of Streaming time series) is depicted in Figure 4. Lines 3-10 describe the update of the clusters, whereas lines 11-17 show how CTCS adapts to the number of clusters. Notice that CTCS does not require new values for all streaming time series to update the clustering. In line 2, only the PLAs of streaming time series that have a new value, are updated.

Additionally, an outlier detection scheme could be applied. A PLA belongs to a cluster, if its distance from the centroid of this cluster is minimized. Let PLA_x be a PLA belonging to cluster C_i. PLA_x will be declared as outlier if the $DPLA$ distance between PLA_x and the centroid of the cluster C_i, is higher than the $DPLA$ distance between the centroids of the cluster C_i and its nearest cluster nC_i. In Figure 4, we can apply the outlier detection before line 6. If PLA_x is an outlier, we insert it into outliers and we continue with the next PLA, omitting the computations of lines 6-9.

Algorithm CTCS
Input
    new values of streaming time series
Output
    set of clusters

---

1. **updC** = ∅ //set of changed clusters
2. update PLAs of streaming time series
3. **for** (each PLA_i)
4. C_k = cluster that PLA_i belongs to
5.   find its new nearest cluster C_j
6.   **if** ($C_j \neq C_k$)
7.   move PLA_i to cluster C_j
8.   insert C_j and C_k to **updC**
9.   **end**
10. **end**
11. **for** (each cluster C_i of **updC**)
12.   remove C_i from **updC**
13.   apply merge criterion
14.   **if** (merge occurs) insert the new cluster to **updC**
15.   apply split criterion
16.   **if** (a split occurs) insert the new clusters to **updC**
17. **end**
18. report the clusters;

Fig. 4. Outline of CTCS algorithm

Table 2. Complexity analysis of CTCS

operation	worst case complexity
PLA update	$O(1)$
stream update	$O(k) + O(C_{old}.n) + O(C_{new}.n)$
split test	$O(C.n)$
split process	$O((C.n)^2) + O(k)$
merge test	$O(1)$
merge process	$O((C_1.n + C_2.n)^2) + O(k)$

Table 2 shows the worst case complexity of the basic operations of CTCS algorithm. These complexities are easily determined by a careful examination of the corresponding operations. Stream update refers to the update of cluster data due to the update of a stream S. k is the current number of clusters, C_{old} is the previous cluster containing S, whereas C_{new} is the new cluster containing S. Notice that, split and merge processes have a quadratic complexity on the number of streams per cluster and therefore, they are more computationally intensive. However, these two operations are executed less frequently than the rest, and thus the overall cost is not affected significantly.

4 Performance Study

In this section, we report the experimental results. We have conducted a series of experiments to evaluate the performance of the proposed method. Algorithm

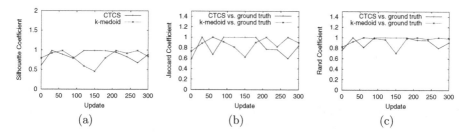

Fig. 5. Quality comparison: a) Silhouette coefficient, b) Jaccard coefficient and c) Rand coefficient for SYNTH

k-medoid is used as a competitor. k-medoid is modified to use the proposed distance function to handle PLAs of streaming time series. Notice that incremental implementations of k-medoid are not applicable due to the evolution of clusters, i.e., the number of clusters varies over time. However, in favor to k-medoid, we assume that the number of clusters is known, and we seed the algorithm with the previously determined medoids. All methods are implemented in C++ and the experiments have been conducted on a Pentium IV system at 3.0GHz, with 1GB of main memory running Windows XP.

We use both real and synthetic data sets. STOCK contains stock prices obtained from http://finance.yahoo.com. The data set consists of 500 time series, and the maximum length of each one is set to 3000. SYNTH is a synthetic data set and it is used in order to evaluate the quality of our method. The data set generator takes as parameters the number of streaming time series, the size of the sliding window and the number of clusters in different time instances. In this way, the number of clusters are varied over time and therefore, we can validate the performance of split and merge criteria.

First, we examine the quality of the results. We use the synthetic data set which consists of 500 streams. The window size is 30. We apply 300 updates and every 30 time instances we measure the silhouette coefficient of the clustering produced by CTCS and k-medoid. Parameter k of k-medoid is set to the actual number of clusters in each update. Figure 5(a) depicts the results. CTCS achieves silhouette coefficients more than 0.6 in all cases. Moreover, we compare the actual clustering with the clusterings of CTCS and k-medoid by using the Jaccard and Rand coefficients [11]. These coefficients range from 0 up to 1. Values close to 1 indicate high correlation between the two clusterings, whereas values close 0 indicate low correlation. Figures 5(b) and (c) depict the results for the Jaccard and Rand coefficient respectively. Jaccard and Rand coefficients are 1 in some cases which means that CTCS gives the actual clustering (ground truth).

To better comprehend the results, we study the number of clusters that CTCS detects. Table 3 shows the number of actual clusters and the number of clusters determined by CTCS. Associating the results of Figure 5 and Table 3, we observe that when CTCS detects the number of clusters, the silhouette coefficient of the clustering is more than 0.85 and the Jaccard and Rand Coefficient is more than 0.8. In cases where CTCS misses one or two clusters, silhouette, Jaccard and

Table 3. Number of clusters over time (SYNTH)

Update	0	30	60	90	120	150	180	210	240	270	300
No. Clusters	6	7	5	6	5	4	6	7	8	7	6
CTCS	3	6	6	4	5	4	6	7	8	4	6

Rand coefficient are good (more than 0.78, 0.67 and 0.81 respectively) which means that CTCS has recognized two clusters as one or the opposite. The results of k-medoid are as good as CTCS but notice that CTCS algorithm automatically detects the number of clusters.

In the next experiment, we examine the quality of the results in a real data set. Figure 6 shows the results with respect to the number of streams. For each run, a number of updates are applied and the average results are given. We set the parameter k of k-medoid equal to the number of clusters of CTCS in each update. As the number of streams increases, the two clusterings have less correlation (Figure 6 (b)). However the silhouette coefficient of CTCS is better than that of k-medoid and it is above 0.6 in all cases which indicates a good clustering.

Next, we study the CPU cost of the proposed method with respect to the number of streaming time series and the window size (Figure 7(a) and (b) respectively). The average CPU cost is given. It is evident that CTCS outperforms k-medoid. Especially, k-medoid is highly affected by the number of streams (the CPU axis is shown in logarithmic scale), whereas CTCS can handle a large number of streams in less than 1 second.

Finally, we examine the scalability of the proposed method. Figure 8(a) depicts the CPU cost with respect to the number of streams for the SYNTH data set. The number of streams varies between 100 and 10000. CTCS outperforms k-medoid in all cases. CTCS algorithm has two basic steps: a) the incremental computation of the PLA of a stream and the update of cluster data that the stream belongs to before and after the update (streams update) and b) the continuous update of the clustering (clusters update). Figure 8(b) shows the CPU

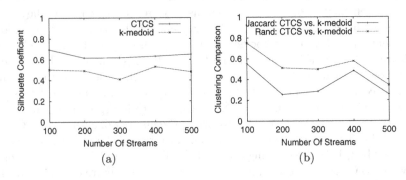

Fig. 6. Quality comparison: a) Silhouette coefficient and b) Clustering comparison vs number of streams for STOCK

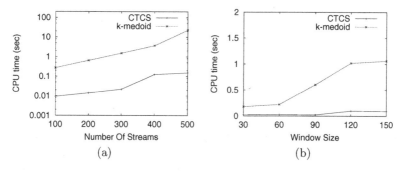

Fig. 7. CPU cost vs a) number of streams and b) window size for STOCK

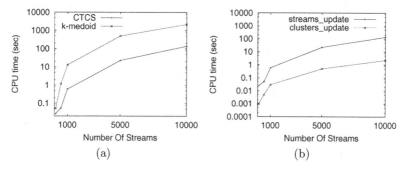

Fig. 8. CPU cost vs number of streams for SYNTH

cost of the two steps separately. It is evident, that the major overhead of the method is the first step, since the clustering update requires 2.5 sec at most. Notice that in each update all the streaming time series are updated and this is the worst scenario, thus the time of the streams update is expected to be smaller in a realistic scenario.

5   Conclusions

In this paper, a novel method has been proposed towards efficient continuous clustering of streaming time series. The proposed algorithm, CTCS, uses the PLAs of the streaming time series in order to achieve a trend-based clustering. Trends are automatically detected and PLAs are updated incrementally. Moreover, a new distance function, DPLA, is proposed. Additionally, CTCS does not require the number of clusters, since split and merge criteria are used to adjust the number of clusters automatically. Performance evaluation results illustrate the superiority of the proposed method against the k-medoid algorithm concerning the CPU cost and the quality of the produced clustering. Moreover, it demonstrates the capability of the proposed method to detect the number of clusters. Future work may include the use of a distance function that obeys the triangular inequality, towards exploiting indexing schemes to improve performance.

References

1. Charikar, M., O'Callaghan, L., Panigrahy, R.: Better Streaming Algorithms for Clustering Problems. In: Proceedings of STOC, pp. 30–39 (2003)
2. Datar, M., Gionis, A., Indyk, P., Motwani, R.: Maintaining stream statistics over sliding windows. In: Proceedings of ACM-SIAM SODA, pp. 635–644 (2002)
3. Domingos, P., Hulten, G.: Mining High-Speed Data Streams. In: Proceedings of ACM SIGKDD, pp. 71–80 (2000)
4. Fung, G.P.C., Yu, J.X., Lam, W.: News Sensitive Stock Trend Prediction. In: Chen, M.-S., Yu, P.S., Liu, B. (eds.) PAKDD 2002. LNCS (LNAI), vol. 2336, pp. 481–493. Springer, Heidelberg (2002)
5. Guha, S., Meyerson, A., Mishra, N., Motwani, R., OCallaghan, L.: Clustering Data Streams: Theory and Practice. IEEE TKDE 15(3), 515–528 (2003)
6. Hulten, G., Spencer, L., Domingos, P.: Mining Time Changing Data Streams. In: Proceedings of ACM KDD, pp. 97–106 (2001)
7. Hutson, J.K.: TRIX - Triple Exponential Smoothing Oscillator. Technical Analysis of Stocks and Commodities, 105–108 (July/August 1983)
8. Kontaki, M., Papadopoulos, A.N., Manolopoulos, Y.: Continuous Trend-Based Classification of Streaming Time Series. In: Eder, J., Haav, H.-M., Kalja, A., Penjam, J. (eds.) ADBIS 2005. LNCS, vol. 3631, pp. 294–308. Springer, Heidelberg (2005)
9. Sacchi, L., Bellazzi, R., Larizza, C., Magni, P., Curk, T., Petrovic, U., Zupan, B.: Clustering and Classifying Gene Expressions Data through Temporal Abstractions. In: Proceedings of IDAMAP, Protaras, Cyprus (2003)
10. Sakurai, Y., Faloutsos, C., Yamamuro, M.: Stream Monitoring Under the Time Warping Distance. In: Proceedings of ICDE, Istanbul, Turkey, pp. 1046–1055 (2007)
11. Tan, P.-N., Steinbach, M., Kumar, V.: Introduction to Data Mining. Addison-Wesley, Reading (2006)
12. Yoon, J.P., Luo, Y., Nam, J.: A Bitmap Approach to Trend Clustering for Prediction in Time-Series Databases. In: Proceedings of Data Mining and Knowledge Discovery: Theory, Tools, and Technology II, Florida, USA (2001)

Mining Multidimensional Sequential Patterns over Data Streams

Chedy Raïssi and Marc Plantevit

LIRMM, University of Montpellier, France
{raissi,plantevi}@lirmm.fr

Abstract. Sequential pattern mining is an active field in the domain of knowledge discovery and has been widely studied for over a decade by data mining researchers. More and more, with the constant progress in hardware and software technologies, real-world applications like network monitoring systems or sensor grids generate huge amount of streaming data. This new data model, seen as a potentially infinite and unbounded flow, calls for new real-time sequence mining algorithms that can handle large volume of information with minimal scans. However, current sequence mining approaches fail to take into account the inherent multi-dimensionality of the streams and all algorithms merely mine correlations between events among only one dimension. Therefore, in this paper, we propose to take multidimensional framework into account in order to detect high-level changes like trends. We show that multidimensional sequential pattern mining over data streams can help detecting interesting high-level variations. We demonstrate with empirical results that our approach is able to extract multidimensional sequential patterns with an approximate support guarantee over data streams.

1 Introduction

Sequential patterns have been studied for more than a decade [1], with substantial research and industrial applications. Sequence pattern mining allows the discovery of frequent sequences and helps identifying relations between itemsets in transactional database. However, sequential pattern mining is a difficult and challenging task as the search space for this problem is huge. To bypass this problem, researchers have developed mining algorithms based on the *Apriori* property [1] or *pattern growth* paradigm [10]. Lately, these approaches have been extended to mine multidimensional sequential patterns [11, 12, 14]. They aim at discovering more interesting patterns that take time into account and involve several analysis dimensions. For instance, in [12], rules like *"A customer who bought a surfboard together with a bag in New York later bought a wetsuit in San Francisco"* are discovered.

With the constant evolutions in hardware and software technologies, it becomes very affordable for companies and organisations to generate and store very large volume of information from various sources: network monitoring with TCP/IP traffic, financial transactions such as credit card customers operations,

I.-Y. Song, J. Eder, and T.M. Nguyen (Eds.): DaWaK 2008, LNCS 5182, pp. 263–272, 2008.

medical records and a wide variety of sensor logs. This large amount of data calls for the study of a new model called data streams, where the data appears as a continuous, high-speed and unbounded flow. Compared to the classical mining approaches from static transaction databases, the problem of mining sequences over data streams is far more challenging. It is indeed often impossible to mine patterns with classical algorithms requiring multiple scans over a database. Consequently, new approaches were proposed to mine itemsets [4, 5, 7, 8]. But, few works focused on sequential patterns extraction over data streams [2][9][13]. In this paper, we propose to consider the intrinsic multidimensionality of the streams for the extraction of more interesting sequential patterns. These patterns help detecting high-level changes like trends or outliers and are by far more advantageous for analysts than low-level abstraction patterns. However, the search space in multidimensional framework is huge. To overcome this difficulty, we only focus on the most specific abstraction level for items instead of mining at all possible levels. Furthermore, we mine time-sensitive data streams using a tilted-time frame approach that is more suitable for pattern mining. As a matter of fact, other models, like the landmark or sliding-window models, are often not appropriate since the set of frequent sequences is time-sensitive and it is often more important to detect changes in the sequences than sequences themselves.

The rest of this paper is organized as follows. Related work is described in Section 2. Preliminary concepts, problem description and a motivating example are introduced in Section 3. Section 4 presents our algorithm. The experiments and their results are described and discussed in Section 5. In the last section we give some conclusions and perspectives for future researches.

2   Related Work

Lately, sequential patterns have been extended to mine multidimensional sequential patterns in [11], [14] and [12]. [11] is the first paper dealing with several dimensions in the framework of sequential patterns. The sequences found by this approach do not contain multiple dimensions since the time dimension only concerns products. In [14], the authors mine for sequential patterns in the framework of Web Usage Mining considering three dimensions (pages, sessions, days), those being very particular since they belong to a single hierarchized dimension.In [12], rules combine several dimensions but also combine over time. In the rule *A customer who bought a surfboard together with a bag in NY later bought a wetsuit in SF*, *NY* appears before *SF*, and *surfboard* appears before *wetsuit*. In [13], the authors propose a new approach, called SPEED (*Sequential Patterns Efficient Extraction in Data streams*), to identify maximal sequential patterns over a data stream. It is the first approach defined for mining sequential patterns in streaming data. The main originality of this mining method is that the authors use a novel data structure to maintain frequent sequential patterns coupled with a fast pruning strategy. At any time, users can issue requests for frequent sequences over an arbitrary time interval. Furthermore, this approach produces an approximate answer with an assurance that it will not bypass user-defined frequency and temporal thresholds. In [9], the authors propose an algorithm based on sequences alignment for mining approximate sequential patterns in Web usage data streams. In [6] is introduced an efficient stream data cubing algorithm which computes only the layers along a popular path and leaves the other cuboids for query driven, on-line computation. Moreover, a tilted-time window model is also used to construct and maintain the cuboid incrementally.

We can denote that the approach defined in [2] is totally different from our proposal even if some notations seem to be similar. Indeed, the authors propose to handle several data streams. However, they only consider one analysis dimension over each data stream.

3   Problem Definition

In this section, we define the problem of mining multidimensional sequential patterns in a database and over a data stream.

Let DB be a set of tuples defined on a set of n dimensions denoted by $\mathcal{D}$. We consider a partitioning of $\mathcal{D}$ into three sets: the set of the analysis dimensions D_A, the set of reference dimensions D_R and the set of temporal dimensions D_t. A *multidimensional item* $a = (d_1, \ldots, d_m)$ is a tuple such that for every $i = 1 \ldots m$, $d_i \in Dom(D_i) \cup \{*\}$, $D_i \in D_A$. The symbol $*$ stands for wild-card value that can be interpreted by ALL.

A *multidimensional itemset* $i = \{a_1, \ldots, a_k\}$ is a non-empty set of multidimensional items such that for all distinct i, j in $\{1 \ldots k\}$, a_i and a_j are incomparable. A *multidimensional sequence* $s = \langle i_1, \ldots, i_l \rangle$ is an ordered list of multidimensional itemsets. Given a table T, the projection over $D_t \cup D_A$ of the set of all tuples in T having the same restriction r over D_R is called a *block*. Thus, each block $\mathcal{B}_r$ identifies a multidimensional data sequences.

A multidimensional data sequence identified by $\mathcal{B}_r$ *supports* a multidimensional sequence $s = \langle i_1, \ldots, i_l \rangle$ if for every item a_i of every itemset i_j, there exists a tuple (t, a'_i) in $\mathcal{B}_r$ such that $a'_i \subseteq a_i$ with respect to the ordered relation (itemset i_1 must be discovered before itemset i_2, etc.). The support of a sequence s is the number of blocks that support s. Given a user-defined minimum support threshold *minsup*, a sequence is said to be *frequent* if its support is greater than or equal to *minsup*.

Given a set of blocks B_{DB,D_R} on a table DB, the problem of mining *multidimensional sequential patterns* is to discover all multidimensional sequences that have a support greater than or equal to the user specified minimum support threshold *minsup* (denoted σ).

Because of wild card value, multidimensional items can be too general. Such items do not describe the data source very well. In other words, they are too general to be useful and meaningful to enhance the decision making process. Moreover, these items combinatorially increase the search space. In this paper, we thus focus on the *most specific* frequent items to generate the multidimensional sequential patterns. For instance, if items $(LA, *, M, *)$ and $(*, *, M, Wii)$

are frequent, we do not consider the frequent items $(LA, *, *, *), (*, *, M, *)$ and $(*, *, M, Wii)$ which are more general than $(LA, *, M, *)$ and $(*, *, M, Wii)$.

Let *data stream* $DS = B_0, B_1, \ldots, B_n$, be an infinite sequence of batches, where each batch is associated with a timestamp t, i.e. B_t, and n is the identifier of the most recent batch B_n. A batch B_i is defined as a set of multidimensional blocks appearing over the stream at the ith time unit, $B_i = \{\mathcal{B}_1, \mathcal{B}_2, \mathcal{B}_3, ..., \mathcal{B}_k\}$. Furthermore, the data model is fixed for the data stream: all batches are defined over the same set of dimensions D. For each block $\mathcal{B}_k$ in B_i we are thus provided with the corresponding list of itemsets. The length L_{DS} of the data stream is defined as $L_{DS} = |B_0| + |B_1| + \ldots + |B_n|$ where $|B_i|$ stands for the cardinality of the set B_i in terms of multidimensional blocks. In our current example, $L_{DS} = |B_{October}| + |B_{November}| = 7$, with $|B_{October}| = 3$ ($\mathcal{B}_{C_1}$, $\mathcal{B}_{C_2}$ and $\mathcal{B}_{C_3}$) and $|B_{November}| = 4$ ($\mathcal{B}_{C_4}$, $\mathcal{B}_{C_5}$, $\mathcal{B}_{C_6}$ and $\mathcal{B}_{C_7}$).

Therefore, given a user-defined minimal support σ, the problem of mining multidimensional sequential patterns over data stream DS is to extract and update frequent sequences S such that: $support(S) \geq \sigma.L_{DS}$.

However, and in order to respect the completeness of our sequential pattern extraction, any data stream mining algorithm should take into account the evolution of sequential patterns support over time: infrequent sequences at an instant t could become frequent later at $t+1$ and a frequent sequence could also not stay such. If a sequence become frequent over time and we did not store its previous support, it will be impossible to compute its correct overall support. Any mining algorithm should thus store necessary informations for frequent sequences, but also for candidate *sub-frequent* patterns that could become frequent [5]. A sequence S is called *sub-frequent* if: $\epsilon \leq Support(S) \leq \sigma$, where ϵ is a user defined support error threshold.

October						November					
CID	Date	Customer Informations				CID	Date	Customer Informations			
C_1	1	NY	Educ.	Middle	CD	C_4	1	NY	Business	Middle	Wii
C_1	1	NY	Educ.	Middle	DVD	C_4	2	NY	Business	Middle	Game
C_1	2	LA	Educ	Middle	CD	C_5	1	LA	Prof.	Middle	Wii
C_2	1	SF	Prof.	Middle	PS3	C_5	1	LA	Prof.	Middle	iPod
C_2	2	SF	Prof.	Middle	xbox	C_6	1	LA	Educ.	Young	PSP
C_3	1	DC	Business	Retired	PS2	C_6	2	LA	Educ.	Young	iPod
C_3	1	LA	Business	Retired	Game	C_7	1	LA	Business	Young	PS2

Fig. 1. Multidimensional tables

3.1   Motivating Example

In order to illustrate our previous definitions, we focus only on the two month tables (also called batches) in Figure 1: October and November. More precisely, these batches describe the customer purchases according to 6 attributes or dimensions: the *Customers id*, the purchase *Date*, the shipment *City*, the customer social group *Customer-group*, the customer age group *Age-group* and the *Product* as shown in Figure 1. Suppose that a company analyst wants to extract multidimensional sequences for the discovery of new marketing rules based on customers purchasing evolutions from October to November.

The analysts would like to get all the sequences that are frequent in at least 50% of the customers in each batch. Note that each row for a customer contains a transaction number and a multidimensional item. From the October batch, the analyst extracts three different sequences: (i) The first sequence: $\langle\{(*,*,M,*)\}\rangle$ is a single item sequence. This states that at least 2 customers out of 3 buying entertainment products are middle aged. (ii) The second sequence: $\langle\{(LA,*,*,*)\}$ is also a single item one. The analyst can infer that at least 2 customers out of 3 asked for their purchases to be shipped in Los Angeles. (iii) The third sequence: $\langle\{(*,*,M,*)\}\{(*,*,M,*)\}\rangle$, is a 2-item sequence. This is an interesting knowledge for the analyst as it means that 2 customers out of 3 are middle aged and that they bought entertainment products twice in October. Then the analyst extracts sequences for November. He finds 9 sequences but 3 are really interesting: (i) $\langle\{(LA,*,Y,*)\}\rangle$, this is an extremely valuable new knowledge as it informs the analyst that 2 out of 4 customers that ask to be shipped in Los Angeles are young people. This is also a specialization of the second sequence from October's batch. (ii) $\langle\{(LA,*,*,iPod)\}\rangle$, this informs the analyst that 2 out of 4 customers that have asked to be shipped in Los Angeles bought the iPod product. Thus, the analyst can use this knowledge to build, for instance, targeted customers offers for the next month. Notice that this sequence is also a specialization of the second sequence from the previous batch. (iii) $\langle\{(*,*,M,Wii)\}\rangle$, this sequence infers that 2 out of 4 customers are middle-aged and bought a brand new Wii console in November. This is also a specialization of the first sequence of October's batch. Plus, this sequence highlights the appearence of a new product on the market: the Wii console.

4   The *MDSDS* Approach

In *MDSDS*, the extraction of multidimensional sequential patterns is the most challenging step. In order to optimize it we divide the process into two different steps:

1. *MDSDS* extracts the most specific multidimensional items. Most specific items are a good alternative to the potential huge set of frequent multidimensional items that can be usually extracted. Indeed, they allow to *factorize* knowledge, as more general patterns can be inferred from them in a post-processing step. Furthermore, mining most specific multidimensional items allows the detection of generalization or specialization of items with wildcard values appearing or disappearing in an item over time.
2. Using the extracted most specific items, we mine sequences containing *only* these items in a classical fashion using PrefixSpan algorithm[10]

When applying this strategy, frequent sequences with too general items are not mined if there exist some more specific ones. However, this is not a disadvantage since these sequences often represent too general knowledge which is useless and uninteresting for the analysts (*e.g.* decision maker). *MDSDS* uses a data structure consisting of a prefix-tree containing items and tilted-time windows tables embedded in each node of the tree to ensure the maintenance of support information for frequent and sub-frequent sequences. We use tilted-time windows table technique in our approach to store sequences supports for every processed batch. Tilted-time windows notion was first introduced in [3] and is based on the fact that people are often interested in recent changes at a fine granularity but long term changes at a coarse one. By matching a tilted-time window for each mined sequence of the stream, we build a history of the support of the sequence over time. The patterns in our approach are divided into three categories: frequent patterns, sub-frequent patterns and infrequent patterns. Sub-frequent patterns may become frequent later (in the next batches from the data stream), thus *MDSDS* has to store and maintain their support count as for the frequent ones. Only the infrequent patterns are not stored in the prefix-tree. The cardinality of the set of sub-frequent patterns that are maintained is decided by the user and called *support error threshold*, denoted ϵ. The updating operations are done after receiving a batch from the data stream: at the prefix-tree level (adding new items and pruning) and at the tilted-time window table level (updating support values). First, multidimensional sequential patterns are extracted from the batch and projected in the prefix-tree structure. Second, the tilted-time windows table for each multidimensional sequential pattern is updated by the support of the sequence in the current batch. The pruning techniques relative to tilted-time windows table are then applied. Generalization and specialization of sequences are detected during the maintenance.

4.1   Algorithm

We now describe in more details the *MDSDS* algorithm. This algorithm is divided in four steps :

1. As previously highlighted, mining most specific multidimensional items is the starting point for our multidimensional sequence extraction. We use a levelwise algorithm to build the frequent multidimensional items having the smallest possible number of wild-card values. Each item on the different analysis dimensions is associated with a unique integer value. We then mine the most specific multidimensional items based on this new mapping.

   For instance, let us consider October's batch from Figure 1. Each value for each analysis dimension will be mapped to an integer value: NY in the *City* dimension will be mapped to 1, LA to 2 and so on until $Game$ in *Product* dimension which will be mapped to the value 15. With this new representation of the batch, the most specific frequent items extracted are: (2) and (8), in multidimensional representation: $(LA, *, *, *)$ and $(*, *, M, *)$. All the multidimensional items are stored in a data structure called mapping array. After this step, we can add an optional processing step in order to detect the appearance of specialization or generalization according to the previous processed batch. Indeed, a specific item which was frequent over the previous batches could become infrequent later. In this case, we consider some more general multidimensional items. In order to detect the specialization or generalization, the subroutine compare (by inclusion) each multidimensional item from the current batch with the multidimensional items in the previous batch. Notice that in order to keep the process fast, the items are compared only with the previous batch.

2. In order to have a consistant mining of frequent sequences we have to consider subfrequent sequences which may become frequent in future batches. Thus, the support is set to ϵ and we use the PrefixSpan algorithm [10] to mine efficiently the multidimensional sequences. The new mined sequences are added to the tree which maintain the set of frequent and subfrequent sequences over the data stream.

3. Finally, a last scan into the pattern tree is done in order to check if each node n was updated when batch B_i was mined. If not, we insert 0 into n's tilted-time window table. Eventually, some pruning is done on the tilted-time window as defined in [5].

5   Performance

In this section, we present the experiments we conducted in order to evaluate the feasability and the performances of the $MDSDS$ approach. Throughout the experiments, we answer the following questions inherent to scalability issues : *Does the algorithm mine and update its data structure before the arrival of the next batch? Does the mining process over a data stream remain bounded in term of memory ?* The experiments were performed on a Core-Duo 2.16 Ghz MacBook Pro with 1GB of main memory, running Mac OS X 10.5.1. The algorithm was written in C++ using a modified Apriori code[1] to mine the most specific multidimensional items and we modified PrefixSpan[2] implementation to enable multidimensional sequence mining. We performed several tests with different real data sets that we gathered from TCP/IP network traffic at the University of Montpellier. The size of the batches is limited to 20000 transactions with an average of 5369 sequences per batch, the time to fill the batch varies w.r.t to the data distribution over the TCP/IP network.

A TCP/IP network can be seen as a very dense multidimensional stream. This is based on the following property: TCP and IP headers contain multiple different informations encapsulated in different formats that can vary drastically depending on the packet destination, its final application or the data it contains. We claim that generalization or specialization of the packets on the network (or even Denial of Services) can be detected by mining multidimensional sequential patterns over a local area network.

$DS1$ is very dense data set composed of 13 analysis dimensions based on the different TCP header options selected for their relevance by a network analyst expert (source port, destination port, time-to-live etc...), the data stream is divided into 204 batches for a total size of 1.58 GB. The results for the different

---

[1] http://www.adrem.ua.ac.be/ goethals/software/

[2] http://illimine.cs.uiuc.edu/

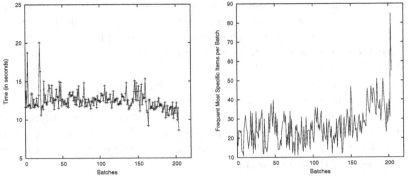

(a) Time needed for each batch process- (b) Number of frequent most specific
ing for data set $DS1$ items per batch for data set $DS1$

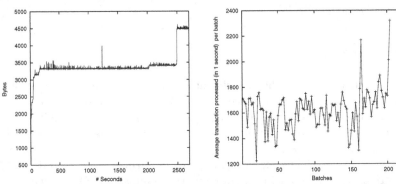

(c) Memory usage for data set $DS1$ (d) Average transactions processed per
                                       second and per batch

Fig. 2. Experiments carried out on TCP/IP network data

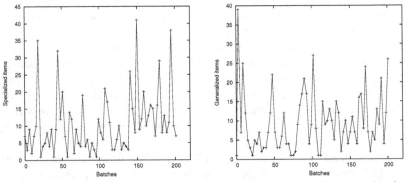

(a) Number of items that specialized per (b) Number of items that generalized per
batch   batch

Fig. 3. Specialization and generalization for the experiments carried out on TCP/IP
network data

experimentations are listed in figures 2(a), 2(b), 2(c) and 2(d). The computation time for a batch over $DS1$ do not exceed 25 seconds, our algorithm never goes beyond this limit when extracting multidimensional sequences, leaving enough time for the data gathering for the next batch. The experimentation total time is 2600 seconds. Figure 2(a) depicts the computation time needed for the batches processing by our algorithm for a support value of $\sigma = 0.15\%$ and an error value of $\epsilon = 0.1\%$. $MDSDS$ is also bounded in term of memory constraint. Figure 2(c) depicts the memory behavior of the algorithm on $DS1$. We see that the space usage is bounded above and stable (less than 6 Mb) with spikes when extracting the sequences and updating the tilted-time windows. The effective processing rate is approximately 1400 transactions per second. Specializations (Figure 3(a)) and generalizations (Figure 3(b)) of the frequent multidimensional items vary between batches but is bounded by a maximum of 41 specializations and 39 generalizations between two batches. From these results, the network analyst expert can deduce some new knowledge on the state of the network traffic, for example between batch 150 and batch 151, the multidimensional item ($VER :$ $4, IPLEN : 5, TOS : 0, PLEN > 21, IPID : [10K, 20K[, [AF], DF : 1, PROT :$ $6, TCPHLEN : 5, SEQ\_NUM : +65K$) which states that most of the packets are based on the TCP protocol with some options on the version, packet length etc gets specialized into ($VER : 4, IPLEN : 5, TOS : 0, PLEN > 21, IPID :$ $[10K, 20K[, [AF], DF : 1, TTL : 128, PROT : 6, TCPHLEN : 5, SEQ\_NUM :$ $+65K$). Notice the appearance of $TTL : 128$, which states that at batch 150 most of the packets had different time-to-live values but starting from batch 151, most of the packets now have a time-to-live to 128. From a network analysis point of view, this means that all the packets are now topologically concentrated in less than 128 hops from the source to the destination host. Several other specializations or generalizations give some other insights on the traffic trends and evolutions. For example, items that specialize with the value $DSTP : [32K[,$ means that the destination ports for the different applications are used for video or music streaming (applications for multimedia streaming usually have their ports number starting from 32771).

6   Conclusion

In this paper, we adress the problem of mining multidimensional sequential patterns in streaming data and propose the first approach called $MDSDS$ for mining such patterns. By considering the most specific multidimensional items, this approach efficiently detects trends. Experiments on real data gathered from TCP/IP network traffic provide compelling evidence that it is possible to obtain accurate and fast results for multidimensional sequential pattern mining. This work can be extended following several directions. For example, we can take hierarchies into account in order to enhance the trend detection and discover trends like *"In october, console sales are frequent whereas a specific console become frequent in November (Wii)"*. Besides, to fit to OLAP framework, we should take approximate values on quantitative dimensions into account with constrained base multidimensional pattern mining.

Finally, our results shows the potential of further work on multidimensional sequential pattern mining and specially in the new challenging data streams model.

References

1. Agrawal, R., Srikant, R.: Mining sequential patterns. In: Proc. 1995 Int. Conf. Data Engineering (ICDE 1995), pp. 3–14 (1995)
2. Chen, G., Wu, X., Zhu, X.: Sequential pattern mining in multiple streams. In: ICDM, pp. 585–588. IEEE Computer Society, Los Alamitos (2005)
3. Chen, Y., Dong, G., Han, J., Wah, B.W., Wang, J.: Multi-dimensional regression analysis of time-series data streams. In: VLDB, pp. 323–334 (2002)
4. Chi, Y., Wang, H., Yu, P.S., Muntz, R.R.: Moment: Maintaining closed frequent itemsets over a stream sliding window. In: Proceedings of the 4th IEEE International Conference on Data Mining (ICDM 2004), pp. 59–66, Brighton, UK (2004)
5. Giannella, G., Han, J., Pei, J., Yan, X., Yu, P.: Mining frequent patterns in data streams at multiple time granularities. In: Kargupta, H., Joshi, A., Sivakumar, K., Yesha, Y. (eds.) Next Generation Data Mining. MIT Press, Cambridge (2003)
6. Han, J., Chen, Y., Dong, G., Pei, J., Wah, B.W., Wang, J., Cai, Y.D.: Stream cube: An architecture for multi-dimensional analysis of data streams. Distributed and Parallel Databases 18(2), 173–197 (2005)
7. Li, H.-F., Lee, S.Y., Shan, M.-K.: An efficient algorithm for mining frequent itemsets over the entire history of data streams. In: Proceedings of the 1st International Workshop on Knowledge Discovery in Data Streams, Pisa, Italy (2004)
8. Manku, G., Motwani, R.: Approximate frequency counts over data streams. In: Proceedings of the 28th International Conference on Very Large Data Bases (VLDB 2002), pp. 346–357, Hong Kong, China (2002)
9. Marascu, A., Masseglia, F.: Mining sequential patterns from data streams: a centroid approach. J. Intell. Inf. Syst. 27(3), 291–307 (2006)
10. Pei, J., Han, J., Mortazavi-Asl, B., Wang, J., Pinto, H., Chen, Q., Dayal, U., Hsu, M.-C.: Mining sequential patterns by pattern-growth: The prefixspan approach. IEEE Transactions on Knowledge and Data Engineering 16(10) (2004)
11. Pinto, H., Han, J., Pei, J., Wang, K., Chen, Q., Dayal, U.: Multi-dimensional sequential pattern mining. In: CIKM, pp. 81–88 (2001)
12. Plantevit, M., Choong, Y.W., Laurent, A., Laurent, D., Teisseire, M.: M^2SP: Mining sequential patterns among several dimensions. In: Jorge, A.M., Torgo, L., Brazdil, P.B., Camacho, R., Gama, J. (eds.) PKDD 2005. LNCS (LNAI), vol. 3721, pp. 205–216. Springer, Heidelberg (2005)
13. Raïssi, C., Poncelet, P., Teisseire, M.: Need for speed: Mining sequential patterns in data streams. In: BDA (2005)
14. Yu, C.-C., Chen, Y.-L.: Mining sequential patterns from multidimensional sequence data. IEEE Transactions on Knowledge and Data Engineering 17(1), 136–140 (2005)

Towards a Model Independent Method for Explaining Classification for Individual Instances

Erik Štrumbelj and Igor Kononenko

University of Ljubljana, Faculty of Computer and Information Science,
Tržaška 25, 1000 Ljubljana, Slovenia
{erik.strumbelj,igor.kononenko}@fri.uni-lj.si

Abstract. Recently, a method for explaining the model's decision for an instance was introduced by Robnik-Šikonja and Kononenko. It is a rare example of a model-independent explanation method. In this paper we make a step towards formalization of the model-independent explanation methods by defining the criteria and a testing environment for such methods. We extensively test the aforementioned method and its variations. The results confirm some of the qualities of the original method as well as expose several of its shortcomings. We propose a new method, based on attribute interactions, that overcomes the shortcomings of the original method and serves as a theoretical framework for further work.

1 Introduction

Numerous different models for classification are being used today and range from those with a transparent decision making process (for example naive Bayes or a simple decision tree) to non-transparent and nearly black-box classifiers such as Support Vector Machines or artificial neural networks. Explaining the prediction of a classifier is sometimes as important as the prediction itself. A good explanation makes the decision easier to understand and easier to trust, which is especially useful when applying machine learning in areas such as medicine. We strive to develop a general model-independent method for instance explanation. Such a method would provide a continuity of explanation even if the model would be replaced. It would also make the explanations from different models easier to compare.

Various methods have been developed for the purpose of decision explanation and most were focused on explaining predictions and extracting rules from artificial neural networks, which [1] is a survey of. Methods for other types of models were developed as well (for SVM [4,6], naive Bayes [8] and additive classifiers [10] to name a few), but each method was limited to a specific classifier or type of classifiers. There have not been any notable attempts at a model independent method. However, recently a method for explanation of instances in classification was developed by Robnik-Šikonja and Kononenko [9]. The advantage of this method is, that it can be applied to any probabilistic classifier as it uses only the classifiers output to generate an explanation. The method defines the i-th attribute's contribution to the classifiers decision $f(x)$ for an instance x as the

I.-Y. Song, J. Eder, and T.M. Nguyen (Eds.): DaWaK 2008, LNCS 5182, pp. 273–282, 2008.
© Springer-Verlag Berlin Heidelberg 2008

difference between the classifiers prediction and the classifiers prediction without the knowledge of the value of attribute A_i:

$$PD_i(x) = f(x) - f(x \backslash A_i) \, . \tag{1}$$

Therefore the explanation assigns a number to each attribute. This number can be either positive (the attribute's value speaks in favor of the class value) or negative (the attribute's value speaks against the class value) and the absolute value indicates the intensity of the contribution.

Contribution and Organization of the Paper. The contribution of this paper is threefold. First, we provide a theoretical framework for testing model independent methods that deal with explaining instances. Second, we extensively test the method proposed by Robnik-Šikonja and Kononenko [9] and several of its variations. Third, we use the findings to provide guidelines for further methods and present a theoretical solution that has the potential to solve all the found weaknesses.

In Section 2 we first briefly describe the method and its variations. We continue in Section 3 with the description of the test environment, classifiers and data sets we used for testing. The results are presented in Section 4. In Section 5 we propose and discuss a new method that is based on attribute interactions and overcomes the shortcomings of the original method. We conclude the paper and set guidelines for further work in Section 6.

2   Application Issues and the Variations of the Method

Evaluating the Prediction Difference. The authors of the original method used three different ways of evaluating the difference (1). We can keep the raw probabilities and leave (1) unchanged or we can transform it either to information (2) or the weight of evidence [3], as shown in (3):

$$PD_i^I(x) = \log_2 f(x) - \log_2 f(x \backslash A_i) \, . \tag{2}$$

$$PD_i^W(x) = \log_2 \left(\frac{f(x)}{1 - f(x)} \right) - \log_2 \left(\frac{f(x \backslash A_i)}{1 - f(x \backslash A_i)} \right) \, . \tag{3}$$

Simulating an Unknown Value. In this paper we will test three different ways of simulating the term $f(x \backslash A_i)$ in the initial decomposition (1). As most classifiers are not able to answer the question *What would your prediction be like, if you didn't know the value of attribute A_i?* the authors of [9] devised a way of simulating such a marginal prediction:

$$f(x \backslash A_i) \dot{=} \sum_{a_s \in val(A_i)} p(A_i = a_s) \, f(x \leftarrow A_i = a_s) \, . \tag{4}$$

Equation (4) approximates the term $p(y|x \backslash A_i)$ with the sum of predictions across all possible values a_s of attribute A_i. Each prediction is weighted with the prior probability of A_i having that particular value. We will refer to this approach with *avg*. They also proposed the use of special unknown values or NA's, where the term is approximated by $p(y|x\backslash A_i)\dot{=}p(y|x \leftarrow A_i = NA)$. However, no sufficient test results for this NA approach were presented. Note that this approach can only be used on classifiers that support the use of unknown values. The third possible way of handling the $p(y|x\backslash A_i)$ term is by removing (*rem*) the attribute from the instance and the learning set and repeating the learning process, which yields a new classifier. Then we use the new classifiers' prediction as an approximation of the marginal prediction. This approach was already mentioned in [9] as a way of approximating the marginal prediction of a single or even multiple attributes, but had not been implemented or tested.

Table 1. A short description of the data sets used for testing the method. The columns N_{imp}, N_{rnd}, Type and TE_i contain, for each data set, the number of important attributes, the number of attributes unrelated to the class, the type of attributes and the true contributions of the important attributes, respectively. The true contributions are positive if the class value is 1 and negative otherwise.

Data Set	N_{imp}	N_{rnd}	Type	TE_i	Description
condInd	4	4	binary	$\pm\frac{i}{10}$	$P(C = I_i) = \frac{5+i}{10}, i \in \{1, 2, 3, 4\}; P(C = 1) = \frac{1}{2}$
xor	3	3	binary	$\pm\frac{1}{3}$	$C = 1 \Leftrightarrow (((I_1 + I_2 + I_3) \bmod 2) = 1)$
group	2	2	continuous	$\pm\frac{1}{3}$	$C = 1 \Leftrightarrow (((\lfloor 3 \cdot I_1 \rfloor + \lfloor 3 \cdot I_2 \rfloor + \lfloor 3 \cdot I_3 \rfloor) \bmod 3) = 1)$
cross	2	4	continuous	$\pm\frac{1}{2}$	$C = 1 \Leftrightarrow ((I_1 - 0.5)(I_2 - 0.5) < 0)$
chess	2	2	continuous	$\pm\frac{1}{2}$	$C = 1 \Leftrightarrow (((\lfloor 4 \cdot I_1 \rfloor + \lfloor 4 \cdot I_2 \rfloor) \bmod 2) = 1)$
random4	0	4	continuous	± 0	$P(C = 1) = \frac{1}{2}$, attributes are unrelated to the class
disjunct5	5	2	binary	$\pm\frac{1}{5}$	$C = 1 \Leftrightarrow (I_1 \wedge I_2 \wedge I_3 \wedge I_4 \wedge I_5)$
sphere	3	2	continuous	$\pm\frac{1}{3}$	$C = 1 \Leftrightarrow (\sqrt{(I_1 - 0.5)^2 + (I_2 - 0.5)^2 + (I_3 - 0.5)^2} < 0.5)$
monk1	3	3	discrete	-	$C = 1 \Leftrightarrow (I_1 = I_2) \vee (I_5 = 1)$
monk2	6	0	discrete	-	$C = 1 \Leftrightarrow$ exactly 2 attributes have the value 1
monk3	3	3	discrete	-	$C = 1 \Leftrightarrow ((I_5 = 3 \wedge I_4 = 1) \vee (I_5 \neq 4 \wedge I_2 \neq 3))$

3   Testing Environment

To systematically evaluate the method of Robnik-Šikonja and Kononenko [9] and any further model independent methods for instance explanation we propose the following two criteria[1]:

− the generated explanations must follow the model
− the generated explanations must be intuitive and easy to understand

To explain the first criteria we use the terms explanation quality and model quality. Ideally, the quality of the generated explanation should always reflect the quality of the model that is being explained - the better the model, the better the explanation and vice-versa. Several ways for evaluating the model's quality

---

[1] These criteria measure the theoretical potential of the method. When applying the method in practice, performance criteria, such as scalability, etc. , have to be taken into account.

exist and we used three of them: prediction accuracy (acc), area under the ROC curve (AUC) and the Brier score ($Brier$). Unlike the former two measures, the latter was not used in [9]. Explanation quality depends on the definition of the explanation method, but every well-defined explanation method can be used to produce for each instance a best possible explanation for that instance. We call this explanation a *true explanation* for that instance. The quality of any other explanation generated for that instance can be measured by how much it differs from the true explanation. Given that an explanation is a set of n attribute contributions, we can use a simple Euclidean distance to compare explanations. Note that this principle can be applied to any instance explanation method (not just those based on attribute contributions), we just have to choose an appropriate distance function.

The second criteria is of a more subjective nature, but when dealing with explanation methods based on attribute contributions, we can set some general guidelines. Irrelevant attribute values should have some sort of a zero contribution, attributes that speak in favor of the decision should be distinguished from those that speak against the decision, and the size of an attribute's contribution should indicate its importance in the decision process.

To summarize both criteria: if the method satisfies the first criterion then its explanations will get closer to the true explanations if the underlaying classifier's prediction quality increases. If the method also satisfies the second criterion then the true explanations will be intuitive and easy to understand and whenever we use the method to explain a prediction of a well-preforming classifier we will get an intuitive explanation that is also close to the true explanation. Note that it is necessary to test such methods on artificial data sets, because true explanations of an instance can only be calculated if we know all the concepts behind the data set.

Classifiers and Data Sets. The method was originally tested on five different classifiers [9]: a naive Bayes classifier (NB), a decision tree based on the C4.5 algorithm ($J48$), a k-nearest neighbors classifier (kNN), a support vector machine with a polynomial kernel (SVM) and a single hidden layer artificial neural network (NN). We expanded the set of classifiers with several variations of the bagging and boosting algorithms (three variations for each of the two algorithms, with either NB, $J48$ or kNN as the base classifier). We also added a classifier based on the Random Forests [2] algorithm ($rForest$).

We use 11 different data sets that are designed to test the explanation method's ability to handle attributes connected with and, xor, or, random attributes, conditionally independent attributes, etc. Each of the first eight data sets consists of 2000 examples, 1000 of which were used for learning and 1000 for testing. With such a high number of examples we can avoid the use of different sampling techniques such as cross-validation. The first five data sets listed in Table 1 were used for testing the original method and were designed so that each data set fits a particular classifier [9]. This enables us to test how the quality of the explanation follows the quality of the model. The next three data sets (*random4*, *disjunct5*, *sphere*) were added, because they introduce concepts that are not present in the first five data sets. The well known *Monk*'s data sets were used to test the intuitiveness of the method's explanations. Note that the method explains an instance with contributions of attribute values to a specific class value (in our testing we always explain the prediction regarding the class value 1). When the class is binary (as in all of our data sets, except *group*) then an attribute's contribution for one class value is exactly the opposite for the other class value.

4   Test Results

To test the original method of Robnik-Šikonja and Kononenko [9] we have to define the true explanations and a distance function. We defined the true explanations in the same way as they did (see Table 1). Due to the symmetrical nature of the important attributes we assign equal contributions to all important attributes on all data sets except *condInd* and *disjunct5*, where the important attributes are not symmetrical. For *disjunct5* attributes with value 1 are assigned a positive contribution and attributes with value 0 are assigned a negative contribution. We also use their modified distance between the true explanation and a generated explanation:

$$d_{\exp}(x) = \left(\sum_{i=1}^{a} \left(\frac{1}{2}(TE_i(x) - \frac{PD_i(x)}{\sum_{i=1}^{a} |PD_i(x)|}) \right)^2 \right)^{\frac{1}{2}} . \tag{5}$$

The distance (5) is a modified Euclidean distance. The contributions are first normalized to sum up to 1 so that they can be more easily compared to the true explanations ($TE_i(x)$ is the i-th attribute's contribution in the true explanation for instance x).

4.1   Explanation Quality and Model Quality

The test show that we get similar results with all three ways of evaluating the prediction difference (see Section 2) as was already mentioned in [9]. However, when using information to evaluate the prediction difference, the prediction differences shift towards the left, due to the asymmetrical nature of information. The similarity of the three approaches suggests that it would be better to use just one approach. No significant difference in explanation results has been found between the *avg*, *rem* and the *NA* approach. The only difference is that the *NA* approach can only be used for certain classifiers.

Table 2 shows the test results using the *avg* approach and the weight of evidence. Quality measures *AUC* and *acc* were omitted to save space. Using the Brier score we get the highest correlation between explanation and model quality, so it is the most appropriate measure of quality. This is no surprise, because it measures the quality of the predictions that the explanation method uses for input. The scatter-plots in Figure 1 show the relation between the Brier score and the average distance across all test instances ($\overline{d_{exp}}$) for two data sets. The Brier score and $\overline{d_{exp}}$ appear almost linearly connected on the *chess* data set

Table 2. The prediction performance measure $Brier$ (upper number) and $\overline{d_{exp}}$ (lower number) by various models and different data sets. For each data set we marked the best performing classifier and explanation (*). The $\overline{d_{exp}}$ values are omitted for the *Monk*'s data sets. These data sets were used for the purpose of testing the intuitiveness of the method and their true explanations were not derived.

	cndInd	xor	group	cross	chess	rnd4	disj5	sphr	monk1	monk2	monk3
NB	*0.075	0.252	0.222	0.258	0.251	*0.251	0.008	0.183	0.187	0.226	0.049
	0.056	0.388	0.465	0.453	0.468	0.278	0.217	0.371	-	-	-
J48	0.081	*0.089	0.222	0.237	0.250	*0.251	*0.000	0.177	0.192	0.240	*0.034
	0.170	*0.008	0.354	0.331	0.354	0.431	0.252	0.383	-	-	-
kNN	0.105	0.094	0.007	0.254	0.190	0.264	0.009	0.184	0.210	0.227	0.162
	0.158	0.101	0.082	0.399	0.326	0.303	0.273	0.380	-	-	-
SVM	0.081	0.251	0.140	*0.007	0.246	0.252	*0.000	0.129	*0.003	0.206	0.067
	0.120	0.376	0.222	*0.039	0.422	0.429	0.226	0.370	-	-	-
NN	0.114	0.094	0.018	0.033	0.119	0.252	*0.000	0.243	0.035	*0.144	0.073
	0.272	0.088	0.020	0.148	*0.156	0.291	0.287	0.354	-	-	-
bagNB	*0.075	0.252	0.222	0.260	0.251	*0.251	0.010	0.183	0.187	0.225	0.053
	*0.055	0.384	0.449	0.443	0.474	*0.275	*0.153	0.371	-	-	-
bagJ48	0.084	*0.089	0.123	0.057	0.250	*0.251	*0.000	0.157	0.112	0.209	0.074
	0.169	*0.009	0.101	0.063	0.361	0.308	0.252	*0.346	-	-	-
bagKNN	0.082	0.100	0.009	0.262	0.193	0.252	0.013	0.181	0.182	0.203	0.123
	0.113	0.094	0.109	0.408	0.326	0.291	0.261	0.379	-	-	-
bstNB	0.090	0.252	0.222	0.258	0.251	*0.251	0.009	*0.128	0.203	0.226	0.039
	0.201	0.459	0.464	0.466	0.468	0.278	0.156	0.375	-	-	-
bstJ48	0.103	0.093	0.222	0.247	0.250	*0.251	*0.000	0.201	0.005	0.245	0.070
	0.192	0.071	0.354	0.345	0.354	0.382	0.252	0.349	-	-	-
bstKNN	0.094	0.094	*0.003	0.263	0.207	0.252	0.024	0.191	0.172	0.225	0.074
	0.238	0.082	*0.011	0.422	0.306	0.290	0.252	0.370	-	-	-
rForest	0.079	0.145	0.026	0.041	*0.096	0.286	0.013	0.151	0.073	0.159	0.049
	0.089	0.111	0.116	0.132	0.165	0.297	0.252	0.367	-	-	-

(see Figure 1(a). In fact, the average correlation coefficient for the first five data sets is 0.88, with *condInd* having the lowest correlation coefficient of 0.72. The correlation coefficients for *sphere*, *random4* and *disjunct5* (see Figure 1(b)) are -0.24, -0.09 and -0.20 respectively, which is a clear sign that there is something wrong with the method's performance on these data sets.

On the *random4* data set all classifiers get the near optimal Brier score for this data set (0.25), yet all the $\overline{d_{exp}}$ are high. The reason for such behavior lies in the equation for d_{exp} (5), where the sum of predicted contributions is normalized to 1, regardless of their size. If all the attributes have a negligible contribution, they get increased and a high d_{exp} is calculated when we compare them with their true explanation of 0 (as is the case with *random4*). Note that normalizing the contributions to 1 was somewhat necessary for the original method. Without the normalization the authors would have to provide a different true explanation for each of the three different evaluations of the prediction difference. Also, without normalization, finding the true explanation becomes less trivial, especially for instances where the attributes do not necessarily have symmetrical contributions.

The cause of the faulty explanation on the *disjunct5* data set is the method's inability to handle disjunctive concepts. The terms $f(x)$ and $f(x \backslash A_i)$ in (1) are equal if A_i is unrelated to the class or if at least one other attribute exists, that contributes in the same way as A_i does. The latter happens when at least two important attributes have the value 1. Removing just one of them does not (significantly) change the initial prediction so all the contributions of attribute

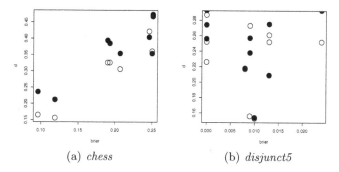

(a) *chess*   (b) *disjunct5*

Fig. 1. The scatter-plots of the relation between a classifiers Brier score and the average quality of its explanations d ($\overline{d_{exp}}$). The color represents the approach used (black - *rem*, white - *avg* approach). In (a) the explanation approaches the true explanation with increasing prediction quality (lower Brier score), while in (b) it does not.

values are subsequently insignificant. The *sphere* data set is similar as we also have disjunction, despite having continuous attributes. It is sometimes sufficient to know just two coordinates to precisely predict whether a point lies inside or outside the sphere. An example of this can be seen in Figure 2(a) where the *SVM* model correctly predicts that the instance with the important attribute values 0.025, 0.031 and 0.901 lies outside the sphere (true class = 0). However, the method assigns zero-contributions to each attribute value. Due to space considerations we omit the detailed results for the *Monk*'s data sets. The method does not produce intuitive explanations for instances from these three data sets, not even on the best performing models. For *monk2* the poor explanations may be due to poor performance of all the models (see Table 2), while on *monk1* and *monk3* the classifiers perform well and poor explanations are due to the data sets' disjunctive concepts.

5   The New Explanation Method Based on the Decomposition of Attribute Interactions

Given a data set with n attributes represented by the set $S = \{A_1, A_2, ..., A_n\}$, an instance x with n attribute values and a classifier, we can choose a class value and obtain a prediction $f(x)$. On the other hand, if we know nothing about an instance then the optimal choice is to predict in accordance with the prior class probability (p_{prior}). The difference $\Delta_S(x) = f(x) - p_{prior}$ between these two predictions is caused by the knowledge about the values of the attributes for x and is exactly what we are trying to explain. This concept can be generalized to an arbitrary subset of attributes:

$$\Delta_Q(x) = f(x \backslash (S - Q)) - p_{prior} \ . \tag{6}$$

The $\Delta_Q(x)$ term in (6) represents the difference between the prediction for x knowing the attribute values of attributes in Q and the prior class probability.

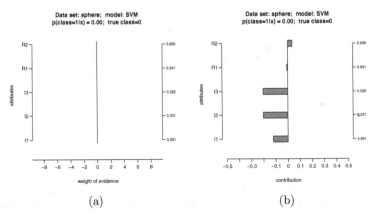

Fig. 2. Two visualizations of the explanation of a *SVM* model decision for an instance from the *sphere* data set. On the left hand side we have the attribute names and on the right hand side their values for this instance. The bars represent the attributes' contributions. In (a) we have the explanation generated by the original method (it assigns zero contributions to every attribute). In (b) we have the explanation generated by the new method It assigns negative contributions to the three important attributes while some noise is captured by the model and results in a very small contribution of the two random attributes.

With this definition we decompose $\Delta_S(x)$ into 2^n parts called interactions (one for each W, where W is a subset of S):

$$\Delta_S = \sum_{W \subseteq S} I_W . \tag{7}$$

By transforming (7) and additionally defining that an interaction of an empty subset is 0, we get our definition of an interaction:

$$I_Q = \Delta_Q - \sum_{W \subset Q} I_W . \tag{8}$$

Equation (8) resembles the inclusion/exclusion principle from the set theory. The study of interactions of attributes also appears in information theory and was initiated with the generalization of mutual information in [7]. Since then, several approaches involving interactions have been taken, including [5], which is also an extensive survey.

We calculate I_Q by taking the sum of all I_W where W is a subset of Q, and by subtracting that sum from the prediction when all attribute values from Q are known. In other words, an interaction I_W is the part of $\Delta_Q(x)$ which represents the contribution of the joint knowledge of all the attribute values in W that can not be found in any subset of W. Unlike the size of a set, an interaction can be either positive or negative. A positive interaction indicates that two (or more) attribute values contribute something extra if they are considered together rather than separately. A negative interaction, on the other hand, indicates that two (or more) attribute values contribute less and implies that there is some redundancy.

Now that we have decomposed the initial prediction difference into interactions, we have to divide the interactions among individual attributes (more precisely: among their values in x):

$$\pi_{A_i}(x) = \sum_{W \subseteq S \wedge A_i \in W} \frac{I_W}{|W|}. \tag{9}$$

As it can be seen in (9), we divided an interaction I_W into $|W|$ equal parts and assigned one part to each attribute in W. The sum of all these values for an attribute is its contribution to the decision.

Preliminary testing of the new method (we used the *rem* approach to generate marginal predictions for every subset) shows that the correlation coefficients for the *sphere, random4* and *disjunct5* increase to 0.78, 0.8 and 0.98, which is a significant increase and an indication that the new explanation method performs much better on these data sets than the original method. The correlation coefficients for the first five data sets do not change significantly. Note that the relation between explanation and model quality might not necessarily be linear, although test show that it is close to linear for both methods.

The main advantage of the new method is that it considers all the interactions between attribute values. The method from [9] is myopic in this respect and fails to capture disjunctive concepts and produces faulty explanations for redundant attributes. A clear example of this can be seen if we compare the explanation's of both methods as shown in 2. Another advantage of the new method is that the difference between the predicted and the prior class probability equals, by definition (8), the sum of all interactions. Interactions are divided among the attributes to form contributions, so the sum of attribute contributions also equals the difference between the predicted and the prior class probability. This implicitly normalizes the contributions, which also resolves the normalization issues we saw on *random4*. The only downside of the new method is that 2^n marginal predictions have to be calculated, where n is the number of attributes, so the method has a high time complexity and will not scale. Note that for any explanation method that does not look at all subsets a worst case instance can be constructed so that the method fails to capture all the contributions. That is why we start with an optimal method, which offers insight into the explanation for individual instances. The framework can later be approximated. Finding a good approximation algorithm extends beyond the scope of this paper, however, it is likely that the approximation algorithm will exploit the fact that interactions of three or more attributes are rare in real life data sets. This heuristic could be used to limit the number of calculated marginal predictions.

6 Conclusion and Further Work

We provided a framework for testing model independent methods for the explanation of a classifiers decision and extensively evaluated the method described in [9]. The method's explanations do not always reflect the quality of the model and its explanations do not always make sense, so we conclude that its performance is not good enough. The weaknesses of the method enabled us to design a new method that eliminates them. The new method's explanations are implicitly normalized, which solves the normalization issues, and the theoretical foundation provides a framework for generating true explanations, which the original method did not provide. Preliminary results show that the new method's explanations are intuitive and closely follow the model on all test data sets.

Theoretical issues, such as the true relation between explanation and model quality, should be further researched. To apply the method in practice we have to develop an approximation method that deals with high time complexity of generating all 2^n Δ terms, either by using heuristics or some other approach. The simulation of unknown values for an arbitrary number of attributes also needs to be addressed.

References

1. Andrews, R., et al.: Survey and Critique of Techniques for Extracting Rules from Trained Artificial Neural Networks. Knowledge-Based Systems 8, 373–389 (1995)
2. Breiman, L.: Random Forests. Machine Learning Journal 42, 5–32 (2001)
3. Good, I.J.: Probability and the Weighing of Evidence. C. Griffin, London (1950)
4. Hamel, L.: Grid Visualization of Support Vector Machines with Unsupervised Learning. In: Proceedings of 2006 IEEE Symposium on Computational Intelligence in Bioinformatics and Computational Biology, Toronto, Canada, pp. 1–8 (2006)
5. Jakulin, A.: Machine Learning Based on Attribute Interactions. PhD Thesis. University of Ljubljana, Faculty of Computer and Information Science (2005)
6. Jakulin, A., et al.: Grid Nomograms for Visualizing Support Vector Machines. In: KDD 2005: Proceeding of the Eleventh ACM SIGKDD International Conference on Knowledge Discovery in Data Mining, Chicago, Illinois, USA, pp. 108–117 (2005)
7. McGill, W.: Multivariate Information Transmission. IEEE Transactions on Information Theory 4, 195–197 (1981)
8. Mozina, M., et al.: Nomograms for Visualization of Naive Bayesian Classifier. In: Boulicaut, J.-F., Esposito, F., Giannotti, F., Pedreschi, D. (eds.) PKDD 2004. LNCS (LNAI), vol. 3202, pp. 337–348. Springer, Heidelberg (2004)
9. Robnik-Šikonja, M., Kononenko, I.: Explaining Classifications for Individual Instances. IEEE TKDE 20(5), 589–600 (2008)
10. Szafron, D., et al.: Visual Explanation of Evidence in Additive Classifiers. In: Proceedings of Innovative Applications of Artificial Intelligence (2006)

Selective Pre-processing of Imbalanced Data for Improving Classification Performance

Jerzy Stefanowski[1] and Szymon Wilk[1,2]

[1] Institute of Computing Science, Poznań University of Technology,
ul. Piotrowo 2, 60–965 Poznań, Poland
jerzy.stefanowski@cs.put.poznan.pl, szymon.wilk@cs.put.poznan.pl
[2] Telfer School of Management, University of Ottawa,
55 Laurier Ave East, K1N 6N5 Ottawa, Canada
wilk@telfer.uottawa.ca

Abstract. In this paper we discuss problems of constructing classifiers from imbalanced data. We describe a new approach to selective pre-processing of imbalanced data which combines local over-sampling of the minority class with filtering difficult examples from the majority classes. In experiments focused on rule-based and tree-based classifiers we compare our approach with two other related pre-processing methods – NCR and SMOTE. The results show that NCR is too strongly biased toward the minority class and leads to deteriorated specificity and overall accuracy, while SMOTE and our approach do not demonstrate such behavior. Analysis of the degree to which the original class distribution has been modified also reveals that our approach does not introduce so extensive changes as SMOTE.

1 Introduction

Discovering classification knowledge from imbalanced data received much research interest in recent years [2,4,11]. A data set is considered to be *imbalanced* if one of the classes (further called a *minority class*) contains much smaller number of examples than the remaining classes (*majority classes*). The minority class is usually of primary interest in a given application. The imbalanced distribution of classes constitutes a difficulty for standard learning algorithms because they are biased toward the majority classes. As a result examples from the majority classes are classified correctly by created classifiers, whereas examples from the minority class tend to be misclassified. As an overall classification accuracy is not appropriate performance measure in this context, such classifiers are evaluated by measures derived from a binary confusion matrix, like *sensitivity* and *specificity*. Sensitivity is defined as the ratio of the number of correctly recognized examples from the minority class (also called positive examples) to the cardinality of this class. On the other hand, specificity corresponds to ability of classifying negative examples of the minority class, so it is defined the ratio of correctly recognized examples from all the majority classes. Receiver Operating Characteristics curve and the area under this curve are also often used to summarize performance of a classifier [2].

I.-Y. Song, J. Eder, and T.M. Nguyen (Eds.): DaWaK 2008, LNCS 5182, pp. 283–292, 2008.

Several methods have been proposed to improve classifiers learned from imbalanced data, for a review see [4,11]. Re-sampling methods that modify the
original class distributions in pre-processing are the most popular approaches.
In particular, methods such as SMOTE, NCR or their combinations, were experimentally shown to work well [1,3,10]. However, some of their properties can be
considered as shortcomings. Focused under-sampling methods, like NCR [6] or
one-side-sampling [5], may remove too many examples from the majority classes.
As a result, improved sensitivity is associated with deteriorated specificity. Random introduction of synthetic examples by SMOTE [3] may be questionable or
difficult to justify in some domains, where it is important to preserve a link
between original data and a constructed classifier (e.g., to justify suggested decisions). Moreover, SMOTE may blindly "over-generalize" the minority area without checking positions of the nearest examples from the majority classes and lead
to overlapping between classes. Finally, the number of synthetic samples has to
be globally parameterized, thus reducing the flexibility of this approach.

Our main research thesis is that focusing on improving sensitivity, which is
typical for many approaches to class imbalance, cannot cause too high decrease of
specificity at the same time. In many problems sufficiently accurate recognition
of the majority classes and preserving the overall accuracy of a classifier at an
acceptable level are still required. Moreover, we hypothesize that it is worth
to develop more flexible approaches based on analyzing local neighborhood of
"difficult" examples rather than using global approaches with fixed parameters.
Following these motivations we introduce our own approach to selective pre-
processing of imbalanced data. It combines filtering of these examples from the
majority classes, which may result in misclassifying some examples from the
minority class, with local over-sampling of examples from the minority class
that are located in "difficult regions" (i.e., surrounded by examples from the
majority classes).

The main aim of this paper is to experimentally evaluate usefulness of our
approach combined with two different learning algorithms. Specifically we use
C4.5 for inducing decision trees and MODLEM [7] for decision rules. We compare
our approach to SMOTE and NCR – two methods that are closely related to
our proposal. The second aim of these experiments is to study how much all
compared methods change the class distribution (the numbers of examples in
the minority and majority classes).

2   Related Works on Focused Re-sampling

Here we discuss only focused re-sampling methods, as they are most related to
our approach and further experiments – for reviews see [2,11]. In [5] one-side-
sampling is used to under-sample the majority classes in a focused way. Noisy and
borderline (i.e., lying on a border between decision classes) examples from the
majority classes are identified using Tomek links and deleted. Another approach
to the focused removal of examples from the majority class is the neighborhood
cleaning rule (NCR) introduced in [6]. It applies the edited nearest neighbor

rule (ENNR) to the majority classes [12]. ENNR first uses the nearest neighbor rule (NNR) to classify examples using a specific number of nearest neighbors (NCR sets it to 3) and then removes incorrectly classified ones. Experiments demonstrated that both above approaches provided better sensitivity than simple random over-sampling. According to [6] NCR performs better than one-side sampling and processes noisy examples more carefully.

The Synthetic Minority Over-sampling Technique (SMOTE) selectively over-samples the minority class by creating new synthetic (artificial) examples [3]. It considers each example from the minority class, finds its k-nearest neighbors from the majority classes, randomly selects j of these neighbors and randomly introduces new artificial examples along the lines joining it with the j-selected neighbors. SMOTE can generate artificial examples with quantitative and qualitative attributes and a number of nearest neighbors depends on how extensive over-sampling is required. Experiments showed that a combination of SMOTE and under-sampling yielded the best AUC among tested techniques [3]. This was confirmed in a comprehensive study [1], where various re-sampling methods were evaluated with different imbalanced data sets. SMOTE was also used in combination with ensemble classifiers as SMOTEBoost [2]. Finally, there are other proposals to focused over-sampling, e.g. Japkowicz used local over-sampling of sub-clusters inside the minority class [4]. Our past research was concerned with the use of rough set theory to detect inconsistent examples in order to remove or relabel them [8]. This technique was combined with rule induction algorithms and experimentally evaluated. Then, in our last paper [9] we preliminary sketched the idea of the selective pre-processing based on ENNR which forms the basis of the approach presented in the next section.

3   An Algorithm for Selective Pre-processing

Our approach uses the the "internal characteristic" of examples to drive their pre-processing. We distinguish between two types of examples – *noisy* and *safe*. Safe examples should be correctly classified by a constructed classifier, while noisy ones are likely to be misclassified and require special processing. We discover the type of an example by applying NNR with the heterogeneous value distance metric (HVDM) [12]. An example is *safe* if it is correctly classified by its k nearest neighbors, otherwise it is *noisy*. We pre-process examples according to their type, and handle noisy examples from the majority classes following the principles of ENNR.

The approach is presented below in details in pseudo-code. We use C for denoting the minority class and O for one majority class (i.e,. for simplicity we group all the majority classes into one). We also use the flags *safe* or *noisy* to indicate appropriate types of examples. Moreover, we introduce two functions: $classify\_knn(x, k)$ and $knn(x, k, c, f)$. The first function classifies x using its k nearest neighbors and returns information whether the classification is correct or not. The second function identifies k nearest neighbors of x and returns a set of those that belong to class c and are flagged as f (the returned set may be empty if none of the k neighbors belongs to c or is flagged as f). Finally, we assume $|\cdot|$ returns the number of items in a set.

```
1: for each x ∈ O ∪ C do
2:     if classify_knn(x, 3) is correct then
3:         flag x as safe
4:     else
5:         flag x as noisy
6: D ← all y ∈ O and flagged as noisy
7: if weak amplification then
8:     for each x ∈ C and flagged as noisy do
9:         amplify x by creating its |knn(x, 3, O, safe)| copies
10: else if weak amplification and relabeling then
11:     for each x ∈ C and flagged as noisy do
12:         amplify x by creating its |knn(x, 3, O, safe)| copies
13:     for each x ∈ C and flagged as noisy do
14:         for each y ∈ knn(x, 3, O, noisy) do
15:             relabel y by changing its class from O to C
16:             D ← D \ {y}
17: else {strong amplification}
18:     for each x ∈ C and flagged as safe do
19:         amplify x by creating its |knn(x, 3, O, safe)| copies
20:     for each x ∈ C and flagged as noisy do
21:         if classify_knn(x, 5) is correct then
22:             amplify x by creating its |knn(x, 3, O, safe)| copies
23:         else
24:             amplify x by creating its |knn(x, 5, O, safe)| copies
25: remove all y ∈ D
```

Our approach consists of two phases. In the first phase (lines 1–5) we identify the type of each example by applying NNR and flagging it accordingly. Following the suggestion from [6] we set the number of the nearest neighbors to 3. Then, in the second phase (lines 6–25) we process examples according to their flags. As we want to preserve all examples from C, we assume only examples from O may be removed (lines 6 and 25, where we apply the principles of ENNR). On the other hand, unlike previously described methods, we want to modify O more carefully, therefore, we preserve all safe examples from this class (NCR removes some of them if they are too close to noisy examples from C). We propose three different techniques for the second phase: *weak amplification, weak amplification and relabeling*, and *strong amplification*. They all involve modification of the minority class, however, the degree and scope of changes varies.

Weak amplification (lines 7–9) is the simplest technique. It focuses on noisy examples from C and amplifies them by adding as many of their copies as there are safe examples from O in their 3-nearest neighborhoods. Thus, the amplification is limited to "difficult" examples from C, surrounded by safe members of O (if there are no such safe neighbors, then an example is not amplified). This increases the "weight" of such difficult examples and enables learning algorithms to capture them, while they could be discarded as noise otherwise.

The second technique – weak amplification and relabeling (lines 10–16) – results from our previous positive experience with changing class labels of selected examples from O [8]. It is also focused on noisy examples from C and extends the first technique with an additional relabeling step. In the first step (lines 11–12) noisy examples from C surrounded by safe examples from O are weakly amplified. In the next step (lines 13–16) noisy examples from O located in the 3-nearest neighborhoods of noisy examples from C are relabeled by changing their class assignment is from O to C (relabeled examples are no longer removed – they are excluded from removal in line 16). Thus, we expand the "cover" around selected noisy examples from C, what further increases their chance of being captured by learned classifiers. Such increasing of density in similar to the technique employed by SMOTE, however, instead of introducing new artificial examples, we use relabeled ones from O.

Strong amplification (lines 17–24) is the most sophisticated technique. It focuses on all examples from C – safe and noisy. First, it processes safe examples from C and amplifies them by adding as many copies as there are safe examples from O in their 3-nearest neighborhoods (lines 17–18). Then, it switches to noisy examples from C (lines 19–23). Each such example is reclassified using an extended neighborhood (i.e., 5 nearest neighbors). If an example is reclassified correctly, it is amplified according to its regular neighborhood (i.e., by adding as many of its copies as there are safe examples from O in its 3-nearest neighborhood), as it should be sufficient to form a "strong" classification pattern. However, if an example is reclassified incorrectly, its amplification is stronger and the number of copies is equal to the number of safe examples from O in the 5-nearest neighborhood. Such more aggressive intervention is caused by the limited number of examples from C in the considered extended neighborhood and it is necessary to strengthen a classification pattern.

4   Experiments

Our approach for selective pre-processing was experimentally compared to NCR and SMOTE. We combined all tested approaches with two learning algorithms – Quinlan's C4.5 for inducing decision trees and MODLEM [7] for decision rules. We focused on these two algorithms because they are both sensitive to the imbalanced distribution of classes. Moreover, MODLEM was introduced by one of the authors and successfully applied to many tasks including our previous research on improving sensitivity of classifiers [8,9].

Both algorithms were run in their unpruned versions to get more precise description of the minority class. To obtain baseline results, we also run them without any prior pre-processing of data. For NCR and our approach the nearest neighborhood was calculated with $k = 3$, as suggested in [6]. Moreover, to find the best over-sampling degree for SMOTE, we tested it with different values from 100% to 600% [3] and selected the best one in terms of obtained sensitivity and

Table 1. Characteristics of evaluated data sets (N – the number of examples, N_A – the number of attributes, C – the minority class, N_C – the number of examples in the minority class, N_O – the number of examples in the majority class, $R_C = N_C/N$ – the ratio of examples in the minority class)

Data set	N	N_A	C	N_C	N_O	R_C
Acl	140	6	with knee injury	40	100	0.29
Breast cancer	286	9	recurrence-events	85	201	0.30
Bupa	345	6	sick	145	200	0.42
Cleveland	303	13	positive	35	268	0.12
Ecoli	336	7	imU	35	301	0.10
Haberman	306	3	died	81	225	0.26
Hepatitis	155	19	die	32	123	0.21
New-thyroid	215	5	hyper	35	180	0.16
Pima	768	8	positive	268	500	0.35

specificity. We implemented MODLEM and all tested pre-processing approaches in WEKA. We also used an implementation of C4.5 available in this environment.

The experiments were carried out on 9 data sets listed in Table 1. They were either downloaded from from the UCI repository or provided by our medical partners (acl). We selected data sets that were characterized by varying degrees of imbalance and that were used in related works (e.g. in [6]). Several data sets originally included more than two classes, however, to simplify calculations we decided to collapse all the majority classes into one.

During experiments we evaluated sensitivity, specificity and overall accuracy – see Tables 2, 3 and 4 respectively (for easier orientation the best result for each data set and classifier is marked with boldface and italics, and the second best with italics). All these measures were estimated in the 10-fold cross validation repeated 5 times. Such a selection of evaluation measures allowed us to observe the degree of trade-off between abilities to recognize the minority and majority classes for the tested pre-processing approaches, what was the primary goal of our experiments. According to it we wanted to examine precisely the decrease of specificity and accuracy at the same time, which is not directly visible in ROC analysis. This is also the reason why we did not report values of AUC. Following the secondary goal of experiments, we observed a degree of changes in class distributions introduced by all approaches – see Table 5. It was evaluated in a single pass of pre-processing. In all tables with results we use *base* for denoting the baseline approach (without any pre-processing), *weak* for weak amplification, *relabel* for weak amplification and relabeling, and *strong* for strong amplification.

In order to compare a performance of pairs of approaches with regard to results on all data sets we used the Wilcoxon Signed Ranks Test (confidence $\alpha = 0.05$). Considering sensitivity (Table 2), the baseline with both learning algorithms was significantly outperformed by all other approaches (only for new-thyroid the baseline with C4.5 performed best). NCR led to the highest gain of sensitivity, especially for haberman (0.386), bupa (0.353) and breast cancer (0.319). NCR improved sensitivity of both learning algorithms, although relative improvements

Table 2. Sensitivity

	MODLEM						C4.5					
Data set	Base	SMOTE	NCR	Weak	Relabel	Strong	Base	SMOTE	NCR	Weak	Relabel	Strong
Acl	0.805	*0.850*	**0.900**	0.830	0.835	0.825	*0.855*	0.840	**0.920**	0.835	0.835	0.850
Breast can.	0.319	0.468	**0.638**	0.437	*0.554*	0.539	0.387	0.463	**0.648**	0.500	*0.576*	0.531
Bupa	0.520	0.737	**0.873**	0.799	*0.838*	0.805	0.491	0.662	**0.755**	0.710	*0.720*	0.700
Cleveland	0.085	*0.245*	**0.343**	0.233	*0.245*	0.235	0.237	0.260	**0.398**	0.343	*0.395*	0.302
Ecoli	0.400	0.632	**0.683**	0.605	*0.643*	0.637	0.580	*0.730*	**0.758**	0.688	0.687	0.690
Haberman	0.240	0.301	**0.626**	0.404	0.468	*0.483*	0.410	0.572	0.608	0.657	**0.694**	*0.660*
Hepatitis	0.383	0.382	**0.455**	0.385	*0.438*	0.437	0.432	0.537	**0.622**	0.513	*0.580*	0.475
New-thyr.	0.812	**0.917**	0.842	0.860	*0.877*	0.865	*0.922*	0.898	0.873	0.897	0.897	*0.913*
Pima	0.485	0.640	**0.793**	0.685	*0.738*	*0.738*	0.601	0.739	**0.768**	0.718	*0.751*	0.715

were better for MODLEM – for C4.5 the opportunity for improvement was limited (the baseline results were better than for MODLEM). Relabeling and strong amplification was the second best after NCR for 7 data sets (all except acl and haberman). Two other variants of our approach – weak and strong amplification – resulted in worse sensitivity than the former one, but still they were better than SMOTE on the majority of data sets. The tested approaches demonstrated a similar performance when combined with C4.5.

Table 3. Specificity

	MODLEM						C4.5					
Data set	Base	SMOTE	NCR	Weak	Relabel	Strong	Base	SMOTE	NCR	Weak	Relabel	Strong
Acl	*0.942*	0.914	0.890	*0.934*	0.922	0.930	*0.940*	0.922	0.898	*0.924*	0.908	0.918
Breast can.	*0.804*	0.657	0.523	*0.710*	0.621	0.606	*0.767*	*0.676*	0.525	0.630	0.609	0.614
Bupa	*0.820*	*0.568*	0.308	0.453	0.473	0.459	*0.775*	*0.611*	0.415	0.524	0.459	0.532
Cleveland	*0.957*	0.887	0.884	*0.934*	0.919	0.927	*0.899*	0.870	0.849	0.877	0.864	*0.887*
Ecoli	*0.969*	0.951	0.924	0.958	0.953	*0.962*	*0.959*	0.921	0.920	0.931	0.916	*0.941*
Haberman	*0.816*	*0.782*	0.658	0.746	0.720	0.713	*0.805*	*0.747*	0.698	0.597	0.565	0.591
Hepatitis	*0.933*	*0.927*	0.894	0.918	0.907	0.908	*0.873*	*0.851*	0.823	0.822	0.807	0.803
New-thyr.	*0.987*	0.986	0.984	**0.990**	*0.990*	0.984	0.973	*0.984*	0.974	0.971	0.972	*0.976*
Pima	*0.856*	*0.778*	0.658	0.774	0.720	0.698	*0.814*	*0.716*	0.656	0.681	0.667	0.687

In case of specificity (Table 3), the baseline for both learning algorithms was significantly better than all other approaches. Specificity attained by NCR was significantly the lowest comparing to other methods. NCR combined with MOD-LEM led to the lowest specificity for all data sets. In particular the highest decreases for occurred for bupa (0.512), breast cancer (0.282), pima (0.200) and haberman (0.152) – these are also the sets for which we noted large increase of sensitivity. Slightly smaller loss of specificity occurred for C4.5. Our approach with weak amplification combined with MODLEM was able to preserve satisfactory specificity for most of the data sets. SMOTE with MODLEM behaved similarly on selected data sets (acl, ecoli, haberman, hepatitis and pima). On the other hand, SMOTE with C4.5 was slightly better than our approach.

Similar observations hold for overall accuracy (Table 4) – the baseline was usually the best, then there were SMOTE and our approach. In particular, the variant with weak amplification combined with MODLEM managed to maintain

Table 4. Overall accuracy [in %]

	MODLEM						C4.5					
Data set	Base	SMOTE	NCR	Weak	Relabel	Strong	Base	SMOTE	NCR	Weak	Relabel	Strong
Acl	90.3	89.6	89.3	*90.4*	89.7	90.0	*91.6*	89.9	*90.4*	89.9	88.7	89.9
Breast can.	*66.0*	60.0	55.6	*62.9*	60.1	58.6	*65.4*	*61.2*	56.1	59.1	59.9	58.9
Bupa	*69.4*	*63.9*	54.5	59.8	57.9	60.4	**65.6**	*63.2*	55.7	60.2	56.8	60.2
Cleveland	**85.6**	81.3	82.1	*85.3*	84.0	84.6	*82.3*	79.9	79.7	81.5	81.0	*81.9*
Ecoli	91.0	91.8	90.0	92.2	92.1	**92.8**	*91.9*	90.1	90.4	90.6	89.2	91.5
Haberman	**66.3**	65.4	64.9	*65.5*	65.2	65.1	*70.1*	70.0	67.4	61.3	59.9	60.9
Hepatitis	**81.9**	*81.5*	80.4	81.0	81.0	81.2	78.5	**78.9**	78.2	75.9	76.2	73.7
New-thyr.	95.8	**97.4**	96.2	96.9	*97.1*	96.5	96.5	**97.0**	95.8	95.9	96.0	*96.6*
Pima	72.7	*73.0*	70.6	**74.3**	72.7	71.2	*74.0*	*72.4*	69.5	69.4	69.6	69.7

high (i.e., the best or second best) accuracy for 6 data sets (acl, breast cancer, cleveland, ecoli, haberman and pima). SMOTE with C4.5 demonstrated similar behavior also for 6 data sets (breast cancer, bupa, haberman, hepatitis, new-thyroid and pima). Finally, overall accuracy achieved by NCR was the worst.

Table 5. Changes in the class distribution (N_C – the number of examples in the minority class, N_O – the number of examples in the majority class N_O, N_R – the number of relabeled examples, N_A – the number of amplified examples)

	SMOTE		NCR		Weak		Relabel				Strong	
Data set	N_C	N_O	N_C	N_O	N_C	N_O	N_C	N_O	N_R	N_A	N_C	N_O
Acl	120	100	40	83	57	98	59	98	2	17	67	98
Breast cancer	255	201	85	101	173	167	197	167	24	88	253	167
Bupa	290	200	145	81	236	145	271	145	35	91	309	145
Cleveland	245	268	35	198	102	255	110	255	8	67	147	255
Ecoli	210	301	35	266	58	288	69	288	11	23	77	288
Haberman	162	225	81	121	162	182	193	182	31	81	223	182
Hepatitis	64	123	32	90	61	113	68	113	7	29	88	113
New-thyroid	175	180	35	174	40	179	40	179	0	5	47	179
Pima	536	500	268	280	430	409	493	409	63	162	573	409

Analysis of changes in class distributions (Table 5), showed that NCR removed the largest number of examples from the majority class, in particular for breast cancer, bupa, haberman and pima it was about 50% of this class. None of the other approaches was such "greedy". On the other hand, SMOTE increased the cardinality of the minority class on average by 250% by introducing new random artificial examples. For cleveland and ecoli it led to the highest increase of cardinality of the minority class (by 600% and 500% respectively). Our approach was in the middle, only the variant with strong amplification increased cardinality of the minority class for 4 data sets (bupa, breast, haberman and pima) to the level similar to SMOTE. For our approach with relabeling, the number of weakly amplified examples was usually higher than the number of relabeled examples. It may signal that difficult noisy examples of the minority class (located inside the majority class) occurred more frequently than noisy examples on the borderline. This was somehow confirmed by introducing many additional examples by the variant with strong amplification as a result of considering a wider neighborhood for amplification. Also when analyzing the changes in the class distribution ratio, we noticed that usually larger changes led to better classification performance (e.g., for cleveland, breast cancer and haberman).

Finally, we would like to note that in our on-going experiments we also consider two additional pre-processing approaches (random under- and over-sampling) and one additional rule learning algorithm (Ripper), however, due to limited space we can only say that the Ripper's performance was between the performance of MODLEM and C4.5, only for cleveland it gave the highest observed sensitivity at the cost of specificity and overall accuracy. Under- and over-sampling demonstrated performance do not outperform the remaining pre-processing approaches.

5   Conclusions

The main research idea in our study focused on improving sensitivity of the minority class while preserving sufficiently accurate recognition of the majority classes. This was our main motivation to introduce the new selective approach for pre-processing imbalanced data and to carry out its experimental comparison with other related methods. The results of experiments clearly showed that although NCR led to the highest increase of sensitivity of induced classifiers, it was obtained at a cost of significantly decreased specificity and consequently deteriorated overall accuracy. Thus, NCR was not able to satisfy our requirements. Our approach and SMOTE did not demonstrate such behavior and both kept specificity and overall accuracy at an acceptable level.

The analysis of the changes in the class distribution showed that NCR tended to remove too many examples from the majority classes – although it could "clean" borders of minority class, it might deteriorate recognition abilities of induced classifiers for the majority classes. Moreover, the results revealed that SMOTE introduced much more extensive changes than our approach, what might have also resulted in swapping the minority and majority classes. Finally, one can notice that our approach tended to introduce more limited changes in the class distribution without sacrificing the performance gain. Additionally, unlike SMOTE, it did not introduce any artificial examples, but replicated existing ones what may be more acceptable in some applications.

To sum up, according to the experimental results, the classification performance of our approach is slightly better or comparable to SMOTE depending on a learning algorithm. Moreover, it does not require tuning the global degree of over-sampling, but in a more flexible way identifies difficult regions in the minority class and modifies only these examples, which could be misclassified. Thus, we claim our approach is a viable alternative to SMOTE.

Selection of a a particular variant in our approach depends on the accepted trade-off between sensitivity and specificity. If a user prefers classifiers characterized by higher sensitivity, then the variant with relabeling is the best choice. When specificity is more important, then the simplest variant with weak amplification is suggested. Finally, if balance between sensitivity and specificity and good overall accuracy are requested, then the variant with strong amplification is preferred. We should however note that relabeling of examples may not be accepted in some specific applications - in such cases users would have to decide between the two amplification variants.

Acknowledgment. This research supported by the grant N N519 3505 33.

References

1. Batista, G., Prati, R., Monard, M.: A study of the behavior of several methods for balancing machine learning training data. ACM SIGKDD Explorations Newsletter 6(1), 20–29 (2004)
2. Chawla, N.: Data mining for imbalanced datasets: An overview. In: Maimon, O., Rokach, L. (eds.) The Data Mining and Knowledge Discovery Handbook, pp. 853–867. Springer, Heidelberg (2005)
3. Chawla, N., Bowyer, K., Hall, L., Kegelmeyer, W.: SMOTE: Synthetic Minority Over-sampling Technique. J. of Artifical Intelligence Research 16, 341–378 (2002)
4. Japkowicz, N., Stephen, S.: The Class Imbalance Problem: A Systematic Study. Intelligent Data Analysis 6(5), 429–450 (2002)
5. Kubat, M., Matwin, S.: Adressing the curse of imbalanced training sets: one-side selection. In: Proc. of the 14th Int. Conf. on Machine Learning, pp. 179–186 (1997)
6. Laurikkala, J.: Improving identification of difficult small classes by balancing class distribution. Tech. Report A-2001-2, University of Tampere (2001)
7. Stefanowski, J.: The rough set based rule induction technique for classification problems. In: Proc. of the 6th European Conference on Intelligent Techniques and Soft Computing EUFIT 1998, Aaachen, pp. 109–113 (1998)
8. Stefanowski, J., Wilk, S.: Rough sets for handling imbalanced data: combining filtering and rule-based classifiers. Fundamenta Informaticae 72, 379–391 (2006)
9. Stefanowski, J., Wilk, S.: Improving Rule Based Classifiers Induced by MODLEM by Selective Pre-processing of Imbalanced Data. In: Proc. of the RSKD Workshop at ECML/PKDD, Warsaw, pp. 54–65 (2007)
10. Van Hulse, J., Khoshgoftarr, T., Napolitano, A.: Experimental perspectives on learning from imbalanced data. In: Proceedings of ICML 2007, pp. 935–942 (2007)
11. Weiss, G.M.: Mining with rarity: a unifying framework. ACM SIGKDD Explorations Newsletter 6(1), 7–19 (2004)
12. Wilson, D.R., Martinez, T.: Reduction techniques for instance-based learning algorithms. Machine Learning Journal 38, 257–286 (2000)

A Parameter-Free Associative Classification Method

Loïc Cerf[1], Dominique Gay[2], Nazha Selmaoui[2], and Jean-François Boulicaut[1]

[1] INSA-Lyon, LIRIS CNRS UMR5205,
F-69621 Villeurbanne, France
{loic.cerf,jean-francois.boulicaut}@liris.cnrs.fr
[2] Université de la Nouvelle-Calédonie, ERIM EA 3791,
98800 Nouméa, Nouvelle-Calédonie
{dominique.gay,nazha.selmaoui}@univ-nc.nc

Abstract. In many application domains, classification tasks have to tackle multiclass imbalanced training sets. We have been looking for a CBA approach (Classification Based on Association rules) in such difficult contexts. Actually, most of the CBA-like methods are one-vs-all approaches (OVA), i.e., selected rules characterize a class with what is relevant for this class and irrelevant for the union of the other classes. Instead, our method considers that a rule has to be relevant for one class and irrelevant for every other class taken separately. Furthermore, a constrained hill climbing strategy spares users tuning parameters and/or spending time in tedious post-processing phases. Our approach is empirically validated on various benchmark data sets.

Keywords: Classification, Association Rules, Parameter Tuning, Multiclass.

1 Introduction

Association rule mining [1] has been applied not only for descriptive tasks but also for supervised classification based on labeled transactional data [2,3,4,5,6,7,8]. An association rule is an implication of the form $X \Rightarrow Y$ where X and Y are different sets of Boolean attributes (also called items). When Y denotes a single class value, it is possible to look at the predictive power of such a rule: when the conjunction X is observed, it is sensible to predict that the class value Y is true. Such a shift between descriptive and predictive tasks needs for careful selection strategies [9]. [2] identified it as an associative classification approach (also denoted CBA-like methods thanks to the name chosen in [2]). The pioneering proposal in [2] is based on the classical objective interestingness measures for association rules – *frequency* and *confidence* – for selecting candidate classification rules. Since then, the selection procedure has been improved leading to various CBA-like methods [2,4,6,8,10]. Unfortunately, support-confidence-based methods show their limits on imbalanced data sets. Indeed, rules with high confidence can also be negatively correlated. [11,12] propose new methods based

I.-Y. Song, J. Eder, and T.M. Nguyen (Eds.): DaWaK 2008, LNCS 5182, pp. 293–304, 2008.
© Springer-Verlag Berlin Heidelberg 2008

on correlation measure to overcome this weakness. However, when considering a n-class imbalanced context, even a correlation measure is not satisfactory: a rule can be positively correlated with two different classes what leads to conflicting rules. The common problem of these approaches is that they are OVA (one-vs-all) methods, i.e., they split the classification task into n two-class classification tasks (positives vs negatives) and, for each sub-task, look for rules that are relevant in the *positive* class and irrelevant for the union of the other classes. Notice also that the popular emerging patterns (EPs introduced in [13]) and the associated EPs-based classifiers (see e.g. [14] for a survey) are following the same principle. Thus, they can lead to conflicting EPs.

In order to improve state-of-the-art approaches for associative classification when considering multiclass imbalanced training sets, our contribution is twofold. First, we propose an OVE (one-vs-each) method that avoids some of the problems observed with typical CBA-like methods. Indeed, we formally characterize the association rules that can be used for classification purposes when considering that a rule has to be relevant for one class and irrelevant for every other class (instead of being irrelevant for their union). Next, we designed a constrained hill climbing technique that automatically tunes the many parameters (frequency thresholds) that are needed. The paper is organized as follows: Section 2 provides the needed definitions. Section 3 discusses the relevancy of the rules extracted thanks to the algorithm presented in Section 4. Section 5 describes how the needed parameters are automatically tuned. Section 6 provides our experimental study on various benchmark data sets. Section 7 briefly concludes.

2   Definitions

Let $\mathcal{C}$ be the set of classes and n its cardinality. Let $\mathcal{A}$ be the set of Boolean attributes. An object o is defined by the subset of attributes that holds for it, i.e., $o \subseteq \mathcal{A}$. The data in Table 1 illustrate the various definitions. It provides 11 classified objects $(o_k)_{k \in 1...11}$. Each of them is described with some of the 6 attributes $(a_l)_{l \in 1...6}$ and belongs to one class $(c_i)_{i \in 1...3}$. This is a toy labeled transactional data set that can be used to learn an associative classifier that may predict the class value among the three possible ones.

2.1   Class Association Rule

A *Class Association Rule* (CAR) is an ordered pair $(X, c) \in 2^{\mathcal{A}} \times \mathcal{C}$. X is the *body* of the CAR and c its *target class*.

Example 1. In Tab. 1, $(\{a_1, a_5\}, c_3)$ is a CAR. $\{a_1, a_5\}$ is the body of this CAR and c_3 its target class.

2.2   Per-class Frequency

Given a class $d \in \mathcal{C}$ and a set $\mathcal{O}_d$ of objects belonging to this class, the *frequency* of a CAR (X, c) in d is $|\{o \in \mathcal{O}_d | X \subseteq o\}|$. Since the frequency of (X, c) in d

Table 1. Eleven classified objects

		a_1	a_2	a_3	a_4	a_5	a_6	c_1	c_2	c_3
$\mathcal{O}_{c_1}$	o_1	•		•	•			•		
	o_2	•	•	•	•	•		•		
	o_3	•	•		•			•		
	o_4	•	•	•		•		•		
	o_5	•	•	•		•		•		
$\mathcal{O}_{c_2}$	o_6		•		•		•		•	
	o_7	•	•		•				•	
	o_8	•	•			•			•	
$\mathcal{O}_{c_3}$	o_9	•				•				•
	o_{10}	•				•				•
	o_{11}					•				•

does not depend on c, it is denoted $f_d(X)$. Given $d \in \mathcal{C}$ and a related frequency threshold $\gamma \in \mathbb{N}$, (X, c) is *frequent* (resp. *infrequent*) in d iff $f_d(X)$ is at least (resp. strictly below) γ.

Example 2. In Tab. 1, the CAR $(\{a_1, a_5\}, c_3)$ has a frequency of 3 in c_1, 1 in c_2 and 2 in c_3. Hence, if a frequency threshold $\gamma = 2$ is associated to c_3, $(\{a_1, a_5\}, c_3)$ is frequent in c_3.

With the same notations, the *relative frequency* of a CAR (X, c) in d is $\frac{f_d(X)}{|\mathcal{O}_d|}$.

2.3 Interesting Class Association Rule

Without any loss of generality, consider that $\mathcal{C} = \{c_i | i \in 1 \ldots n\}$. Given $i \in 1 \ldots n$ and $(\gamma_{i,j})_{j \in 1 \ldots n} \in \mathbb{N}^n$ (n per-class frequency thresholds pertaining to each of the n classes), a CAR (X, c_i) is said *interesting* iff:

1. it is frequent in c_i, i.e., $f_{c_i}(X) \geq \gamma_{i,i}$
2. it is infrequent in every other class, i.e., $\forall j \neq i, f_{c_j}(X) < \gamma_{i,j}$
3. any more general CAR is frequent in at least one class different from c_i, i.e., $\forall Y \subset X, \exists j \neq i | f_{c_j}(Y) \geq \gamma_{i,j}$ (*minimal body* constraint).

Example 3. In Tab. 1, assume that the frequency thresholds $\gamma_{3,1} = 4$, $\gamma_{3,2} = 2$, and $\gamma_{3,3} = 2$ are respectively associated to c_1, c_2, and c_3. Although it is frequent in c_3 and infrequent in both c_1 and c_2, $(\{a_1, a_5\}, c_3)$ is not an interesting CAR since $\{a_5\} \subset \{a_1, a_5\}$ and $(\{a_5\}, c_3)$ is neither frequent in c_1 nor in c_2.

3 Relevancy of the Interesting Class Association Rules

3.1 Selecting Better Rules

Constructing a CAR-based classifier means selecting relevant CARs for classification purposes. Hence, the space of CARs is to be split into two: the *relevant*

CARs and the *irrelevant* ones. Furthermore, if $\preceq$ denotes a relevancy (possibly partial) order on the CARs, there should not be a rule r from the relevant CARs and a rule s from the irrelevant CARs s.t. $r \preceq s$. If this never happens, we say that the frontier between relevant and irrelevant CARs is *sound*. Notice that [3] uses the same kind of argument but conserves a one-vs-all perspective.

Using the "Global Frequency + Confidence" Order. The influential work from [2] has been based on a frontier derived from the conjunction of a global frequency (sum of the per-class frequencies for all classes) threshold and a confidence (ratio between the per-class frequency in the target class and the global frequency) threshold. Let us consider the following partial order $\preceq_1$: $\forall(X,Y) \in (2^A)^2, \forall c \in C,$

$$(X,c) \preceq_1 (Y,c) \Leftrightarrow \forall d \in C, \begin{cases} f_c(X) \leq f_d(Y) & \text{if } c = d \\ f_c(X) \geq f_d(Y) & \text{otherwise.} \end{cases}$$

Obviously, $\preceq_1$ is a sensible relevancy order. However, as emphasized in the example below, the frontier drawn by the conjunction of a global frequency threshold and a confidence threshold is not sound w.r.t. $\preceq_1$.

Example 4. Assume a global frequency threshold of 5 and a confidence threshold of $\frac{3}{5}$. In Tab. 1, the CAR $(\{a_3\}, c_1)$ is not (globally) frequent. Thus it is on the irrelevant side of the frontier. At the same time, $(\{a_4\}, c_1)$ is both frequent and with a high enough confidence. It is on the relevant side of the frontier. However, $(\{a_3\}, c_1)$ correctly classifies more objects of O_1 than $(\{a_4\}, c_1)$ and it applies on less objects outside O_1. So $(\{a_4\}, c_1) \preceq_1 (\{a_3\}, c_1)$.

Using the "Emergence" Order. *Emerging patterns* have been introduced in [13]. Here, the frontier between relevancy and irrelevancy relies on a growth rate threshold (ratio between the relative frequency in the target class and the relative frequency in the union of all other classes). As emphasized in the example below, the low number of parameters (one growth rate threshold for each of the n classes) does not support a fine tuning of this frontier.

Example 5. Assume a growth rate threshold of $\frac{8}{5}$. In Tab. 1, the CAR $(\{a_1\}, c_1)$ has a growth rate of $\frac{3}{2}$. Thus it is on the irrelevant side on the frontier. At the same time, $(\{a_2\}, c_1)$ has a growth rate of $\frac{8}{5}$. It is on the relevant side of the frontier. However $(\{a_1\}, c_1)$ correctly classifies more objects of O_1 than $(\{a_2\}, c_1)$ and more clearly differentiates objects in O_1 from those in O_2.

Using the "Interesting" Order. The frontier drawn by the growth rates is sound w.r.t. $\preceq_1$. So is the one related to the so-called interesting CARs. Nevertheless, the latter can be more finely tuned so that the differentiation between two classes is better performed. Indeed, the set of interesting CARs whose target class is c_i is parametrized by n thresholds $(\gamma_{i,j})_{j \in 1...n}$: one frequency threshold $\gamma_{i,i}$ and $n-1$ infrequency thresholds for each of the $n-1$ other classes (instead of one for all of them). Hence, to define a set of interesting CARs targeting every class, n^2 parameters enable to finely draw the frontier between relevancy and irrelevancy.

In practice, this quadratic growth of the number of parameters can be seen as a drawback for the experimenter. Indeed, in the classical approaches presented above, this growth is linear and finding the proper parameters already appears as a dark art. This issue will be solved in Section 5 thanks to an automatic tuning of the frequency thresholds.

3.2   Preferring General Class Association Rules

The minimal body constraint avoids redundancy in the set of interesting CARs. Indeed, it can easily be shown that, for every CAR (X, c), frequent in c and infrequent in every other class, it exists a body $Y \subseteq X$ s.t. (Y, c) is interesting and $\forall Z \subset Y, (Z, c)$ is not. Preferring shorter bodies means focusing on more general CARs. Hence, the interesting CARs are more prone to be applicable to new unclassified objects. Notice that the added-value of the minimal body constraint has been well studied in previous approaches for associative classification (see, e.g., [5,7]).

4   Computing and Using the Interesting Class Association Rules

Let us consider n classes $(c_i)_{i \in 1...n}$ and let us assume that Γ denotes a $n \times n$ matrix of frequency thresholds. The ith line of Γ pertains to the subset of interesting CARs whose target class is c_i. The jth column of Γ pertains to the frequency thresholds in c_j. Given Γ and a set of classified objects, we discuss how to efficiently compute the complete set of interesting CARs.

4.1   Enumeration

The complete extraction of the interesting CARs is performed one target class after another. Given a class $c_i \in \mathcal{C}$, the enumeration strategy of the candidate CARs targetting c_i is critical for performance issues. The search space of the CAR bodies, partially ordered by $\subseteq$, has a lattice structure. It is traversed in a breadth-first way. The two following properties enable to explore only a small part of it without missing any interesting CAR:

1. If (Y, c_i) is not frequent in c_i, neither is any (X, c_i) with $Y \subseteq X$.
2. If (Y, c_i) is an interesting CAR, any (X, c_i) with $Y \subset X$ does not have a minimal body.

Such CAR bodies Y are collected into a prefix tree. When constructing the next level of the lattice, every CAR body in the current level is enlarged s.t. it does not become a superset of a body in the prefix tree. In this way, entire sublattices, which cannot contain bodies of interesting CARs, are ignored.

4.2   Algorithm

Algorithm 1 details how the extraction is performed. `parents` denotes the current level of the lattice (i.e., a list of CAR bodies). `futureParents` is the next level. `forbiddenPrefixes` is the prefix tree of forbidden subsets from which is computed `forbiddenAtts`, the list of attributes that are not allowed to enlarge `parent` (a CAR body in the current level) to give birth to its `children` (bodies in the next level).

```
forbiddenPrefixes ← ∅
parents ← [∅]
while parents ≠ [] do
    futureParents ← ∅
    for all parent ∈ parents do
        forbiddenAtts ← FORBIDDENATTS(forbiddenPrefixes, parent)
        for all attribute > LASTATTRIBUTE(parent) do
            if attribute ∉ forbiddenAtts then
                child ← CONSTRUCTCHILD(parent, attribute)
                if f_{c_i}(child) ≥ γ_{i,i} then
                    if INTERESTING(child) then
                        output (child, c_i)
                        INSERT(child, forbiddenPrefixes)
                    else
                        futureParents ← futureParents ∪ {child}
                else
                    INSERT(child, forbiddenPrefixes)
        parents ← parents \ {parent}
    parents ← futureParents
```

Algorithm 1. EXTRACT(c_i: target class)

4.3   Simultaneously Enforcing Frequency Thresholds in All Classes

Notice that, along the extraction of the interesting CARs targeting c_i, all frequency thresholds $(\gamma_{i,j})_{j \in 1...n}$ are simultaneously enforced. To do so, every CAR body in the lattice is bound to n bitsets related to the $(\mathcal{O}_{c_i})_{i \in 1...n}$. Thus, every bit stands for the match ('1') or the mismatch ('0') of an object and bitwise ANDs enables an incremental and efficient computation of `children`'s bitsets.

Alternatively, the interesting CARs targeting c_i could be obtained by computing the $n-1$ sets of emerging patterns between c_i and every other class c_j (with $\frac{\gamma_{i,i}|\mathcal{O}_j|}{\gamma_{i,j}|\mathcal{O}_i|}$ as a growth rate), one by one, and intersecting them. However, the time complexity of $n^2 - n$ extractions of loosely constrained CARs is far worse than ours (n extractions of tightly constrained CARs). When n is large, it prevents from automatically tuning the parameters with a hill climbing technique.

4.4   Classification

When an interesting CAR is output, we can output its vector of relative frequencies in all classes at no computational cost. Then, for a given unclassified object $o \subseteq \mathcal{A}$, its likeliness to be in the class c_i is quantifiable by $l(o, c_i)$ which is the sum of the relative frequencies in c_i of all interesting CARs applicable to o:

$$l(o, c_i) = \sum_{c \in \mathcal{C}} \left(\sum_{\text{interesting } (X,c) \text{ s.t. } X \subseteq o} \left(\frac{f_{c_i}(X)}{|\mathcal{O}_{c_i}|} \right) \right)$$

Notice that the target class of an interesting CAR does not hide the exceptions it may have in the other classes. The class $c_{\max}$ related to the greatest likeliness value $l(o, c_{\max})$ is where to classify o. $\min_{i \neq \max} \left(\frac{l(o, c_{\max})}{l(o, c_i)} \right)$ quantifies the certainty of the classification of o in the class $c_{\max}$ rather than c_i (the other class with which the confusion is the greatest). This "certainty measure" may be very valuable in cost-sensitive applications.

5 Automatic Parameter Tuning

It is often considered that manually tuning the parameters of an associative classification method, like our CAR-based algorithm, borders the dark arts. Indeed, our algorithm from Sec. 4 requires a n-by-n matrix Γ of input parameters. Fortunately, analyzing the way the interesting CARs apply to the learning set, directly indicates what frequency threshold in Γ should be modified to probably improve the classification. We now describe how to algorithmically tune Γ to obtain a set of interesting CARs that is well adapted to classification purposes. Due to space limitations, the pseudo-code of this algorithm, called fitcare[1], is only available in an associated technical report [15].

5.1 Hill Climbing

The fitcare algorithm tunes Γ following a hill climbing strategy.

Maximizing the Minimal Global Growth Rate. Section 4.4 mentioned the advantages of not restricting the output of a CAR to its target class (its frequencies in every class are valuable as well). With the same argument applied to the global set of CARs, the hill climbing technique, embedded within fitcare, maximizes *global growth rates* instead of other measures (e.g., the number of correctly classified objects) where the loss of information is greater.

Given two classes $(c_i, c_j) \in \mathcal{C}^2$ s.t. $i \neq j$, the *global growth rate* $g(c_i, c_j)$ quantifies, when classifying the objects from $\mathcal{O}_{c_i}$, the confusion with the class c_j. The greater it is, the less confusion made. We define it as follows:

$$g(c_i, c_j) = \frac{\sum_{o \in \mathcal{O}_{c_i}} l(o, c_i)}{\sum_{o \in \mathcal{O}_{c_i}} l(o, c_j)}$$

From a set of interesting CARs, fitcare computes all $n^2 - n$ global growth rates. The maximization of the minimal global growth rate drives the hill climbing, i.e., fitcare tunes Γ so that this rate increases. When no improvement can

---

[1] fitcare is the recursive acronym for fitcare is the class association rule extractor.

be achieved on the smallest global growth rate, `fitcare` attempts to increase the second smallest (while not decreasing the smallest), etc. `fitcare` terminates when a maximum is reached.

Choosing One $\gamma_{i,j}$ to Lower. Instead of a random initialization of the parameters (a common practice in hill climbing techniques), Γ is initialized with high frequency thresholds. The hill climbing procedure only lowers these parameters, one at a time, and by decrements of 1. However, we will see, in Sec. 5.2, that such a modification leads to lowering other frequency thresholds if Γ enters an undesirable state.

The choice of the parameter $\gamma_{i,j}$ to lower depends on the global growth rate $g(c_i, c_j)$ to increase. Indeed, when classifying the objects from $\mathcal{O}_{c_i}$, different causes lead to a confusion with c_j. To discern the primary cause, every class at the denominator of $g(c_i, c_j)$ is evaluated separately:

$$
\begin{pmatrix}
\sum_{o \in \mathcal{O}_{c_i}} \sum_{\text{interesting } (X, c_1) \text{ s.t. } X \subseteq o} \left(\frac{f_{c_j}(X)}{|\mathcal{O}_{c_j}|} \right) \\
\sum_{o \in \mathcal{O}_{c_i}} \sum_{\text{interesting } (X, c_2) \text{ s.t. } X \subseteq o} \left(\frac{f_{c_j}(X)}{|\mathcal{O}_{c_j}|} \right) \\
\vdots \\
\sum_{o \in \mathcal{O}_{c_i}} \sum_{\text{interesting } (X, c_n) \text{ s.t. } X \subseteq o} \left(\frac{f_{c_j}(X)}{|\mathcal{O}_{c_j}|} \right)
\end{pmatrix}
$$

The greatest term is taken as the primary cause for $g(c_i, c_j)$ to be small. Usually it is either the ith term (the interesting CARs targeting c_i are too frequent in c_j) or the jth one (the interesting CARs targeting c_j are too frequent in c_i). This term directly indicates what frequency threshold in Γ should be preferably lowered. Thus, if the ith (resp. j^{th}) term is the greatest, $\gamma_{i,j}$ (resp. $\gamma_{j,i}$) is lowered. Once Γ modified and the new interesting CARs extracted, if $g(c_i, c_j)$ increased, the new Γ is committed. If not, Γ is rolled-back to its previous value and the second most promising $\gamma_{i,j}$ is decremented, etc.

5.2   Avoiding Undesirable Parts of the Parameter Space

Some values for Γ are obviously bad. Furthermore, the hill climbing technique cannot properly work if too few or too many CARs are interesting. Hence, `fitcare` avoids these parts of the parameter space.

Sensible Constraints on Γ. The relative frequency of an interesting CAR targeting c_i should obviously be strictly greater in c_i than in any other class:

$$
\forall i \in 1 \dots n, \forall j \neq i, \frac{\gamma_{i,j}}{|\mathcal{O}_{c_j}|} < \frac{\gamma_{i,i}}{|\mathcal{O}_{c_i}|}
$$

Furthermore, the set of interesting CARs should be *conflictless*, i.e., if it contains (X, c_i), it must not contain (Y, c_j) if $Y \subseteq X$. Thus, an interesting CAR targeting c_i must be strictly more frequent in c_i than any interesting CAR whose target class is not c_i:

$$
\forall i \in 1 \dots n, \forall j \neq i, \gamma_{i,j} < \gamma_{j,j}
$$

Whenever a modification of Γ violates one of these two constraints, every $\gamma_{i,j}$ $(i \neq j)$ in cause is lowered s.t. Γ reaches another sensible state. Then, the extraction of the interesting CARs is performed.

Minimal Positive Cover Rate Constraint. Given a class $c \in \mathcal{C}$, the *positive cover rate* of c is the proportion of objects in $\mathcal{O}_c$ that are covered by at least one interesting CAR targeting c, i.e., $\frac{|\{o \in \mathcal{O}_c | \exists \text{ interesting } (X,c) \text{ s.t. } X \subseteq o\}|}{|\mathcal{O}_c|}$. Obviously, the smaller the positive cover rate of c, the worse the classification in c.

By default, `fitcare` forces the positive cover rates of every class to be 1 (every object is positively covered). Thus, whenever interesting CARs, with c_i as a target class, are extracted, the positive cover rate of c_i is returned. If it is not 1, $\gamma_{i,i}$ is lowered by 1 and the interesting CARs are extracted again.

Notice that `fitcare` lowers $\gamma_{i,i}$ until $\mathcal{O}_{c_i}$ is entirely covered but not more. Indeed, this could bring a disequilibrium between the average number of interesting CARs applying to the objects in the different classes. If this average in $\mathcal{O}_{c_i}$ is much higher than that of $\mathcal{O}_{c_j}$, $g(c_i, c_j)$ would be artificially high and $g(c_j, c_i)$ artificially low. Hence, the hill climbing strategy would be biased.

On some difficult data sets (e.g., containing misclassified objects), it may be impossible to entirely cover some class c_i while verifying $\forall i \in 1 \ldots n, \forall j \neq i, \frac{\gamma_{i,j}}{|\mathcal{O}_{c_j}|} < \frac{\gamma_{i,i}}{|\mathcal{O}_{c_i}|}$. That is why, while initializing Γ, a looser minimal positive cover rate constraint may be decided.

Here is how the frequency thresholds in the ith line of Γ are initialized (every line being independently initialized):

$$\forall j \in 1 \ldots n, \gamma_{i,j} = \begin{cases} |\mathcal{O}_{c_j}| \text{ if } i = j \\ |\mathcal{O}_{c_j}| - 1 \text{ otherwise} \end{cases}$$

The interesting CARs targeting c_i are collected with EXTRACT(c_i). Most of the time, the frequency constraint in c_i is too high for the interesting CARs to entirely cover $\mathcal{O}_{c_i}$. Hence `fitcare` lowers $\gamma_{i,i}$ (and the $(\gamma_{i,j})_{j \in 1 \ldots n}$ s.t. $\forall i \in 1 \ldots n, \forall j \neq i, \frac{\gamma_{i,j}}{|\mathcal{O}_{c_j}|} < \frac{\gamma_{i,i}}{|\mathcal{O}_{c_i}|}$) until $\mathcal{O}_{c_i}$ is entirely covered. If $\gamma_{i,i}$ reaches 0 but the positive cover rate of c_i never was 1, the minimal positive cover rate constraint is loosened to the greatest rate encountered so far. The frequency thresholds related to this greatest rate constitute the ith line of Γ when the hill climbing procedure starts.

6   Experimental Results

The `fitcare` algorithm has been implemented in C++. We performed an empirical validation of its added-value on various benchmark data sets. The LUCS-KDD software library [16] provided the discretized versions of the UCI data sets [17] and a Java implementation of CPAR. Notice that we name the data sets according to Coenen's notation, e.g., the data set "breast.D20.N699.C2" gathers 699 objects described by 20 Boolean attributes and organized in 2 classes. To put the focus on imbalanced data sets, the repartition of the objects into the classes is mentioned as well. Bold faced numbers of objects indicate minor classes, i.e., classes having, at most, half the cardinality of the largest class.

The global and the per-class accuracies of `fitcare` are compared to that of CPAR, one of the best CBA-like methods designed so far. The results, reported in Tab. 2, were obtained after 10-fold stratified cross validations.

Table 2. Experimental results of `fitcare` and comparison with CPAR

Data Sets	Global		Per-class (True Positive rates)	
	fitcare	CPAR	fitcare	CPAR
anneal.D73.N898.C6 8/99/684/0/67/40	92.09	**94.99**	**87.5**/46.46/98.09/-/100/90	17/90.24/99.44/-/100/**96.25**
breast.D20.N699.C2 458/241	82.11	**92.95**	73.36/98.75	98.58/84.68
car.D25.N1728.C4 1210/**384/69/65**	**91.03**	80.79	98.67/**73.43/66.66/78.46**	92.25/58.74/46.03/23.67
congres.D34.N435.C2 267/168	88.96	**95.19**	89.13/88.69	97.36/92.31
cylBands.D124.N540.C2 228/312	**68.7**	68.33	30.7/92.94	61.99/79.99
dermatology.D49.N366.C6 72/112/61/**52/49/20**	77.86	**80.8**	80.55/82.14/62.29/**78.84**/79.59/**85**	80.65/88.86/67.71/77.94/**96.67**/46
glass.D48.N214.C7 70/76/**17/0/13/9/29**	**72.89**	64.1	80/68.42/**23.52**/-/76.92/100/86.2	54.49/65.71/0/-/45/30/90
heart.D52.N303.C5 164/**55/36/35/13**	**55.44**	55.03	81.7/**21.81/19.44/34.28/23.07**	78.68/14.86/**23.26**/23.79/10
hepatitis.D56.N155.C2 **32/123**	**85.16**	74.34	**50**/95.93	45.05/94.37
horsecolic.D85.N368.C2 232/136	81.25	**81.57**	81.46/80.88	85.69/76.74
iris.D19.N150.C3 50/50/50	**95.33**	**95.33**	100/94/92	100/91.57/96.57
nursery.D32.N12960.C5 4320/2/**328**/4266/4044	**98.07**	78.59	100/-/**81.7**/96.78/99.45	77.64/-/21.24/73.53/98.74
pima.D42.N768.C2 500/268	72.78	**75.65**	84.2/51.49	78.52/69.03
ticTacToe.D29.N958.C2 626/332	65.76	**71.43**	63.73/69.57	76.33/63
waveform.D101.N5000.C3 1657/1647/1696	**77.94**	70.66	59.56/88.4/85.73	72.87/69.13/71.67
wine.D68.N178.C3 59/71/48	**95.5**	88.03	96.61/94.36/95.83	85.38/87.26/94.67
Arithmetic Means	**81.3**	79.24	**76.57**	69.08

6.1   2-Class vs. Multiclass Problem

2-class Problem. Five of the seven data sets where CPAR outperforms `fitcare` correspond to well-balanced 2-class problems, where the minimal positive cover constraint has to be loosened for one of the classes (see Sec. 5.2). On the two remaining 2-class data sets, which do not raise this issue (cylBands and hepatitis), `fitcare` has a better accuracy than CPAR.

Multiclass Problem. `fitcare` significantly outperforms CPAR on all the nine multiclass data sets but two – anneal and dermatology – on which `fitcare` lies slightly behind CPAR. On the nursery data, the improvement in terms of global accuracy even reaches 25% w.r.t. CPAR.

6.2   True Positive Rates in Minor Classes

When considering imbalanced data sets, True Positive rates (TPr) are known to better evaluate classification performances. When focusing on the TPr in the minor classes, `fitcare` clearly outperforms `CPAR` in 14 minor classes out of the 20 (bold values). Observe also that the 2-class data sets with a partial positive cover of the largest class have a poor global accuracy but the TPr of the smallest classes often are greater than `CPAR`'s (see breast, horsecolic, ticTacToe).

Compared to `CPAR`, `fitcare` presents better arithmetic means in both the global and the per-class accuracies. However, the difference is much greater with the latter measure. Indeed, as detailed in Sec. 5.1, `fitcare` is driven by the minimization of the confusion between every pair of classes (whatever their sizes). As a consequence, `fitcare` optimizes the True Positive rates. In the opposite, `CPAR` (and all one-vs-all approaches), focusing only on the global accuracy, tends to over-classify in the major classes.

7 Conclusion

Association rules have been extensively studied along the past decade. The CBA proposal has been the first associative classification technique based on a "support-confidence" ranking criterion [2]. Since then, many other CBA-like approaches have been designed. Even if suitable for typical two-class problems, it appears that support and confidence constraints are inadequate for selecting rules in multiclass imbalanced training data sets. Other approaches (see, e.g., [11, 14, 12]) address the problem of imbalanced data sets but show their limits when considering more than 2 classes. We analyzed the limits of all these approaches, suggesting that a common weakness relies on their one-vs-all principle. We proposed a solution to these problems: our associative classification method extracts the so-called interesting class association rules w.r.t. a one-vs-each principle. It computes class association rules that are frequent in the positive class and infrequent in every other class taken separately (instead of their union). Tuning the large number of parameters required by this approach may appear as a bottleneck. Therefore, we designed an automatic tuning method that relies on a hill-climbing strategy. Empirical results have confirmed that our proposal is quite promising for multiclass imbalanced data sets.

Acknowledgments. This work is partly funded by EU contract IST-FET IQ FP6-516169 and by the French contract ANR ANR-07-MDCO-014 Bingo2. We would like to thank an anonymous reviewer for its useful concerns regarding the relevancy of our approach. Unfortunately, because of space restrictions, we could not address them all in this article.

References

1. Agrawal, R., Imielinski, T., Swami, A.N.: Mining Association Rules Between Sets of Items in Large Databases. In: Proceedings of the 1993 ACM SIGMOD International Conference on Management of Data, pp. 207–216. ACM Press, New York (1993)
2. Liu, B., Hsu, W., Ma, Y.: Integrating Classification and Association Rule Mining. In: Proceedings of the Fourth ACM SIGKDD International Conference on Knowledge Discovery and Data Mining, pp. 80–86. AAAI Press, Menlo Park (1998)

3. Bayardo, R., Agrawal, R.: Mining the Most Interesting Rules. In: Proceedings of the Fifth ACM SIGKDD International Conference on Knowledge Discovery and Data Mining, pp. 145–154. ACM Press, New York (1999)

4. Li, W., Han, J., Pei, J.: CMAR: Accurate and Efficient Classification Based on Multiple Class-Association Rules. In: Proceedings of the First IEEE International Conference on Data Mining, pp. 369–376. IEEE Computer Society, Los Alamitos (2001)

5. Boulicaut, J.F., Crémilleux, B.: Simplest Rules Characterizing Classes Generated by Delta-free Sets. In: Proceedings of the Twenty-Second Annual International Conference Knowledge Based Systems and Applied Artificial Intelligence, pp. 33–46. Springer, Heidelberg (2002)

6. Yin, X., Han, J.: CPAR: Classification Based on Predictive Association Rules. In: Proceedings of the Third SIAM International Conference on Data Mining, pp. 369–376. SIAM, Philadelphia (2003)

7. Baralis, E., Chiusano, S.: Essential Classification Rule Sets. ACM Transactions on Database Systems 29(4), 635–674 (2004)

8. Bouzouita, I., Elloumi, S., Yahia, S.B.: GARC: A New Associative Classification Approach. In: Proceedings of the Eight International Conference on Data Warehousing and Knowledge Discovery, pp. 554–565. Springer, Heidelberg (2006)

9. Freitas, A.A.: Understanding the Crucial Differences Between Classification and Discovery of Association Rules – A Position Paper. SIGKDD Explorations 2(1), 65–69 (2000)

10. Wang, J., Karypis, G.: HARMONY: Efficiently Mining the Best Rules for Classification. In: Proceedings of the Fifth SIAM International Conference on Data Mining, pp. 34–43. SIAM, Philadelphia (2005)

11. Arunasalam, B., Chawla, S.: CCCS: A Top-down Associative Classifier for Imbalanced Class Distribution. In: Proceedings of the Twelveth ACM SIGKDD International Conference on Knowledge Discovery and Data Mining, pp. 517–522. ACM Press, New York (2006)

12. Verhein, F., Chawla, S.: Using Significant Positively Associated and Relatively Class Correlated Rules for Associative Classification of Imbalanced Datasets. In: Proceedings of the Seventh IEEE International Conference on Data Mining, pp. 679–684. IEEE Computer Society Press, Los Alamitos

13. Dong, G., Li, J.: Efficient Mining of Emerging Patterns: Discovering Trends and Differences. In: Proceedings of the Fifth ACM SIGKDD International Conference on Knowledge Discovery and Data Mining, pp. 43–52. ACM Press, New York (1999)

14. Ramamohanarao, K., Fan, H.: Patterns Based Classifiers. World Wide Web 10(1), 71–83 (2007)

15. Cerf, L., Gay, D., Selmaoui, N., Boulicaut, J.F.: Technical Notes on fitcare's Implementation. Technical report, LIRIS (April 2008)

16. Coenen, F.: The LUCS-KDD software library (2004), http://www.csc.liv.ac.uk/~frans/KDD/Software/.

17. Newman, D., Hettich, S., Blake, C., Merz, C.: UCI Repository of Machine Learning Databases (1998), http://www.ics.uci.edu/~mlearn/MLRepository.html

The Evaluation of Sentence Similarity Measures

Palakorn Achananuparp[1], Xiaohua Hu[1,2], and Xiajiong Shen[2]

[1] College of Information Science and Technology
Drexel University, Philadelphia, PA 19104
[2] College of Computer and Information Engineering, Hehan University, Henan, China
pkorn@drexel.edu, thu@cis.drexel.edu, shenxj@henu.edu.cn

Abstract. The ability to accurately judge the similarity between natural language sentences is critical to the performance of several applications such as text mining, question answering, and text summarization. Given two sentences, an effective similarity measure should be able to determine whether the sentences are semantically equivalent or not, taking into account the variability of natural language expression. That is, the correct similarity judgment should be made even if the sentences do not share similar surface form. In this work, we evaluate fourteen existing text similarity measures which have been used to calculate similarity score between sentences in many text applications. The evaluation is conducted on three different data sets, TREC9 question variants, Microsoft Research paraphrase corpus, and the third recognizing textual entailment data set.

Keywords: Sentence similarity, Paraphrase Recognition, Textual Entailment Recognition.

1 Introduction

Determining the similarity between sentences is one of the crucial tasks which have a wide impact in many text applications. In information retrieval, similarity measure is used to assign a ranking score between a query and texts in a corpus. Question answering application requires similarity identification between a question-answer or question-question pair [1]. Furthermore, graph-based summarization also relies on similarity measures in its edge weighting mechanism. Yet, computing sentence similarity is not a trivial task. The variability of natural language expression makes it difficult to determine semantically equivalent sentences. While many applications have employed certain similarity functions to evaluate sentence similarity, most approaches only compare sentences based on their surface form. As a result, they fail to recognize equivalent sentences at the semantic level. Another issue pertains to the notions of similarity underlying sentence judgment. Since sentences convey more specific information than documents, a general notion of topicality employed in document similarity might not be appropriate for this task. As Murdock [16] and Metzler et al. [14] point out, there are multiple categories of sentence similarity based on topical specificity. Furthermore, specific notions such as paraphrase or entailment might be needed for certain applications. In this work, we investigate the performance of three classes of measures: word overlap, TF-IDF, and linguistic measures. Each

I.-Y. Song, J. Eder, and T.M. Nguyen (Eds.): DaWaK 2008, LNCS 5182, pp. 305–316, 2008.
© Springer-Verlag Berlin Heidelberg 2008

sentence pair is judged based on the notion that they have identical meaning. For example, two sentences are considered to be similar if they are a paraphrase of each other, that is, they talk about the same event or idea judging from the common principal actors and actions. Next, two sentences are similar if one sentence is a superset of the other. Note that this is also a notion used in textual entailment judgment where directional inference between two sentences is made.

The paper is organized as follows. First, we review the work related to our study. Next, we briefly describe fourteen similarity measures used in the evaluation. In section 4, we explain the experimental evaluation, including evaluation metrics and data sets, used in this study. We discuss about the result and conclude the paper in section 5 and 6, respectively.

2   Related Work

Previous works have been done to evaluate different approaches to measure similarity between short text segments [15]. Specifically, many studies have focused on a comparison between probabilistic approaches and the existing text similarity measures in a sentence retrieval experiment [14][16][3]. For example, Metzler et al. [14] evaluate the performance of statistical translation models in identifying topically related sentences compared to several simplistic approaches such as word overlap, document fingerprinting, and TF-IDF measures. In [15], the effectiveness of lexical matching, language model, and hybrid measures, in computing the similarity between two short queries are investigated. Next, Balasubramanian et al. [3] compare the performance of nine language modeling techniques in sentence retrieval task. Despite their superiority in coping with vocabulary mismatch problem, most probabilistic measures do not significantly outperform existing measures in sentence retrieval task [17]. Although we share the same goal of comparing the performance of sentence similarity measures, there are a few key differences in this study. First, our focus is to evaluate the effectiveness of measures in identifying the similarity between two arbitrary sentences. That is, we perform a pair-wise comparison on a set of sentence pairs. In contrast, sentence retrieval evaluation concentrates on estimating the similarity between the reference query or sentence and the top-N retrieved sentences. Second, the text unit in the previous research is a short text segment such as a short query while we are interested in a syntactically well-formed sentence. Lastly, we conduct the comparative evaluation on public data sets which contain different notions of text similarity, e.g. paraphrase and textual entailment, whereas the prior studies evaluate the effectiveness of measures based on the notion of topical relevance.

3   Sentence Similarity Measures

We describe three classes of measures that can be used for identifying the similarity between sentences. The similarity score produces by these measures has a normalized real-number value from 0 to 1.

3.1 Word Overlap Measures

Word overlap measures is a family of combinatorial similarity measure that compute similarity score based on a number of words shared by two sentences. In this work, we consider four word overlap measures: Jaccard similarity coefficient, simple word overlap, IDF overlap, and phrasal overlap.

3.1.1 Jaccard Similarity Coefficient
Jaccard similarity coefficient is a similarity measure that compares the similarity between two feature sets. When applying to sentence similarity task, it is defined as the size of the intersection of the words in the two sentences compared to the size of the union of the words in the two sentences.

3.1.2 Simple Word Overlap and IDF Overlap Measures
Metzler et al. [14] defined two baseline word overlap measures to compute the similarity between sentence pairs. Simple word overlap fraction ($sim_{overlap}$) is defined as the proportion of words that appear in both sentences normalized by the sentence's length, while IDF overlap ($sim_{overlap,IDF}$) is defined as the proportion of words that appear in both sentences weighted by their inverse document frequency.

3.1.3 Phrasal Overlap Measure
Banerjee and Pedersen [4] introduced the overlap measure based on the Zipfian relationship between the length of phrases and their frequencies in a text collection. Their motivation stems from the fact that a traditional word overlap measure simply treats sentences as a bag of words and does not take into account the differences between single words and multi-word phrases. Since a phrasal n-word overlap is much rarer to find than a single word overlap, thus a phrasal overlap calculation for m phrasal n-word overlaps is defined as a non-linear function displayed in equation 1 below.

$$overlap_{phrase}(s_1, s_2) = \sum_{i=1}^{n} \sum_m i^2 \tag{1}$$

where m is a number of i-word phrases that appear in sentence pairs. Ponzetto and Strube [19] normalized equation 1 by the sum of sentences' length and apply the hyperbolic tangent function to minimize the effect of the outliers. The normalized phrasal overlap similarity measure is defined in equation 2.

$$sim_{overlap,phrase}(s_1, s_2) = \tanh\left(\frac{overlap_{phrase}(s_1, s_2)}{|s_1| + |s_2|}\right) \tag{2}$$

3.2 TF-IDF Measures

Three variations of measures that compute sentence similarity based on term frequency-inverse document frequency (TF-IDF) are considered in this study.

3.2.1 TF-IDF Vector Similarity
Standard vector-space model represents a document as a vector whose feature set consists of indexing words. Term weights are computed from TF-IDF score. For

sentence similarity task, we adopt the standard vector-space approach to compare the similarity between sentence pairs by computing a cosine similarity between the vector representations of the two sentences. A slight modification is made for sentence representation. Instead of using indexing words from a text collection, a set of words that appear in the sentence pair is used as a feature set. This is done to reduce the degree of data sparseness in sentence representation. The standard TF-IDF similarity ($sim_{TFIDF,vector}$) is defined as cosine similarity between vector representation of two sentences. For a baseline comparison, we also include $sim_{TF,vector}$ which utilizes term frequencies as the basic term weights.

3.2.2 Novelty Detection and Identity Measure

Allan et al. [2] proposed TF-IDF measure ($sim_{TFIDF,nov}$) for detecting topically similar sentences in TREC novelty track experiment. The formulation is based on the sum of the product of term frequency and inverse document frequency of words that appear in both sentences. Identity measure ($sim_{identity}$) [9] is another variation of TF-IDF similarity measure originally proposed as a measure for identifying plagiarized documents or co-derivation. It has been shown to perform effectively for such application. Essentially, the identity score is derived from the sum of inverse document frequency of the words that appear in both sentences normalized by the overall lengths of the sentences and the relative frequency of a word between the two sentences. The formulation of the two measures can be found in [14].

3.3   Linguistic Measures

Linguistic measures utilize linguistic knowledge such as semantic relations between words and their syntactic composition, to determine the similarity of sentences. Three major linguistic approaches are evaluated in this work. Note that there are several approaches that utilize word semantic similarity scores to determine similarity between sentences. For a comprehensive comparison of word similarity measures, we recommend the readers to the work done by Budanitsky and Hirst [5]. In this work, we use Lin' universal similarity [12] to compute word similarity scores.

3.3.1 Sentence Semantic Similarity Measures

Li et al. [10] suggest a semantic-vector approach to compute sentence similarity. Sentences are transformed into feature vectors having words from sentence pair as a feature set. Term weights are derived from the maximum semantic similarity score between words in the feature vector and words in a corresponding sentence. In addition, we simplify Li et al.'s measure by only using word similarity scores as term weights. Moreover, we only compute semantic similarity of words within the same part-of-speech class. Then, semantic similarity between sentence pair (sim_{ssv}) is defined as a cosine similarity between semantic vectors of the two sentences.

Another semantic measure, proposed by Mihalcea et al. [16], also combines word semantic similarity scores with word specificity scores. Given two sentences s_1 and s_2, the sentence similarity calculation begins by finding the maximum word similarity score for each word in s_1 with words in the same part of speech class in s_2. Then, apply the same procedure for each word in s_2 with words in the same part of speech class in s_1. The derived word similarity scores are weighted with *idf* scores that belong to the corresponding word. Finally, the sentence similarity formulation is defined in equation 3.

$$sim_{sem,IDF}(s_1,s_2) = \frac{1}{2}\left(\frac{\sum_{w\in\{s_1\}}(\max Sim(w,s_2)\times idf(w))}{\sum_{w\in\{s_1\}} idf(w)} + \frac{\sum_{w\in\{s_2\}}(\max Sim(w,s_1)\times idf(w))}{\sum_{w\in\{s_2\}} idf(w)}\right) \quad (3)$$

where $maxSim(w,s_i)$ is the maximum semantic similarity score of w and words in s_i that belong to the same part-of-speech as w while $idf(w)$ is an inverse document frequency of w. The reason for computing the semantic similarity scores only between words in the same part of speech class is that most WordNet-based measures are unable to compute semantic similarity of cross-part-of-speech words.

Malik et al. [13] have proposed a simplified variation of semantic similarity measure (sim_{sem}) by determining sentence similarity based on the sum of maximum word similarity scores of words in the same part-of-speech class normalized by the sum of sentence's lengths.

3.3.2 Word Order Similarity

Apart from lexical semantics, word composition also plays a role in sentence understanding. Basic syntactic information, such as word order, can provide useful information to distinguish the meaning of two sentences. This is particularly important in many similarity measures where a single word token was used as a basic lexical unit when computing similarity of sentences. Without syntactic information, it is impossible to discriminate sentences that share the similar bag-of-word representations. For example, "the sale manager hits the office worker" and "the office manager hits the sale worker" will be judged as identical sentences because they have the same surface text. However, their meanings are very different.

To utilize word order in similarity calculation, Li et al. [10] defines word order similarity measure as the normalized difference of word order between the two sentences. The formulation for word order similarity is described in equation 4 below:

$$sim_{wo}(s_1,s_2) = 1 - \frac{\|r_1 - r_2\|}{\|r_1 + r_2\|} \quad (4)$$

where r_1 and r_2 is a word order vector of sentence s_1 and s_2, respectively. Word order vector is a feature vector whose feature set comes from words that appear in a sentence pair. The index position of the words in the corresponding sentence are used as term weights for the given word features. That is, each entry in the word order vector r_i is derived from computing a word similarity score between a word feature w with all the words in the sentence s_i. An index position of the word in s_i that gives the maximum word similarity score to w is selected as w's term weight.

3.3.3 The Combined Semantic and Syntactic Measures

Using the notion that both semantic and syntactic information contribute to the understanding of a sentence, Li et al. [10] defined a sentence similarity measure as a linear combination of semantic vector similarity and word order similarity (equation 5). The relative contribution of semantic and syntactic measures is controlled by a coefficient alpha. It has been empirically proved [10][1] that a sentence similarity measure performs the best when semantic measure is weighted more than syntactic measure (alpha = 0.8). This follows the conclusion from a psychological experiment conducted by [10] which emphasizes the role of semantic information over syntactic information in passage understanding. In this study, we also introduce a minor variation of the combined sentence similarity formulation by substituting the semantic vector similarity measure in equation 5 with Malik et al.'s measure (equation 6). The same semantic coefficient value (alpha = 0.8) is applied.

$$sim_{ssv+wo}(s_1,s_2) = \alpha sim_{ssv}(s_1,s_2) + (1-\alpha)sim_{wo}(s_1,s_2) \tag{5}$$

$$sim_{sem+wo}(s_1,s_2) = \alpha sim_{sem}(s_1,s_2) + (1-\alpha)sim_{wo}(s_1,s_2) \tag{6}$$

4   Experimental Evaluation

4.1   Evaluation Criteria

We define six evaluation metrics based on the general notion of positive and negative judgments in information retrieval and text classification as follows.

  Recall is a proportion of correctly predicted similar sentences compared to all similar sentences. *Precision* is a proportion of correctly predicted similar sentences compared to all predicted similar sentences. *Rejection* is a proportion of correctly predicted dissimilar sentences compared to all dissimilar sentences. *Accuracy* is a proportion of all correctly predicted sentences compared to all sentences. F_1 is a uniform harmonic mean of precision and recall. Lastly, we define f_1 as a uniform harmonic mean of rejection and recall. A scoring threshold for similar pairs is defined at 0.5. In this work, we include rejection and f_1 metrics in addition to the standard precision-recall based metrics as it presents another aspect of the performance based on the tradeoff between true positive and true negative judgments.

4.2   Data Sets

Three publicly-available sentence pair data sets are used to evaluate the performance of the sentence similarity measures. The data sets are TREC9 question variants key (*TREC9*) [1], Microsoft Research paraphrase corpus (*MSRP*) [7], and the third recognising textual entailment challenge (*RTE3*) data set [6].

  TREC9 comprises 193 paraphrased pairs used in TREC9 Question Answering experiment. The original questions were taken from a query log of user submitted questions while the paraphrased questions were manually constructed by human assessors. For this study, we randomly pair original questions with non-paraphrased questions to create additional 193 pairs of dissimilar questions. Despite its semi-artificial nature, the data set contains adequate complexity to reflect the variability of nature language expression judging from its various compositions of paraphrasing categories [21].

  MSRP contains 1,725 test pairs automatically constructed from various web new sources. Each sentence pair is judged by two human assessors whether they are semantically equivalent or not. Overall, 67% of the total sentence pairs are judged to be the positive examples. Semantically equivalent sentences may contain either identical information or the same information with minor differences in detail according to the principal agents and the associated actions in the sentences. Sentence that describes the same event but is a superset of the other is considered to be a dissimilar pair. Note that this rule is similar to the one used in text entailment task.

RTE3 consists of 800 sentence pairs in the test set. Each pair comprises two small text segments, which are referred to as *text* and *hypothesis*. The text-hypothesis pairs are collected by human assessors and can be decomposed into four subsets corresponding to the application domains: information retrieval, multi-document summarization, question answering, and information extraction. Similarity judgment between sentence pairs is based on directional inference between text and hypothesis. If the hypothesis can be entailed by the text, then that pair is considered to be a positive example.

From the complexity standpoint, we consider TREC9 to be the lowest complexity data set for its smallest vocabulary space and relatively simple sentence construction. On the other hand, MSRP and RTE3 are considered to be higher complexity data sets due to larger vocabulary space and longer sentence lengths and differences.

Table 1. Summary of three sentence pair data sets used in the experiment

Summary	TREC9	MSRP	RTE3
Number of sentence pairs	386	1,725	800
Number of unique words	252	8,256	5,700
Percentage of unique words covered by WordNet	84.5%	64.5%	70.1%
Average sentence length (in characters)	39.35	115.30	227.87
Average difference in length between two comparing sentences (in characters)	4.32	9.68	132.81
Linguistic complexity	Low	High	High

4.3 Preprocessing

For each data set, we perform a part-of-speech tagging on a sentence using LingPipe libraries (http://alias-i.com/lingpipe/). Next, single word tokens in the sentences are extracted. Then, we remove functional words, such as articles, pronouns, prepositions, conjunctions, auxiliary verbs, modal verbs, and punctuations from the sentence since they do not carry semantic content, but keep the cardinal numbers. Stemming is not applied in the case of linguistic measures to preserve the original meaning of the words. Information about word relations is obtained from WordNet.

5 Results and Discussion

5.1 Question Paraphrase Identification

Table 2 displays the performance of sentence similarity measures on TREC 9 data set. Overall, linguistic measure is the best performer according to F_1, f_1, and accuracy metrics. Within this class of measures, sentence semantic similarity (sim_{sem}) and combined similarity measures (sim_{sem+wo}) perform significantly better than other measures at $p < 0.05$. Phrasal overlap measure is the best performer in word overlap category and standard TF-IDF vector and identity measure perform equally well in TF-IDF measures. Most word order measures and TF-IDF measures exhibit a strong rejection rate. This is to be expected, as the dissimilar pairs in TREC9 contain a relatively small number of word overlaps.

Table 2. Comparison of the performance of sentence similarity measures on TREC9 data set. Results with * indicate that the differences are not statistically significant.

Sentence Similarity Measures	Prec.	Rec.	Rej.	F_1	f_1	Acc.
$sim_{jaccard}$	1	0.383	1	0.554	0.554	0.691
$sim_{overlap}$	0.99	0.362	0.995	0.53	0.532	0.679
$sim_{overlap,IDF}$	0.978	0.233	0.995	0.377	0.378	0.614
$sim_{overlap,phrase}$	1	0.637	1	**0.778**	**0.778**	**0.819**
$sim_{TF,vector}$	0.993	0.689	0.995	0.813	0.814	0.842
$sim_{TFIDF,vector}$	1	0.762	1	**0.865***	**0.865***	**0.881***
$sim_{TFIDF,nov}$	1	0.192	1	0.322	0.322	0.6
$sim_{identity}$	0.98	0.767	0.984	**0.86***	**0.862***	**0.876***
sim_{ssv}	0.67	0.969	0.523	0.79	0.68	0.746
sim_{sem}	0.983	0.912	0.984	**0.946***	**0.947***	**0.948***
$sim_{simsem,IDF}$	0.949	0.575	0.969	0.716	0.722	0.772
sim_{wo}	0.644	0.487	0.731	0.555	0.584	0.609
sim_{ssv+wo}	0.68	0.979	0.539	0.803	0.695	0.759
sim_{sem+wo}	0.963	0.933	0.964	**0.948***	**0.948***	**0.948***

5.2 Paraphrase Recognition

Similar to TREC9 result, linguistic measure is also the overall best performer according to F_1 metric on MSRP data set. Many best linguistic measures perform at an equal F_1 score of 80%. Word overlap and TF-IDF measures perform at a lower F_1 score but the performance gap is very minimal. The performance on f_1 metric, on the other hand, is different from that of TREC9. Due to the fact that most linguistic measures have a very low rejection rate compared to word overlap and TF-IDF measures, they perform poorly on f_1 metric. In this case, the best performer in f_1 category is Jaccard similarity coefficient ($sim_{jaccard}$). A further analysis has shown that several false positive cases are in a "difficult" subset which requires entailment judgment. For example, the following non paraphrase pair produces an average 85% similarity score from the linguistics measures which results in a false positive judgment:

> **Sentence 1:** Russian stocks fell after the arrest last Saturday of Mikhail Khodorkovsky, chief executive of Yukos Oil, on charges of fraud and tax evasion.
> **Sentence 2:** The weekend arrest of Russia's richest man, Mikhail Khodorkovsky, chief executive of oil major YUKOS, on charges of fraud and tax evasion unnerved financial markets.

According to the above example, sentence 1 and sentence 2 describe a parallel event with slightly different detail (generic vs. specific information). Moreover, it requires a semantic inference to relate the two phrases "Russian stocks fell" and "unnerved financial markets." In the cases of superset-subset relationship, all classes of similarity measures fail to make a correct prediction, for example:

> **Sentence 1:** He said the attackers left behind leaflets urging staff at the Ishtar Sheraton to stop working at the hotel and demanding U.S. forces leave Iraq.
> **Sentence 2:** He said the attackers left behind leaflets urging workers at the Ishtar Sheraton to stop working at the hotel.

Table 3. Comparison of the performance of sentence similarity measures on MSRP data set. Results with * indicate that the differences are not statistically significant.

Sentence Similarity Measures	Prec.	Rec.	Rej.	F_1	f_1	Acc.
$sim_{jaccard}$	0.835	0.603	0.763	0.7	**0.674**	0.657
$sim_{overlap}$	0.76	0.678	0.574	0.717	0.622	0.643
$sim_{overlap,IDF}$	0.829	0.325	0.867	0.467	0.473	0.507
$sim_{overlap,phrase}$	0.7	0.892	0.244	**0.785**	0.383	**0.675**
$sim_{TF,vector}$	0.713	0.881	0.298	0.789	0.445	**0.686***
$sim_{TFIDF,vector}$	0.734	0.836	0.398	0.782	**0.539**	**0.69***
$sim_{TFIDF,nov}$	0.858	0.283	0.907	0.426	0.431	0.492
$sim_{identity}$	0.665	1	0	**0.798**	0.01	0.664
sim_{ssv}	0.669	0.989	0.031	**0.798***	0.06	0.668
sim_{sem}	0.674	0.99	0.052	**0.802***	0.099	**0.675***
$sim_{simsem,IDF}$	0.714	0.835	0.337	0.77	0.48	0.668
sim_{wo}	0.681	0.619	0.424	0.648	**0.503**	0.554
sim_{ssv+wo}	0.673	0.983	0.052	**0.799***	0.099	**0.671***
sim_{sem+wo}	0.674	0.977	0.064	**0.8***	0.12	**0.671***

5.3 Textual Entailment Recognition

The performance comparison of sentence similarity measures on RTE3 data is shown in table 4. Overall, linguistic measures outperform other classes of measures in F_1, f_1, and accuracy metrics. Most linguistic measures perform equally well on F_1 metric while the combined sentence semantic and word order measure (sim_{sen+wo}) significantly outperforms other linguistic measures on f_1 and accuracy. Word overlap measures other than phrasal overlap are not viable for text entailment task at all due to low F_1 and f_1 scores. Since sentence length in RTE3 is relatively long compared to the other two data sets, and text length is much greater than hypothesis length, measures that rely on the proportion of word overlap or word distribution are penalized by the unequal sentence lengths. Like MSRP result, linguistic measures produce a significantly lower rejection rate than word overlap and TF-IDF measures. The example of false positive judgment, where no similarity measures are able to correctly reject the above sentence pair, is as follow:

> **Sentence 1 (text):** It's very difficult to get <u>teams from China</u> the right to <u>stay</u> here for a longer period of time.
> **Sentence 2 (hypothesis):** It is difficult to get the right to <u>stay</u> in <u>China</u> for a long period of time.

5.4 The Effect of Word Specificity

There are no clear advantages of word specificity measure such as IDF on the overall performance of sentence similarity measures. Apart from the result of TREC9 evaluation, where an IDF measure, $sim_{TFIDF,vector}$, performs significantly better across all evaluation metrics compared to its non-IDF counterpart, $sim_{TF,vector}$, other IDF-based measures perform poorer on recall, accuracy, F_1, and f_1 metrics. Note that IDF does help improve precision and rejection scores in most measures. This indicates its relative effectiveness in handling false positive cases. However, the loss in recall far

Table 4. Comparison of the performance of sentence similarity measures on RTE3 data set. Results with * indicate that the differences are not statistically significant.

Sentence Similarity Measures	Prec.	Rec.	Rej.	F_1	f_1	Acc.
$sim_{jaccard}$	0.579	0.027	0.979	0.051	0.052	0.491
$sim_{overlap}$	0.565	0.032	0.974	0.06	0.061	0.491
$sim_{overlap,IDF}$	0.6	0.007	0.995	0.014	0.015	0.489
$sim_{overlap,phrase}$	0.638	0.417	0.751	**0.504**	**0.536**	**0.58**
$sim_{TF,vector}$	0.652	0.324	0.812	0.433	0.465	**0.565**
$sim_{TFIDF,vector}$	0.644	0.283	0.836	0.393	0.423	0.553
$sim_{TFIDF,nov}$	0.69	0.141	0.933	0.235	0.246	0.528
$sim_{identity}$	0.539	0.471	0.577	**0.503**	**0.518**	0.523
sim_{ssv}	0.52	0.893	0.133	**0.657***	0.232	0.523
sim_{sem}	0.592	0.727	0.474	**0.653***	0.574	0.604
$sim_{simsem,IDF}$	0.602	0.585	0.592	0.593	0.589	0.589
sim_{wo}	0.569	0.424	0.661	0.486	0.517	0.54
sim_{ssv+wo}	0.532	0.863	0.203	**0.659***	0.328	0.541
sim_{sem+wo}	0.614	0.695	0.541	**0.652***	**0.608**	**0.62**

outweighs the gain in precision and rejection. The results offer a contradicting implication to the previous work [18] where IDF has been empirically proven to be an optimal weight for document retrieval and reinforce the challenge of sentence similarity task. The inclusion of word specificity into the similarity calculation might provide a significant improvement to the task of identifying topically related documents. However, it does have the same effect in the case of paraphrase recognition and entailment identification.

5.5   WordNet Coverage and Linguistic Measures

The effectiveness of linguistic measures depends on a heuristic to compute semantic similarity between words as well as the comprehensiveness of the lexical resource. As WordNet is used as a primary lexical resource in this study, its comprehensiveness is determined by the proportion of words in the text collections that are covered by its knowledge base. In general, a major criticism of WordNet-based similarity measures is in its limited word coverage to handle a large text collection, particularly on the named entities coverage. As indicated in table 1, the percentage of word coverage in WordNet decreases as the size of test collection and vocabulary space increases. Thus, the effectiveness of linguistic measures is likely to be effected because word-to-word similarity calculation will inevitably produce many "misses". One solution is to resort to approaches that utilize other knowledge resources, such as Wikipedia [19] or web search results [20], to derive semantic similarity between words.

6   Conclusions

We have investigated the performance of several classes of sentence similarity measures on multiple sentence pair data sets. In a low-complexity data set, linguistic measures are superior in identifying paraphrases than word overlap and TF-IDF measures.

They are also the best performer in the higher-complexity data sets but the perform-ance gap between measures diminishes depending on the characteristics of the test data. Several factors influence the result. First, MSRP data set contains a high degree of word overlap. Therefore, overlap-based measures are able to produce a reasonable result. Second, linguistics measures perform relatively poor in judging dissimilar pairs in high-complexity data sets. Thus, it adversely affects the overall accuracy. Keep in mind that word overlap and TF-IDF measures tend to reject many dissimilar sentence pairs since their proportion of overlap or the word occurrence is likely to be smaller in high-complexity data sets due to the difference in sentence pair lengths. For "harder" test pairs, such as those in RTE3 or part of MSRP, which require even more specific judgment such as textual entailment, most sentence similarity measures do not pro-duce a satisfactory result.

We are aware of other factors apart from the similarity measure itself which con-tribute to the application performance. Many of which are considered in our future work. For example, instead of representing a sentence as a bag of words, a graph-based representation can be used. Next, different lexical unit that is more meaningful, such as multi-word phrase, can be used as opposed to a single word. Different heuris-tics to compute semantic similarity between words and different lexical resources can be used, etc. Nevertheless, we strongly believe that the comparative evaluation of sentence similarity in this study offers an interesting and useful insight into the per-formance of these similarity measures which are crucial to any sentence-level text applications.

Acknowledgments. This work is supported in part by NSF Career grant (NSF IIS 0448023), NSF CCF 0514679, PA Dept of Health Tobacco Settlement Formula Grant (No. 240205 and No. 240196) and PA Dept of Health Grant (No. 239667).

References

1. Achananuparp, P., Hu, X., Zhou, X., Zhang, X.: Utilizing Sentence Similarity and Ques-tion Type Similarity to Response to Similar Questions in Knowledge-Sharing Community. In: Proceedings of QAWeb 2008 Workshop, Beijing, China (to appear, 2008)
2. Allan, J., Bolivar, A., Wade, C.: Retrieval and novelty detection at the sentence level. In: Proceedings of SIGIR 2003, pp. 314–321 (2003)
3. Balasubramanian, N., Allan, J., Croft, W.B.: A comparison of sentence retrieval tech-niques. In: Proceedings of SIGIR 2007, Amsterdam, The Netherlands, pp. 813–814 (2007)
4. Banerjee, S., Pedersen, T.: Extended gloss overlap as a measure of semantic relatedness. In: Proceedings of IJCAI 2003, Acapulco, Mexico, pp. 805–810 (2003)
5. Budanitsky, A., Hirst, G.: Evaluating WordNet-based measures of semantic distance. Computational Linguistics 32(1), 13–47 (2006)
6. Dagan, I., Glickman, O., Magnini, B.: The PASCAL recognising textual entailment chal-lenge. In: Proceedings of the PASCAL Workshop (2005)
7. Dolan, W., Quirk, C., Brockett, C.: Unsupervised construction of large paraphrase corpora: Exploiting massively parallel new sources. In: Proceedings of the 20th International Con-ference on Computational Linguistics (2004)
8. Fellbaum, C.: WordNet: An Electronic Lexical Database. MIT Press, Cambridge (1998)

9. Hoad, T., Zobel, J.: Methods for identifying versioned and plagiarized documents. Journal of the American Society of Information Science and Technology 54(3), 203–215 (2003)
10. Landauer, T.K., Laham, D., Rehder, B., Schreiner, M.E.: How Well Can Passage Meaning Be Derived without Using Word Order? A Comparison of Latent Semantic Analysis and Humans. In: Proc. 19th Ann. Meeting of the Cognitive Science Soc., pp. 412–417 (1997)
11. Li, Y., McLean, D., Bandar, Z.A., O'Shea, J.D., Crockett, K.: Sentence Similarity Based on Semantic Nets and Corpus Statistics. IEEE Transactions on Knowledge and Data Engineering 18(8), 1138–1150 (2006)
12. Lin, D.: An Information-Theoretic Definition of Similarity. In: Proceedings of the Fifteenth international Conference on Machine Learning, San Francisco, CA, pp. 296–304 (1998)
13. Malik, R., Subramaniam, V., Kaushik, S.: Automatically Selecting Answer Templates to Respond to Customer Emails. In: Proceedings of IJCAI 2007, Hyderabad, India, pp. 1659–1664 (2007)
14. Metzler, D., Bernstein, Y., Croft, W., Moffat, A., Zobel, J.: Similarity measures for tracking information flow. In: Proceedings of CIKM, pp. 517–524 (2005)
15. Metzler, D., Dumais, S.T., Meek, C.: Similarity Measures for Short Segments of Text. In: Amati, G., Carpineto, C., Romano, G. (eds.) ECIR 2007. LNCS, vol. 4425, pp. 16–27. Springer, Heidelberg (2007)
16. Mihalcea, R., Corley, C., Strapparava, C.: Corpus-based and Knowledge-based Measures of Text Semantic Similarity. In: Proceedings of AAAI 2006, Boston (July 2006)
17. Murdock, V.: Aspects of sentence retrieval. Ph.D. Thesis, University of Massachusetts (2006)
18. Papineni, K.: Why inverse document frequency? In: Proceeding of the North American Chapter of the Association for Computational Linguistics, pp. 25–32 (2001)
19. Ponzetto, S.P., Strube, M.: Knowledge Derived From Wikipedia for Computing Semantic Relatedness. Journal of Artificial Intelligence Research 30, 181–212 (2007)
20. Sahami, M., Heilman, T.D.: A web-based kernel function for measuring the similarity of short text snippets. In: Proceedings of WWW 2006, Edinburgh, Scotland, pp. 377–386 (2006)
21. Tomuro, N.: Interrogative Reformulation Patterns and Acquisition of Question Paraphrases. In: Proceedings of the 2nd international Workshop on Paraphrasing, pp. 33–40 (2003)

Labeling Nodes of Automatically Generated Taxonomy for Multi-type Relational Datasets

Tao Li and Sarabjot S. Anand

Department of Computer Science, University of Warwick
Coventry, United Kingdom
{li.tao,s.s.anand}@warwick.ac.uk

Abstract. Automatic Taxonomy Generation organizes a large dataset into a hierarchical structure so as to facilitate people's navigation and browsing actions. To better summarize the content of each node as well as to reflect the distinctiveness between sibling ones, meaningful labels need to be assigned to all the nodes within a derived taxonomy. Current research only focuses on labeling taxonomies that are built from a corpora of textual documents. In this paper we address the problem of labeling taxonomies built for multi-type relational datasets. A novel measure is proposed to quantitatively evaluate the homogeneity of each node and the heterogeneity of its sibling nodes using information-theoretical techniques, based on which the labels of taxonomic nodes are determined. We perform some experiments on a real dataset to prove the effectiveness of our method.

Keywords: Taxonomy, Multi-type, Relational, Hierarchical Clustering.

1 Introduction

Automatic Taxonomy Generation (ATG) is a promising approach to organizing a large dataset into a hierarchical structure so as to facilitate people's navigation and browsing actions in a more efficient way [1]. After the taxonomy has been constructed, meaningful node labels need to be assigned to best summarize the content of each node as well as to reflect its distinctiveness from sibling ones [2]. Most of the current research focuses on labeling taxonomies generated from a large corpora of textual documents. Techniques of Natural Language Processing, Information Retrieval or Computational Linguistics are often applied to pre-process the text and extract keywords/concepts, among which the node labels are selected according to the frequency of keywords or the correlation between concepts [3][4][5]. Nevertheless, to the best of our knowledge, there is no systematic way of labeling taxonomies built for multi-type relational datasets.

Multi-type relational datasets pertain to domains with a number of data types and a multitude of relationships between them [6]. Relational objects, being defined by a set of attributes and links related to other objects, are usually stored in multiple tables of a relational database. Fig. 1 is the simplified schema of a commercial movie database. Each of the movie objects has attributes Title,

I.-Y. Song, J. Eder, and T.M. Nguyen (Eds.): DaWaK 2008, LNCS 5182, pp. 317–326, 2008.
© Springer-Verlag Berlin Heidelberg 2008

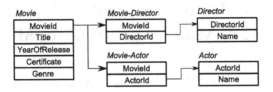

Fig. 1. Example Movie Database Schema

YearOfRelease, Certificate and Genre that are of types *text, numeric, categorical,* and *taxonomy-based* respectively. Moreover, these movies are linked to objects of types Director and Actor via relationships "Movie-Director" and "Movie-Actor". All the directors/actors have the attribute Name of type *text* and are in turn linked to multiple movies. Given the task of generating taxonomy for movie objects, some automated approaches can be employed to (heuristically) search the space of possible taxonomies and find the optimal one. The search may be conducted in a single step as in the case of CobWeb[7] or in two steps where a hierarchical clustering is performed on the dataset followed by a post-processing of the dendrogram, merging nodes in the dendrogram, to derive a multi-split tree structure [4][8]. Once the taxonomy has been learned, labels of the derived taxonomic nodes need to be selected carefully. In our example, the attributes and related objects of movies are usually of different importance in discriminating the nodes within the derived taxonomy: When the taxonomic nodes mainly consists of action movies, users often care about the leading Actors (e.g. Arnold Schwarzenegger, Sylvester Stallone, Jackie Chan); for taxonomic nodes of ethical films, the Director (e.g. Ingmar Bergman, Federico Fellini, Krzysztof Kieslowski) may better summarize the movies allocated to these nodes; while the attribute YearOfRelease and keywords appearing in the Title are good discriminant indicators for the taxonomic nodes consisting of documentaries.

In this paper, we try to answer the following question: *Given a taxonomy that is automatically generated for relational datasets using some unsupervised learning techniques, how can the taxonomic nodes be labeled appropriately so as to summarize the content of each node and reflect the distinctiveness between sibling nodes?* We develop a novel approach to selecting the labels of taxonomic nodes by quantitatively evaluating the homogeneity of each node and the heterogeneity of its sibling nodes. Experiments conducted on a real dataset prove the effectiveness of our method.

The paper is organized as follows: In Section 2 we review some related works. Our new algorithm is explained in Section 3. Experimental results are provided in Section 4. Finally we summarize some conclusions and future works.

2   Related Works

In order to more effectively organize a large document collection and discover knowledge within it, many algorithms for automatically generating taxonomies upon document collections have been developed [3][9][5]. Usually documents are first pre-processed by utilizing techniques in Natural Language Processing, Information Retrieval or Statistics to extract their linguistic features such as keywords or concepts [10]. These keywords or concepts are then used as the labels of the taxonomic nodes to make the taxonomy more understandable. Lawrie and Croft proposed to use a set of topical summary terms as the labels of taxonomic nodes. These topical terms are selected by maximizing the joint probability of their topicality and predictiveness, which is estimated by the statistical language models of the document collection [11]. Kummamuru et al. developed an incremental learning algorithm DisCover to maximize the coverage as well as the distinctiveness of the taxonomy. They used meaningful nouns, adjectives and noun phrases (with necessary pre-processing such as stemming, stop-word elimination or morphological generalization) extracted from the document set as the labels of the derived taxonomic nodes [2]. Krishnapuram and Kummamuru concluded that the set of words that have a high frequency of occurrence within the nodes can be used as the labels [12]. However, no research has been reported on how the taxonomic nodes built from relational datasets can be labeled.

Unlike supervised learning where the classes of data instances are known prior to the learning procedure, the automatic taxonomy generation is more akin to unsupervised learning where the data classes are not available. Labeling taxonomic nodes from structured (propositional or relational) datasets can be viewed as a procedure of learning the features (attributes or related objects) that best discriminate the content of a node from its siblings and hence bears similarity with the idea of choosing decision attributes within Decision Tree Induction [13][14][15]. The pure Information Gain measure based on the node entropy prefers to use attributes with many values as the decision attribute. To reduce such bias, the Gain Ratio was introduced which incorporates the split information to penalize the above over-fitting problem [13][16].

3   Algorithm for Labeling Taxonomic Nodes

In this section, we first investigate the applicability of Kullback-Leibler Divergence within a taxonomy built for relational datasets. The bias of Kullback-Leibler Divergence is then analyzed, leading to the development of a new synthesized criterion. Finally we propose two strategies to determine the node labels using KL Divergence.

The aim of labeling taxonomic nodes is to choose some predictive labels that best summarize the content of each node as well as highlight its distinctiveness from siblings. As shown in Fig. 2, given a set of relational objects $O = \{o_i\}$ ($1 \leq i \leq N$) organized by a taxonomy T and t_s a non-leaf node in T with K child nodes $\{t_{sk}\}$ ($1 \leq k \leq K$), we try to determine labels for each child node t_{sk} based on the attributes and linkages of all relational objects contained in t_{sk} and its sibling nodes. The Kullback-Leibler Divergence [17], which is widely used in probability theory and information theory to measure the divergence of one probability distribution from another, will be adopted in our approach.

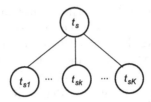

Fig. 2. Example Taxonomy Sub-Tree

We first consider the propositional case where all the objects are only defined by a set of attributes $\{A_r\}$. The attribute-value pairs that best distinguish the objects contained in the current node from those in its sibling nodes are chosen as the node labels. Given an attribute A_r, assuming p_{kr} and q_r are the probability distributions defined over the domain of A_r for objects allocated to nodes t_{sk} and t_s respectively, the KL-divergence of p_{kr} from q_r is defined as:

$$KL_r(p_{kr}, q_r) = \sum_{v_j \in \mathrm{Dom}(A_r)} p_{kr}(v_j) \log \frac{p_{kr}(v_j)}{q_r(v_j)} \tag{1}$$

where $\mathrm{Dom}(A_r)$ is the domain of A_r. The distribution q_r is defined by the weighted sum of its child-node distributions: $q_r = \sum_{k=1}^{K} \frac{N_k}{N} p_{kr}$, where N_k is the number of objects contained in child-node t_{sk} and N is the number of objects contained in parent-node t_s. Because $KL_r(p_{kr}, q_r)$ measures how important the attribute A_r is in distinguishing the members of node t_{sk} from the members of its sibling nodes, the attribute that has the maximum value of KL-divergence provides the basis for naming the node t_{sk}. Although the KL-divergence metric is generally unbounded according to Eq. 1, we can estimate its upper bound in the context of the taxonomy labeling task, as $\log \frac{N}{N_k}$. It is interesting that this upper bound is independent of the attribute domain from which the KL-divergence is computed. More details about the proof are provided in the Appendix A.1.

More generally in the relational case where objects are defined by both attributes and relational links to other objects, the above procedure can be easily extended to calculate the KL-divergences for all attributes as well as relationships. For example, each movie object is related to a set of actors via the relationship *actedBy* (through the table "Movie-Actor"). Given a set of movie objects contained in t_{sk}, the size of the related actor objects is $N_{k.actor}$, which is usually greater than N_k. Then the range of KL-divergence for relationship *actedBy* will be $\left[0, \log \frac{N.actor}{N_k.actor}\right]$ (we ignore the proof here for brevity). Similarly for relationship *directedBy*, the corresponding range is $\left[0, \log \frac{N.director}{N_k.director}\right]$. Since the KL-divergences for different relationships have different ranges, they should be normalized to the interval $[0, 1]$ before being compared.

As shown in Section 4, the KL-divergence, when being used as the criterion of determining node labels, is biased towards the preference of attributes with more unique values over those with fewer values. This phenomenon is similar to the bias in the Information Gain metric used in the procedure of Decision

Tree induction [16]. To address this problem, we introduce the KL-Ratio (KLR) metric as:

$$KLR_r(p_{kr}, q_r) = \frac{KL_r(p_{kr}, q_r)}{Entropy_r(t_s)} \qquad (2)$$

where $Entropy_r(t_s) = -\sum_j \left(\frac{n_j}{N} \log \frac{n_j}{N} \right)$ is the entropy of the attribute A_r within the parent node t_s. Experimental results presented in Section 4 show that the normalization and the entropy-based adjustment together can effectively reduce the bias of the original KL-divergence metric.

With the KLR criterion defined in Eq. 2, we develop two strategies for labeling taxonomic nodes:

- *Strategy 1*: All the sibling nodes use the same attribute to construct their labels, solely differing in the attribute values. The selected attribute has the maximum weighted sum of KLR values across all child-nodes:

$$\underset{A_r}{\arg\max} \sum_k \frac{N_k}{N} KLR_r(p_{kr}, q_r)$$

It is interesting that this strategy is mathematically equivalent to the use of Gain Ratio as the criterion to determine the split attribute in Decision Tree Induction. The detailed proof is provided in the Appendix A.2. However, we must point out the fundamental difference between Decision Tree Induction and our ATG-based approach: the former belongs to supervised learning, i.e. the class information for all training data are already known before the learning procedure; in contrast, our approach is based on the ATG algorithms, which is in essential unsupervised and hence has no prior knowledge about the class information. In summary, Decision Tree Induction and our ATG-based labeling aim at solving similar problems under different motivations and prerequisites.

- *Strategy 2*: Each of the child nodes can be assigned an independent attribute-value pair as its label, which might be different from its siblings. In this case the attribute used to label a child node would be the one with the maximum KLR value for that node:

$$\underset{A_r}{\arg\max} KLR_r(p_{kr}, q_r)$$

In comparison to the Decision Tree Induction, each branch within the derived taxonomy could be labeled using a different attribute.

When being applied in practice, the first strategy is more suitable for users to navigate the taxonomy in a top-down fashion, because all the child nodes that belong to the same parent node use the same attributes with different split values as their labels, so users can easily understand the distinctiveness between these sibling nodes. While the second one might more accurately reflect the content of each node in the taxonomy, because the most representative attribute (with the maximum KLR value) is selected as the node labels.

Table 1. Results of Simulation Experiment ($PopMovieSet = 100$)

Attribute	Number of Unique Values	KL-divergence	Normalized KL-divergence	Entropy	KLR
Title	542.480±16.770	1.132 ± 0.076	0.713 ± 0.028	8.774 ± 0.060	0.081 ± 0.003
Year	16.180± 0.997	0.082 ± 0.027	0.052 ± 0.017	3.243 ± 0.078	0.016 ± 0.005
Certificate	9.580± 0.698	0.046 ± 0.019	0.029 ± 0.012	2.647 ± 0.064	0.011 ± 0.005
Genre	35.720± 2.810	0.195 ± 0.037	0.123 ± 0.023	3.969 ± 0.118	0.031 ± 0.006
Director	225.460±13.237	1.031 ± 0.065	0.650 ± 0.030	6.404 ± 0.218	0.101 ± 0.003
Actor	1064.960±55.599	1.338 ± 0.100	0.842 ± 0.018	9.736 ± 0.107	0.086 ± 0.002

4   Experimental Results

Some experiments were conducted to evaluate our algorithms presented in Section 3. We first measure the bias within the original KL-divergence metric and show to what extent the KLR metric can reduce such bias. Then two strategies for labeling nodes are compared through the simulation of a user locating a given set of movies in the taxonomy. A real movie dataset used by an online DVD retailer, of which the database schema has been shown in Fig. 1, is used in our experiments. After data pre-processing, there are 62,955 movies, 40,826 actors and 9,189 directors. The dataset also includes a genre taxonomy of 186 genres. Additionally, we have 542,738 browsing records from 10,151 users. Based on the user visits, we select 10,000 most popular movies for our analysis.

4.1   Bias within the KL Metric

To determine whether the use of the original KL-divergence metric for selecting the most informative attribute is biased or not, we conducted the following experiments: three sets of 100 movies were randomly chosen to form sibling nodes t_{s1}, t_{s2} and t_{s3}, which shared the common parent node t_s. For each subnode composed of sampled movies, we calculate the KL-divergences for all the attributes. This experiment was repeated 50 times with different random seeds.

Table 1 shows the average number of unique values for each attribute contained in the parent node t_s as well as the mean and the standard deviation of using different labeling criteria with respect to each attribute. The movie titles were processed to extract meaningful nouns, verb and adjectives using WordNet (http://wordnet.princeton.edu/). In Tables 1, the original KL-divergence values for attributes Title, Director and Actor are greater than 1 while the others are far less than 1, which proves the necessity for normalizing the KL-Divergence as suggested in Section 3. Furthermore, the number of unique values for different attributes vary greatly and the original KL-divergence is proportional to this number. The entropy, which acts as the penalty factor in our approach, is also impacted by that because the attributes with more unique values also have higher entropy. As can be seen from the last column of Table 1, KLR that synthesizes the KL-Divergence and the entropy can effectively reduce the bias.

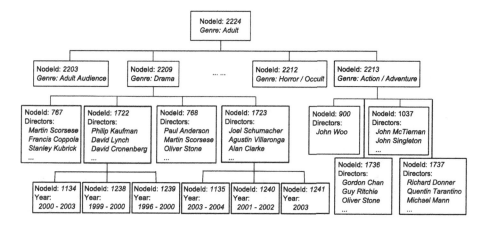

Fig. 3. Labeling Taxonomy for Movie Dataset (Strategy 1)

4.2 Evaluating Labeling Strategies

Figures 3 and 4 show parts of the movie taxonomy that was labeled using two strategies introduced in Section 3. In Fig. 3 all the sibling nodes have the same decision attribute with different values as their labels, while in Fig. 4 each of the sibling nodes may use a different attribute as its own label. It is worth noting that, depending on the utilized techniques of taxonomy generation, some data objects belonging to two different taxonomic nodes might share the same attribute values, which makes the derived node labels overlapped, e.g. both Node 767 and Node 768 in Figures 3 use the director "Martin Scorsese" in their labels. In such case, other attribute values in the node labels can provide useful information to distinguish the content the nodes.

To quantitatively evaluate the goodness of the derived labels, a robot was developed to simulate a user navigating the taxonomy. For a set of randomly selected movies, the robot navigated the taxonomic structure in a top-down fashion to locate the corresponding leaf nodes that contain the given movies. When examining a non-leaf node, the robot uses the node's label to determine which sub-node should be explored in the next iteration. If the target movie object matches the labels of more than one sub-node, all the matched sub-nodes will be explored in a best-match-first order. We use a criterion $Cost$ to measure the time spent in the above exploration procedure, which is defined by the average number of attribute values within the labels of the taxonomic nodes that have been examined by the robot before finding the correct leaf nodes. General speaking, a smaller value of $Cost$ means that the corresponding labeling strategy is preferred. In our experiment, 100 randomly selected movies were used in each run and the experiment was repeated 10 times. The mean and standard deviation of $Cost$ values for two strategies are 378.31 ± 26.97 and 399.88 ± 26.14 respectively. They are statistically significant different, meaning the first labeling strategy is more effective to locate the target objects in the derived taxonomy.

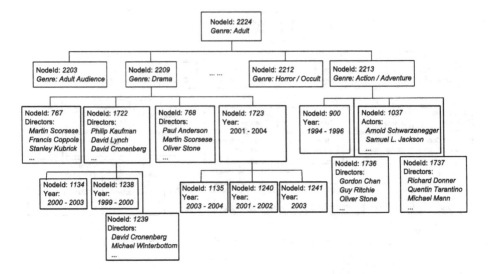

Fig. 4. Labeling Taxonomy for Movie Dataset (Strategy 2)

5   Conclusions and Future Works

ATG techniques can efficiently organize large datasets into hierarchical structures. Usually a bundle of labels will be assigned to taxonomic nodes in order to summarize their respective content and to reflect their distinctiveness among siblings. In this paper we propose a novel approach, based on the evaluation of the homogeneity of each node and heterogeneity of its siblings, to label taxonomies built for multi-type relational datasets. Moreover, we effectively remove the induction bias within the original KL-Divergence metric.

In the future, we will continue investigating other approaches to labeling the automatically generated taxonomy for relational dataset and compare their effectiveness and efficiency. Furthermore, we will study the possibility of using multiple attributes to label the taxonomic nodes.

References

1. Zhao, Y., Karypis, G.: Evaluation of hierarchical clustering algorithms for document datasets. In: Proceedings of ACM CIKM 2002, USA (2002)
2. Kummamuru, K., Lotlikar, R., Roy, S., Singal, K., Krishnapuram, R.: A hierarchical monothetic document clustering algorithm for summarization and browsing search results. In: Proceedings of WWW 2004 (2004)
3. Muller, A., Dorre, J., Gerstl, P., Seiffert, R.: The TaxGen framework: Automating the generation of a taxonomy for a large document collection. In: Proceedings of the 32nd Hawaii International Conference on System Sciences (1999)
4. Chuang, S.L., Chien, L.F.: Towards automatic generation of query taxonomy: A hierarchical query clustering approach. In: Proceedings of ICDM 2002, Washington, USA, pp. 75–82. IEEE Computer Society, Los Alamitos (2002)

5. Cimiano, P., Hotho, A., Staab, S.: Comparing conceptual, partitional and agglomerative clustering for learning taxonomies from text. In: Proceedings of ECAI 2004,
pp. 435–439. IOS Press, Amsterdam (2004)
6. Džeroski, S.: Multi-relational data mining: An introduction. SIGKDD Explorations
Newsletter 5(1), 1–16 (2003)
7. Clerkin, P., Cunningham, P., Hayes, C.: Ontology discovery for the semantic web
using hierarchical clustering. In: Proceedings of Semantic Web Mining Workshop
co-located with ECML/PKDD, Freiburg, Germany (September 2001)
8. Cheng, P.J., Chien, L.F.: Auto-generation of topic hierarchies for web images from
users' perspectives. In: Proceedings of ACM CIKM 2003 (2003)
9. Lawrie, D., Croft, W.B., Rosenberg, A.: Finding topic words for hierarchical summarization. In: Proceedings of ACM SIGIR 2001, pp. 349–357. ACM, New York (2001)
10. Baeza-Yates, R.A., Ribeiro-Neto, B.A.: Modern Information Retrieval. ACM Press
/ Addison-Wesley (1999)
11. Lawrie, D.J., Croft, W.B.: Generating hierarchical summaries for web searches. In:
Proceedings of ACM SIGIR 2003, pp. 457–458. ACM, New York (2003)
12. Krishnapuram, R., Kummamuru, K.: Automatic taxonomy generation: Issues and
possibilities. In: De Baets, B., Kaynak, O., Bilgiç, T. (eds.) IFSA 2003. LNCS,
vol. 2715, pp. 52–63. Springer, Heidelberg (2003)
13. Quinlan, J.R.: Induction of decision trees, vol. 1, pp. 81–106. Kluwer Academic
Publishers, Hingham
14. Quinlan, J.R.: C4.5: programs for machine learning. Morgan Kaufmann Publishers
Inc., San Francisco (1993)
15. Kramer, S., Widmer, G.: Inducing classification and regression trees in first order
logic. In: Relational Data Mining, September 2001, pp. 140–160. Springer, Heidelberg (2001)
16. Mitchell, T.M.: Machine Learning. McGraw-Hill, New York (1997)
17. Kullback, S.: The kullback-leibler distance. The American Statistician 41, 340–341
(1987)

A Appendix

A.1 Proof of the KL-Divergence Bounds

Given the symbols N, N_k, A_r, p_{kr} and q_r defined as in Section 3, it is easy
to see that the KL-divergence for attribute A_r will be maximized when the
distributions p_{kr} and $q_{\backslash k,r}$ have non-zero probabilities for disjoint subsets of
values in A_r, where $q_{\backslash k,r} = \sum_{l \neq k} \frac{N_l}{N-N_k} p_{lr}$. The upper bound is then:

$$\max KL_r(p_{kr}, q_r) = \sum_{v_j \in \mathrm{Dom}(A_r)} p_{kr}(v_j) \log \frac{p_{kr}(v_j)}{q_r(v_j)}$$

$$= \sum_j \left(\frac{n_j}{N_k} \log \frac{\frac{n_j}{N_k}}{\frac{n_j}{N}} \right)$$

$$= \sum_j \left(\frac{n_j}{N_k} \log \frac{N}{N_k} \right)$$

$$= \log \frac{N}{N_k}$$

where $\frac{n_j}{N_k}$ $\left(\sum_j \frac{n_j}{N_k} = 1\right)$ is the frequency that the j-th value of A_r occurs in the objects contained in child-node t_{sk}. Therefore, the range of KL-divergence for node t_{sk} is $\left[0, \log \frac{N}{N_k}\right]$, which is not impacted by the pre-specified attribute A_r.

A.2   Proof of the Equivalence between Information Gain and KL-Based Strategy

Proof. The Information Gain used in the Decision Tree Induction is defined as:

$$
\begin{aligned}
&InfoGain_r(t_s) \\
&= Entropy(t_s) - \sum_k \left(\frac{N_k}{N} \cdot Entropy(t_{sk})\right) \\
&= -\sum_j \left(\frac{n_j}{N} \log \frac{n_j}{N}\right) + \sum_k \left[\frac{N_k}{N} \cdot \sum_j \left(\frac{n_{kj}}{N_k} \log \frac{n_{kj}}{N_k}\right)\right] \\
&= -\sum_j \sum_k \left(\frac{n_{kj}}{N} \log \frac{n_j}{N}\right) + \sum_k \sum_j \left(\frac{n_{kj}}{N} \log \frac{n_{kj}}{N_k}\right) \\
&= \sum_k \sum_j \left(-\frac{n_{kj}}{N} \log \frac{n_j}{N} + \frac{n_{kj}}{N} \log \frac{n_{kj}}{N_k}\right) \\
&= \sum_k \sum_j \left(\frac{n_{kj}}{N} \log \frac{\frac{n_{kj}}{N_k}}{\frac{n_j}{N}}\right) \\
&= \sum_k \sum_j \left(\frac{N_k}{N} \cdot p_{kr}(v_j) \log \frac{p_{kr}(v_j)}{q_r(v_j)}\right) \\
&= \sum_k \left(\frac{N_k}{N} \cdot KL_r(p_{kr}, q_r)\right)
\end{aligned}
$$

The Gain Ratio is defined as the ratio of Information Gain and the entropy of parent node with respect to the attribute A_r [16], which gives:

$$
\begin{aligned}
GainRatio_r(t_s) &= \frac{InfoGain_r(t_s)}{Entropy_r(t_s)} \\
&= \sum_k \left(\frac{N_k}{N} \cdot \frac{KL_r(p_{kr}, q_r)}{Entropy_r(t_s)}\right) \\
&= \sum_k \frac{N_k}{N} KLR_r(p_{kr}, q_r)
\end{aligned}
$$

This completes the proof.   □

Towards the Automatic Construction of Conceptual Taxonomies

Dino Ienco and Rosa Meo

Dipartimento di Informatica, Università di Torino, Italy

Abstract. In this paper we investigate the possibility of an automatic construction of conceptual taxonomies and evaluate the achievable results. The hierarchy is performed by Ward algorithm, guided by Goodman-Kruskal τ as proximity measure. Then, we provide a concise description of each cluster by a keyword representative selected by PageRank.

The obtained hierarchy has the same advantages - both descriptive and operative - of indices on keywords which partition a set of documents with respect to their content.

We performed experiments in a real case - the abstracts of the papers published in ACM TODS in which the papers have been manually classified into the ACM Computing Taxonomy (CT). We evaluated objectively the generated hierarchy by two methods: Jaccard measure and entropy. We obtained good results by both the methods. Finally we evaluated the capability to classify in the categories of the two taxonomies showing that KH provides a greater facility than CT.

1 Introduction

We propose a method based on data mining techniques to detect the significant topics in a document, i.e., a method for determining which concepts in the document are relatively important. The need to determine the importance of a particular concept within a document is motivated by several applications, including information retrieval, authorship automatic determination, similarity metrics for cross-document clustering, automatic indexing and document summarization.

The objective of this paper is to build a hierarchical organization of terms from a set of documents. We want to build this hierarchy in a completely automatic way. On the contrary many approaches, such as in semantic web and in creation of ontologies in a Web Ontology Language like in OWL, rely heavily on human intervention. In our meaning, a concept is represented by a keyword contained and extracted from a text corpus. A conceptual taxonomy is then a hierarchical organization of the keywords (Keyword Hierarchy, KH) such that the keywords at the higher hierarchy levels are representatives of a higher number of other keywords and as a consequence there is a higher number of documents that contain them. Our method has the following merits. It provides automatically at the same time: 1) a concise description of the content area which the documents belongs to and 2) it produces a keyword hierarchy that can be used as an index facility to browse and partition the collection of documents. However, it has the problem of polysemy and synonymy of keywords representing concepts.

I.-Y. Song, J. Eder, and T.M. Nguyen (Eds.): DaWaK 2008, LNCS 5182, pp. 327–336, 2008.

Using keywords to represent concepts is often adopted in literature [4,5,6,8,9,16] because keywords have solid statistics and algorithms are already well-established and efficient.

A different approach [14] induces the description of a concept not in terms of keywords, but by induction of a schematic pattern, describing the concept in terms of a language. Our proposed method will also provide a description of the concepts. As we will see later, we adopt PageRank algorithm [13], which determines the list of frequently occurring terms ordered by their authoritative score.

Much of the literature on this field is related to text categorization, often referred to as automated indexing or authority control and it is the assignment of texts to one or more of a pre-existing set of categories [11,3,1]. In Section 5 we review some of the related works in the field of document categorization and topic taxonomy generation. In this work we focus instead on an automated indexing in a set of categories which are not predefined but are determined automatically by the documents content (the recurrent documents terms).

1.1   A Summary of the Proposed Method

We produced the terms hierarchy by application of Ward hierarchical clustering algorithm to the contained keywords. In order to represent a concept, we extract from documents the recurrent keywords (after Porter Stemming preprocessing), cluster them by Ward algorithm, obtaining a hierarchical keyword tree (KH). Ward algorithm is guided by an objective function that evaluates the best elements (in our case the feature sets) to be agglomerated in a cluster. We coupled Ward algorithm with a measure of distance between features derived by Goodman-Kruskal τ, an associative measure often used in statistics to determine the ability of one category of predict another category. Coupling Ward with τ is an original combination that already produced good results in other domains [10]. This is another contribution of the present method.

We then determine from the group of terms in any hierarchy node the authoritative term that represents the group. In this step we apply PageRank [13] which has been brought to success by its adoption in the popular Google search engine and has been already applied to many domains to detect the authoritative elements in a large collection. Section 3 discusses the proposed algorithm. In Section 4 we report the experiments for the validation of the obtained hierarchy.

We performed experiments in a real case - the abstracts of the papers published in the last 8 years in ACM Transactions on Database Systems Journal in which the papers have been manually classified into the ACM Computing Taxonomy (CT) whose categories were created by computer science experts. We used the ACM CT to validate the automatic term hierarchy produced. ACM CT was created by computer science experts to organize the collection of papers and concisely represents by abstract keywords the computer science knowledge. We succeeded to perform a comparison between ACM CT and KH since the papers whose abstracts we have used were manually classified by the authors themselves into the ACM CT categories. We applied three different measures to perform the comparison: Jaccard measure, Entropy and classification accuracy into the taxonomy categories by a set of classifiers. We obtained considerably superior results by all the classifiers into our KH taxonomy, that means that the discovered taxonomy is much more consistent than ACM CT with the document textual content.

2 Background Knowledge

In this Section, we briefly describe the components on which our framework is based: Goodman-Kruskal τ, Ward algorithm and PageRank.

2.1 Goodman-Kruskal τ as a Cluster Evaluation Measure

Goodman-Kruskal τ can be described as a measure of proportional reduction in the error prediction [7]. It describes the association between two categorical variables - a dependent and an independent one - in terms of the increase in the probability of correct prediction of the category of the dependent variable when information is supplied about the category of the other, independent variable. τ is intended for use with any categorical variable while other measures (eg., γ) are intended for ordinal ones. In the present work we applied τ to boolean features each corresponding to a keyword observed in documents with a frequency higher than a minimum threshold (chosen to eliminate noise).

In order to present the meaning of τ consider the table in Figure 1 whose cells at the intersection of the row I_1 (resp. I_0) with the column D_1 (resp. D_0) contain the number of documents with (resp. without) keyword I in a minimum number of occurrences together with (resp. without) keyword D in a minimum number of occurrences.

	D_1	D_0	Total
I_1	n_{11}	n_{10}	I_1 Total
I_0	n_{01}	n_{00}	I_0 Total
Total	D_1 Total	D_0 Total	Matrix Total

Fig. 1. Cross-classification table of two categorical features I and D

τ determines the predictive power of a keyword I for the prediction of the presence of a keyword D as a function of the prediction error. The error is first computed when we do not have any knowledge on the variable I. This error is denoted by E_D and considers the prediction of category D_j with the relative frequency determined by $\frac{D_j Total}{Total}$ (which reduces as much as possible the error without knowledge on I).

$$E_D = \sum_{j=0,1} (\frac{Total - D_j Total}{Total} \cdot D_j Total) \tag{1}$$

The error in the prediction of D when we know the value of I in any database example is denoted by $E_D|I$ and considers instead the prediction of D_j with the relative frequency $\frac{n_{ij}}{I_i Total}$ (it preserves the class distribution).

$$E_D|I = \sum_{i=0,1} \sum_{j=0,1} (\frac{I_i Total - n_{ij}}{I_i Total} \cdot n_{ij}) \tag{2}$$

The proportional reduction in the prediction error of D given I, here called $\tau_{I \rightsquigarrow D}$, is computed by:

$$\tau_{I \rightsquigarrow D} = \frac{E_D - E_D|I}{E_D} \tag{3}$$

It has an operational meaning and corresponds to the relative reduction in the prediction error in a way that preserves the class distribution. This measure is not symmetrical: $\tau_{I \leadsto D} \neq \tau_{D \leadsto I}$. This could be a problem for the adoption of τ as measure of proximity between keywords to be later used in distance calculations. We solve the problem by using a function of τ as measure of distance $dist$. Taken $d_{I,D} = 1 - \tau_{I \leadsto D}$, the distance between two keywords I, D is defined by $dist(I, D) = \max\{d_{I,D}, d_{D,I}\}$.

Note that the domain of $dist$ is in $(0,1)$ but it increases as the association between the two keywords decreases. Furthermore, it is $dist(D, D) = 0$ for any D.

2.2 Ward's Hierarchical Clustering Method

Ward's hierarchical method is one of the hierarchical clustering methods most used in literature [2]. It is a greedy, agglomerative hierarchical method, that determines a diagram - the dendrogram - that records the sequence of fusions of clusters into larger clusters. At the end of the process, a unique cluster - with all the population - is usually determined. The final number of clusters could be any desired number of cluster k<N (with N the total number of elements).

Ward's method is iterative. An objective function determines the best candidate clusters that will be merged in a new cluster at each iteration of the algorithm (corresponding to a dendrogram level). The objective function is based on the computation of an overall, global measure of goodness of each candidate clustering solution. In turn, the global measure of each solution is given by a summation, over all clusters of the solution, of a measure of cluster cohesion. (Let call it W_i for cluster i.) When two clusters are merged, in the agglomerative hierarchical clustering, the overall global measure increases. All the candidate solutions are computed but the best one is determined by the minimum increase in the objective function. Ward's algorithm, like other clustering hierarchical algorithms, takes in input a matrix containing the distances (`dist`) between any pair of elements (in our case the keywords).

At the core of the method lies the computation of the cluster cohesion of each new cluster and the update of the matrix of distances for the inclusion of the new cluster. The new cluster cohesion is computed on the basis of the cohesions of the clusters that are considered for the merge, as follows

$$W_{ir} = \frac{(|i| + |p|)W_{ip} + (|i| + |q|)W_{iq} - |i|W_{pq}}{|i| + |r|} \tag{4}$$

where p and q represent two clusters, r represents the new cluster formed by the fusion of clusters p and q, while i represents any other cluster. Notation $|i|$ is for the cardinality (number of elements) of cluster i while W_{ij} denotes the cohesion of the cluster that would be obtained if cluster i and j were merged. For the initial clusters composed by I and D single keywords, W_{ID} is initialized by $dist(I, D)$.

2.3 PageRank

In our approach PageRank algorithm is used for ranking the keywords inside a specific cluster. PageRank [13] is a graph-based ranking algorithm already used in the Google search engine and in a great number of unsupervised applications. A good definition of

PageRank and of some of its applications is given in [12]. The basic idea that supports the adoption of a graph-based ranking algorithm is that of voting or recommendation: when a first vertex is connected to a second vertex by a weighted edge the first vertex basically votes for the second one proportionally to the edge weight connecting them. The higher is the sum of the weights obtained by the second vertex by the other vertices the higher is the importance of that vertex in the graph. Furthermore, the importance of a vertex determines the importance of its votes. A Random Walk on a graph describes the probability of moving between the graph vertices. In our case, being graph vertices the keywords, it describes the probability of finding a document with the keywords. Random Walks search the stationary state in the Markov Chain and this situation assigns to each state in the Markov Chain a probability that is the probability of being in that state after an infinite walk on the graph guided by the transition probabilities. Through Random Walks on the graph PageRank determines the authority of a graph vertex, essentially by an iterative algorithm, i.e. collectively by aggregation of the transition probabilities between all the graph vertices. PageRank produces for each graph vertex a score (according to formula 5 that will be discussed below) and orders them by the score value. As a conclusion, it finds a ranking between the keywords.

In our problem, we have an undirected and weighted graph $G = (V, E)$, where V is the set of vertices corresponding to the keywords in a cluster, $E \subseteq V \times V$ is the set of edges between two vertices corresponding to the probability of having a document with both the keywords. For an edge connecting vertices V_a and $V_b \in V$ there is a weight denoted by w_{ab}. PageRank assigns a score to any keyword corresponding to a vertex of the graph: the score at vertex V_a is as greater as is the importance of the vertex. The importance is determined by the vertices to which V_a is connected. PageRank determines iteratively the score for each vertex V_a in the graph as a weighted contribution of all the scores assigned to the vertices V_b connected to V_a, as follows:

$$\mathbf{WP}(V_a) = (1 - d) + d \cdot [\sum_{V_b, V_b \neq V_a} \frac{w_{ba}}{\sum_{V_c, V_c \neq V_b} w_{bc}} \mathbf{WP}(V_b)] \tag{5}$$

where d is a parameter that is set between 0 and 1 (setted to 0.85, the usual value). **WP** is the resulting score vector, whose i-th component is the score associated to vertex V_i. The greater is the score, the greater is the importance of the vertex according to its similarity with the other vertices to which it is connected. This algorithm is used in many applications, particularly in NLP tasks such as Word Sense Disambiguation [12].

3   Algorithm

The algorithm for the automatic generation of the topic taxonomy is reported. *createTaxonomy* takes in input the collection of documents, pre-processes them by function *preProcessData* which performs the following tasks: 1) it selects the subset of documents which are present in CT categories having at least a minimum number of documents and 2) it obtains the list of relevant keywords present in these documents in at least a minimum number of occurrences.

Algorithm 1. createTaxonomy(C)

1: Input: **C**: a collection of text (abstracts) with category in CT.
2: Output: a term taxonomy **KH** with labelled nodes
3: TDlist = preProcessData(C)
4: M = BuildDistanceMatrix(TDlist)
5: – Obtain Term Taxonomy T –
6: T = HCL-Ward(M)
7: KH is initialized with the same structure and nodes of T
8: **for all** Node $n \in$ T **do**
9:     – Label Taxonomy nodes –
10:    term label=LabelNodeByPageRank(n)
11:    assign-label(KH,n,label)
12: **end for**
13: **return** KH

Then, the keywords distance matrix is prepared (by *BuildDistanceMatrix* function) and given to Ward clustering algorithm (*HCL-Ward* function) which produces KH. Finally *LabelNodeByPageRank* calls PageRank algorithm to determine the authoritative score for each keyword in any KH node the top of which is selected as the node label.

4   Evaluation of the Obtained KH Taxonomy

For the evaluation of the proposed method, we run it on short texts - the abstract of 126 papers submitted in the last 8 years to ACM Transactions of Database Systems. In Figure 2 we show a snapshot of the KH term hierarchy automatically generated and shown by the implemented system. An evident, subjective meaning of the hierarchical organization of terms obtained by KH, can immediately be highlighted by humans. In fact, at lower levels of the hierarchy there are specific and very related terms, such as *nearest* and *neighbours* while at higher levels there are more general terms such as *thread, process, distributed, language, relation*. A brief overview of this hierarchy is shown in Figure 2. This hierarchy is functional also to document search and organization. In fact, under the concepts corresponding to nodes at higher hierarchy levels, more documents can be found. As soon as we go deeper in the tree, a lower number of documents is retrieved instead. It can be noticed that some terms, such as *target* were selected repeatedly by PageRank as the representative term in many nested clusters that are in an ancestor-descendant relationship: this means that the concept is a strong authoritative one, chosen as representative from a large set of related terms, contained in nearby clusters. Our term organization presents also a description of any single concept. A concept placed in a node of the hierarchy is explained in more precise terms by the list of related and frequently occurring terms that are shown in the left frame of Figure 2. This list is ordered by the authoritative score of the terms determined by PageRank. For instance, the node labelled by *target* is related to terms *source, exchange, universal schema*.

Now we want to evaluate also objectively the obtained KH taxonomy. We objectively evaluated KH in three ways.

Algorithm 2. preProcessData(C, minWfreq, minCfreq)

Input: **C**: collection of text (abstracts) with category in CT.
Input: **minWfreq**: minimum word frequency in texts.
Input: **minCfreq**: minimum category frequency in terms of texts.
– Each text is identified by a doc-id, and has a category id –
Output: **TDlist**: structure with term and list of document ids containing the term.
C' = SELECT * FROM C WHERE category IN
    (SELECT category FROM C
    GROUP BY category HAVING COUNT(doc-id) $>=$ minCfreq)
– Compute the frequency in C' of each word –
TDlist = $\emptyset$
for all document $d \in C'$ **do**
  for all word w $\in d$ **do**
    if not(w $\in$ StopWordList OR freq(w) $<$ minWfreq) **then**
      k = Stemming(w)
      if k $\in$ TDlist **then**
        add(TDlist[k],d.doc-id)
      else
        append(TDlist,k)
        add(TDlist[k],d.doc-id)
      end if
    end if
  end for
end for
return TDlist

First method. It is the computation of the Jaccard measure for the similarity between each node in CT and each node in KH in terms of the overlapping between the documents sets under the two category nodes (D_{CT} and D_{KH}): $\frac{D_{CT} \cap D_{KH}}{D_{CT} \cup D_{KH}}$

We have observed that Jaccard measure is high for categories that are low in the taxonomies and are very specific in meaning. As soon as we consider upper nodes of the taxonomies, categories get more general and more documents are found in the categories. As a consequence, relatively less common documents are present in both of the nodes and Jaccard measure decreases. In our system (see Figure 2), when the user has the focus on any given node of the KH taxonomy (in case of Figure 2 it is the concept *target*), the lower frame of the window shows the list of CT categories that are relevant for the current KH category. (Each CT category is represented by four fields corresponding to the values of the nodes in a path of the hierarchy of ACM CT). For each CT category, the last column shows the computed Jaccard value.

Second method. It is the entropy measure that evaluates the impurity of the CT label of documents clustered under any node in KH. We observed a similar situation as for Jaccard measure. Higher entropy clusters are the nodes at higher levels of the taxonomy KH. In the lower frame of our system the entropy value is reported for the KH category with the focus.

Third method. It is the augmented ease of a classifier to correctly classify documents in the leaf nodes of the hierarchy. We considered the nodes at the same level - fourth - in

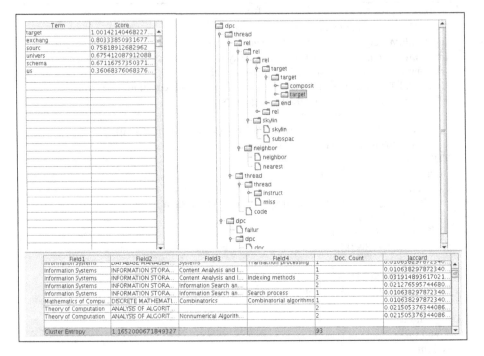

Fig. 2. Overview of the system

Classifier	ACM Category Tree (CT)	Cluster KH Taxonomy
NB	26,98%	**71,83%**
J48	11,90%	**67,24%**
KNN7	28,57%	**70,68%**

Fig. 3. Accuracy of three different classifiers on the ACM Categories and on clusters of KH

both the hierarchies. We thus compared the ability of a classifier to classify documents (according to their content only) into the categories of KH, our generated taxonomy, and CT, the human generated categories taxonomy of ACM.

We used three different families of classifiers implemented in Weka (available at http://www.cs.waikato.ac.nz/~ml/index.html: Naive Bayes (NB) which maximizes the likelihood of the class given the observations, Decision Trees (J48) and K-Nearest Neighbours (KNN7, with K equal to 7).In Figure 3 it is evident that the accuracy obtained by all the three classifiers is very low for CT and considerably higher for KH. This allows us to establish the superiority of KH on CT, at least regarding the document categorization task. This is due to the fact that the papers were placed into a category in CT by several different persons - the papers authors - and this allows us to foresee a non completely coherent way of assigning the papers to a category label.

5   Previous Works

A rich set of works based on document clustering or performing a hierarchical organization of topics of documents exist. They make use very often of an already existing set of categories or NLP domain knowledge either to learn a classifier or to generate topic taxonomies. [17] clusters documents in the result of web search engines to improve user browsing. It intersects the sets of words contained in documents to obtain the document clusters, whose cohesion is defined as the number of common words in the cluster documents. In [15] salient words and phrases are extracted from documents and are organized hierarchically using a type of co-occurrence derived from subsumption. [8] investigated the adoption of hierarchical clustering methods in the context of document management. It adopted a statistical model based on log-linear regression for the evaluation of similarity between information. In [9,16] EM clustering algorithm is adopted instead, once the words distribution conditioned to the classes is given. [5] studies a spherical k-means algorithm for clustering very sparse document vectors. [4] identifies recurrent topics in a text corpus by a combination of NLP (to detect named entities) and data mining techniques, to detect clusters of named entities frequently occurring together in documents. Here clustering partitions the hypergraph resulting from the relationships between frequent itemsets. [6] presents a system for the construction of taxonomies which yield high accuracies with automated categorization systems on the web.

The proposed technique is similarly based on clustering. Differently to previous works, it is guided by τ predictive measure and exploits PageRank. Thus, it succeeds to be domain- and language-independent and exploits little knowledge on NLP.

6   Conclusions

This paper presents a data mining technique based on clustering to generate an unsupervised concept taxonomy. We exploited Goodman-Kruskal τ, a predictive measure for categorical attributes, and PageRank to make the keywords which are the most recommended by others in terms of co-occurrence, to emerge from a set of keywords and represent each cluster.

We validated the proposed approach by comparison with ACM CT. We adopted three different measures: Jaccard, Entropy and classification accuracy. We showed that our taxonomy is better suited to text categorization. In addition, it has been generated automatically, and it is domain and language-independent. Furthermore, the hierarchical keyword organization works as an index which partitions a large document collection with respect to the search space of the contents. In future work we plan to integrate in the system a document retrieval interface and add more sophisticated NLP (like NER) in order to identify named entities to label the hierarchy concepts.

References

1. Aggarwal, C.C., Gates, S.C., Yu, P.S.: On the merits of building categorization systems by supervised clustering. In: Proc. of 5th ACM Int. Conf. on Knowledge Discovery and Data Mining, San Diego, US, pp. 352–356 (1999)

2. Anderberg, M.R.: Cluster analysis for applications, 2nd edn. Academic (1973)
3. Chakrabarti, S., Dom, B., Agrawal, R., Raghavan, P.: Scalable feature selection, classification and signature generation for organizing large text databases into hierarchical topic taxonomies. VLDB Journal 7(3), 163–178 (1998)
4. Clifton, C., Cooley, R., Rennie, J.: Topcat: Data mining for topic identification in a text corpus. IEEE Trans. Knowledge and Data Engineering 16(8), 949–964 (2004)
5. Dhillon, I.S., Modha, D.S.: Concept decompositions for large sparse text data using clustering. Mach. Learn. 42(1/2), 143–175 (2001)
6. Gates, S.C., Teiken, W., Cheng, K.-S.F.: Taxonomies by the numbers: building high-performance taxonomies. In: ACM CIKM 2005: Proc. of the 14th ACM international conference on Information and knowledge management, pp. 568–577 (2005)
7. Goodman, L.A., Kruskal, W.H.: Measures of association for cross classifications. Journal American Statistical Association 49(268), 732–764 (1954)
8. Hatzivassiloglou, V., Gravano, L., Maganti, A.: An investigation of linguistic features and clustering algorithms for topical document clustering. In: ACM SIGIR 2000, pp. 224–231 (2000)
9. Hofmann, T.: The cluster-abstraction model: Unsupervised learning of topic hierarchies from text data. In: IJCAI, pp. 682–687 (1999)
10. Ienco, D., Meo, R.: Exploration and reduction of the feature space by hierarchical clustering. In: SDM 2008 (2008)
11. Lewis, D.D.: Evaluating text categorization. In: Proc. Speech and Natural Language Workshop, HLT (1991)
12. Mihalcea, R.: Unsupervised large-vocabulary word sense disambiguation with graph-based algorithms for sequence data labeling. In: HLT/EMNLP 2005 (2005)
13. Brin, S., Page, L.: The anatomy of a large-scale hypertextual web search engine. Computer Networks and ISDN Systems 30 (1998)
14. Michalski, R.S., Stepp, R.E.: Learning from observation: Conceptual clustering. Machine Learning: An Artificial Intelligence Approach, 331–363 (1983)
15. Sanderson, M., Croft, W.B.: Deriving concept hierarchies from text. In: Research and Development in Information Retrieval, pp. 206–213 (1999)
16. Segal, E., Koller, D., Ormoneit, D.: Probabilistic abstraction hierarchies. In: Proc. NIPS 2001 (2001)
17. Zamir, O., Etzioni, O., Madani, O., Karp, R.M.: Fast and intuitive clustering of web documents. In: SIGACM KDD Conference, pp. 287–290 (1997)

Adapting LDA Model to Discover Author-Topic Relations for Email Analysis

Liqiang Geng[1], Hao Wang[1], Xin Wang[2], and Larry Korba[1]

[1] Institute of Information Technology, National Research Council of Canada
Fredericton, New Brunswick, Canada
{liqiang.geng,hao.wang,larry.korba}nrc-cnrc.gc.ca
[2] Department of Geomatics Engineering, University of Calgary
Calgary, Alberta, Canada
xcwang@ucalgary.ca

Abstract. Analyzing the author and topic relations in email corpus is an important issue in both social network analysis and text mining. The Author-Topic model is a statistical model that identifies the author-topic relations. However, in its inference process, it ignores the information at the document level, i.e., the co-occurrence of words within documents are not taken into account in deriving topics. This may not be suitable for email analysis. We propose to adapt the Latent Dirichlet Allocation model for analyzing email corpus. This method takes into account both the author-document relations and the document-topic relations. We use the Author-Topic model as the baseline method and propose measures to compare our method against the Author-Topic model. We did empirical analysis based on experimental results on both simulated data sets and the real Enron email data set to show that our method obtains better performance than the Author-Topic model.

1 Introduction

Identifying topics and author-topic relations in emails is an important issue in social network analysis. It adds semantics to social network analysis and provides additional perspectives for role analysis. Both supervised and unsupervised text mining techniques have been used for topic identification in emails.

When supervised learning methods are applied to identify email topics, email messages need to be labeled before the classification model is built [2, 6]. This is not a trivial task, especially without domain knowledge and context. Also generally speaking, email messages can involve any topics and it is very difficult to predefine the email topics. Clustering on "a bag of words" representation is an unsupervised learning method and thus does not require labeled training data sets. However, it only assigns one email into one cluster or topic [5, 7]. Furthermore, none of the above-mentioned methods can identify topics and author-topic relations at the same time.

Statistical models for document modeling have attracted a lot of attentions in the recent years. Latent Dirichlet Allocation (LDA) was first proposed to extract topics from large text corpora [1]. LDA is a generative model that represents each document as a mixture of probabilistic topics and represents each topic as a probabilistic distribution

I.-Y. Song, J. Eder, and T.M. Nguyen (Eds.): DaWaK 2008, LNCS 5182, pp. 337–346, 2008.

over words. One of the advantages of the LDA model is that this generative probabilistic model can be scaled up to introduce more levels of structure for inference [1]. Author-Topic (AT) model can be considered as an extension of the LDA model by incorporating a layer of authors [9, 10]. It is the first probabilistic model to identify the topics and author-topic relations simultaneously. To tackle the efficiency issues of LDA and AT models, Gibbs sampling was proposed to estimate the parameters of the models. However, in the Gibbs sampling process for the AT model, the relations between documents and the words are not taken into account. This results in some information loss, i.e., the co-occurrence of words in the same document will be ignored in the algorithm. This is especially true when each document only involves one or very few topics, which is common in email messages. For example, if an author wrote two emails each consisting of two words as follows.

Email 1: *Computer Science*
Email 2: *Civil Engineering*

In the Gibbs sampling algorithm, these two documents will be mixed together. Co-occurrence between *computer* and *science* and that between *civil* and *engineering* will be ignored.

In this paper, we propose to adapt the LDA model in a different way to identify the author-topic relations for email analysis. The idea is that we adopt the LDA model to derive document-topic relations and then aggregate the results on authors to obtain the author-topic relations. In this way, both document-topic and author-topic relations are taken into account. We also propose evaluation criteria for comparing the LDA and the AT models. The rest of the paper is organized as follows. Section 2 introduces the LDA and AT models and presents the adapted LDA model. In Section 3, we propose the evaluation criteria for comparing the AT model and the adapted LDA model. Section 4 presents the experimental results on both simulated data sets and a real data set. Section 5 concludes the paper and discusses possible future work.

2 Adapted LDA Model for Email Analysis

LDA is a generative statistical model that describes how words in a document might be generated on the basis of latent random variables. It assumes that a document is a multinomial distribution over topics and that a topic is a multinomial distribution over words. In the generation process, LDA first chooses a topic in terms of the probabilities of a document over topics. Then it chooses a word according to the chosen topic and the probability distribution of the topic over words. The process is repeated until the corpus is generated.

The probability of choosing a word token w_i in a particular document is

$$P(w_i) = \sum_{j=1}^{T} P(w_i \mid z_i = j)P(z_i = j),$$

(1)

where $P(z_i = j)$ is the probability of topic j being sampled for word token w_i in this document. $P(w_i \mid z_i = j)$ is the probability of word w_i under topic j. T is the number of topics. This model specifies the probability distribution over words within a document.

Let $\phi^{(j)} = P(w \mid z = j)$ refer to the multinomial distribution over words for topic j and $\theta^{(d)} = P(z)$ refer to the multinomial distribution over topics for document d. The parameters ϕ and θ indicate which words are important for a given topic and which topics are important for a particular document, respectively.

Given a document collection, the topic identification problem becomes the model fitting that finds the best estimate of the parameters ϕ and θ, i.e., the topic-word distributions and the document-topic distributions. Gibbs sampling is an efficient method to solve this model fitting problem. Gibbs sampling simulates a high-dimensional distribution by sampling on lower-dimensional subsets of variables where each subset is conditioned on the values of all other variables. The sampling is done sequentially and proceeds until the sampled values approximate the target distribution [3].

For the LDA model, the Gibbs sampling procedure considers each word token in the document collection in turn, and estimates the probability of assigning the current word token to each topic, conditioned on the topic assignments of all other word tokens. The conditioned probability is written as:

$$P(z_i = j \mid w_i, z_{-i}, w_{-i}, d_i, \ldots) \propto \frac{C_{w_i j}^{WT} + \beta}{\sum_{w=1}^{W} C_{wj}^{WT} + W \cdot \beta} \cdot \frac{C_{d_i j}^{DT} + \alpha}{\sum_{t=1}^{T} C_{dt}^{DT} + T \cdot \alpha} \qquad (2)$$

where $z_i = j$ represents the topic assignment of token w_i to topic j, z_{-i} refers to the topic assignments of all other word tokens, and "..." refers to all other known or observed information. T is the number of the topics, W is the number of word tokens, D is the number of documents, and α and β are prior parameters that need to be specified before the sampling process. Empirical guidelines for choosing the appropriate values for α and β are discussed in [1, 4]. A word-topic matrix C^{WT} and a topic-document matrix C^{DT} are maintained in the Gibbs sampling process to calculate the probabilities according to equation (2).

$$C^{WT} = \begin{bmatrix} a_{11} & a_{12} & \cdots & a_{1T} \\ a_{21} & a_{22} & \cdots & a_{2T} \\ \cdots & \cdots & & \cdots \\ a_{W1} & a_{W2} & \cdots & a_{WT} \end{bmatrix}_{W \times T} \qquad C^{DT} = \begin{bmatrix} b_{11} & b_{12} & \cdots & b_{1T} \\ b_{21} & b_{22} & \cdots & b_{2T} \\ \cdots & \cdots & & \cdots \\ b_{D1} & b_{D2} & \cdots & b_{DT} \end{bmatrix}_{D \times T}$$

Word-Topic matrix   Topic-Document matrix

The word-topic matrix C^{WT} contains the number of times w_i is assigned to topic j, not including the current token of w_i; the topic-document matrix C^{DT} contains the number of times topic j is assigned to some word token in document d, not including the current instance w_i.

After the sampling process, the estimate of parameters ϕ and θ could be obtained from the word-topic matrix and the topic-document matrix with equations (3) and (4).

$$\hat{\phi}_i^{(j)} = \frac{C_{w_i j}^{WT} + \beta}{\sum_{w=1}^{W} C_{wj}^{WT} + W \cdot \beta} \tag{3}$$

$$\hat{\theta}_j^{(d)} = \cdot \frac{C_{d_i j}^{DT} + \alpha}{\sum_{t=1}^{T} C_{dt}^{DT} + T \cdot \alpha} \tag{4}$$

The Gibbs sampling procedure is an iterative process as follows.

1. Initialize C^{WT} and C^{DT}
2. For $i = 1$ to N do // N is the number of Gibbs sampling iterations
3.   Randomly read a word token w from documents
3.   Calculate the probabilities of assigning w to topics based on equation 2.
4.   Sample a topic in terms of the estimated probabilities obtained in step 3
5.   Update the matrix C^{WT} and C^{DT} with new sampling results
6.   Go to step 3 until all of word tokens have been scanned.
7. Endfor

The AT model is an extension of the LDA model by substituting the variable *author* for variable *document*, which means each author is associated with a multinomial distribution over topics. In the AT model each word w in a document is associated with two latent parameters: an author x, and a topic z [9].

In general, one document can have more than one author. However, for email collections, usually there is only one author for one email (except for forwarded emails, where the forwarder may add more content or modify the content of the original sender). Therefore, we simplify the Gibbs sampling process for the AT model on emails by ignoring the author sampling. In this case, it is straightforward that the Gibbs sampling procedure for the AT model is equivalent to first aggregating documents on authors and then applying the LDA model on the aggregated documents.

We can see that while the AT model tries to identify the relationship between authors and topics, the relationship between documents and topics is ignored, i.e., the information about co-occurrence of words within a document is ignored. The author-topic relations are obtained by integrating document-topic relations at the beginning of the Gibbs sampling process. For documents like email messages, each of which may only involve one or a few topics, the ignorance of document-topic relation may deteriorate the results. We propose an adapted LDA model to derive the author and topic relationship. The idea is straightforward. First, Gibbs sampling algorithm for LDA is used to derive the document-topic relations. Then the author-topic relations are obtained by aggregating the document-topic matrix using the following SQL sentence:

Select *author*, sum(*Topic$_1$*),..., sum(*Topic$_T$*) from Table(C^{DT}) inner join Table(*AD*) on Table(C^{DT}).*document* = Table(*AD*).*document* group by *author*.

where Table(C^{DT}) denotes the table corresponding to the document-topic matrix and Table(AD) denotes the table representing the relationship between documents and authors.

It should be noted that in the adapted LDA model, the author variable is isolated from other factors, and thus is not involved in the inference process. Therefore, the inference process in the model is identical to that of the LDA model, i.e., only the document-topic relation is taken into account in the inference process. The author-document relation is used in the aggregation process after the inference process to derive the author-topic relation. This is different from the AT model in that the AT model directly uses the author-topic structure in the inference process.

3 Evaluation Criteria

To evaluate our model and compare it with the AT model, we did experiments on both simulated data and real data. The simulated data is generated by the predefined probability distributions, and therefore, the number of the topics and the probability distributions of authors over topics and those of topics over words are known. The evaluation can be done straightforwardly by comparing the degree of match between the real distributions and the discovered distributions.

Since the AT model and the adapted LDA model are unsupervised methods and the discovered topics are randomly ordered, we used a greedy algorithm to compare the discovered topics and the originally assigned topics to determine the degree of match between them. The algorithm first compares the discovered topics and the actual topics in a pair-wise fashion. The pair with the minimum distance will be matched if the distance is below a threshold. Then this pair of topics is removed from the topic lists and the next round begins until the current minimum distance is above a threshold, which means that the rest of the real topics and the discovered topics do not match any more. Here we used $1 - \text{cosine}(T, T')$ as the distance measure. Figure 1 shows the procedure.

Based on the degree of match between the real and the discovered topics, we also evaluated the degree of match between real and discovered author distribution over topics with similar experiments. The difference here is that when we match real and discovered author distributions over topics, we need to consider the discovered topics and the real topics that do not match. For example, suppose we have three real topics t_1, t_2, t_3 and we discovered three topics t_1', t_2', t_3'. If t_1 matches t_1', t_2 matches t_2', but t_3 does not matches t_3', we need to calculate the distance based on distribution over t_1, t_2, t_3, and t_3'.

For the real data sets, we do not know the real topics as we do for the simulated data. Therefore, we cannot use the degree of match to evaluate the models. We use different measures to compare our method with the AT model. Perplexity can be used as a measure to indicate the prediction power of the AT and LDA models [9], but here we focus on the quality of the topics in terms of the clustering results rather than the prediction power.

To measure the intra-topic quality, we use entropy to evaluate the correlation among the words of each topic. A uniform distribution of topics over words conveys no meaning to users and thus a topic of high entropy value will be considered as low

---

Function *TopicMatch*

Input: $T[1, n]$, $T'[1, n]$ //Real topics and identified topics as probability distribution over words, n is the number of topics.

    Count = 0   //number of topics matched

    For $i = 1$ to n,

        $(k, l) = \text{argmin}_{(i,j)} (Dist(T[i], T'[j]))$ //find the currently best matched topics

        if dist($T[k]$, $T'[l]$) < *threshold*   // It is a match

            remove $T[k]$ and $T'[l]$ from arrays respectively

            count++;

        else   //It is not a match

            break;

    endfor

    degreeOfMatch = *count/n*

Fig. 1. Algorithm that calculates the degree of topic match

quality. On the other hand, when a topic concentrates on a small group of words, which results in a lower entropy value, we say it is a topic with higher quality. The average entropy value for a topic distribution over words is defined as

$$Entropy = \frac{1}{n} \sum_{i=1}^{n} \sum_{j=1}^{m} p_{ij} \log_2 p_{ij}$$, where p_{ij} denotes the probability of topic i

taking word j.

To measure the inter-topic quality, we use the average minimum Kullback–Leibler divergence (or KL-distance) to evaluate how close the discovered topics are to each other. The average minimum KL distance is defined as

$$KL = \frac{1}{n} \sum_{i=1}^{n} \min_{j=1,n, j \neq i} kl(p_i, p_j)$$, where $kl(p_i, p_j)$ denotes the symmetric KL-

distance between topic i and topic j and is defined as

$$kl(p_i, p_j) = \frac{1}{2} (\sum_{k} p_i(k) \log \frac{p_i(k)}{q_i(k)} + \sum_{k} q_i(k) \log \frac{q_i(k)}{p_i(k)}) .$$

A greater KL-distance value means the topics are far away from each other and thus are desired.

4   Experimental Results

We first did experiments on simulated data to compare the adapted LDA model and the AT model. To simplify the generating process and facilitate the comparison of the results, we assume that different topics do not share any common words and all topics have uniform distributions over the words within that topic. We assume two types of documents in terms of the document-topic structure: single-topic documents and multi-topic documents. A single-topic document is generated from the words within a single topic, while a multi-topic document is generated from words from more than one topic. We also assume two types of author-topic structures: separated-author-topic structure and the mixed-author-topic structure. The separated-author-topic structure requires that any two authors either share all the topics that they are involved in or share no topics at all. The mixed-author-topic structure allows authors to share some of their topics with other authors.

We get four combinations based on the author-topic and document-topic structures: single-topic documents with separated author-topic structure, single-topic documents with mixed author-topic structure, multi-topic documents with separated author-topic structure, and multi-topic documents with mixed author-topic structure. The multi-topic documents with separated author-topic structure can be converted to single-topic documents with separated-author-topic structure if we combine the topics under the same author to one topic. Therefore, we only consider the three other cases.

We generated three data sets to simulate the three cases respectively. Each data set consists of 5000 emails with a vocabulary of 200 words. We set 20 topics with each topic consisting of 10 words. We set 20 authors and each author has two topics. When running the AT and the adapted LDA models, we set the number of topics to 20, which means that we already know the number of topics in advance. This facilitates the comparison of the results (Some principles for choosing the appropriate number of the topics were discussed in [4]). We follow the suggestions from [9] and set $\alpha = 50/T$, and $\beta = 0.01$.

In Figure 2, the X-axis denotes the distance threshold for match. If the distance value of two topics is less than this threshold, we say there is a match between the two topics. Otherwise, there is no match between the two topics. The Y-axis denotes the degree of match between a set of real topics and a set of discovered topics, as defined in Figure 1. Figure 2 shows that in all three cases, the LDA model outperforms the AT model when the distance threshold is less than 0.3. Figure 2(a) shows that for the single-topic document with separated author-topic structure, The LDA model performs much better than the AT model. This is because the AT model mixed the documents of a single author together, and therefore co-occurrence information is totally lost. Figure 2(b) shows that for the single-topic document and mixed author-topic structure, LDA still has better performance than the AT model, but not as significant as in the first case. This is because in the aggregating process in the AT model, although some co-occurrence information within a document is lost, some co-occurrence can still be embodied in the author-topic structure. Figure 2(c) shows that for the multiple-topic document with mixed author-topic structure, LDA just performs slightly better than the AT model. This is because in aggregation process in the AT algorithm, the loss of co-occurrence information is very limited.

We also evaluated the results based on F1 measure and observed the very similar trends for all three cases. Here the *precision* is defined as the ratio between the number of the discovered topics whose minimum distance to the actual topics are below a threshold and the number of the discovered topics. The *recall* is defined as the ratio between the number of the actual topics whose minimum distance to the discovered topics are below a threshold and the number of the actual topics.

Based on the degree of match between the real and the discovered topics, we also evaluated the degree of match between real and discovered authors' distribution over topics with similar experiments. We set the distance threshold for both the topic-word distributions and the author-topic distributions to 0.3 and obtained the results as shown in Table 1.

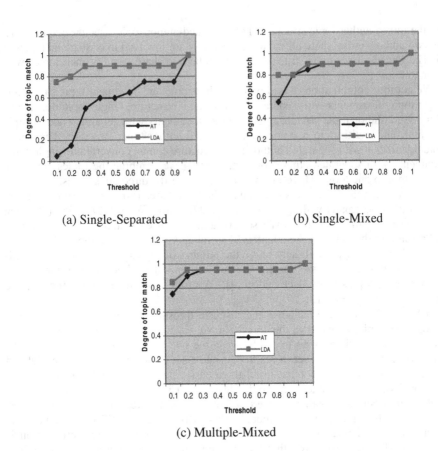

(a) Single-Separated   (b) Single-Mixed

(c) Multiple-Mixed

Fig. 2. Degree of match for topic-word relation for three simulated data sets

Table 1. Degree of match for author-topic relation for three simulated data sets

%	Single-Separated	Single-Mixed	Multiple-Mixed
LDA	90	90	90
AT	55	80	90

We then did experiments on the Enron email data set [11] to compare the LDA and the AT models.

We variated the number of the topics from 20 to 200 and recorded the entropy and KL-distance measures in Figure 3. Figure 3(a) shows that when the specified number of topics increases, the average entropy of the results generated from the AT model increases, while the entropy of the results generated from the LDA model remain stable. This means that the intra-topic quality of the LDA model is relative stable to the specified number of topic and the AT model produce deteriorated results when the number of topics increases. We also observed that the LDA model consistently produces better results than the AT model in terms of intra-topic quality.

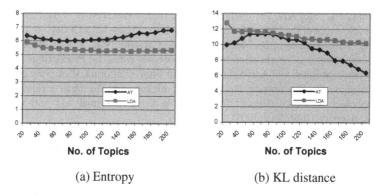

(a) Entropy   (b) KL distance

Fig. 3. Comparison of the AT and LDA models on Enron email data set

Figure 3(b) shows that the LDA model consistently attains greater KL-distance values than the AT model regardless of the number of specified topics. This means that the LDA model produces results with higher inter-topic quality. Also when the number of the specified topics increases, the difference between the KL values from the two models increases. This means that when the number of topics increases, the inter-topic quality of the AT model deteriorates dramatically, while that of the LDA model remains stable.

5 Conclusions and Future Work

We proposed a method to find topics and author-topic relations in emails based on the LDA model. Compared with the AT model, our method takes into account the word co-occurrence information within documents. Experimental results on both synthetic and real data sets show that the adapted LDA method obtains better results than the AT model for email corpus where each document has one author and involves only one or a few topics.

We will extend our work to identify author-recipient-topic relations based on the adapted LDA method and compare the results with the Author-Recipient-Topic model [8]. Another approach we are interested in involves taking into account the threading information in our method to facilitate the discovery of topics and author-topic relations.

Reference

[1] Blei, D.M., Ng, A.Y., Jordan, M.I.: Latent Dirichlet allocation. Journal of Machine Learning Research 3, 993–1022 (2003)
[2] Dredze, M., Lau, T.A., Kushmerick, N.: Automatically classifying emails into activities. In: Intelligent User Interfaces, Sydney, Australia, January, 2006, pp. 70–77 (2006)
[3] Gilks, W., Richardson, S., Spiegelhalter, D.: Markov Chain Monte Carlo in Practice. Chapman & Hall, New York (1996)

[4] Griffiths, T.L., Steyvers, M.: Finding scientific topics. Proceedings of the National Academy of Sciences of the United States of America 101, 5228–5235 (2004)

[5] Huang, Y., Govindaraju, D., Mitchell, T.M., de Carvalho, V.R., Cohen, W.W.: Inferring ongoing activities of workstation users by clustering email. In: Proceedings of the First Conference on Email and Anti-Spam, Mountain View, California, USA (July 2004)

[6] Khoussainov, R., Kushmerick, N.: Email task management: An iterative relational learning approach. In: Proceedings of the Second Conference on Email and Anti-Spam. Stanford University, California (2005)

[7] Li, H., Shen, D., Zhang, B., Chen, A., Yang, Q.: Adding semantics to email clustering. In: Proceedings of the 6th IEEE International Conference on Data Mining, Hong Kong, China, pp. 938–942 (2006)

[8] McCallum, A., Wang, X., Corrada-Emmanuel, A.: Topic and role discovery in social networks with experiments on Enron and academic email. Journal of Artificial Intelligence Research 30, 249–272 (2007)

[9] Rosen-Zvi, M., Griggiths, T.L., Smyth, P., Steyvers, M.: Learning author topic models from text corpora,
http://citeseer.ist.psu.edu/rosen-zvi05learning.html

[10] Steyvers, M., Smyth, P., Rosen-Zvi, M., Griffiths, T.L.: Probabilistic author-topic models for information discovery. In: Proceedings of the Tenth ACM SIGKDD International Conference on Knowledge Discovery and Data Mining, Seattle, USA, August, 2004, pp. 306–315 (2004)

[11] Enron email data set,
http://www.isi.edu/~adibi/Enron/Enron.htm

A New Semantic Representation for Short Texts

M.J. Martín-Bautista[1], S. Martínez-Folgoso[2], and M.A. Vila[1]

[1] Intelligent Databases and Information Systems Research Group
Dept. of Computer Science and Artificial Intelligence
University of Granada, 18071 - Granada (Spain)
[2] Dept. of Computer Science
University of Camagüey, 74650 - Camagüey (Cuba)
mbautis@decsai.ugr.es, smartinez@inf.reduc.edu.cu, vila@decsai.ugr.es
http://idbis.ugr.es

Abstract. The aim of this work is to present a new semantic structure of knowledge representation for textual fields. The purpose of this
structure is to allow us to handle this kind of textual fields as the rest
of the fields in the database in processes such as OLAP, datawarehousing, semantic querying, etc. The architecture of the system is described
in the work as well as a detailed description of the mathematical formalism of the structure. The mechanism to carry out the transformation
is given together with an experimental example with a medical database.

Keywords: Textual fields, AP-Sets, AP-Structure, Apriori property, frequent itemsets.

1   Introduction

The lack of structure of textual data makes difficult the automatic processes to
handle them in a massive way. This happens not only in textual repositories
but also in text fields in relational databases such as the medical ones [?], [8]
polls, e-mail databases, and so on. There are several researching areas such as
Processing Natural Language, Text Mining, Information Retrieval and so on,
that treat with this problem and try to solve it from different points of view.
Most of the solutions come from finding some kind of structure in the text and/or
transforming it into a what is called intermediate form [4].

In this paper, we try to solve this problem by transforming the text fields
in a database into an intermediate structure [4] based on frequent itemsets.
This structure is called AP-Set (it comes from sets obtained via the Apriori
algorithm [1] asserting the Apriori property [1]). This intermediate structure can
be implemented as an ADT (Abstract Data Type) with its associated methods,
in such a way that the text fields become into new structured representations.
These new fields can be managed together with the other non-textual ones in
the database for querying, analysis, warehousing or mining purposes.

I.-Y. Song, J. Eder, and T.M. Nguyen (Eds.): DaWaK 2008, LNCS 5182, pp. 347–356, 2008.

For this aim, we include some related work in next section and our hypothesis and system architecture in section 3. The AP-Set concept, some operations to manage it, as well as the concept of AP-Structure and its associated operations are presented in Section 4. The mechanism to apply such concepts to text fields can be found in Section 5 by using an experimental example with a medical database. Finally, some conclusions and future work are given in Section 6.

2   Related Work

There are some approaches in the literature integrating information management systems with structured, semi-structured and unstructured data [9]. In all of them, the architecture of the primitive system is modified in some aspects to support the management of other type of data. Three information management systems are considered for this purpose: Semi-Structured Text Retrieval Systems, Text Retrieval Relational/Object Database Management Systems and Semi-structured (XML) Database Management Systems [9].

From these systems, we work in a framework of a Semi-structured Database Management System in an Object Oriented Model. But unlike most of the systems of this type in the literature [11],[5], [2], [3], we do not obtain an XML representation of the attributes in the database. We just transform textual attributes in order to manage them in the same way as the rest of the attributes of the database. For this purpose, we obtain an intermediate form via mining techniques, so the new representation of the textual attribute in the database can be obtained automatically by an ADT.

3   System Description

The developed system deals with short texts composed of some words or phrases in a certain domain. They can be not only in textual repositories but also in databases (Relational, Object Oriented Relational, etc...). The operations over these fields are the classical ones in databases for text attributes such as to ask for the content of the field, or to search for the fields containing a certain word. However, the lack of a defined domain and a structure in these attributes prevents them from being treated as the rest of the attributes in the database. The problem can be worst when processes such as Data Warehousing, Data Mining or just a semantic querying have to be carried out over these text attributes. Besides to count the number of words repeated in the text fields asserting some conditions, it is not possible to resume them with the traditional relational database representation/operators.

Our hypothesis is based on the fact that there is an underlying semantic on a certain textual attribute of a database. Although the domain is very complex to manage, since it is composed of natural language elements, the semantic and even the vocabulary in the fields are, in some way, restricted. Therefore, we can find sets of terms related semantically that appear repeatedly in the analyzed attribute. For this purpose, we use an intermediate form based on frequent itemsets obtained via the Apriori algorithm with the Apriori property [1]. Such intermediate form is called AP-Set. With all the AP-Sets obtained in the database, a global intermediate structure called AP-Structure is formed. This structure reflects the semantic of the textual attribute in the database.

3.1   System Architecture

The system is basically composed of an initial database, a text management module, a modified database and some query modules. Since the main objective of this architecture is the management of textual attributes as the rest of the attributes in the database, we could have a datawarehouse server or an OLAP server to take advantages of the system, as it is shown in figure 1. Any of these modules are explained in the following:

– **Initial database:** The database stores different types of data, including textual fields. These textual fields can be transformed in order to make possible to query the database including these fields. Although we are managing an OOR (Object-Oriented Relational) database, any other kind of database can be considered.

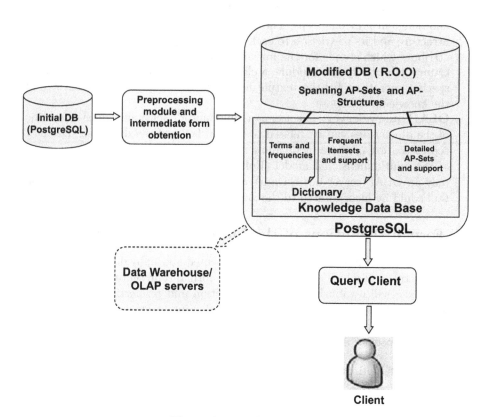

Fig. 1. System Architecture

- **Text management module:** In this module, the textual fields of the database are processed in order to get an intermediate form. The text management module performs a mining process to obtain the whole semantic structure of the text attribute, (in some way "the attribute semantic domain"), as well as its particular restriction to each tuple value. The new knowledge structures will allow us to query all the fields of the database and to carry out any process of summarization, datawarehousing and so on. The details of the mining process are presented in the next section.
- **Modified database:** This database stores the new knowledge structures corresponding to the textual fields in the initial database. These structures are based on sets of terms obtained by the Apriori algorithm. That is the reason why they are called AP-Sets and AP-Structures. This database can be used to update the modified database as well as to complete the answers of the system to the query client by suggesting new query terms, for instance. Different intermediate structures such as a dictionary are obtained in the process to get these knowledge structures. The dictionary is composed of all the relevant terms of each textual field (once removed stop words, preprocessed acronyms, etc.) and by the frequent itemsets and their support. This dictionary and the expanded AP-Sets with their support are stored in the Knowledge database. It should be remarked that the modified database includes the obtained knowledge in previous steps. The semantic domain appears as a global AP-Structure and its restriction to each tuple gives us the tuple value for a new attribute which stands for the initial textual one.
- **Query client:** By this module, a client is used to query the database, with specific operations for the textual fields such as to query the structures of the knowledge database.
- **OLAP query server:** The user can build multidimensional cubes where one of the dimensions can be related to textual fields of the database. The new representation of these fields allow us to performance the usual operations of the multidimensional model (Roll up, Drill Down, Slice and Dice). Not only an OLAP server but a DataWarehousing or any other summarization tool could be incorporated to the architecture.

4   Formal Definitions of the Mathematical Structures

In this section we will formalize the ideas presented above, by defining the mathematical structures which are the basis of the formal representation of the data.

Firstly, we will establish the formal definition and properties of the sets of subsets with the Apriori property [1] (AP-Sets). Finally, we will give the formal definition and properties of the structure underlying in the texts (AP-Structure) consisting of a set of AP-Sets.

4.1   AP-Set Definition and Properties

Definition 1. AP-Set
Let be $X = \{x_1...x_n\}$ a finite set of items and $\mathcal{R} \subseteq \mathcal{P}(X)$ a set of frequent itemsets, being $\mathcal{P}(X)$ the set of parts of X. We will say that $\mathcal{R}$ is an AP-Set if and only if:

1. $\forall Z \in \mathcal{R} \Rightarrow \mathcal{P}(Z) \subseteq \mathcal{R}$
2. $\exists Y \in \mathcal{R}$ *such that* :
   (a) $card(Y) = max_{Z \in \mathcal{R}}(card(Z))$ *and   not exists* $Y' \in \mathcal{R}$ *such that* $card(Y') = card(Y)$
   (b) $\forall Z \in \mathcal{R}; Z \subseteq Y$

The first condition of the above definition guaranties that any AP-Set verifies de Apriori property. The second one assures us the existence of an unique set called Y of maximal cardinality *spanning set of* $\mathcal{R}$, which characterizes the AP-Set. We will denote $\mathcal{R} = g(Y)$, that is $g(Y)$ is an AP-Set with spanning set Y. Let us remark that $g(Y)$ is also the power-set of Y.

We will call *Level of* $g(Y)$ to the cardinal of Y. Obviously, AP-Set of level 1 is composed of the elements of X. We will consider the empty set $\emptyset$ as the AP-Set of level zero.

Example 1. Let be $X = \{1, 2, 3, ..., 10\}$ and
$\mathcal{R} = \{\{1\}, \{3\}, \{5\}, \{1,3\}, \{1,5\}, \{3,5\}, \{1,3,5\}\}$, the spanning set is $Y = \{1,3,5\}$

Let us remark that the definition 1 implies that any AP-Set $g(Y)$ is in fact the reticulum of $\mathcal{P}(Y)$ with respect to the set inclusion.

4.2   AP-Structure Definition and Properties

Once we have established the AP-Set concept, we will use it to define the information structures which appear when frequent itemsets are computed. It should be considered that such structures are obtained in a constructive way, by initially generating itemsets with cardinal equal to 1, next these ones are combined to obtain those of cardinal equal to 2, and by continuing until getting itemsets of maximal cardinal, with a fixed minimal support. Therefore the final structure is that of a set of AP-Sets, which formally is defined as follows.

Definition 2. AP-Structure
Let be $X = \{x_1...x_n\}$ *any referential and*
$S = \{A, B, ...\} \subseteq \mathcal{P}(X)$ *such that:*

$$\forall A, B \in S \, ; \, A \not\subseteq B \, , \, B \not\subseteq A$$

We will call AP-Structure of spanning S,
$\mathcal{T} = g(A, B, ...)$, *to the set of AP-Set whose spanning sets are* $A, B, ...$

Let us remark that any AP-Structure is a reticulum of subsets whose upper extremes are their spanning sets.

There are some other definitions and properties related to the AP-Sets and AP-Structures concepts which, due to a lack of space, are not included in this work, but they can be seen in [6].

5   From Text Fields to AP-Structures: A Mechanism to Obtain the ADT

In this section, we present a methodology that is applied to the textual attributes of a database with the aim to obtain a new structured representation that keeps the semantics of the texts. This representation must not be static but be a mathematical model with operations over the data, that is, be an Abstract Data Type (ADT). In this way, we can implement the structure in an Object Relational Data Base System (ORDBS) as well as its operations in order to query the database by terms and to carry out semantic querying, mining, datawarehousing or OLAP processes.

The mechanism to get the AP-Sets from the original data and to form the AP-Structure and the ADT is described as follows. Let be R a relation with attributes $\{A_1, A_2, A_T, ..., A_n\}$, where A_T is a textual attribute without a predictable structure. To obtain the ADT of this attribute, the tasks to perform are the following [7]:

1. To obtain a data dictionary consisting in a list of the terms removing stop-words [10].
2. To transform the data into a transactional database, where the attributes are the different terms of the data dictionary, and each tuple corresponds to a record.
3. To obtain the frequent itemsets of the transactional database following an Apriori like algorithm [1].
4. The maximal itemsets form the reticular structure, that is, the AP-Sets, following the Apriori property.

In order to illustrate this mechanism, we present an experimental example, both for clarifying the ideas and for showing how the processes have been carried out in a real case. The experimental data set consists of textual attributes from a surgical operation database of the University Clinic Hospital San Cecilio of Granada in Spain. From 24527 records of data, we have extracted the textual attributes corresponding to surgery descriptions and medical diagnosis. These fields are short (from one up to fifteen terms), and they can include one or more sentences. The first column of the table 1 presents an example of these original texts, for the surgery attribute description. Since we have data from a Spanish Hospital, they are in Spanish.

5.1   Data Cleaning

In this first phase, different cleaning processes can be performed based on the characteristics of the textual data. The number of sentences in the field must be determined, as well as the separators. The substitution of acronyms by their complete words and the elimination of stop-words [10] are other actions to carry out in this phase.

In our real case, this task has been specially hard, because of the wide variety of acronyms and synonymous used in the medical vocabulary, even using expert knowledge.

Table 1. Example of an original text field with the cleaned form and the associated ADT

Original Text	Cleaned Text	Associated ADT
CURA DE ABCESO PERIANAL	CURA ABSCESO PERIANAL	{CURA, ABSCESO, PERIANAL}
HISTERECTOMIA ABDOMINAL	HISTERECTOMIA ABDOMINAL	{ABDOMINAL, HISTERECTOMIA}
EMBOLECTOMIA BYPASS AXILO-FEMORAL IZDO	EMBOLECTOMIA BY-PASS AXILO-FEMORAL IZQUIERDO	{AXILO-FEMORAL, BY-PASS, EMBOLECTOMIA, IZQUIERDO}
EECC + L.I.O. O.I.(20.0)	EECC + LIO OI(20.0)	{EECC}, {LIO, OI}
ARTROPLASTIA DE RODILLA DCHA	ARTROPLASTIA RODILLA DERECHA	{ARTROPLASTIA, RODILLA}, {DERECHA }
CURA DE PIE IZQ	CURA PIE IZQUIERDO	{CURA, IZQUIERDO,PIE}
ABDUCION DE YESO	ABDUCION YESO	{YESO}, {ABDUCION}
REDUCCION INCRUENTA DE YESO	REDUCCION INCRUENTA YESO	{INCRUENTA,REDUCCION}, {REDUCCION,YESO}

5.2   Global Knowledge AP-Structure Obtaining

As we mentioned in section 3, our basic hypothesis is that the knowledge included in a free text attribute of any database can be obtained by means of a simple mining process. We assume that the frequent itemset structure [1] of the data dictionary involves the major part of the semantics of the considered attribute. We consider that, once a support is fixed, frequent itemsets are the most frequent sets of terms that appear in the cleaned data. Therefore, they can be viewed as a representation of the most frequent sentences included in the data. Furthermore, frequent itemsets have the Apriori property, so they form an AP-Structure whose spanning sets are those frequent itemsets non included in any other, i.e. the maximal itemsets. Consequently:

We can get a global semantic structure of the text attribute by computing the AP-Structure formed by frequent itemsets of the data dictionary. Maximal frequent itemsets are spanning sets of such global AP-Structure.

For this purpose, once fixed a minimal support (minsupp), the Apriori algorithm is executed over the transactional database with the textual data to get the frequent itemsets. Then, the global AP-Structure is obtained by considering the maximal ones.

Let us remark that this process is strongly dependent on both the minimal support and the maximal cardinal of the maximal frequent itemset. In fact, these two parameters are related since a high minsupp produces a low maximal cardinal.

Regarding our real case, different tests of the Apriori algorithm have been carried out using different values of minsupp. The spanning sets form the global AP-Structures corresponding to the operation description attribute and the medical diagnosis for 24000 tuples. They have been obtained with a minsupp low value of 0.001 per cent. These results show that in a controlled short text context, we obtain the following:

- The vocabulary, that is, the amount of different used terms, is not too extended. In our example, there are less than 1900 terms for the surgery attribute and 1300 for the medical diagnosis one.
- The amount of possible maximal sentences is even smaller. In fact, there are no maximal sentences with more than 7 terms for the operation description field and with more than 5 terms for the medical diagnosis one.

This experimental analysis reinforces our hypothesis about how the global semantics of a textual attribute is covered by the global AP-Structure.

5.3   Particular AP-Structures Obtaining

Let be R a relation with attributes $\{A_1, A_2, A_T, ..., A_n\}$, where A_T is a textual attribute without a predictable structure. At this point of the process, we will assume that the attribute A_T is cleaned. In our example, they are in the second column of table 1.

As we have established above, the global AP-Structure obtained by a mining process covers the semantics of the attribute A_T and so, it will provide us with the domain for the ADT which will replace the mentioned attribute. Let be $T = g(A, B, ...)$ this AP-Structure.

Let us consider now a tuple $x \in R$. To obtain the instance of the ADT for such a tuple, it suffices to get the AP-Substructure of T induced by the A_T value for x, that is, $x[A_T]$. If we denote by A_{TN} this new attribute, we have:

$$\forall x \in R \; x[A_{TN}] = T \bigwedge x[A_T]$$

It is clear that every value of this new attribute is an AP-Substructure of the global one. In this sense, we consider the global AP-Structure as the "value domain" of this attribute. Both global AP-Structure and the induced AP-Substructures are of the same ADT. The resulting new attribute is also stored in a new column in the modified database.

Regarding our real case, we have computed the induced AP-Substructures corresponding to each tuple in the database by designing an algorithm *ad hoc*. The third column in table 1 shows the spanning sets of the AP-Structures induced by the values of the second column which includes examples of AP-Substructures. It is possible that some terms in any tuple value can be missed, since by computing the induced AP-Substructure we restrict the term sets. An example of this case appears in the fourth row of the table.

In order to measure the reduction of data by replacing the term sets with the AP-Structures, the proportion of missed terms in the database after the restriction process is computed. We assume that this measuring is strongly dependent of the minsupp values used to get the global AP-Structures. To confirm this assumption, we have computed this proportion using different minsupp values and taking different amount of tuples of the database. The graphics of figure 2 show the obtained results. It is clear that the missing information depends on the minsupp, but the most interesting result is that this measure is almost independent of the amount of tuples considered. In our opinion, this result reinforces

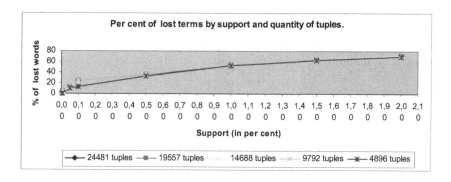

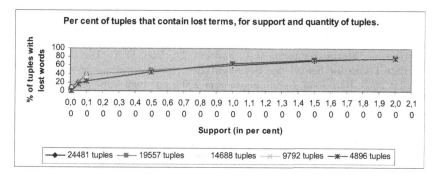

Fig. 2. Results of experiments with lost terms

our assumptions that the global AP-Structure actually captures the semantics of the attribute. Moreover, it is possible to get the semantics underlying in a textual attribute without using the whole database; that is, the mining process is scalable. New experiments with other types of databases have been designed for testing this initial analysis.

6   Conclusions and Future Work

The methodology presented in the paper gives us a way to manage textual fields in databases (Relational, OOR, etc...). With a minimal preprocessing of the text and the obtaining of the frequent itemsets, a structure following the Apriori property can be constructed. This structure has a mathematical model which allows us to consider it as an ADT easily implementable.

We have applied this methodology to a real case and the results have reinforced our initial assumptions and suggested new issues to tackle in the future. Some of them are:

– To carry out new experiments with different kinds of text attributes to confirm the experimental results.

- To complete the query process implementation by using the mathematical result presented in [6,7]
- To define a second level mining process by using the AP-Structure as the basic data domain.

References

1. Agrawal, R., Srikant, R.: Fast Algorithms for mining Association rules. In: Proceedings of VLDB, Santiago, Chile (September 1994)
2. Fernandez, M., Tan, W.C., Suciu, D.: SilkRoute: Trading between relations and XML. In: Proceedings of the 9th Intl. World Wide Web Conference, Amsterdam, Holland, May 2000, pp. 723–745 (2000)
3. Hiemstra, D., de Vries, A.P., Blok, H.E., van Keulen, M., Jonker, W., Kersten, M.L.: CIRQUID: Complex Information Retrieval Queries in a Database. In: Proceedings of the VLDB PhD Workshop, Berlin, Germany (September 2003)
4. Justicia, C., Martín-Bautista, M.J., Sánchez, D., Vila, M.A.: Text Mining: Intermediate Forms for Knowledge Representation. In: Proceedings of the conference Eusflat 2005, Barcelona, Spain, September 2005, pp. 1082–1087 (2005)
5. Lahiri, T., Abiteboul, S., Widom, J.: Ozone: Integrating structured and semistructured data. In: Connor, R.C.H., Mendelzon, A.O. (eds.) DBPL 1999. LNCS, vol. 1949, pp. 297–323. Springer, Heidelberg (2000)
6. Marín, N., Martín-Bautista, M.J., Prados, M., Vila, M.A.: Enhancing short text retrieval in databases. In: Proceedings of FQAS, Milan, Italy (June 2006)
7. Martín-Bautista, M.J., Prados, M., Vila, M.A., Martínez-Folgoso, S.: A knowledge representation for short texts based on frequent itemsets. In: Proceedings of IPMU, Paris, France (July 2006)
8. Prados, M., Peña, C., Prados-Suárez, B., Vila, M.A.: Generation and use of one ontology for intelligent information retrieval from electronic health medical record. In: 8th International Conference on Enterprise Information Systems, Cyprus (May 2006)
9. Raghavan, S., Garcia-Molina, H.: Integrating diverse information management systems: a brief survey. IEEE Data Engineering Bulletin 24(4), 44–52 (2001)
10. Salton, G., McGill, M.J.: Introduction to Modern Information Retrieval. McGraw-Hill, New York (1983)
11. SQL-MM Standards: SQL Multimedia and Application Packages. Part 2: Full-Text. ISO/IEC FDIS 13249-2:2000 (E) (2000), http://www.wiscorp.com/sqlfulltext.zip

Document-Base Extraction for Single-Label Text Classification

Yanbo J. Wang[*,**], Robert Sanderson, Frans Coenen, and Paul Leng

Department of Computer Science, The University of Liverpool
Ashton Building, Ashton Street, Liverpool, L69 3BX, UK
{jwang,azaroth,frans,phl}@ csc.liv.ac.uk
wangya@cs.man.ac.uk

Abstract. Many text mining applications, especially when investigating Text Classification (TC), require experiments to be performed using common text-collections, such that results can be compared with alternative approaches. With regard to single-label TC, most text-collections (textual data-sources) in their original form have at least one of the following limitations: the overall volume of textual data is too large for ease of experimentation; there are many predefined classes; most of the classes consist of only a very few documents; some documents are labeled with a single class whereas others have multiple classes; and there are documents found with little or no actual text-content. In this paper, we propose a standard approach to automatically extract "*qualified*" document-bases from a given textual data-source that can be used more effectively and reliably in single-label TC experiments. The experimental results demonstrate that document-bases extracted based on our approach can be used effectively in single-label TC experiments.

Keywords: Textual Data Preparation, Document-base Extraction, Knowledge Discovery in Databases, (Single-label) Text Classification, Textual Data Sources, Text Mining.

1 Introduction

The increasing number of electronic documents that are available to be explored online has led to text mining becoming a promising field of current research in Knowledge Discovery in Databases (KDD). It "*aims at disclosing the concealed information by means of methods which on the one hand are able to cope with the large number of words and structures in natural language and on the other hand allow to handle vagueness, uncertainty and fuzziness*" [9]. One important aspect of text mining is Text Classification (TC) — "*the task of assigning one or more predefined categories to natural language text documents, based on their contents*" [6]. Early studies of TC can be dated back to the early 1960s (see for instance [13]). During the last decade,

---

[*] Corresponding author.
[**] Who has recently started his postdoctoral position in the School of Computer Science & National Centre for Text Mining at the University of Manchester, UK.

I.-Y. Song, J. Eder, and T.M. Nguyen (Eds.): DaWaK 2008, LNCS 5182, pp. 357–367, 2008.
© Springer-Verlag Berlin Heidelberg 2008

TC has been well investigated as an intersection of research into KDD (e.g. [1]) and machine learning (e.g. [14]).

In a general context, the TC problem can be separated into two significant divisions: (1) assigning only one predefined category to each "unseen" natural language text document as in [3] and often defined as the non-overlapping or **single-label TC** task; and (2) assigning more than one predefined category to an "unseen" document as in [5] and often defined as the overlapping or **multi-label TC** task. "*A special case of single-label TC is binary TC*" [14], which in particular assigns either a predefined category or its complement to an "unseen" document. Many studies have addressed this approach in the past, i.e. [10], [14], [15], etc. In contrast, single-label TC tasks other than the binary approach are recognized as *multi-class* approaches, and simultaneously deal with all given categories and assign the most appropriate category to each "unseen" document. Individual studies under this heading include [2], [7], and [16]. When handling a set of textual data with more than two predefined categories, a sufficient set of binary TC tasks will implement a multi-class TC task with a possibly better accuracy of classification, but a drawback in terms of processing efficiency.

One important facet of developing TC approaches is being able to show a set of experimental results using common textual datasets. There are many such datasets, e.g. Reuters-21578[1], Usenet Articles[2], MedLine-OHSUMED[3], etc. With regard to single-label TC, most datasets, in their original form, have at least one of the following limitations: (i) the overall volume of textual data is too large for ease of experimentation; (ii) there are many predefined classes involved; (iii) most of the classes consist of only a very few documents; (iv) some documents are labeled with a single class whereas others have multiple classes; and (v) there are documents found without any actual textual content, i.e. a document containing less than δ recognized words (a recognized word is further defined in section 2.1), where δ is usually a small constant. Hence it is difficult to run TC experiments using a textual dataset in its original form, especially when dealing with multi-label datasets while trying to perform experiments in a single-label TC environment. When comparing the performance among alternative TC approaches, it is often necessary to extract sub datasets (which we call document-bases) from the original data source. In this paper, we investigate the textual data preparation problem, and propose a standard document-base extraction approach for single-label TC, which automatically generates "*qualified*" document-bases (such document-bases contain "*qualified*" documents only, further defined in section 3) from a given textual data source that can be used more effectively and more reliably in single-label TC experiments.

The rest of this paper is organized as follows. Section 2 describes some previous work in document-base extraction for TC. In section 3, we propose a five-state document-base extraction approach for single-label TC. The results are presented in section 4, where one document-base (RE.D6643.C8) is generated from the Reuters-21578 collection; two document-bases (NG.D9482.C10 & NG.D9614.C10) are from Usenet Articles; and another document-base (OH.D6855.C10) is extracted from MedLine-OHSUMED. We show these document-bases can be used effectively in single-label TC experiments. Finally our conclusions and open issues for further research are given in section 5.

---

[1] http://www.daviddlewis.com/resources/testcollections/reuters21578/

[2] http://www.cs.cmu.edu/afs/cs/project/theo-11/www/naive-bayes/20_newsgroups.tar.gz

[3] http://trec.nist.gov/data/filtering/

2 Previous Work

2.1 Reuters-21578

Reuters-21578 is a popular text collection widely applied in text mining research. It comprises 21,578 documents collected from the Reuters newswire with 135 predefined classes. Within the entire collection, 13,476 documents are labeled with at least one class, 7,059 are clearly not marked with any class and 1,043 documents have their class-label as *"bypass"* (which, at least in our study, is not considered to be a proper class-label). Within these 13,476 classified documents, 2,308 appear to have a class but on further investigation that class turns out to be spurious. This leaves 11,168 documents, of which 9,338 are single-labeled and 1,830 are multi-labeled.

There are in total 135 classes. However, many TC studies (see for example [12] and [17]) have used only the 10 most populous classes for their experiments and evaluations. There are 68 classes that consist of fewer than 10 documents, and many others consist of fewer than 100 documents. The extracted document-base, suggested in [12] and [17], can be referred to as RE.D10247.C10 and comprises 10,247 documents with 10 classes. However RE.D10247.C10 includes multi-labeled documents that are inappropriate for a single-label TC environment.

In [4] Deng et al. introduce the Reuters_100 document-base that comprises 8,786 documents with 10 classes. Deng et al. assign *"one document (to) one category and adopt categories that contain training documents (of) more than 100"*. Unfortunately which 10 of the 135 classes had been chosen was not specified, but it can be assumed that they are close to or identical with the classes included in RE.D10247.C10 where many documents were in fact found without a *"proper"* text-content — the document contains less than δ recognized words, where δ is usually a small constant (20 in our study). Herein, a recognized word can be defined as a text-unit, separated by punctuation marks, white space or wild card characters within paragraphs, which belongs to one of the known languages (e.g. English, French, Chinese, etc.) and does not associate with any non-language component (i.e. numbers, symbols, etc.). Filtering away such non-text documents from the extracted document-base is suggested, which ensures that document-base quality is maintained.

2.2 Usenet Articles

The Usenet Articles is another well-known textual data source. It was compiled by Lang [11] from 20 different newsgroups and is sometimes referred to as the "20 Newsgroups" text collection. Each newsgroup represents a predefined class. There are exactly 1,000 documents per class with an exception — the class "soc.religion.christian" contains 997 documents only. In comparison with other common text collections (e.g. Reuters-21578), the structure of the "20 Newsgroups" collection is relatively consistent — every document within this collection is labeled with one class only and almost all documents (higher than 95% of all documents) have a proper text-content ($\delta \geq 20$). Previous TC studies have used this text collection in various ways. For example: (*i*) in [4] the entire "20 Newsgroups" was randomly divided into two non-overlapping and (almost) equally sized document-bases covering 10 classes each: NG.D10000.C10 and NG.D9997.C10; and (*ii*) in [15] four smaller document-bases were extracted from the collection and used in evaluations: NG.Comp.D5000.C5, NG.Rec.D4000.C4, NG.Sci.D4000.C4, and NG.Talk.D4000.C4. Note here that of the total 19,997 documents, 901 of them fall into our non-text category — each document contains less than 20 recognized words ($\delta < 20$). This may weaken the overall quality for these above listed ("20 Newsgroups" based) document-bases.

2.3 MedLine-OHSUMED

The MedLine-OHSUMED text collection, collected by Hersh et al. [8], consists of 348,566 records relating to 14,631 predefined MeSH (Medical Subject Headings) categories. The OHSUMED collection accounts for a subset of the MedLine text collection[4] for 1987 to 1991. Characteristics of OHSUMED include: (1) many multi-labeled documents; (2) the total 14,631 classes are named (and also considered to be arranged) in hierarchies (e.g. classes "male" and "female" can be assumed as subclasses of the class "human"; classes "adult" and "child" can be assumed as subclasses of "male" and/or "female"); and (3) the text-content of each document comprises either a title on its own (without a text-content), or a "*title-plus-abstract*" (with a text-content) from various medical journals.

With the goal of investigating the multi-label TC problem, Joachims [10] uses the first 10,000 title-plus-abstracts texts of the 50,216 documents for 1991 as the training instances, and the second 10,000 such documents as the test instances. This defines the OH.D20000.C23 document-base, in which the classes are 23 MeSH "diseases" categories. Since each record within this document-base may be labeled with more than one class, it does not satisfy the single-label TC investigation. This is also the case for the OH.Maximal document-base [17], which consists of all OHSUMED classes incorporating all 233,445 title-plus-abstract documents.

3 Proposed Document-Base Extraction

It is claimed that common textual data sources in their original form are not usually suitable to be directly employed in TC experiments. In this section, we propose a standard textual data preparation approach that automatically extracts qualified document-bases from a given large textual data source (text collection). The entire process of the proposed document-base extraction approach is illustrated graphically in Fig. 1. It consists of five component-functions (states).

1. **Top-k Class Identification:** Given a large text collection D, it is possible to find hundreds (sometimes thousands or even more) predefined classes there. However, many of them are assigned to only one or very few (usually less than 10) documents. Hence, it is considered necessary to identify the k most populous (top-k) classes with their associated documents in D. To fulfill this, we introduce the Top-k_Class_Extraction function (see Algorithm 1).

---

[4] http://medline.cos.com/

2. **Target Class Determination:** Given a collection of documents D', based on the k most populous classes (either collected originally or identified by applying the Top-k_Class_Extraction function), some classes may be within a taxonomy-like form (sharing a super-and-sub class-relationship). Note that all documents, that are included in a predefined (sub) class, are considered to be also involved in its super-class. Hence, retaining both super-and-sub classes within a created document-base would cause a conflict when running a single-label TC experiment using this document-base — each single document should not be assigned more than one class. With regard to this super-and-sub class-relationship problem, a smaller group of $k*$ target-classes are suggested to be further extracted from D', where $k* \leq k$ and $k*$ is suggested to be chosen as a non-prime integer (which has some positive divisors that can be further used in the next state). To fulfill this, we introduce the Target-$k*$_Class_Extraction function (see Algorithm 2), which takes a tree |structure representing the taxonomy-like class-relationship(s) among the top-k classes as the input.

3. **Class-group Allocation:** Given a collection of documents D'', based on the $k*$ target-classes (either collected originally or determined by applying the Top-k_Class_Extraction function and/or the Target-$k*$_Class_Extraction function), we then equally and randomly allocate these $k*$ target-classes into g non-overlapping class-groups, where g is a small constant (integer) defined by the user, usually as a positive divisor of $k*$ and $1 \leq g \leq k*/2$ with a consideration — each class-group contains at least 2 target-classes. In this state, we introduce the Class-Group_Allocation function (see Algorithm 3).

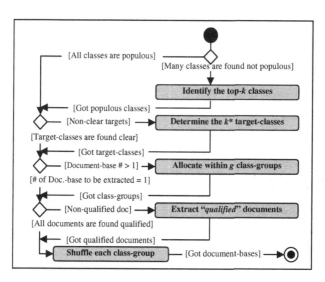

Fig. 1. A state-chart diagram for the proposed document-base extraction approach

Function Top-k_Class_Extraction
input:	(i) a given large text collection D;
	(ii) an integer k (usually ≤ 100);
output:	a collection of documents D', based on the k-most populous classes;

```
(1)  begin
(2)    Set D' ← ∅;
(3)    for each document dⱼ ∈ D do
(4)      catch its class-label(s) C'ⱼ;
(5)      for each single class-label cᵢ ∈ C' do
(6)        Set Kᵢ ← ∅;
(7)        if (Kᵢ ∈ D') then
(8)          Kᵢ ← get it from D'ⱼ;
(9)          add dⱼ into Kᵢ;
(10)         add Kᵢ into D'ⱼ;
(11)     end for
(12)   end for
(13)   sort descending all elements (classes) in D' based on
         their size (num. of contained documents);
(14)   remain the top-k elements in D'ⱼ;
(15)   return (D');
(16) end begin
```

Algorithm 1. The top-k class extraction function

Function Target-k*_Class_Extraction
input:	(i) a given collection of documents D', based on the top-k classes;
	(ii) a tree structure that represents the taxonomy-like class-relationship(s) $Tree$;
output:	a collection of documents D'', based on the k^* target-classes;

```
(1)  begin
(2)    Set D'' ← ∅;
(3)    for each class based document-set Kᵢ ∈ D' do
(4)      catch its class-label cᵢ;
(5)      if (cᵢ is found as a leaf-node ∈ Tree) then
(6)        add Kᵢ into D'';
(7)    end for
(8)    if (|D''| is a prime-number) then
(9)      remove the minimum sized element from D'';
(10)   return (D'');
(11) end begin
```

Algorithm 2. The target-k* class extraction function

Function Class-group Allocation
input:	(i) a given collection of documents D'', based on the k^* target-classes;
	(ii) an integer g ($1 ≤ g ≤ k^*/2$, and as a positive divisor of k^*);
output:	a set of g-many (equally sized) class-groups G, where each class-group is a collection of documents, based on at least 2 target-classes;

```
(1)  begin
(2)    Set G ← ∅;
(3)    Set G_temp ← ∅;
(4)    for l = 0 to g-1 do
(5)      Set Gₗ ← ∅;
(6)      add Gₗ into G_temp;
(7)    end for
(8)    for each class based document-set Kᵢ ∈ D'' do
(9)      r ← get a random decimal between 0 and 1;
(10)     l ← ⌊r × |G_temp|⌋; // '⌊⌋' gives a floor integer
(11)     catch class-group Gₗ ∈ G_temp;
(12)     add Kᵢ into Gₗ;
(13)     if (|Gₗ| = |D''| / g) then
(14)       add Gₗ into G;
(15)       remove Gₗ from G_temp;
(16)   end for
(17)   return (G);
(18) end begin
```

Algorithm 3. The class-group allocation function

Function Qualified-Document_Extraction_1
input:	a given collection of documents G, presented as a class-group;
output:	a refined collection of documents G', where each document labels to one class only;

```
(1)  begin
(2)    Set G' ← ∅;
(3)    for i = 0 to |G|-2 do
```

```
(4)      catch the class based document-set Kᵢ ∈ G;
(5)      for each document dₐ ∈ Kᵢ do
(6)        Boolean delete ← false;
(7)        for j = i+1 to |G|-1 do
(8)          catch the class based document-set Kⱼ ∈ G;
(9)          if (dₐ ∈ Kⱼ) then
(10)           delete ← true;
(11)           remove dₐ from Kⱼ;
(12)         end for
(13)         if delete then
(14)           remove dₐ from Kᵢ;
(15)       end for
(16)       add Kᵢ into G';
(17)   end for
(18)   add K_{|G|-1} into G';
(19)   return (G');
(20) end begin
```

Algorithm 4. The qualified-document extraction function (Part 1)

Function Qualified-Document_Extraction_2
input:	a given collection of documents G', presented as a class-group (each document is single-labeled);
output:	a further refined collection of documents G'', where each single-labeled document contains more than δ recognized words;

```
(1)  begin
(2)    Set G'' ← ∅;
(3)    for each class based document-set Kᵢ ∈ G' do
(4)      for each document dₐ ∈ Kᵢ do
(5)        if ((num. of recognized words in dₐ) < δ) then
(6)          remove dₐ from Kᵢ;
(7)      end for
(8)      add Kᵢ into G'';
(9)    end for
(10)   return (G'');
(11) end begin
```

Algorithm 5. The qualified-document extraction function (Part 2)

Function Document_Shuffle
input:	an ordered set of qualified documents G'', presented as a sufficiently refined class-group;
output:	a (shuffled) document-base $Ð$;

```
(1)  begin
(2)    Set Ð ← ∅;
(3)    σ ← find the minimum |Kᵢ| ∈ G'';
           // Kᵢ is a class based (qualified) document-set
(4)    Set S ← ∅;
(5)    for u = 0 to σ-1 do
(6)      Set Sᵤ ← ∅;
(7)      add Sᵤ into S;
(8)    end for
(9)    for each class based document-set Kᵢ ∈ G'' do
(10)     w ← ⌊|Kᵢ| / σ⌋; // '⌊⌋' gives a floor integer
(11)     for a = 0 to |Kᵢ|-1 do
(12)       if (a ≤ w × σ) then
(13)         v ← ⌊a / w⌋; // '⌊⌋' gives a floor integer
(14)         catch Sᵥ ∈ S;
(15)         add (document dₐ ∈ Kᵢ) into Sᵥ;
(16)         mark dₐ as a removable document in Kᵢ;
(17)     end for
(18)     remove all removable documents from Kᵢ;
(19)   end for
(20)   remove empty Kᵢ from G'';
(21)   z ← 0;
(22)   for each class based document-set Kⱼ ∈ G'' do
(23)     for each document d_b ∈ Kⱼ do
(24)       catch Sᵤ ∈ S;
(25)       add (document d_b ∈ Kⱼ) into Sᵤ;
(26)       z ← z + 1;
(27)       if (z = σ) then
(28)         z ← 0;
(29)     end for
(30)   end for
(31)   for each Sᵧ ∈ S;
(32)     for each document dₑ ∈ Sᵧ do
(33)       add dₑ into Ð;
(34)     end for
(35)   end for
(36)   return (Ð);
(37) end begin
```

Algorithm 6. The document shuffle function

4. **Qualified Document Extraction:** For each class-group G (either collected originally as a text collection or generated from ⟨state(s) 1, 2 and/or 3⟩), we now extract all *"qualified"* documents from G. We define a qualified document as a document that (*i*) belongs to only one predefined class; and (*ii*) consists of at least δ recognized words. Regarding (*i*), it is possible to discover single documents that are simultaneously labeled with two classes although they do not share a super-and-sub class-relationship (as per state 2). To solve this problem, we provide the Qualified-Document_Extraction_1 function (see Algorithm 4). Regarding (*ii*), a further refined document-base will be generated — at least δ recognized words are ensured within each extracted document. Hence, multi-word (phrases, quasi phrases and/or single-word combinations) are more likely to be discovered. This addresses a diversified feature selection approach (i.e. "bag of phrases" vs. "bag of words") in a further document-base preprocessing phase. The Qualified-Document_Extraction_2 function, aiming to filter away such non-text documents from the output of Qualified-Document_Extraction_1, is provided (see Algorithm 5).

5. **Document Shuffle:** Given an ordered set of documents G'', presented as a class-group with qualified documents only, we finally shuffle these documents, and construct a document-base $Đ$. Note that when investigating single-label TC, especially the multi-class problem, the cross-validation procedure is suggested to be addressed in a further training-and-test experimental phase. Employing the cross-validation procedure in a TC experiment requires (*i*) dividing the given document-base into f-fold (normally $f = 10$); (*ii*) in each of the f runs, treating the ith-fold as a test set (of instances) whereas the rest folds as the training dataset; and (*iii*) calculating the average of f-run TC results (accuracies). The cross-validation procedure requires inputting a sufficiently shuffled document-base, where documents sharing a common predefined class should be evenly and dispersedly distributed within the entire document-base. This ensures that when randomly picking up a fraction of the document-base having its minimum size $\approx \sigma$, where σ represents the size of the smallest class (containing the least documents) in G'', a sufficient number of documents are found within each predefined class. In this state, we introduce the Document_Shuffle function (see Algorithm 6).

4 Results

In this section, we show four extracted document-bases[5] regarding the case of single-label multi-class TC, where one is generated from Reuters-21578, two from "20 Newsgroups", and another one from MedLine-OHSUMED.

* **The Reuters-21578 based Document-base:** Given Reuters-21578 in its original form, we first of all identified the Top-10 populous classes by applying the Top-k_Class_Extraction function, which confirm the 10 most populous classes, suggested in [12] and [17]. Since super-and-sub class-relationships were not found within the Top-10 classes, we skipped the state of determination of the $k*$

---

[5] The four extracted document-bases may be obtained from http://www.csc.liv.ac.uk/~jwang/

target-classes. We treated the Top-10 classes as a unique class-group that ensures only one document-base would be extracted from this data source. After running an implementation of both Qualified-Document_Extraction_1 and Qualified-Document_Extraction_2 (with $\delta = 20$) functions, we found that the class "wheat" contains only one qualified document, and no qualified document was contained in class "corn". Hence, the final document-base, namely RE.D6643.C8, omitted these classes of "wheat" and "corn", leaving a total of 6,643 documents in 8 classes. To complete the document-base extraction, we fairly shuffled these 6,643 documents finally. A description of this document-base is given in Table 1.

- **Two "20 Newsgroups" based Document-bases:** When generating document-bases from "20 Newsgroups", the first and second states of our proposed approach were skipped because (*i*) all of the 20 given classes are equally populous and (*ii*) there is not a hierarchy of class relationships within the 20 classes. We decided to adopt the approach of Deng et al. [4] and randomly split the entire data source, by applying the Class-Group_Allocation function, into two class-groups covering 10 classes each.

  o Focusing on the first class-group, we then checked the qualification of each document. Since all documents are known to be single-labeled, we skipped to the Qualified-Document_Extraction_1 function. Having $\delta = 20$, we refined this class-group by using the Qualified-Document_Extraction_2 function. A total of 518 non-text documents were filtered away. We finally shuffled this class-group and created the NG.D9482.C10 document-base. Table 2(a) shows the detail of NG.D9482.C10.

  o Focusing on the second class-group, the qualification of each document was then verified. Again, since all "20 Newsgroups" based documents are single-labeled, we skipped the Qualified-Document_Extraction_1 function. Having $\delta = 20$, we refined this class-group by applying the Qualified-Document_Extraction_2 function. A total of 383 non-text documents were filtered away. We finally shuffled this class-group and created the NG.D9614.C10 document-base. A description of NG.D9614.C10 is provided in Table 2(b).

The OHSUMED based Document-base: When generating document-bases from MedLine-OHSUMED, we first of all identified the Top-100 populous classes by applying the Top-k_Class_Extraction function. It is obvious that some of the Top-k classes are originally named in hierarchies (as previously described in section 2.3). Hence we assume that the super-and-sub class-relationships exist among these classes. Due to the difficulty of obtaining a precise tree structure that describes all possible taxonomy-like class-relationships within the Top-100 classes, instead of applying the Target-k^*_Class_Extraction function, we simply selected 10 target-classes from these classes by hand, so as to exclude obvious super-and-sub class-relationships. We simply treated the Top-10 classes as a unique class-group that ensures only one

- document-base would be extracted from this data source. We then checked the qualification of each document. Since a document may be multi-labeled, we called the Qualified-Document_Extraction_1 function to remove the documents that do not label to exactly 1 of the 10 target-classes. Having $\delta = 20$, we further refined this class-group by applying the **Qualified-Document_Extraction_2** function. As a consequence 6,855 documents within 10 classes were comprised in the refined form of this class-group. We finally shuffled it and created the OH.D6855.C10 document-base. Table 3 shows the detail of this document-base.

Table 1. Document-base description (RE.D6643.C8)

Class	# of documents	Class	# of documents
acq	2,108	interest	216
crude	444	money	432
earn	2,736	ship	174
grain	108	trade	425

Table 2. Document-base description (NG.D9482.C10 & NG.D9614.C10)

(a) NG.D9482.C10		(b) NG.D9614.C10	
Class	# of documents	Class	# of documents
comp.windows.x	940	comp.graphics	919
rec.motorcycles	959	comp.sys.mac.hardware	958
talk.religion.misc	966	rec.sport.hockey	965
sci.electronics	953	sci.crypt	980
alt.atheism	976	sci.space	977
misc.forsale	861	talk.politics.guns	976
sci.med	974	comp.os.ms-windows.misc	928
talk.politics.mideast	966	rec.autos	961
comp.sys.ibm.pc.hardware	955	talk.politics.misc	980
rec.sport.baseball	932	soc.religion.christian	970

Table 3. Document-base description (OH.D6855.C10)

Class	# of documents	Class	# of documents
amino_acid_sequence	333	kidney	871
blood_pressure	635	rats	1,596
body_weight	192	smoking	222
brain	667	tomography,_x-ray_computed	657
dna	944	united_states	738

These four (extracted) document-bases were further evaluated in a single-label TC environment. All evaluations described here were conducted using the TFPC (Total From Partial Classification) associative text classifier[6] [18]; although any other classifier could equally well have been used. All algorithms involved in the evaluation were implemented using the standard Java programming language. The experiments were

---

[6] TFPC associative text classier may be obtained from http://www.csc.liv.ac.uk/~frans/KDD /Software/TextMiningDemo/textMining.html

run on a 1.87 GHz Intel(R) Core(TM)2 CPU with 2.00 GB of RAM running under Windows Command Processor.

In the preprocessing of each document-base, we first of all treated these very common and rare words (with a document-base frequency > 20% or < 0.2%) as the noise words and eliminated them from the document-base. For the rest of words, we simply employed the *mutual information* feature selection approach [14] to identify these *key* words that significantly serve to distinguish between classes. Finally the top 100 words (based on their mutual information score) were decided to be remained in each class. With a support threshold value of 0.1% and a confidence threshold value of 35% (as suggested in [18]), we identified (using Ten-fold Cross Validation): the classification accuracy generated using the RE.D6643.C8 document-base was 86.23%, whereas NG.D9482.C10 and NG.D9614.C10 produced the accuracies of 77.49% and 81.26%, and 79.27% was given by using the OH.D6855.C10 document-base. We expect better TC results, based on these extracted document-bases, when applying improved textual data preprocessing and/or classification approaches.

5 Conclusion

When investigating text mining and its applications, especially when dealing with different TC problems, being able to show a set of experimental results using common text collections is required. Due to a list of major limitations (see section 1), we indicate that most text collections (textual data sources), in their original form, are not suggested to be directly addressed in TC experiments. In this paper, we investigated the problem of textual data preparation, and introduced a standard document-base extraction approach for single-label TC. Based on three well-known textual data sources (Reuters-21578, Usenet Articles, and MedLine-OHSUMED), we extracted four document-bases and tested them (with a simple preprocessing approach and an associative classifier) in a single-label TC environment. The experimental results demonstrate the effectiveness of our approach. Further single-label TC related studies are invited to utilize our proposed document-base extraction approach or directly make use of our generated document-bases (RE.D6643.C8, NG.D9482.C10, NG.D9614.C10, and OH.D6855.C10) in their result and evaluation part. In the future, many further textual data preparation approaches can be proposed for a variety of text mining applications. One possible task is to extract qualified document-bases from a large textual data source for multi-label TC experiments.

References

1. Antonie, M.-L., Zaïane, O.R.: Text Document Categorization by Term Association. In: Proceedings of the 2002 IEEE International Conference on Data Mining, Maebashi City, Japan, December 2002, pp. 19–26. IEEE, Los Alamitos (2002)
2. Berger, H., Merkl, D.: A Comparison of Text-Categorization Methods applied to N-Gram Frequency Statistics. In: Proceedings of the 17th Australian Joint Conference on Artificial Intelligence, Cairns, Australia, December 2004, pp. 998–1003. Springer, Heidelberg (2004)
3. Cardoso-Cachopo, A.: Improving Methods for Single-label Text Categorization. Ph.D. Thesis, Instituto Superior Técnico – Universidade Ténica de Lisboa / INESC-ID, Portugal

4. Deng, Z.-H., Tang, S.-W., Yang, D.-Q., Zhang, M., Wu, X.-B., Yang, M.: Two Odds-radio-based Text Classification Algorithms. In: Proceedings of the Third International Conference on Web Information Systems Engineering Workshop, Singapore, December 2002, pp. 223–231. IEEE, Los Alamitos (2002)

5. Feng, Y., Wu, Z., Zhou, Z.: Multi-label Text Categorization using K-Nearest Neighbor Approach with M-Similarity. In: Proceedings of the 12th International Conference on String Processing and Information Retrieval, Buenos Aires, Argentina, November 2005, pp. 155–160. Springer, Heidelberg (2005)

6. Fragoudis, D., Meretaskis, D., Likothanassis, S.: Best Terms: An Efficient Feature-Selection Algorithm for Text Categorization. Knowledge and Information Systems 8(1), 16–33 (2005)

7. Giorgetti, D., Sebastiani, F.: Multiclass Text Categorization for Automated Survey Coding. In: Proceedings of the 2003 ACM Symposium on Applied Computing, Melbourne, FL, USA, March 2003, pp. 798–802. ACM Press, New York (2003)

8. Hersh, W.R., Buckley, C., Leone, T.J., Hickman, D.H.: OHSUMED: An Interactive Retrieval Evaluation and New Large Test Collection for Research. In: Proceedings of the 17th Annual International ACM SIGIR Conference on Research and Development in Information Retrieval, Dublin, Ireland, July 1994, pp. 192–201. ACM/Springer (1994)

9. Hotho, A., Nürnberger, A., Paaß, G.: A Brief Survey of Text Mining. LDV Forum – GLDV Journal for Computational Linguistics and Language Technology 20(1), 19–62 (2005)

10. Joachims, T.: Text Categorization with Support Vector Machines: Learning with Many Relevant Features. LS-8 Report 23 – Research Reports of the Unit no. VIII (AI), Computer Science Department, University of Dortmund, Germany

11. Lang, K.: NewsWeeder: Learning to Filter Netnews. In: Proceedings of the Twelfth International Conference on Machine Learning, Tahoe City, CA, USA, July 1995, pp. 331–339. Morgan Kaufmann Publishers, San Francisco (1995)

12. Li, X., Liu, B.: Learning to Classify Texts using Positive and Unlabeled Data. In: Proceedings of the Eighteenth International Joint Conference on Artificial Intelligence, Acapulco, Mexico, August 2003, pp. 587–594. Morgan Kaufmann Publishers, San Francisco (2003)

13. Maron, M.E.: Automatic Indexing: An Experimental Inquiry. Journal of the ACM (JACM) 8(3), 404–417 (1961)

14. Sebastiani, F.: Machine Learning in Automated Text Categorization. ACM Computing Surveys 34(1), 1–47 (2002)

15. Wu, H., Phang, T.H., Liu, B., Li, X.: A Refinement Approach to Handling Model Misfit in Text Categorization. In: Proceedings of the Eighth ACM SIGKDD International Conference on Knowledge Discovery and Data Mining, Edmonton, Alberta, Canada, July 2002, pp. 207–215. ACM Press, New York (2002)

16. Wu, K., Lu, B.-L., Uchiyama, M., Isahara, H.: A Probabilistic Approach to Feature Selection for Multi-class Text Categorization. In: Proceedings of the 4th International Symposium on Neural Networks, Nanjing, China, June 2007, pp. 1310–1317. Springer, Heidelberg (2007)

17. Zaïane, O.R., Antonie, M.-L.: Classifying Text Documents by Associating Terms with Text Categories. In: Proceedings of the 13th Australasian Database Conference, Melbourne, Victoria, Australia, January-February 2002, pp. 215–222. CRPIT 5 Australian Computer Society (2002)

18. Coenen, F., Leng, P., Sanderson, R., Wang, Y.J.: Statistical Identification of Key Phrases for Text Classification. In: Proceedings of the 5th International Conference on Machine Learning and Data Mining, Leipzig, Germany, July 2007, pp. 838–853. Springer, Heidelberg (2007)

How an Ensemble Method Can Compute a Comprehensible Model

José L. Triviño-Rodriguez, Amparo Ruiz-Sepúlveda, and Rafael Morales-Bueno

Department of Computer Science and Artificial Intelligence
University of Málaga, Málaga (Spain)
{trivino,amparo,morales}@lcc.uma.es
http://www.lcc.uma.es

Abstract. Ensemble machine learning methods have been developed as an easy way to improve accuracy in theoretical and practical machine learning problems. However, hypotheses computed by these methods are often considered difficult to understand. This could be an important drawback in fields such as data mining and knowledge discovery where comprehensibility is a main criterion. This paper aims to explore the area of trade-offs between accuracy and comprehensibility in ensemble machine learning methods by proposing a learning method that combines the accuracy of boosting algorithms with the comprehensibility of decision trees. The approach described in this paper avoids the voting scheme of boosting by computing simple classification rules from the boosting learning approach while achieving the accuracy of AdaBoost learning algorithm in a set of UCI datasets. The comprehensibility of the hypothesis is thus enhanced without any loss of accuracy.

1 Introduction

For a long time developments in machine learning have been focused on the improvement of models accuracy. This research has given rise to a wide range of representations of induced hypotheses that allow modelling many kinds of problems with some accuracy. For example, there are learning algorithms that represent their hypotheses as decision trees [1,2], decision lists [3] inference rules [4], neural networks [5], hidden Markov models [6], Bayesian networks [7] and stored lists of examples [8]

Some of these hypothesis models improve the accuracy at the expense of comprehensibility. Neural networks, for example, provide good predictive accuracy in a wide variety of problem domains, but produce models that are notoriously difficult to understand.

Ensemble methods have been another means of improving model accuracy [9]. These methods combine several hypotheses in order to obtain a meta hypothesis with higher accuracy than any of the combined hypotheses on their own. Decision tree learning (either propositional or relational) is especially benefited by ensemble methods [10,11]. Well known techniques for generating and combining hypotheses are boosting [12,10], bagging [13,10], randomization [14], stacking

I.-Y. Song, J. Eder, and T.M. Nguyen (Eds.): DaWaK 2008, LNCS 5182, pp. 368–378, 2008.

[15], bayesian averaging [16] and windowing [17]. Ensemble methods have been found to be quite effective in practice [10] and also has substantial theoretical foundations.

However, from the point of view of knowledge acquisition, the learning hypothesis generated by these methods cannot be easily understood [18]. For example, while a single decision tree can easily be understood if it is not too large, fifty such trees, even if individually simple, are beyond the capacity of even the most patient user.

Comprehensibility is a criterion that cannot be easily overlooked. Its importance is argued by Michalski [4] in his comprehensibility postulate:

> The results of computer induction should be symbolic descriptions of given entities, semantically and structurally similar to those a human expert might produce observing the same entities. Components of these descriptions should be comprehensible as single *chunks* of information, directly interpretable in natural language, and should relate quantitative and qualitative concepts in an integrated fashion. (pg. 122)

Recently, the use of machine learning methods for data mining and knowledge discovery has set high value on algorithms that produce comprehensible output [19]. There are also practical reasons for this. In many applications, it is not enough for a learner model to be accurate; it also needs to be understood by its users, before they can trust and accept it.

Focusing on comprehensibility without losing too much accuracy has been the main goal of several developments in machine learning in the last years. Two different approaches have been followed in order to achieve this goal with ensemble methods.

The first approach [20] considers the ensemble hypothesis as an oracle that would allow us to measure the similarity of each single hypothesis in relation to this oracle. Once the ensemble hypothesis has been computed, this approach searches for the single hypothesis most similar (semantically) to the oracle (ensemble hypothesis). The search task applies a measure of hypothesis similarity to compare every single hypothesis with the oracle. This introduces a new heuristic parameter in the learning model with a strong effect on the accuracy of the selected hypothesis.

Instead of measuring the similarity of each single hypothesis in relation to this oracle, several researchers, including [21,22], generate artificial data based on the ensemble to build a new comprehensible model.

A second approach called SLIPPER [23] computes directly a comprehensible hypothesis using metrics based on the formal analysis of boosting algorithm. It tries to combine the boosting accuracy with a well known user readable hypothesis model as rulesets. Although the ruleset used as hypothesis by SLIPPER is more comprehensible than a boosting hypothesis, the label assigned to an instance depends on the vote of several rules. So, it conforms to a voting approach. It is therefore more difficult to understand than a simple decision tree where every instance is classified based on only one disjoint rule.

Following this approach, Freund et. al. [24] suggest a new combination of decision trees with boosting that generate classifications rules. They call it *alternating decision trees* (ADTrees). This representation generalizes both voted-stumps and decision trees. However, unlike decision trees, the classification that is associated with a path in the tree is not the leaf label, instead, it is the sign of the sum of the predictions along the path. Thus ADTrees are a scoring classification model that does not compute disjoint rules to classify a sample. It therefore retains the same drawback in comprehensibility as SLIPPER.

This paper attempts to explore the potential of trade-offs between accuracy and comprehensibility by proposing a learning method that combines the accuracy of boosting algorithm with the comprehensibility of decision trees. Boosting [12] algorithm has been shown to be a good accuracy ensemble method and is well-understood formally.

On the other hand, decision trees are a hypothesis model easy to understand. They can be transformed into a set of disjoint rules. So the classification of an instance only depends on one rule. They therefore avoid the main drawback of ensemble methods. Moreover it has been proved that Top-Down Induction Decision Tree Learning (TDIDTL) can be viewed as a boosting algorithm [25].

The approach described in the present paper substitutes the linear classifier of the AdaBoost [12] algorithm with a decision tree learning algorithm like C4.5. The voting scheme of the boosting algorithm is therefore substituted by a set of disjoint rules over the classification done by the weak learners algorithms. If the hypothesis computed by weak learners is comprehensible, then the rules computed by the decision tree meta learner should be easy to understand.

In section 2 of this paper an approach that combines boosting with decision trees is described. Then in section 3 this model will be empirically evaluated with regard to the accuracy, stability and size of the model in relation to ADTrees and Boosting.

2   Boosting with Decision Trees

In [25], Kearns et al. examine top-down induction decision tree (TDIDT) learning algorithms in the model of *weak learning* [26], that is, like a boosting algorithm. They proved that the standard TDIDT algorithms are in fact boosting algorithms.

From this point of view, the class of functions labeling the decision tree nodes acts as the set of weak hypothesis in the boosting algorithm. Thus, splitting a leaf in a top-down decision tree is equivalent to adding a new weak hypothesis in a boosting algorithm. A sample is classified by a decision tree by means of the vote of functions in the nodes of the tree. Every vote represents a step in a path from the root node of the tree to a leaf of the tree. A sample is classified with the label of the leaf at the end of the path defined by the votes of the functions in the internal nodes of the tree.

However, Dieterich et al. [27] have shown that top-down decision trees and Adaboost have qualitatively different behavior. Briefly, they find that although decision trees and Adaboost are both boosting algorithm, Adaboost creates successively *"harder"* filtered distributions, while decision trees create successively *"easier"* ones. In other words, while Adaboost concentrates the probability distribution over the training sample on samples misclassified by weak hypothesis, focusing the learning task more and more on the hard part of the learning problem; top-down decision trees form increasingly refined partitions of the input space. These partitions eventually become sufficiently fine that very simple functions can approximate the target function well within a cell. Thus, while boosting algorithm uses all the learning sample to train every weak learner, the learning sample on a decision tree is distributed on the leaves of the tree. And so the number of samples on every leaf decreases successively with the depth of the tree. A small number of samples on a leaf could produce overfitting. In order to avoid overfitting in top-down decision tree learning, several methods and heuristics such as pruning have been developed.

Nonetheless, both approaches could be combined in a learning method that has the accuracy and stability of boosting algorithms and the comprehensibility of decision trees hypothesis. In algorithm 1, the AdaBoost learning algorithm is shown. It is composed of a main loop that computes a set of weak hypothesis $H = \{h_t | 1 \leq t \leq T\}$ and a last statement that computes the final hypothesis by means of equation 1.

Algorithm 1. The AdaBoost algorithm for binary classification problems returning $h_{final} : X \rightarrow Y$ [12]

Let:

- A training E set of m samples $E = \{(x_1, y_1), \ldots, (x_m, y_m)\}$ with $x_i \in X = \{$vectors of attribute values$\}$ and labels $y_i \in Y = \{-1, 1\}$
- A weak learning algorithm $WeakLearner$
- $T \in \mathbb{N}$ specifiying number of iterations

1: Let $D_1(i) = 1/m$ for all $1 \leq i \leq m$
2: **for** $t = 1$ to T **do** {default}
3:    Call $WeakLearner$ providing it with the distribution D_t
4:    Get back a weak hypothesis $h_t : X \rightarrow Y$
5:    Calculate error of h_t: $\epsilon_t = \sum_{i:h_t(x_i) \neq y_i} D_t(i)$
6:    Let $\alpha_t = \frac{1}{2} \ln \frac{1-\epsilon_t}{\epsilon_t}$
7:    Update distribution D_t:

$$D_{t+1}(i) = \frac{D_t(i)}{Z_t} * \begin{cases} e^{-\alpha_t} & \text{if } h_t(x_i) = y_i \\ e^{\alpha_t} & \text{otherwise} \end{cases}$$

where Z_t is a normalization constant (chosen so that D_{t+1} will be a distribution)
8: **end for**
9: Final hypothesis:

$$h_{final} = sign \sum_{t=1}^{T} (\alpha_t * h_t(x)) \tag{1}$$

In AdaBoost algorithm, given a sample x, $h_t(x)$ could be interpreted as the coordinate t of x in an instance space $F = \{\langle h_1(x), h_2(x), \ldots, h_T(x)\rangle | x \in X\} \sqsubseteq Y^T$. So F is the set of all points of Y^T that are generated by the set of weak hypotheses $H = \{h_t | 1 \le t \le T\}$. Thus, we could think of the main loop as a mapping from the input space X into a new space F. From this point of view, the final hypothesis h_{final} computed by equation 1, which is a weighted majority vote of the T weak hypotheses where α_t is the weight assigned to h_t, defines a linear classifier over the space F. In other words, we could think of AdaBoost in terms of a mapping from an input space X to an instance space F and a linear classifier over the instance space F.

If we assume that the weak hypothesis learner is simple enough, for example a decision stump, then we could assume that all the weak hypotheses are comprehensible and that the difficulty in understanding the final hypothesis lies in the linear classifier over a space defined by weak learners.

As we have seen previously, the top-down decision tree can be a good substitute for the linear classifier of AdaBoost algorithm. It can be regarded as a boosting algorithm and it computes comprehensible hypotheses.

Thus, we could substitute the linear classifier of AdaBoost computed by equation 1 in the algorithm shown in algorithm 1 with a decision tree. This proposed approach is shown in algorithm 2. Although this approach could be seen as a kind of stacking, there are several features that make it closer to an AdaBoost approach than to the stacking approach described by Wolpert [15]: stacking approach splits the training dataset into several subsets each one used to train an independent weak learner in order to map every training sample to the space F, while each weak learner in AdaBoost is trained with the entire dataset. Moreover, Adaboost modifies the weight of samples of the learning dataset throughout the learning task. Finally, the accuracy of each weak learner is used to modify the weight of samples of the learning dataset in order to train the next weak learner; so each weak learner is influenced by the accuracy of the previous one, while stacking only considers the accuracy of the weak learners at the meta learning level.

Moreover, since decision stumps are classifiers that consider only a single variable of the input space, they are equivalent to the class of single variable projection functions usually used in the top-down decision tree learners [17]. If then the linear classifier of AdaBoost over decision stumps is substituted by a top-down decision tree learner, then the final hypothesis will be compounded by a tree with a query about a single variable of the input space in the internal nodes. Thus, this final hypothesis will be equivalent to a decision tree computed directly over the input space. However, as we shall see in the next section, the mapping of the input space improves the accuracy and stability of the learned decision tree.

Of course, we could have taken into account more complex classifiers as weak learners in Boosting such as, for example, bayesian classifiers or neural networks. But their comprehensibility is lower than the comprehensibility of decision stumps and they will not take into account in this paper.

Algorithm 2. The Decision Tree Boosting algorithm for binary classification problems returning $h_{final} : X \rightarrow Y$

Let:

- A training set E of m samples $E = \{(x_1, y_1), \ldots, (x_m, y_m)\}$ with $x_i \in X = \{$vectors of attribute values$\}$ and labels $y_i \in Y = \{-1, 1\}$
- A weak learning algorithm $WeakLearner$
- $T \in \mathbb{N}$ specifying number of iterations

1: Compute a set H of T weak hypothesis following steps 1 to 8 of the AdaBoost.M1 algorithm shown in algorithm 1
2: Map every instance of E to the instance space $F = \{h_t(x)|x \in X\} \sqsubseteq Y^T$:

$$\bar{E} = \{(< h_1(x_i), \ldots, h_T(x_i) >, y_i)|(x_i, y_i) \in E\}$$

3: Let h_{final} be the decision tree learned from the training dataset $\bar{E}$

---

3 Empirical Evaluation

In this section, we present an experimental evaluation of our approach. The accuracy, stability and size of boosting with decision trees (DTB) is compared with AdaBoost, Stacking and ADTrees.

We have used several datasets from the UCI dataset repository [28]. In table 1 the datasets used to compare these methods are shown. It displays the dataset name, the size in number of examples and the number of attributes including the class attribute.

In order to implement Boosting with Decision Trees (DTB), a Top-Down Decision Tree learning algorithm and a weak learner are needed. In order to make use of previous implementation of this algorithm, Boosting with Decision Trees method has been implemented and integrated into the Weka software package [29]. In the experiments then we will use the J48 implementation of the

Table 1. Information about UCI datasets used in the experiments

Dataset	Size	Attributes	Dataset	Size	Attributes
Breast-cancer	286	10	Kr-vs-kp	3196	37
Wisconsin-breast-cancer	699	10	Labor	57	17
Horse-colic	368	23	Liver-disorders	345	7
Credit-rating	690	16	Molecular-biology-p.	106	59
German-credit	1000	21	Mushroom	8124	23
Cylinder-bands	540	40	Sick	3772	30
Pima-diabetes	768	9	Sonar	208	61
Haberman	306	4	Spambase	4601	58
Heart-statlog	270	14	Tic-tac-toe	958	10
Hepatitis	155	20	Vote	435	17
Ionosphere	351	35			

C4.5 algorithm as the Top-Down Decision Tree learning algorithm. Moreover, in order to compare our method with Boosting and Alternate Decision Trees [24], several tests have been carried out using the AdaBoostM1 implementation of Boosting in Weka and the ADTree classifier. Finally, the Decision-Stump learning algorithm of Weka has been taken as the weak learner algorithm.

All of the tested learning models have a learning parameter that defines the number of weak learners computed throughout the learning task (AdaBoost, ADTrees) or the number of folds to split the training sample (Stacking). Before comparing the results of these learning models over DTB, we optimized this parameter for every dataset and learning algorithm. So the results shown in this paper are the most accurate obtained from these learning algorithms after optimizing this learning parameter. However, this parameter has not been fully optimized for DTB and only two values for this parameter (10 and 100 weak learners) have been taken into account in order to avoid unfair advantage to DTB.

Once the number of weak learners has been set, the accuracy of the DTB, AdaBoost, Stacking and ADTrees over the datasets has been computed by means of a ten-fold cross validation. The results of this test and the standard deviation of the accuracy throughout the ten iterations of this test are shown in table 2.

Table 2 shows that DTB has the same accuracy or even better accuracy than Boosting and Stacking. With the same weak learner (Decision Stump), DTB achieves significantly the same accuracy as Boosting and it improves it in 4 of the 21 datasets. It is necessary to apply Boosting to a weak learner like C4.5 in order to improve the accuracy achieved by DTB, but it can only be improved in 4 of the 21 datasets.

DTB performs better than Stacking using Decision Stumps in 12 of the 21 datasets and only in one dataset is the accuracy of Stacking using C4.5 greater than the accuracy of DTB.

Table 2. Average accuracies (Acc.) and their standard deviations (SD). Bold face notes that the accuracy is significantly lower than DTB using a t-Test with significance 0.01. '*' notes that the accuracy is significantly higher than DTB.

Dataset	AdaBoost DS		AdaBoost C4.5		Stacking DS		Stacking C4.5		ADTree		DTB DS	
	Acc.	SD	Acc.	SD	Acc.	SD	Acc.	SD	Acc.	SD	Acc.	SD
Breast-cancer	71.69	7.14	**66.89**	7.33	**68.86**	5.15	72.33	5.34	72.00	7.04	75.02	6.10
Wisconsin-breast-cancer	95.55	2.35	96.68	2.14	**92.33**	3.27	94.85	2.88	95.78	2.45	95.08	2.35
Horse-colic	82.53	5.43	81.74	5.66	81.52	5.82	85.13	5.89	84.10	5.90	83.12	5.70
Credit-rating	86.17	4.22	86.38	3.85	85.51	3.96	84.45	3.80	85.38	4.14	84.42	3.91
German-credit	74.63	3.64	74.40	3.23	**70.00**	0.00	70.56	2.90	72.69	3.91	73.25	3.37
Cylinder-bands	76.33	5.27	**57.78**	0.74	68.57	3.93	**57.78**	0.74	77.50	5.47	73.28	5.51
Pima-diabetes	75.45	4.39	73.78	4.56	71.60	5.68	73.84	5.14	74.04	4.61	74.92	5.93
Haberman	74.10	6.03	69.82	6.91	73.11	2.21	72.77	2.63	72.17	6.27	74.92	5.93
Heart-statlog	81.63	6.66	80.04	6.53	**71.67**	8.31	78.07	7.68	79.52	7.13	81.52	7.44
Hepatitis	82.40	8.30	84.74	7.48	79.05	3.31	78.28	7.45	82.55	8.59	79.10	8.77
Ionosphere	92.63	4.16	93.93	3.81	**82.57**	4.88	89.72	4.42	91.92	3.86	93.19	4.09
Kr-vs-kp	**95.15**	1.26	99.60*	0.31	**66.05**	1.72	99.44*	0.37	99.58*	0.34	98.08	0.69
Labor	91.40	12.39	88.90	14.11	80.07	14.82	75.90	17.37	86.73	15.58	85.90	13.50
Liver-disorders	73.59	6.87	70.08	7.14	**57.52**	4.66	65.10	7.22	67.34	7.54	67.33	8.487
Molecular-biology-promote	92.08	8.01	94.54	7.01	**70.34**	10.83	75.66	12.65	90.81	9.27	86.41	9.47
Mushroom	**99.90**	0.11	100.00	0.00	**88.68**	1.11	100.00	0.00	100.00	0.00	99.99	0.05
Sick	97.29	0.77	99.06*	0.45	96.55	0.91	98.60	0.57	99.16*	0.49	98.04	0.71
Sonar	84.77	7.39	85.25	7.65	72.25	9.44	72.84	9.62	72.78	7.25	76.85	9.40
Spambase	93.44	1.22	95.34*	0.87	**77.68**	3.29	92.64	1.16	94.59*	1.14	93.01	1.22
Tic-tac-toe	**88.60**	2.54	99.04*	1.01	**69.21**	4.34	**85.35**	3.30	97.83*	1.64	95.09	2.31
Vote	96.25	2.92	95.51	3.05	95.43	3.07	96.55	2.60	96.46	2.75	95.70	2.81
Average	85.98	4.81	85.40	4.47	77.07	4.80	81.90	4.94	85.85	5.02	84.98	5.05

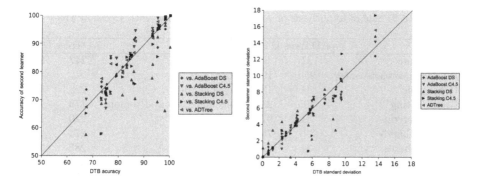

Fig. 1. Left: Ratio between accuracy of other classifiers (Y axis) and Boosting with Decision Trees accuracy (X axis). Points below the lines $x = y$ correspond to datasets for which DTB performs better than some second learner. Right: Ratio between boosting with decision trees standard deviation of accuracy on a ten cross validation (X axis) and other classifiers standard deviation (Y axis). Points over the lines $x = y$ correspond to datasets for which DTB has a lower standard deviation than a second classifier.

Best results have been obtained with ADTrees. ADTrees improve significantly the accuracy of DTB in 4 of the 21 datasets. In the rest of the datasets, DTB achieves the same accuracy as ADTrees. However, we should take into account that ADTrees achieve this accuracy with a hypothesis significantly bigger than DTB hypothesis as shown in table 3.

Figure 1 shows how DTB improves the accuracy of other learners. For example, the accuracy of Boosting using Decision Stump in the Breast-cancer dataset is 71.19% and the accuracy of DTB is 75.02%. So, DTB improves the accuracy of Boosting using Decision Stumps by 3.33%. This figure shows the accuracy of Boosting, Stacking and ADTrees (Y axis) relative to the accuracy of DTB (X axis). Points below the lines $x = y$ correspond to datasets for which DTB performs better than any second learner. Visual inspection confirms that DTB's performance is almost always close to other learners.

Another important feature of a learning model is consistency. This feature has been widely studied in machine learning in order to find robust to noise machine learning methods, see [30]. Neumann [31] describes the consistency of a rule extraction method such as the ability of the algorithm to extract rules, under various training sessions, with the same degree of accuracy. Standard deviation over the ten-fold cross evaluation is a good measure of consistency. Figure 1 shows the standard deviation of DTB (X axis) relative to Boosting, Stacking and ADTrees (Y axis) for each dataset.

In figure 1, points over the lines $x = y$ correspond to datasets for which DTB has a standard deviation lower than some second classifier. This figure shows that DTB has the same consistency as Boosting in most of the datasets. Furthermore, DTB has a higher stability than Stacking for several datasets.

However, the accuracy and consistency of DTB is not as good as it could be to justify the practical application of this model and the cost of computing a

Table 3. Size of hypothesis of ADTrees and Boosting with Decision Trees

Dataset	ADTrees	DTB	Dataset	ADTrees	DTB
Breast-cancer	31	5	Kr-vs-kp	247	27
Wisconsin-breast-cancer	250	19	Labor	82	11
Horse-colic	31	13	Liver-disorders	298	47
Credit-rating	295	23	Molecular-Biology-Promote	175	17
German-credit	31	97	Mushroom	220	19
Cylinder-bands	301	75	Sick	274	29
Pima-diabetes	31	49	Sonar	283	29
Haberman	31	7	Spambase	301	151
Heart-statlog	31	15	Tic-tac-toe	259	77
Hepatitis	238	23	Vote	31	11
Ionosphere	271	21			

Top-down Decision Tree over several weak learners instead of computing directly a Boosting voting model or an ADTree. So the main advantage of DTB over Boosting is that DTB can compute or improve a weak learner just as Boosting can but DTB computes a comprehensible hypothesis.

In order to show this, the size of the DTB hypothesis has been compared with the size of the ADTrees hypothesis. Table 3 shows the size[1] of the tree computed by ADTrees in the datasets and the size of the tree computed by DTB. It is important to note that the TDIDT computed by DTB has not necessarily included all the weak learners.

Results in table 3 are very interesting, and reflect the main goal achieved by this paper, since they show that DTB can achieve the same accuracy as Boosting or ADTrees with a size of hypothesis smaller than the size of the hypothesis of ADTrees and without a scoring classification scheme. Of course, the size of the hypothesis of a DTB is smaller and more comprehensible than the hypothesis of Boosting because the size of the latter is the number of the weak learners that is greater than 10 at best and usually greater than 100 weak hypotheses.

4 Conclusion

Knowledge discovery has increased the value in comprehensibility in machine learning models. Since that knowledge has to be understood by human experts, the comprehensibility of the model is a feature as important as its accuracy. So a lot of research has been carried out in order to find comprehensible models without loss of accuracy.

Ensemble machine learning methods have been developed as an easy way to improve accuracy in theoretical and practical machine learning problems. However, hypotheses computed by these methods are often considered difficult to understand.

---

[1] The table shows the size of the hypothesis computed by running on all the data, not the average of the cross-validation runs.

This paper has shown that an ensemble method like boosting can be combined with an easily understood hypothesis like decision tree without loss of accuracy. Furthermore, the accuracy can be improved in several datasets. This paper compares the accuracy of DTB with boosting over Decision Stumps and Boosting over C4.5. Of course, we could have taken into account more complex classifiers as weak learners in Boosting such as, for example, bayesian classifiers or neural networks. But their comprehensibility is lower than the comprehensibility of decision trees and we would have had to show in the first instance that more complex classifiers compute an accuracy significantly higher than Boosting over C4.5. However, that was beyond the scope of this paper.

The accuracy and consistency of this mixed model has been computed in an empirical way over several datasets. This empirical evaluation has shown that the mixed model achieves an accuracy as high as other machine learning models like Stacking, Boosting and ADTrees. Moreover, the size of the computed hypothesis is lower than hypotheses computed by ADTrees.

References

1. Breiman, L., Friedman, J., Richard, O., Charles, S.: Classification and regression trees. Wadsworth and Brooks (1984)
2. Quinlan, J.: Induction of decision trees. Machine Learning 1, 81–106 (1986)
3. Clark, P., Niblett, T.: The CN2 induction algorithm. Machine Learning 3, 261–284 (1989)
4. Michalski, R.: A theory and methodology of inductive learning. Artificial Intelligence 20, 111–161 (1983)
5. Rumelhart, D., Hinton, G., Williams, R.: Learning internal representations by error propagation. Parallel Distributed Processing: Explorations in the microstructure of cognition, vol. 1. MIT Press, Cambridge (1986)
6. Rabiner, L., Juang, B.: An introduction to hidden Markov models. IEEE Acoustics, Speech & Signal Processing Magazine 3, 4–16 (1986)
7. Smyth, P., Heckerman, D., Jordan, M.: Probabilistic independence networks for hidden markov probability models. Technical report, Microsoft Research (June 1996)
8. Aha, D., Kibler, D., Albert, M.: Instance-based learning algorithms. Machine Learning 6, 37–66 (1991)
9. Dietterich, T.G.: Ensemble methods in machine learning. In: First International Workshop on Multiple Classifier Systems, pp. 1–15 (2000)
10. Quinlan, J.: Bagging, boosting, and c4.5. In: Proc. of the 13th Nat. Conf. on A.I. and the 8th Innovate Applications of A.I. Conf., pp. 725–730. AAAI/MIT Press (1996)
11. Morales, R., Ramos, G., del Campo-Ávila, J.: Fe-cidim: Fast ensemble of cidim classifiers. International Journal of Systems Science 37(13), 939–947 (2006)
12. Freund, Y., Schapire, R.E.: Experiments with a new boosting algorithm. In: International Conference on Machine Learning, pp. 148–156 (1996)
13. Breiman, L.: Bagging predictors. Machine Learning 24(2), 123–140 (1996)
14. Dietterich, T.G.: An experimental comparison of three methods for constructing ensembles of decision trees: Bagging, boosting, and randomization. Machine Learning 40(2), 139–157 (2000)

15. Wolpert, D.H.: Stacked generalization. Neural Networks 5(2), 241–259 (1992)
16. Buntine, W.: A Theory of Learning Classification Rules. PhD thesis, University of Technology, Sydney, Australia (1990)
17. Quinlan, J.R.: C4.5: Programs for Machine Learning. Morgan Kaufmann, San Francisco (1993)
18. Kohavi, D., Sommerfield, D.: Targetting business users with decision table classifiers. In: 4th Internationals Conference on Knowledge Discovery in Databases, pp. 249–253 (1998)
19. Ridgeway, G., Madigan, D., Richardson, T., O'Kane, J.: Interpetable boostednaive bayes classification. In: 4th International Conference on Knowledge Discovery in Databases, pp. 101–104 (1998)
20. Ferri, C., Hernández, J., Ramírez, M.: From ensemble methods to comprehensible models. In: 5th Int. Conf. on Discovery Science (2002)
21. Domingos, P.: Knowledge discovery via multiple models. Intelligent Data Analysis 2(3) (1998)
22. Johansson, U., Niklasson, L., König, R.: Accuracy vs. comprehensibility in data mining models. In: Proceedings of the Seventh International Conference on Information Fusion, June 2004, pp. 295–300 (2004)
23. Cohen, W., Singer, Y.: A simple, fast, and effective rule learner. In: Sixteenth National Conference on Artificial Intelligence, pp. 335–342. AAAI Press, Menlo Park (1999)
24. Freund, Y., Mason, L.: The alternating decision tree learning algorithm. In: Proc. 16th International Conf. on Machine Learning, pp. 124–133. Morgan Kaufmann, San Francisco (1999)
25. Kearns, M., Mansour, Y.: On the boosting ability of top-down decision tree learning algorithms. In: Twenty-eighth annual ACM symposium on Theory of computing, Philadelphia, Pennsylvania, United State, pp. 459–468 (1996)
26. Schapire, R.: The streng of weak learnability. Machine Learning 5(2), 197–227 (1990)
27. Dietterich, T., Kearns, M., Mansour, Y.: Applying the weak learning framework to understand and improve C4.5. In: Proc. 13th International Conference on Machine Learning, pp. 96–104. Morgan Kaufmann, San Francisco (1996)
28. Newman, D., Merz, C., Hettich, S., Blake, C.: UCI repository of machine learning databases (1998)
29. Witten, I., Frank, E.: Data Mining: Practical machine learning tools and techniques, 2nd edn. Morgan Kaufmann, San Francisco (2005)
30. Bousquet, O., Elisseeff, A.: Stability and generalization. Journal of Machine Learning Research 2, 499–526 (2002)
31. Neumann, J.: Classification and Evaluation of Algorithms for Rule Extraction from Artificial Neural Networks. PhD thesis, University of Edinburgh (August 1998)

Empirical Analysis of Reliability Estimates for Individual Regression Predictions

Zoran Bosnić and Igor Kononenko

University of Ljubljana,
Faculty of Computer and Information Science,
Tržaška 25, Ljubljana, Slovenia
{zoran.bosnic,igor.kononenko}@fri.uni-lj.si

Abstract. In machine learning, the reliability estimates for individual predictions provide more information about individual prediction error than the average accuracy of predictive model (e.g. relative mean squared error). Such reliability estimates may represent a decisive information in the risk-sensitive applications of machine learning (e.g. medicine, engineering, business), where they enable the users to distinguish between better and worse predictions. In this paper, we compare the sensitivity-based reliability estimates, developed in our previous work, with four other approaches, proposed or inspired by the ideas from the related work. The results, obtained using 5 regression models, indicate the potentials for the usage of the sensitivity-based and the local modeling approach, especially with the regression trees.

Keywords: Regression, reliability, reliability estimate, sensitivity analysis, prediction accuracy, prediction error.

1 Introduction

When using supervised learning for modeling data, we aim to achieve the best possible prediction accuracy for the unseen examples which were not included in the learning process [1]. For evaluation of the prediction accuracies, the averaged accuracy measures are most commonly used, such as the mean squared error (MSE) and the relative mean squared error (RMSE). Although these estimates evaluate the model performance by summarizing the error contributions of all test examples, they provide no local information about the expected error of individual prediction for a given unseen example. Having information about single prediction reliability [2] at disposal might present an important advantage in the risk-sensitive areas, where acting upon predictions may have financial or medical consequences (e.g. medical diagnosis, stock market, navigation, control applications). For example, in the medical diagnosis, physicians are not interested only in the average accuracy of the predictor, but expect from a system to be able to provide the estimate of prediction reliability besides the bare prediction itself. Since the averaged accuracy measures do not fulfill this requirement, the reliability measures for individual predictions are needed in this area.

I.-Y. Song, J. Eder, and T.M. Nguyen (Eds.): DaWaK 2008, LNCS 5182, pp. 379–388, 2008.

To enable the users of classification and regression models to gain more insight into the reliability of individual predictions, various methods aiming at this task were developed in the past. Some of these methods were focused on extending formalizations of the existing classification and regression models, enabling them to make predictions with their adjoined reliability estimates. The other group of methods focused on the development of model-independent approaches, which are more general, but harder to analytically evaluate with individual models.

In this paper we compare five approaches to model-independent reliability estimation for individual examples in regression. We mostly focus on comparing the sensitivity-based estimates, developed in our previous work, to the other traditional approaches. We evaluate the performance of selected reliability estimates using 5 regression models (regression trees, linear regression, neural networks, support vector machines and locally weighted regression) on 15 testing domains.

The paper is organized as follows. Section 2 summarizes the previous work from related areas of individual prediction reliability estimation and Sect. 3 presents and proposes the tested reliability estimates. We describe the experiments, testing protocol and interpret the results in Sect. 4. Section 5 provides conclusions and ideas for the further work.

2   Related Work

The idea of the reliability estimation for individual predictions originated in statistics, where confidence values and intervals are used to express the reliability of estimates. In machine learning, statistical properties of predictive models were utilized to extend the predictions with the adjoined reliability estimates, e.g. with support vector machines [3,4], ridge regression [5], and multilayer perceptron [6]. Since these approaches are bound to a particular model formalism, their reliability estimates can be probabilistically interpretable, thus being the *confidence measures* (0 represents the confidence of the most inaccurate prediction and 1 the confidence of the most accurate one). However, since not all approaches offer probabilistic interpretation, we use more general term *reliability estimate* to name a measure that provides information about trust in the accuracy of the individual prediction.

In contrast to the previous group of methods, the second group is more general and model-independent (not bound to particular model). These methods utilize model-independent approaches to found their reliability estimates, e.g. local modeling of prediction error based on input space properties and local learning [7,8], using the variance of bagged models [9], or by meta-predicting the leave-one-out error of a single example [10]. Inspired by transductive reasoning [11], Kukar and Kononenko [12] proposed a method for estimation of classification reliability. Their work introduced a set of reliability measures which successfully separate correct and incorrect classifications and are independent of the learning algorithm. We later adapted this approach to regression [13,14].

The work presented here extends the initial work of Bosnić and Kononenko [13] and compares the performance of their proposed estimates to four other approaches. The previously developed sensitivity-based estimates were compared either to existing estimates, proposed in related work, or adapted from their ideas. They are summarized in the following section.

3   Reliability Estimates

3.1   Sensitivity Analysis-Based Reliability Estimates

In our previous work [13] we used the sensitivity analysis approach to develop two reliability estimates (RE_1 and RE_3), which estimate the local variance and local bias for a given unlabeled example. For greater clarity, in this paper we re-name these two estimates to $SAvar$ (Sensitivity Analysis - variance) and $SAbias$ (Sensitivity Analysis - bias). Both estimates are based on observing the change in the output prediction if a small and controlled change is induced in the learning set, i.e. by expanding the initial learning set with an additional learning example. The predictions of the *sensitivity models* (i.e. models which are generated on the modified learning sets) are afterwards used to compose the reliability estimates, which are defined as:

$$SAvar = \frac{\sum_{\varepsilon \in E}(K_\varepsilon - K_{-\varepsilon})}{|E|} \tag{1}$$

and

$$SAbias = \frac{\sum_{\varepsilon \in E}(K_\varepsilon - K) + (K_{-\varepsilon} - K)}{2|E|} . \tag{2}$$

In both of the above estimates, K represents the prediction of the initial regression model, whereas K_ε and $K_{-\varepsilon}$ denote predictions of two sensitivity models which are obtained using positive and negative value of parameter ε (chosen in advance). The chosen ε determines the label value of the additional learning example (which equals K, modified by some $f(\varepsilon)$) and therefore indirectly defines the magnitude of the induced change in the initial learning set. It is defined relative to the interval of the learning examples' label values, hence its values are domain-independent. To widen the observation windows in the local problem space and make the measures robust to local anomalies, the reliability measures use predictions from sensitivity models, gained and averaged using various parameters $\varepsilon \in E$. In our experiments we used $E = \{0.01, 0.1, 0.5, 1.0, 2.0\}$. For more details, see [13].

3.2   Variance of a Bagged Model

In related work, the variance of predictions in the bagged aggregate [15] of artificial neural networks has been used to indirectly estimate the reliability of the aggregated prediction [9]. Since an arbitrary regression model can be used with the bagging technique, we evaluate the proposed reliability estimate in combination with other learning algorithms.

Given a bagged aggregate of m predictive models, where each of the models yields a prediction K_m, the prediction of an example is therefore computed by averaging the individual predictions:

$$K = \frac{\sum_{i=1}^{m} K_i}{m} \qquad (3)$$

and reliability estimate $BAGV$ as the prediction variance

$$BAGV = \frac{1}{m} \sum_{i=1}^{m} (K_i - K)^2 . \qquad (4)$$

In our experimental work the testing bagged aggregates contained $m = 50$ model instances.

3.3   Local Cross-Validation Reliability Estimate

Approaches in the related work have demonstrated a potential for applying the general cross-validation procedure locally [7,8,16] and use it for the local estimation of prediction reliability.

Analogously, we implemented the LCV (Local Cross-Validation) reliability estimate, which is computed using the local leave-one-out (LOO) procedure. For a given unlabeled example, in each iteration (for every nearest neighbor) we generate a leave-one-out (LOO) model, excluding one of the neighbors. By making the LOO prediction K_i for each of the neighbors $i = 1, \ldots, k$ and using its true label C_i, we then compute absolute LOO prediction error $|C_i - K_i|$. The reliability of unlabeled example is therefore defined as weighted average of the absolute LOO prediction errors of its nearest neighbors (weighted by distance):

$$LCV(x) = \frac{\sum\limits_{(x_i, C_i) \in N} d(x_i, x) \cdot |C_i - K_i|}{\sum\limits_{(x_i, C_i) \in N} d(x_i, x)} \qquad (5)$$

where N denotes the set of the k nearest neighbors $N = \{(x_1, C_1), \ldots, (x_k, C_k)\}$ of an example $(x, \_)$, $d()$ denotes a distance function, and K_i denotes a LOO prediction for neighbor (x_i, C_i) computed on a learning set $N \backslash (x_i, C_i)$. In our work, we implemented the LCV algorithm to be adaptive to the size of the neighborhood with respect to the number of examples in the learning set. The parameter k was therefore dynamically assigned as $\lceil \frac{1}{20} \times L \rceil$, where L denotes the number of learning examples.

3.4   Density-Based Reliability Estimate

One of the traditional approaches to estimation of the prediction reliability is based on the distribution of learning examples in the input space. The density-based estimation of the prediction error assumes that the error is lower for predictions which are made in *denser* problem subspaces (a local part of the input space with higher number of learning examples), and higher for predictions which are made in *sparser* subspaces (a local part of the input space with fewer learning examples).

A typical use of this approach is, for example, with decision and regression trees, where we trust each prediction according to the proportion of learning examples that fall in the same leaf of a tree as the predicted example. But although this approach considers quantity of available information, it also has a disadvantage that it does not consider the learning examples' labels. This causes the method to perform poorly in noisy data and when the distinct examples are dense but not clearly separable. Namely, although it is obvious that such cases present a modeling challenge, the density estimate would characterize such regions of the input space as highly reliable.

We define the reliability estimate *DENS* as the value of the estimated probability density function for a given unlabeled example. To estimate the density we use the Parzen windows [17] with the Gaussian kernel. We reduced the problem of calculating the multidimensional Gaussian kernel to calculation of two-dimensional kernel using a distance function applied to the pairs of example vectors. Given the learning set $L = ((x_1, y_1), \ldots, (x_n, y_n))$, the density estimate for an unlabeled example $(x, \_)$ is therefore defined as:

$$p(x) = \frac{\sum_{i=1}^{n} \kappa(D(x, x_i))}{n} \tag{6}$$

where D denotes a distance function and κ a kernel function (in our case the Gaussian). Since we expect the prediction error to be higher in the cases where the value of density is lower, this means that $p(x)$ would correlate negatively with the prediction error. To establish the positive correlation with prediction error as with the other reliability estimates (which would improve the comparability), we need to invert $p(x)$, defining *DENS* as:

$$DENS(x) = \max_{i=1..n} (p(x_i)) - p(x) \; . \tag{7}$$

3.5   Local Modeling of Prediction Error

For the comparison with the previous reliability estimates we also propose an approach to the local estimation of prediction reliability using the nearest neighbors' labels. If we are given a set of nearest neighbors $N = \{(x_1, C_1), \ldots, (x_k, C_k)\}$, we define the estimate *CNK* ($C_{Neighbors} - K$) for an unlabeled example $(x, \_)$ as a difference between average label of the nearest neighbors and the example's prediction K (using the model that was generated on all learning examples):

$$CNK = \frac{\sum_{i=1}^{k} C_i}{k} - K \; . \tag{8}$$

In our experiments we computed the estimate *CNK* using 5 nearest neighbors.

4   Experimental Results

We tested and compared the estimates *SAvar* and *SAbias*, developed in our previous work, to other estimates, presented in Sect. 3. The first phase of the testing consisted of calculating the prediction errors and six reliability estimates for every example, computing them using the leave-one-out cross-validation procedure. In the evaluation phase of the testing, the correlation coefficients between the errors and each estimate were computed. The significance of the correlation coefficients was statistically evaluated using the t-test for correlation coefficients with the maximal significance level $\alpha = 0.05$.

Note that all of the estimates are expected to correlate positively with the prediction error. This means that all the estimates are designed so that their higher absolute values represent less reliable predictions and their lower absolute values represent more reliable predictions (the value 0 represents the reliability of the most reliable prediction). Also note, that all of the estimates except *SAbias* and *CNK* can take only positive values. Besides the absolute magnitude of these two estimates, which we interpret as the prediction reliability, the estimates also provide the additional information about the error direction (whether the value of prediction was too high or too low), which holds a potential for the further work in the correcting of the initial predictions using these two estimates. We therefore performed the experiments by correlating only the magnitudes of estimates (absolute values of *SAbias* and *CNK*) to the absolute prediction error of test examples. For estimates *SAbias* and *CNK*, we also correlated their signed values to the signed prediction error of test examples. In this way we actually tested the performance of eight estimates: *SAvar*, *SAbias-s* (signed), *SAbias-a* (absolute), *BAGV*, *LCV*, *DENS*, *CNK-s* (signed) and *CNK-a* (absolute).

The performance of reliability estimates was tested using five regression models, implemented in statistical package R [18]: regression trees (RT), linear regression (LR), neural networks (NN), support vector machines (SVM) and locally weighted regression (LWR). For testing, 15 standard benchmark data sets were used, each data set representing a regression problem. The application domains varied from medical, ecological, technical to mathematical and physical. Most of the data sets are available from UCI Machine Learning Repository [19] and from StatLib DataSets Archive [20]. All data sets are available from authors on request. A brief description of data sets is given in Table 1.

The results of the experiments are shown in Tables 2 and 3. The data in Table 2 indicate the percent of domains with the significant positive/negative correlation between the reliability estimates and the prediction error. As another perspective of the achieved results, Table 3 shows the correlation coefficients for the individual domains, averaged across five used regression models. The results confirm our expectation that the reliability estimates should positively correlate with the prediction error. Namely, we can see that the number of positive (desired) correlations dominates the number of negative (non-desired) correlations in all regression model/reliability estimate pairs. The last line in Table 2 and Fig. 1 presents the results of the reliability estimates, averaged across all five testing regression models. We can see, that the best results were achieved

Table 1. Basic characteristics of testing data sets

Data set	# examples	# disc.attr.	#cont.attr.
baskball	96	0	4
brainsize	20	0	8
cloud	108	2	4
cpu	209	0	6
diabetes	43	0	2
elusage	55	1	1
fruitfly	125	2	2
grv	123	0	3
mbagrade	61	1	1
pwlinear	200	0	10
pyrim	74	0	27
servo	167	2	2
sleep	58	0	7
transplant	131	0	2
tumor	86	0	4

using the estimates $BAGV$, $SAvar$, CNK-a and LCV (in the decreasing order with respect to the percent of the significant positive correlations). The estimate $SAbias$-a achieved the worst average results.

By observing the detailed results we can see that the performance of the estimate CNK-s with the regression trees stands out the most. Namely, the estimate significantly positively correlated with the prediction error in 80% of experiments and did not negatively correlate with the prediction error in any experiment. Estimate $SAbias$-s achieved similar performance, it positively correlated with the prediction error in 73% of tests with the regression trees and negatively in 0% of tests.

Table 2. Percentage of experiments exhibiting significant positive/negative correlations between *reliability estimates* and *prediction error*

model	$SAvar$ +/−	$SAbias$-s +/−	$SAbias$-a +/−	$BAGV$ +/−	LCV +/−	$DENS$ +/−	CNK-s +/−	CNK-a +/−
RT	33/0	73/0	40/0	53/0	33/0	33/0	80/0	60/0
LR	53/0	7/0	7/0	53/0	40/0	27/7	33/0	47/0
NN	47/0	20/7	27/7	53/0	33/0	27/7	20/7	33/7
SVM	40/7	40/7	27/0	47/0	60/0	33/7	27/13	27/0
LWR	40/13	0/13	13/13	47/0	33/0	33/7	27/20	33/0
average	**43/4**	**28/5**	**23/4**	**51/0**	**40/0**	**31/6**	**37/8**	**40/1**

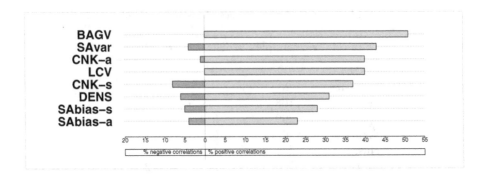

Fig. 1. Ranking of reliability estimates by the average percent of significant positive and negative correlations with the prediction error

Table 3. Average correlation coefficients, achieved by the reliability estimates in the individual domains. Statistically significant values are denoted with the boldface. Significant negative values are additionally underlined.

domain	SAvar	SAbias-s	SAbias-a	BAGV	LCV	DENS	CNK-s	CNK-a
baskball	0.053	0.078	-0.017	0.083	0.064	0.071	**0.150**	0.035
brainsize	-0.099	0.049	-0.160	-0.168	-0.139	-0.160	0.108	-0.077
cloud	**0.510**	0.110	0.148	**0.394**	**0.212**	**0.426**	0.155	**0.323**
cpu	**0.606**	**0.226**	**0.339**	**0.620**	**0.513**	**0.370**	**0.195**	**0.684**
diabetes	0.289	0.037	-0.023	0.200	0.057	**0.422**	0.272	0.012
elusage	0.222	0.083	0.192	**0.351**	**0.300**	0.202	0.197	0.046
fruitfly	-0.052	0.054	0.001	-0.014	-0.072	-0.049	-0.138	0.023
grv	**0.180**	0.122	0.096	**0.187**	0.169	0.018	**0.185**	0.118
mbagrade	0.009	-0.049	-0.039	0.023	0.061	-0.061	-0.068	0.037
pwlinear	**0.240**	0.069	0.009	**0.278**	0.027	**0.155**	0.132	**0.177**
pyrim	**0.410**	0.130	0.001	**0.337**	**0.276**	0.105	**0.340**	**0.598**
servo	0.059	0.079	**0.158**	**0.520**	**0.536**	**-0.275**	-0.122	**0.206**
sleep	0.059	**0.276**	0.062	**0.263**	0.160	-0.032	0.135	0.185
transplant	**0.320**	0.162	**0.228**	**0.436**	**0.484**	**0.471**	**0.289**	**0.344**
tumor	-0.097	0.169	0.109	0.086	0.108	-0.107	0.156	-0.007

Results, achieved by estimates *SAbias-a* and *SAbias-s*, also confirm our expectations and analysis from the previous work [13] that the sensitivity-analysis based estimates are not appropriate for models which do not partition input space (in our case, linear regression and locally weighted regression). Namely, from Table 2 we can see that these two estimates achieved the worst performance with linear regression and locally weighted regression. Also, no other reliability estimate performed worse, indicating that the other tested estimates present a better choice for the usage with linear regression and locally weighted regression.

5   Conclusion and Further Work

In this paper we compared the sensitivity-based reliability estimates, proposed in our previous work, with four other approaches to estimating the reliability of individual predictions: using variance of bagged predictors, local cross-validation, density-based estimation and local error estimation. The results, obtained using 15 testing domains and 5 regression models, indicated the promising results for the usage of estimates *SAbias-s* and *CNK-s* with the regression trees, where these two estimates significantly positively correlated with the signed prediction error in 73% and 80% of tests, respectively. The best average performance was achieved by estimate *BAGV*, which turned out to be the best choice for the usage with linear regression, neural networks and locally weighted regression. With linear regression, the estimate *SAvar* performed comparably to *BAGV*, preceding the performance of other tested approaches.

The summarized results therefore show the potential of using the sensitivity-based reliability estimates for the estimation of prediction error with selected regression predictors. They also show the favorable performance of the newly proposed local error modeling estimate *CNK* to the other estimates. The achieved results and the ideas, arising from this work, offer the challenges for the further work, which includes:

- A good performance of the signed reliability estimates (*SAbias-s* and *CNK-s*) with the signed prediction error implies the potential for the usage of the reliability estimates for the correction of regression predictions. We shall therefore explore whether these two reliability estimates can be utilized to significantly reduce the error of regression predictions.
- Different magnitudes of correlation coefficients in different testing domains (see Table 3) indicate that the potential for estimation of prediction reliability is in some domains more feasible than in the others. The domain characteristics, which lead to a good performance of reliability estimates, shall be analyzed.
- It shall be tested, whether the performance of individual estimate can be improved by combining two or more estimates, which achieve their best results with complementary sets of regression models.
- Since different estimates perform with different success in different domain-model pairs, the approach for selection of the best performing reliability estimate for a given domain and model shall be developed.

References

1. Kononenko, I., Kukar, M.: Machine Learning and Data Mining: Introduction to Principles and Algorithms. Horwood Publishing Limited, UK (2007)
2. Crowder, M.J., Kimber, A.C., Smith, R.L., Sweeting, T.J.: Statistical concepts in reliability. Statistical Analysis of Reliability Data. Chapman and Hall, London (1991)

3. Gammerman, A., Vovk, V., Vapnik, V.: Learning by transduction. In: Proceedings of the 14th Conference on Uncertainty in Artificial Intelligence, Madison, Wisconsin, pp. 148–155 (1998)

4. Saunders, C., Gammerman, A., Vovk, V.: Transduction with confidence and credibility. In: Proceedings of IJCAI 1999, vol. 2, pp. 722–726 (1999)

5. Nouretdinov, I., Melluish, T., Vovk, V.: Ridge regressioon confidence machine. In: Proc. 18th International Conf. on Machine Learning, pp. 385–392. Morgan Kaufmann, San Francisco (2001)

6. Weigend, A., Nix, D.: Predictions with confidence intervals (local error bars). In: Proceedings of the International Conference on Neural Information Processing (ICONIP 1994), Seoul, Korea, pp. 847–852 (1994)

7. Birattari, M., Bontempi, H., Bersini, H.: Local learning for data analysis. In: Proceedings of the 8th Belgian-Dutch Conference on Machine Learning, pp. 55–61 (1998)

8. Giacinto, G., Roli, F.: Dynamic classifier selection based on multiple classifier behaviour. Pattern Recognition 34(9), 1879–1881 (2001)

9. Heskes, T.: Practical confidence and prediction intervals. In: Mozer, M.C., Jordan, M.I., Petsche, T. (eds.) Advances in Neural Information Processing Systems, vol. 9, pp. 176–182. MIT Press, Cambridge (1997)

10. Tsuda, K., Rätsch, G., Mika, S., Müller, K.: Learning to predict the leave-one-out error of kernel based classifiers, p. 331. Springer, Heidelberg (2001)

11. Vapnik, V.: The Nature of Statistical Learning Theory. Springer, Heidelberg (1995)

12. Kukar, M., Kononenko, I.: Reliable classifications with machine learning. In: Elomaa, T., Mannila, H., Toivonen, H. (eds.) ECML 2002. LNCS (LNAI), vol. 2430, pp. 219–231. Springer, Heidelberg (2002)

13. Bosnić, Z., Kononenko, I.: Estimation of individual prediction reliability using the local sensitivity analysis. Applied Intelligence, 1–17 (in press, 2007), http://www.springerlink.com/content/e27p2584387532g8/

14. Bosnić, Z., Kononenko, I., Robnik-Šikonja, M., Kukar, M.: Evaluation of prediction reliability in regression using the transduction principle. In: Zajc, B., Tkalčič, M. (eds.) Proceedings of Eurocon 2003, Ljubljana, pp. 99–103 (2003)

15. Breiman, L.: Bagging predictors. Machine Learning 24(2), 123–140 (1996)

16. Schaal, S., Atkeson, C.G.: Assessing the quality of learned local models. In: Cowan, J.D., Tesauro, G., Alspector, J. (eds.) Advances in Neural Information Processing Systems, vol. 6, pp. 160–167. Morgan Kaufmann Publishers, Inc., San Francisco (1994)

17. Silverman, B.W.: Density Estimation for Statistics and Data Analysis. Monographs on Statistics and Applied Probability. Chapman and Hall, London (1986)

18. R Development Core Team: A language and environment for statistical computing. R Foundation for Statistical Computing,Vienna, Austria (2006)

19. Asuncion, A., Newman, D.J.: UCI machine learning repository (2007)

20. Department of Statistics at Carnegie Mellon University: Statlib – data, software and news from the statistics community (2005), http://lib.stat.cmu.edu/

User Defined Partitioning - Group Data Based on Computation Model

Qiming Chen and Meichun Hsu

HP Labs
Palo Alto, California, USA
Hewlett Packard Co.
{qiming.chen,meichun.hsu}@hp.com

Abstract. A technical trend in supporting large scale scientific applications is converging data intensive computation and data management for fast data access and reduced data flow. In a combined cluster platform, co-locating computation and data is the key to efficiency and scalability; and to make it happen, data must be partitioned in a way consistent with the computation model. However, with the current parallel database technology, data partitioning is primarily used to support *flat* parallel computing, and based on existing partition key values; for a given application, when the data scopes of function executions are determined by a high-level concept that is related to the application semantics but not presented in the original data, there would be no appropriate partition keys for grouping data.

Aiming at making application-aware data partitioning, we introduce the notion of User Defined Data Partitioning (UDP). UDP differs from the usual data partitioning methods in that it does not rely on existing partition key values, but extracts or generates them from the original data in a *labeling* process. The novelty of UDP is allowing data partitioning to be based on application level concepts for matching the data access scoping of the targeted computation model, and for supporting data dependency graph based parallel computing.

We applied this approach to architect a hydro-informatics system, for supporting periodical, near-real-time, data-intensive hydrologic computation on a database cluster. Our experimental results reveal its power in tightly coupling data partitioning with "pipelined" parallel computing in the presence of data processing dependencies.

1 Introduction

Dynamic data warehousing and operational BI applications involve large-scale data intensive computations in multiple stages from information extraction, modeling, analysis to prediction [8,17,20,22]. To support such applications, two disciplines are required: High-Performance Computation (HPC) [3-6] and scalable data warehousing [9-12]; nowadays both are based on computer cluster technology, and based on partitioning tasks and data for parallel processing. In such an environment, co-locating data and computation for reduced data move and fast data access is the key to efficiency and scalability.

I.-Y. Song, J. Eder, and T.M. Nguyen (Eds.): DaWaK 2008, LNCS 5182, pp. 389–401, 2008.
© Springer-Verlag Berlin Heidelberg 2008

1.1 The Problem

While both parallel computing and parallel data management had significant progress along with the advance of cluster technology [3,17,18], they were previously studied separately [6]. For scientific applications, data are stored in separate repositories and brought in for computation; for database, applications are no more than external clients. Very often, a task and the data to be applied by it are not co-located, causing significant overhead of data flow. Such locality mismatch is considered as the major performance bottleneck.

The usual hash, range and list partitioning mechanisms do not address this issue as they focus on general-purpose parallel data access but without taking into account the application-level semantics. These methods map rows of a table to partitions based on existing partition key values presented in the original data. If the data grouping and partitioning need to be driven by certain application-level concept not presented in the original data, there would be no appropriate partition keys to be used; should that happen, "moving computation to data" is hard to realize.

Some "flat" parallel computing schemes, such as Map-Reduce characterized by applying one function to multiple data objects, does not catch the order dependency of data processing.

In summary, we have identified the need for partitioning data based on high-level concepts not presented in the original data, and determined that the current data partitioning approaches lack such capability.

1.2 Related Work

With the cluster technology, rapid advancement has been seen in scalable data management, e.g. parallel databases, multi-databases, large file systems such as GFS, HDFS, and in HPC e.g. MPI, MapReduce (or Hadoop) [13,14,17,18]. Although the importance of co-locating data and computation has been widely recognized, it remains a significant challenge.

We share the same view as the DISC project (Data Intensive Super Computing) [6]: moving data is often more expensive than moving programs, thus computation should be data-driven [1,2,15]. However, this locality related goal can be achieved only if data partitioning and allocation are driven by the computation model. The hydrologic application described in this paper provides a convincible example; based on hydrologic fundamentals a watershed computation must be made region by region from upstream to downstream in a river drainage network, therefore the data must be partitioned accordingly for computation efficiency. This kind of dependencies has not been systematically addressed in data partitioning.

Map-Reduce on Google File System (GFS), or Hadoop on Hadoop File System (HDFS) [13,14,17,18] has become increasingly popular in parallel computing, which is characterized by applying a single operator to multiple data objects in one-level (flat) parallel. Many applications in content search, etc, belong to this simple case, and many other applications belong to more complex cases. For more complicated computations where the data-dependency graph based control flow of applying a function to multiple data objects presents, Map-Reduce cannot act properly.

Parallel database systems [15] such as Teradata [21] and Neoview [20], and multi-database systems such as Terraserver [3], do provide data partitioning mechanisms for parallel query processing. However the standard hash, range, list partitioning mechanisms provided by these systems deal with data partition at the data access level rather than at the data intensive computation level, thus may not be able to group and partition data in the way suitable for the computation model.

1.3 Our Solution – User Defined Partitioning (UDP)

We propose the notion of User Defined Data Partitioning (UDP) for grouping data based on the semantics at the data intensive computing level. This allows data partitioning to be consistent with the data access scoping of the computation model, which underlies the co-location of data partitions and task executions.

Unlike the conventional hash or range partitioning method which maps rows of a table to partitions based on the existing partition key values, with UDP, the partition keys are generated or learnt from the original data by a *labeling process* based on the application level semantics and computation model, representing certain high-level concepts.

Further, unlike the conventional data partitioning that is primarily used to support *flat* parallel computing, i.e. applying a function to independent data objects, the UDP partitions data by taking into account the control flow in parallel computing based on data dependency graph [12].

Therefore, the novelty of the proposed UDP methodology consists in supporting computation-model aware data partitioning, for tightly incorporating parallel data management with data intensive computation and for accommodating the order dependency in multi-steps parallel data processing.

The rest of this paper is organized as follows: Section 2 outlines the watershed computation on river drainage network, for providing an application context of our discussion. Section 3 introduces the notion of User Defined Partitioning (UDP). Section 4 discusses implementation issues. Section 5 concludes.

2 River Drainage Network Modeling Example

A Hydro-informatics System (HIS), like most Earth information systems, is a HPC system that carries out a class of space-time oriented data intensive hydrologic computations periodically with near-real-time response. The computation results are stored in the underlying databases to be retrieved for analysis, mash-up and visualization. To the efficiency of such data intensive computation, the locality match of parallel computing and parallel data management becomes a major challenge.

The majority of data held in a HIS are location sensitive geographic information. Specifically, a river drainage network is modeled as an unbalanced binary tree as shown in Figure 1, where river segments are named by binary string codification. The river segments binary tree is stored in a table, where each row represents a river segment, or a tree node. Among others, at minimum, the table contains attributes *node_id, left_child_id, right_child_id, node_type* (RR if it is the root of a region; or RN otherwise), and *region_id* to be explained later.

In the watershed computation, a kind of hydrologic applications, the river segments should be grouped to regions and be processed from upstream regions to downstream regions.

Regions also form a tree but not necessary a binary tree. Each region is represented by a node in the region tree, and viewed as a partition of the river segments tree. A region has the following properties

- *region_id*, that takes the value of the root *node_id*,
- *region_level*, as the length of its longest descendant path counted by region, bottom-up from the leaves of the region tree,
- *parent_region_id*, the *region_id* of the parent region.

Note that the notion of *region* is not an original property of river segments; this notion represents the results of a data *labeling* process; the generated *region_id* instances from that process serve as the partition keys of the river segments table. The river-segment table is partitioned by region across multiple server nodes to be accessed in parallel.

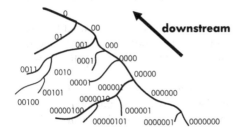
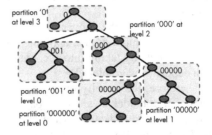

Fig. 1. Model river drainage network as binary treeing

Fig. 2. Bottom-up binary tree partition

Common to many geo-aware scientific applications, in watershed computation, the same function is applied, in a certain order, to multiple data partitions corresponding to geographic regions. The computation on a region needs retrieve the updated information of the root nodes of its child regions. The results of local executions are communicated through database access, using either permanent tables or temporary tables.

The watershed computation is made region by region from upstream to downstream, which means the region tree is *post-order traversed* - the root is visited *last*. This is a kind of "data dependency graph" based parallel processing, as

- geographically dependent regions must be processed in certain order, but
- the parallelism opportunities exist for the regions which can be computed in any order; for instance, regions at different tree branches may be processed in parallel.

3   User Defined Data Partitioning (UDP)

3.1   Data Partitioning

Data partitioning addresses a key issue in supporting very large tables and indexes by decomposing them into smaller and more manageable pieces called *partitions*. Each partition of a table or index must have the same logical attributes, such as column names, data types, and constraints, but each partition can have separate physical properties. Distribute table partitions over multiple nodes of a parallel database or multiple databases is essential for parallel query processing.

The usual partitioning methods map rows to partitions by the *range, hash* or *discrete* values of the selected *partition key* attribute, referred to as range partitioning, hash partitioning and list partitioning respectively. In hash partitioning, for example, the hash value of a data object determines its membership in data grouping. Assuming there are four partitions numbered by 0, 1, 2, 3, the hash function could return a value from 0 to 3. Composite partitioning allows for certain combinations of the above partitioning schemes; for example, first applying a range partitioning and then a hash partitioning.

However, the focus of the "general-purpose" data partitioning methods is placed on accelerating query processing through increased data access throughput. These methods may not be well coped with data intensive computations at the application level. In order to support data-parallel computing, the computing models should be taken into account in partitioning the data. This is our standing point of studying UDP.

3.2 User Defined Partitioning

As mentioned above, the usual hash, range and list partitioning methods rely on existing partition key values to group data. For many applications data must be grouped based on the criteria presented at an aggregate or summarized level, and there is no partition keys preexist in the original data for such grouping.

To solve the above problem we introduce the notion of User Defined Partitioning (UDP), which is characterized by partitioning data based on certain higher-level concepts reflecting the application semantics. In UDP, partition key values may not present in the original data, but instead they are generated or learnt by a *labeling process*.

In general, a UDP for partitioning a table T includes the following methods.

- a *labeling* method to mark rows of T for representing their group memberships, i.e. to generate partition keys;
- an *allocating* method to distribute data groups (partitions) to multiple server nodes; and
- a *retrieving* method for accessing data records of an already partitioned table.

The labeling, allocating and retrieving methods are often data model oriented. Below we will illustrate those using the river drainage tree model and the corresponding watershed computation.

3.3 UDP Example - Partition River Drainage Tree

As watershed computation is applied to river segments regions from upstream to downstream, we group the river segments into regions and allocate the regions over multiple databases. A region contains a binary tree of river segments. The regions themselves also form a tree but not necessarily a binary tree. The partitioning is also made bottom-up from upstream (child) to downstream (parent) of the river, to be consistent with the geographic dependency of hydrologic computation.

The river segments tree is partitioned based on the following criterion. Counted bottom-up in the river segments tree, every sub-tree of a given height forms a region, which is counted from either leaf nodes or the root nodes of its child regions. In order to capture the geographic dependency between regions, the notion of *region level* is introduced as the *partition level* of a region that is counted bottom-up from its farthest leave region, thus represents the length of its longest descendant path on the region tree. As illustrated in Figure 2, the levels between a pair of parent/child regions may not be consecutive. The computation independence (i.e. parallelizability) of the regions at the same level is statically guaranteed.

Labeling. Labeling aims at grouping the nodes of the river segments tree into regions and then assigning a *region_id* to each tree node. Labeling is made bottom-up from leaves. Each region spans k levels in the river-segment tree, where k is referred to as *partition_depth*, and for a region, counted from either leaf nodes river segments tree or the root nodes of its child regions. The top-level region may span the remainder levels smaller than k. Other variables are explained below.

- The depth of a node is its distance from the root; the depth of a binary tree is the depth of its deepest node; the *height* of a node is defined as the depth of the binary three rooted by this node. The height of a leave node is 0.
- The *node_type* of a node is assigned to either RR or RN after its group is determined during the labeling process. This variable also indicates whether a node is already labeled or not.
- CRR is used to abbreviate the Closest RR nodes beneath a node t where each of these RR nodes can be identified by checking the *parent_region_id* value of the region it roots, as either the *region_id* of t, or un-assigned yet. Correspondingly, we abbreviate the Closest Descendant Regions beneath a node as its CDR.

The following functions on a tree node, t, are defined.

- *is-root()* returns True if t is the root of the whole binary tree.
- *cdr()* returns the CDR regions beneath t.
- *adj-height()* returns 0 if the node type of t is RR, otherwise as the height of the binary tree beneath t where all the CRR nodes, and the sub-trees beneath them, are ignored.
- *adj-desc()* returns the list of descendant nodes of t where all the CRR nodes, and the sub-trees beneath them, are exclusive.
- *max-cdr-level()* returns the maximal *region_level* value of t's CRR (or CDR).

The labeling algorithm generates *region_id* for each tree node as its label, or partition key; as well as the information about partitioned regions, including the *id, level, parent region* for each region. The labeling algorithm is outlined below.

Allocating. After labeling, the river segments are partitioned into regions. Regions form a tree. Counted from the leaves of the region tree and in the bottom-up order, each region has a region-level as its longest path. Figure 3 shows a region tree with region levels in data partitioning.

The allocation method addresses how to map labeled river regions to multiple databases. As the river regions at the same region level have no geographic dependency

Algorithm PostorderTreeNodeLabeling *(bt, k)*
Input: (1) BinaryTree *bt*
           (2) int *k* as partition depth
Output: (1) region_id of each node (label)
           (2) id, level, parent of each region
Procedure
 1: **if** *bt* = ∅ **then**
 2:    return
 3: **if** *bt*.node_type ≠ UNDEF **then**
 4:    return
 5: **if** *bt*.left_child ≠ ∅ && *bt*.left_child.adj-height() ≥ *k* **then**
 6:                                PostorderTreeNodeLabeling   *(bt*.left_child)
 7: **if** *bt*.right_child≠∅ && *bt*.right_child.adj-height() ≥ *k* **then**
 8:                                PostorderTreeNodeLabelping   *(bt*.right_child)
 9: **if** bt.is_root() ‖ bt.adj-height() = *k* **then**
 10:    Region *p* = new Region(*bt*.node_id)
 11:    *bt*.region_id = *p*.get-id() // optionally as *bt*.node_id
 12:    *bt*.region_level = *bt*.max-cdr-level() + 1
 13:    *bt*.node_type = RR
 14:    List cdr = *bt*.cdr()
 15:    **for each** n_{cdr} in cdr **do**
 16:        n_{cdr}.parent_region_id = *bt*.region_id
 17:    List members = *bt*.adj-desc()
 18:    **for each** n_m in members **do**
 19:        n_m.region_id = *bt*.region_id
 20:        n_m.node_type = RN

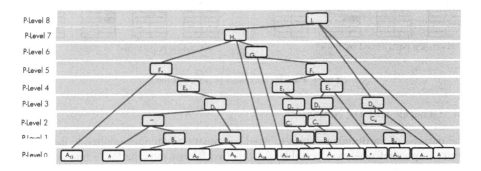

Fig. 3. Partition-levels of regions

thus can be processed in parallel, we start with the most conservative way to distribute regions, in the following steps.

- Step 1: *generate region-hash from region_id*;
- Step 2: *map the region-hash values to the keys of a mapping table* that is independent of the cluster configuration; then *distribute regions to server-nodes* based on

that mapping table. The separation of logical partition and physical allocation makes the data partitioning independent of the underlying infrastructure.

- Step 3: balance load, i.e. maximally evening the number of regions over the server nodes level by level in the bottom-up order along the region hierarchy.
- Step 4: *record the distribution of regions* and make it visible to all server nodes.

Note that we focus on static data allocation for all applications, rather than static task partitioning for one particular application.

Retrieving. *After* data partitioning, to locate the region of a river segment given in a query can be very different from searching the usual hash partitioned or range partitioned data, in case the partition keys are generated through labeling but not given in the "unlabeled" query inputs; the general mechanism is based on "ALL-NODES" parallel search as shown on the left-hand of Figure 4.

Another possibility is to create "*partition indices*", i.e. to have region_ids indexed by river segment_ids and to hash partition the indices. In this case, the full records of river segments are partitioned by region, and in addition, the river segment_ids for indexing regions are partitioned by hash. Then querying a river segment given its id but without region, is a two steps search as shown on the right hand of Figure 4: first, based on the hash value of the river segment id, only one node will be identified for indexing its region, and second, based on the hash value of the region, the node containing the full record of the river segment is identified for data retrieval. As the full record size of a river segment is really long, the storage overhead of preparing "partition indices" is worth to pay for this application.

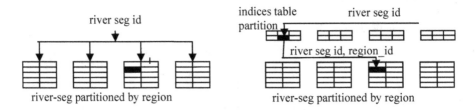

Fig. 4. Parallel access of partitioned data

3.4 Generalized UDP Notion

The purpose of partitioning data is to have computation functions applied to data partitions in parallel whenever possible; for this two factors must be taken into account: the scope of data grouping must match the domain of the computation function, and the order dependency of function applications must be enforced.

A *flat* data-parallel processing falls in one of the following typical cases:

- apply a function to multiple objects, i.e. $f : \langle x_1, \ldots, x_n \rangle = \langle f{:}x_1, \ldots, f{:}x_n \rangle$
- apply multiple functions to an object, i.e. $[f_1, \ldots, f_n] : x = \langle f_1{:}x, \ldots, f_n{:}x \rangle$

More generally a computation job is parallelized based on a *data dependency graph*, where the above flat-data parallel execution plans are combined in processing data partitions in sequential, parallel or branching. Here we focus on *embarrassing* parallel computing without in-task communication but with retrieval of previous computation results through database accessing.

The conventional data partitioning methods expect to group data objects based on existing partition key values, which may not be feasible if there are no key values suitable for the application preexist. The UDP is characterized by partitioning data based on the high-level concept relating to the computation model, which are extracted from the original data and serve as the generated partition keys. In the watershed computation example, we partition data based on the concept *region* whose values are not pre-associated with the original river segment data, but generated in the labeling process. Below let us discuss step by step how the UDP notion can be generalized.

- UDP aims at partitioning data objects to *regions* and distribute data belonging to different regions over a number K of server nodes.
- In the watershed computation, a region is a geographic area in the river drainage network. In other sciences, the notion of region is domain specific; but in general a region means a multidimensional space.
- We view an object with attributes, or features, $x_1, \ldots x_n$ as a vector $X = \{x_1, \ldots x_n\}$ that in general does not contain a partition key thus UDP is used to generate or even learn a *label* on X, and eventually maps the label to a number in $\{0, \ldots, K\}$ for allocating X to a server node numbered by k ($0 \leq k \leq K\text{-}1$).
- *Labeling* is a mapping, possibly with probabilistic measures.
  - It is a mapping from a feature space (e.g. medical computer tomography (CT) features, molecular properties features) $X = \{x_1, \ldots x_n\}$ to a label space $Y = \{Y_1, \ldots Y_m\}$ where Y_i is a vector in the label space;
  - A labeling mapping potentially yields a *confident* ranging over 0 to 1.
  - The labeling algorithm is used to find the appropriate or best-fit mappings $X \rightarrow Y_i$ for each i.
- *Allocating* is a mapping from the above label space to an integer; i.e. map a label vector with probabilistic measures to a number that represents a server node. We have suggested making this mapping in two steps.
  - In the first step, a label vector is mapped to a *logical* partition id called *region-hash* (e.g. 1-1024) independent of the actual number (e.g. 1-128) of server node.
  - In the second step that *region-hash* is mapped to a physical partition id such as a server node number by a hash-map.

The method for generating label-hash is domain specific. As an example, ignoring the confident measures, a mapping from a multidimensional vector to a unique single value can be done using spatial filing curves [23] that turn a multidimensional vector to an integer, and then such an integer can be hash mapped to a label hash value. Methods taking into account of confidence of labels are also domain specific, e.g. in computer tomography interpretation.

4   Implementation Issues

We applied the UDP approach to the hydro-informatics system for
- converging parallel data management and parallel computing and,
- managing data dependency graph based parallel computations.

For the watershed computation,
- the river segments data are divided into partitions based on the watershed computation model and allocated to multiple servers for parallel processing;
- the same function is applied to multiple data partitions (representing geographic regions) with order dependencies (i.e. from upstream regions to downstream regions;
- the data processing on one region retrieves and updates its local data, where accessing a small amount of neighborhood information from upstream regions is required;
- data communication is made through database access.

4.1 Convergent Cluster Platform

We have the cluster platforms of parallel data management and parallel computing converged, for shared resource utilization, for reduced data movement between database and applications, and for mutually optimized performance.

For parallel data management, we have options between using a parallel database or multiple individual databases, and we choose the latter for this project. We rely on a single cluster of server machines for both parallel data management and parallel computing. The cluster can contain 4, 16, 128, 256, ... nodes interconnected by high-bandwidth Ethernet or Infiniband (IB). The clustered server nodes run individual share-nothing Oracle DBMS; data are partitioned to multiple databases based on their domain specific properties, allowing the data access throughput to increase linearly along with the increase of server nodes. The server nodes form one or more cliques in data accessing, allowing a data partition to be visible to multiple nodes, and a node to access multiple data partitions. This arrangement is necessary for simplifying inter-node messaging and for tolerating faults (as mentioned early, the computation on a region needs retrieve the updated information of the root nodes of its child regions). The architecture is outlined in Figure 5.

The computation functions are implemented as database User Defined Functions (UDFs), which represents our effort in the synthesis of data intensive computation and data management.

While employing multiple server nodes and running multiple DBMSs, our platform offers application a single system image transparent to data partitioning and execution parallelization, This is accomplished by building a Virtual Software Layer (VSL) on top of DBMS that provides Virtual Data Management (VDM) for dealing with data access from multiple underlying databases, and Virtual Task Management (VTM) for handling task partition and scheduling.

In the current design, the VSL resides at each server node, all server nodes are treated equally: every server node holds partitions of data, as well as the meta-data describing data partitioning; has VDM capability as well as VTM capability. The locations of data partitions and function executions are consistent but transparent from applications.

4.2 Task Scheduling

The parallel computation opportunities exist *statically* in processing the geographically independent regions either at the same level or not, and *dynamically* in processing the regions with all their children regions have been processed. These two kinds of opportunities will be interpreted and realized by the system layer.

The computation functions, i.e. UDFs are made available on all the server nodes. The participating server nodes also know the partition of regions and their locations, the connectivity of regions, particular computation models, UDF settings and default values. Further, Each VTM is provided with a **UDF invoker** and an ODBC connector.

A computation job can be task-partitioned among multiple server nodes to be executed in parallel. Task scheduling is data-driven, based on the locality and geo-dependency of the statically partitioned data. UDFs are scheduled to run at the server nodes where the applied data partitions reside. Local execution results are stored in databases, and communicated through database access. The computation results from multiple server nodes may be assembled if necessary.

In more detail, task scheduling is based on the master-slave architecture. Each server node can act as either master or slave, and can have both of them.

- The VTM-master is responsible for scheduling tasks based on the location of data partitions, their processing dependencies, and the execution status. It determines the parallel processing opportunities for the UDF applications without static and dynamic dependencies, send task requests together with parameters to the VTM-slaves where the data to be computed on reside, monitors execution status, re-executes tasks upon failure, etc. Currently, the resembling of local results is handled directly by the VTM-master module.
- Upon receipt of task execution requests and parameters from the VTM-master, the VTM-slaves execute their tasks through UDF invokers.

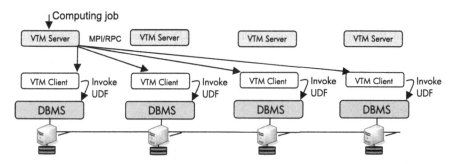

Fig. 5. Tasks distributed to meet data partitions, and executed through UDF/SP invocation

For messaging, the MPI protocol is currently utilized where VTM master and slaves serve as MPI masters and slaves. Although the data from *master* to *slave* may include static inputs associated with a new region, processes on different regions pass information through database access.

5 Conclusions

In a convergent cluster platform for data intensive application and data management, partitioning data over the cluster nodes is the major mechanism for parallel processing.

However, the conventional data partitioning methods do not take into account the application level semantics thus may not be able to partition data properly to fit in the computation model. These partitioning methods are primarily used to support *flat* parallel computing, and based on the existing partition key values, but the criterion of partitioning data could relate to a concept presented at the application level rather than in the original data; should that happen, there would be no appropriate partition keys identifiable.

We have proposed the notion of User Defined Data Partitioning (UDP) for correlating data partitioning and application semantics. With UDP, partition key values are not expected to pre-exist, but generated or learnt in a *labeling process* based on certain *higher level concept* extracted from the original data, which relates to the computation model, and especially the "complex" parallel computing scheme based on data dependency graphs.

Introducing UDP represents an initial step to support computation model aware data partitioning, and to correlate data analysis, machine learning to parallel data management.

We applied this approach to architect a hydro-informatics system, for supporting periodical, near-real-time, data-intensive hydrologic computation on a database cluster. Our experimental results reveal its power in tightly coupling data partitioning with "complex" parallel computing in the presence of data processing dependencies. In the future we plan to apply UDP to partition content data based on feature extraction, and to embed UDP in the extended SQL framework with user defined functions.

References

1. Agrawal, R., Shim, K.: Developing Tightly-Coupled Data Mining Applications on a Relational Database System. In: Proceedings Second KDD Int. Conf. (1996)
2. Asanovic, K., Bodik, R., Catanzo, B.C., Gebis, J.J., Husbands, P., Keutzer, K., Patterson, D.A., Plishker, W.L., Shalf, J., Williams, S.W., Yelick, K.A.: The landscape of parallel computing research: A view from Berkeley, Tech Rep EECS-2006-183, U.C.Berkeley (2006)
3. Barclay, T., Gray, J., Chong, W.: TerraServer Bricks – A High Availability Cluster Alternative, Technical Report, MSR-TR-2004-107 (October 2004)
4. Barroso, L.A., Dean, J., H"olze, U.: Web search for a planet: The Google cluster architecture. IEEE Micro 23(2), 22–28 (2003)
5. Brewer, E.A.: Delivering high availability for Inktomi search engines. In: Haas, L.M., Tiwary, A. (eds.) ACM SIGMOD Conf. (1998)
6. Bryant, R.E.: Data-Intensive Supercomputing: The case for DISC, CMU-CS-07-128 (2007)
7. Dayal, U., Chen, Q., Hsu, M.: Dynamic Data Warehousing. In: Mohania, M., Tjoa, A.M. (eds.) DaWaK 1999. LNCS, vol. 1676. Springer, Heidelberg (1999)
8. Chen, Q., Dayal, U., Hsu, M.: An OLAP-based Scalable Web Access Analysis Engine. In: Kambayashi, Y., Mohania, M., Tjoa, A.M. (eds.) DaWaK 2000. LNCS, vol. 1874. Springer, Heidelberg (2000)
9. Chen, Q., Hsu, M., Dayal, U.: A Data Warehouse/OLAP Framework for Scalable Telecommunication Tandem Traffic Analysis. In: Proc. of 16th ICDE Conf. (2000)

10. Chen, Q., Dayal, U., Hsu, M.: A Distributed OLAP Infrastructure for E-Commerce. In: Proc. Fourth IFCIS CoopIS Conference, UK (1999)
11. Chen, Q., Dayal, U., Hsu, M.: OLAP-based Scalable Profiling of Customer Behavior. In: Mohania, M., Tjoa, A.M. (eds.) DaWaK 1999. LNCS, vol. 1676. Springer, Heidelberg (1999)
12. Chen, Q., Kambayashi, Y.: Nested Relation Based Database Knowledge Representation. In: ACM-SIGMOD Conference (1991)
13. Dean, J.: Experiences with MapReduce, an abstraction for large-scale computation. In: Int. Conf. on Parallel Architecture and Compilation Techniques. ACM, New York (2006)
14. Dean, J., Ghemawat, S.: MapReduce: Simplified data processing on large clusters. In: Operating Systems Design and Implementation (2004)
15. DeWitt, D., Gray, J.: Parallel Database Systems: the Future of High Performance Database Systems. CACM 35(6) (June 1992)
16. Gray, J., Liu, D.T., Nieto-Santisteban, M.A., Szalay, A.S., Heber, G., DeWitt, D.: Scientific Data Management in the Coming Decade. SIGMOD Record 34(4) (2005)
17. Ghemawat, S., Gobioff, H., Leung, S.T.: The Google file system. In: Symposium on Operating Systems Principles, pp. 29–43. ACM, New York (2003)
18. HDFS: http://hdf.ncsa.uiuc.edu/HDF5/
19. Hsu, M., Xiong, Y.: Building a Scalable Web Query System. In: Bhalla, S. (ed.) DNIS 2007. LNCS, vol. 4777. Springer, Heidelberg (2007)
20. HP Neoview enterprise datawarehousing platform, http://h71028.www7.hp.com/ERC/downloads/4AA0-7932ENW.pdf
21. O'Connell, et al.: A Teradata Content-Based Multimedia Object Manager for Massively Parallel Architectures. In: ACM-SIGMOD Conf., Canada (1996)
22. Saarenvirta, G.: Operational Data Mining. DB2 Magazine 6 (2001)
23. Sagan, H.: Space-Filling Curves. Springer, Heidelberg (1994)

Workload-Aware Histograms for Remote Applications

Tanu Malik and Randal Burns

[1] Cyber Center
Purdue University
[2] Department of Computer Science
Johns Hopkins University
tmalik@cs.purdue.edu, randal@cs.jhu.edu

Abstract. Recently several database-based applications have emerged that are remote from data sources and need accurate histograms for query cardinality estimation. Traditional approaches for constructing histograms require complete access to data and are I/O and network intensive, and therefore no longer apply to these applications. Recent approaches use queries and their feedback to construct and maintain "workload aware" histograms. However, these approaches either employ heuristics, thereby providing no guarantees on the overall histogram accuracy, or rely on detailed query feedbacks, thus making them too expensive to use. In this paper, we propose a novel, incremental method for constructing histograms that uses minimum feedback and guarantees minimum overall residual error. Experiments on real, high dimensional data shows 30-40% higher estimation accuracy over currently known heuristic approaches, which translates to significant performance improvement of remote applications.

1 Introduction

Recently several applications have emerged that interact with databases over the network and need estimates of query cardinalities. Examples include replica maintenance [1], proxy caching [2], and query schedulers in federated systems [3]. Accurate estimates of query cardinality are required for core optimization decisions: pushing updates versus pulling data in replica systems, caching versus evicting objects in proxy caches, choosing a distributed query execution schedule in a federated system. Optimal decisions improve performance of these applications.

We are interested in the performance of Bypass Yield (BY) cache, a proxy caching framework for the National Virtual Observatory (NVO), a federation of astronomy databases. BY caching reduces the bandwidth requirements of the NVO by a factor of five. It achieves this reduction by replicating database objects, such as columns (attributes), tables, or views, near clients so that queries to the database may be served locally, reducing network bandwidth requirements. BY caches load and evict the database objects based on their expected yield: the size of the query results against that objecti.The five-fold reduction in bandwidth is an upper bound, realized when the cache has perfect, a priori knowledge of query result sizes. In practice, a cache must estimate yield or equivalently the cardinality of the query.

Query cardinality estimation has long been employed by query optimizers to evaluate various query execution plans. For this, the most popular data structure is the

I.-Y. Song, J. Eder, and T.M. Nguyen (Eds.): DaWaK 2008, LNCS 5182, pp. 402–412, 2008.
© Springer-Verlag Berlin Heidelberg 2008

histogram. However, traditional approaches for constructing histograms do not translate to proxy caches. Traditional approaches [4,5] require complete access to data, and involve expensive data scans. Distributed applications such, as proxy caches, however, are remote from data sources and are severely constrained in the data they have access to or resources they can expend on query cardinality estimation [6]. Resource constraints include lack of access to source data, or limitations in storage space, processing time, or amount of network interactions with the remote data sources.

Distributed applications benefit from "workload-aware" histograms that are constructed by exploiting workload information and query feedback. The key idea is to use queries and the corresponding feedback to construct a data distribution over heavily accessed regions while allowing for inaccuracies in the rest of the regions. However, previous approaches [7,8,9] for constructing workload aware histograms either require extremely detailed query feedback or lack in consistency and accuracy. Approaches such as STHoles [7] and ISOMER [8] require sub-query cardinality feedback at each level of a query execution plan to construct a consistent and accurate histogram. These approaches are practical for applications that are co-located with the data source. For remote applications, such detailed feedback not only comes at a high cost, it is also redundant: The applications only need a top level query cardinality estimate; estimating sub-query cardinality is an over-kill [10]. STGrid [9] is seemingly suitable for distributed applications as it requires only the query and its feedback to construct a histogram. However, it uses heuristics to refine the histogram leading to inconsistencies and inaccuracies in the constructed histogram.

In this paper, we propose a novel technique for histogram construction that is suitable for distributed applications such as the Bypass-Yield proxy cache. It uses queries and only the top-level cardinality feedback to learn the data distribution of a cached object. The histogram is maintained incrementally in that query feedbacks refine the bucket frequencies. Unlike previous approaches [9] in which the refinement is done heuristically, our technique uses recursive least squares technique to gradually "regress" the distribution. This incremental regression technique takes into account all previous query feedbacks and thus produces a consistent histogram. Histogram accuracy is maintained as the refinement procedure avoids inclusion of erroneous assumptions and heuristics into the histogram. In addition, the overhead of our technique has a negligible cost in constant time for each query feedback, making it extremely suitable for resource-constrained distributed applications.

Extensive experimental results show that refining a histogram consistently and in a principled manner leads to 30-40% higher accuracy. We conduct our experiments on the Sloan Digital Sky Survey (SDSS), which is the largest federating site of the NVO. We generate synthetic workload whose access patterns mimic the workload logs of the SDSS When the cardinality is known apriori, BY caches reduce network traffic requirements of SDSS by a factor of five. Our workload-aware histograms reduce network requirements by a factor of 5.5 in addition to being computationally and space efficient.

2   Related Work

There has been intensive research in constructing accurate histograms for cardinality estimation in query optimizers. Chen and Roussoupoulos [11] first introduced the idea of constructing a data distribution using queries and their feedback. However, they approximate a data distribution of a relational attribute as a pre-chosen model function and not a histogram. By virtue of using histograms our approach easily extends to multi-dimensional queries, which are not discussed by Chen et. al. and left as future work.

Using queries and its cardinality, STGrid [9] progressively learns a histogram over a data distribution. A histogram is refined by heuristically distributing the error of the *current* query amongst histogram buckets that overlap with the selectivity range of the query. While this gives a very fast algorithm for refining the histogram, it lacks in consistency. It favors in reducing error for the current query but does not take account feedback from previous queries over the same histogram.

In STHoles [7] and ISOMER [8], a superior histogram structure is proposed, which improves accuracy, but at the expense of additional feedback [10]. This additional feedback comes either by constructing queries with artificial predicates and obtaining their cardinality from the database, or by closely monitoring the query execution plan of each query and obtaining feedbacks at every level of the plan. This makes these approaches suitable for query optimizers, which have closer proximity to data, and can monitor and create such feedbacks with low overhead. However, they are an overkill for distributed applications which are oblivious to query execution plans, resource constrained and need only a top-level cardinality estimate [10]. In this paper, we have introduced consistent and efficient workload-aware histograms that can be used for cardinality estimation in remote, distributed applications such as the Bypass Yield proxy cache to make core optimization decisions.

3   Cardinality Estimation in BY Caches

In this section, we briefly describe the core caching decision made by the bypass yield application, how cardinality estimates are required and used in cache replacement algorithms, and metrics which define the network performance of the system. We then introduce workload-aware histograms which can be used in Bypass caches for cardinality estimation.

3.1   The BY Cache Application

The bypass-yield cache [2] is a proxy cache, situated close to clients. For each user query, the BY cache evaluates whether to service the query locally, loading data into the cache, versus shipping each query to be evaluated at the database server. By electing to evaluate some queries at the server, caches minimize the total network cost of servicing all the queries.

In order to minimize the WAN traffic, a BY cache management algorithm makes an economic decision analogous to the rent-or-buy problem [12]. Algorithms choose between loading (buying) an object and servicing queries for that object in the cache versus bypassing the cache (renting) and sending queries to be evaluated at sites in the federation. The BY cache, caches objects such as tables, columns or views. At the heart of a cache algorithm is byte yield hit rate (BYHR), a savings rate, which helps the cache make the load versus bypass decision. BYHR is maintained on all objects in the system regardless of whether they are in cache or not. BYHR is defined as

$$\text{BYHR} = \sum_j \frac{p_{i,j} y_{i,j} f_i}{s_i^2} \tag{1}$$

for an object o_i of size s_i and fetch cost f_i accessed by queries Q_i with each query $q_{i,j} \in Q_i$ occurring with probability $p_{i,j}$ and yielding $y_{i,j}$ bytes. Intuitively, BYHR prefers objects for which the workload yields more bytes per unit of cache space. BYHR measures the utility of caching an object (table, column, or view) and measures the rate of network savings that would be realized from caching that object. BY cache performance depends on accurate BYHR, which, in turn, requires accurate yield estimates of incoming queries. (The per object yield $y_{i,j}$ is a function of the yield of the incoming query.) Since the yield of the incoming query is a function of the number of tuples, we estimate cardinality for each query.

3.2 Learning Histograms

Histograms are the most popular and general model used in databases for cardinality estimation. Histograms model the data distribution as piece-wise constant functions. In our workload-aware technique, histograms are initialized as a constant function (uniformity assumption), with a fixed number of buckets. With each query and its feedback, the current function is updated to obtain a new piecewise constant function, the latter providing a better approximation of the underlying data distribution. This function update is not arbitrary, but is based on past queries and their feedback. Specifically, that function is obtained which minimizes the sum of square of residual errors, i.e., the squared error between the actual and estimated cardinality over all queries. For this, cardinality estimation error over all queries need not be maintained and our technique uses the well-known recursive-least-square (RLS) algorithm to incrementally minimize the sum of residual errors. Periodically bucket boundaries are readjusted to give a new function which further improves accuracy. For ease of presentation, we describe the technique by constructing histograms using queries with single range clauses. We later revise the technique for queries with multiple range clauses. Finally, we show how bucket boundaries are periodically organized.

Single-Dimensional Histograms. Let A be an attribute of a relation R, and let its *domain* be the range $D = [A_{\min}, A_{\max}]$ (currently assuming numerical domains only). Let f_A be the actual distribution of A, and F_A be the corresponding cumulative distribution function (CDF). A *range* query on attribute A, $\sigma_{l \le R.A \le h}(R)$, where $l \le h$ is denoted as $q(l, h)$. The cardinality s of query q defined as $s_q = f_A([l, h])$, is the number of tuples in the query result, and is equal to

$$s = \sum_l^h f(x) = F_A(h) - F_A(l)$$

The query feedback is then defined as $\tau = (l, h, s)$. Given several such feedbacks, the goal is to learn $\hat{f}_A$ (or equivalently $\hat{F}_A$), an approximation of f_A that gives the minimum sum of squared residuals.

We model $\hat{f}_A$ as a histogram of B buckets. The corresponding $\hat{F}_A$ is a piece-wise linear function with a linear piece in each bucket of the histogram.

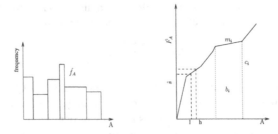

Fig. 1. The cumulative frequency distribution on attribute A can be learned to estimate query cardinality

To model (compute) $\hat{F}_A(\cdot)$, we need to maintain the following data: For each bucket i, the values $i_{\min}, i_{\max}$ are the boundaries of the bucket m_i the slope of the linear segment within bucket i, and c_i is the y-intercept of the linear segment. Thus $\hat{F}_A(x)$ is computed as: $\hat{F}_A(x) = m_i x + c_i, i \in B$, and $i_{\min} \le x < i_{\max}$.

Using this form of approximation the estimated cardinality, $\hat{s}$ for a query $q = (l, h)$ is computed as follows:

$$\hat{s} = \hat{F}_A(h) - \hat{F}_A(l) = m_u h + c_u - (m_v l + c_v), \tag{2}$$

in which $u_{\min} \le h < u_{\max}$ and $v_{\min} \le l < v_{\max}$. Figure 1 shows a histogram of 6 buckets constructed over attribute A. The adjacent figure shows the corresponding $\hat{F}_A$ and how it can be used to estimate cardinality of queries. However, to estimate the cardinality precisely we need to approximate the parameters m_i and c_i in each bucket of $\hat{F}_A$. For this we first cast our parameters m_i and c_i as vectors. This gives a concise representation and a neat expression that can then be estimated using RLS.

$\mathbf{M} = (m_1, \ldots, m_B)$ is a $B \times 1$ vector in which each element is the slope of the bucket. Similarly, $\mathbf{C} = (c_1, \ldots, c_B)$ is a $B \times 1$ vector in which each entry is the y-intercept. To obtain $\hat{s}$, we cast the range values h and l in a $1 \times B$ vector in which the uth element is h, the vth element is $-l$ and the rest of the elements are 0. For the intercept, we define a $1 \times B$ unit vector in which the uth element is 1, the vth element is -1 and the rest of the elements are 0. This gives a concise vector notation for $\hat{s}$:

$$\hat{s} = \left((0, .., h_u, .., -l_v, .., 0)(0, .., 1_u, .., -1_v, .., 0)\right) \begin{pmatrix} \mathbf{M} \\ \mathbf{C} \end{pmatrix}, \tag{3}$$

The vectors $\mathbf{M}$ and $\mathbf{C}$ are estimated after a sequence of query feedbacks $\tau_1, \ldots, \tau_n$ have been collected, where $n \ge B$. A standard criterion [13] to find the optimal $\mathbf{M}$ and $\mathbf{C}$ is to minimize the sum of the squares of the estimation error or the least square error:

$$\sum_n (\hat{s} - s)^2 = \sum_n \left(((0, .., h_u, .., -l_v, .., 0)(0, .., 1_u, .., -1_v, .., 0)) \begin{pmatrix} \mathbf{M} \\ \mathbf{C} \end{pmatrix} - s\right)^2. \tag{4}$$

Alternatively, above problem can be reformulated as:

$$minimize \parallel \mathbf{X}\hat{\theta} - \mathbf{Y} \parallel^2, given \tag{5}$$

$$\mathbf{X}_{n \times 2B} = \begin{pmatrix} (0,..,h_u,..,-l_v,..,0)(0,..,1_{u1},..,-1_{v1},..,0) \\ (0,..,h_{u2},..,-l_{v2},..,0)(0,..,1_{u2},..,-1_{v2},..,0) \\ ... \\ (0,..,h_{uN},..,-l_{vN},..,0)(0,..,1_{uN},..,-1_{vN},..,0) \end{pmatrix},$$

$$\mathbf{Y}_{n \times 1} = \big(s_1, s_2, \ldots, s_n\big)^T, \text{ and}$$

$$\hat{\theta}_{2B \times 1} = (\mathbf{M}, \mathbf{C})^T = (m_1, \ldots, m_B, c_1, \ldots, c_B)^T.$$

If $\mathbf{X}^T$ is the transpose of $\mathbf{X}$ then the solution to above equation is

$$\hat{\theta}_{LS} = (\mathbf{X}^T\mathbf{X})^{-1}\mathbf{X}^T\mathbf{Y} \tag{6}$$

The above formulation assumes that all the feedbacks $\tau_1 \ldots \tau_n$ are known and thus $\hat{\theta}_{LS}$ can be computed by computing the inverse of X^TX. However, feedbacks come one at a time and we use recursive least square (RLS) [13] method to calculate $\hat{\theta}_{LS}$ incrementally. The RLS algorithm can be started by collecting 2B samples and solving 6 to obtain an initial value of $\hat{\theta}_{LS}$. Each incremental feedback now takes $O(B^2)$ time to compute.

The obtained $\hat{\theta}_{LS}$ is our $\hat{F}_A(\cdot)$. Since there is a one-to-one mapping between $\hat{F}_A(\cdot)$ and $\hat{f}_A(\cdot)$, the corresponding histogram can be easily obtained.

Input: Initialize $\mathbf{P}_N = [\mathbf{X}^T\mathbf{X}]^{-1}$ and $\hat{\theta}_N$ from an initial collection of N examples using
(4) and (5)
foreach *subsequent example* $(\mathbf{x}_i, y_i)$ *compute* **do**
$\hat{\theta}_i = \hat{\theta}_{i-1} + \frac{\mathbf{P}_{i-1}\mathbf{x}_i(y_i - \mathbf{x}_i^T\hat{\theta}_{i-1})}{1+\mathbf{x}_i^T P_{i-1}x_i}$
$\mathbf{P}_i = \mathbf{P}_{i-1} - \frac{\mathbf{P}_{i-1}\mathbf{x}_i\mathbf{x}_i^T\mathbf{P}_{i-1}}{1+\mathbf{x}_i^T\mathbf{P}_{i-1}\mathbf{x}_i}$
end

Algorithm 1. Recursive least squares (RLS) algorithm

Multi-dimensional Range Queries. We show how the above formulation can be extended to multi-dimensional (m-d) range queries. For m-d range queries we need to learn f_A the actual distribution over the attribute set $A = A_1, \ldots, A_k$ of R. The corresponding range query on A is $\sigma_{\bigwedge_d\ l_d \le R.A_d \le h_d}(R)$, in which $1 \le d \le k$ and $l_d \le h_d$. The cardinality of q is

$$s = \sum_{x_1=l_1}^{h_1} \cdots \sum_{x_k=l_k}^{h_k} f(x_1, \ldots, x_k) = F_A(h_1, \ldots, h_k) - F_A(l_1, \ldots, l_k)$$

We model $\hat{f}_A$ as a histogram of $B = B_1 \cdots B_k$ buckets. The corresponding $\hat{F}_A$ is a piece-wise linear function with a planar segment in each bucket of the histogram. To model (compute) $\hat{F}_A(\cdot)$, we need to maintain the following data: For each bucket i, and each attribute A_d the values $i_{d_{\min}}, i_{d_{\max}}$ are the boundaries of the bucket, m_{i_d} the slope of the planar segment along A_d within bucket i, and c_i is the intercept of the planar segment. Thus $\hat{F}_A(x_1, \ldots, x_k)$ is computed as: $\hat{F}_A(x_1, \ldots, x_k) = \sum_d m_{i_d}x_d + c_i$, where i is the bucket such that $i_{d_{\min}} \le x_d < i_{d_{\max}}$ for each attribute A_d.

Using this form of approximation the estimated cardinality $\hat{s}$ for a query $q = q((l_1, h_1), \ldots, (l_k, h_k)$ $q = (\{(l_d, h_d)\})$ is computed as follows:

$$\hat{s} = \hat{F}_A(h_1, \ldots, h_k) - \hat{F}_A(l_1, \ldots, l_k) \tag{7}$$

Given a sequence of feedbacks, an algebraic form similar to equation 3, and finally equation 5 can be obtained which can be minimized recursively using RLS.

In the above, we have maintained $\hat{F}(\cdot)$ corresponding to the sequence S of query-feedbacks we have received at any stage. For each query-feedback $\tau \in S$ there is a corresponding error squared wrt to $\hat{F}(\cdot)$, which is $(\hat{F}(h_1, \ldots, h_k) - \hat{F}(l_1, \ldots, l_k) - s)^2$. Based on the formulation of $\hat{F}(\cdot)$ and the RLS algorithm [13], we can prove the following theorem.

Theorem 1. *For any set of query-feedbacks S, the corresponding function $\hat{F}(\cdot)$ has the minimum sum of squared error with respect to S over all piece-wise linear functions defined over the set of buckets used by $\hat{F}(\cdot)$.*

4 Bucket Restructuring

In the above section, a histogram of B buckets is chosen to approximate the data distribution. However, the choice of B is arbitrary and a lower B can result in even lower error over the same workload. Bucket restructuring refers to the iteration process in which bucket boundaries are collapsed or expanded to obtain a histogram of B' buckets in which $B' \neq B$ such that the overall error is minimized. The restructuring process is not a learning process and does not depend on query feedback. Thus any good restructuring process suffices. In order to test the accuracy of our approach, in this paper, we have used the same restructuring process as described in [9]. Specifically, high frequency buckets are split into several buckets. Splitting induces the separation of high frequency and low frequency values into different buckets, and the frequency refinement process later adjusts the frequencies of these new buckets. In order to ensure that the number of buckets assigned to the histogram does not increase due to splitting, buckets with similar frequencies are reclaimed. The restructuring process is performed periodically.

5 Experiments

The goal of our experiments is two-fold: Firstly to test the accuracy of an incremental, self-tuning approach in learning histograms and secondly to measure the impact of the learned histogram on the performance of the BY cache. We first describe our experimental setup and then report our results.

Datasets. For both the experiments we use a synthetic workload over a real astronomy database, the Sloan Digital Sky Survey. The Sloan Digital Sky Survey database consists of over 15TB of high dimensional data. These dimensions consist of various properties of the astronomical bodies. Over here we show results for one, two and three dimensional datasets. The datasets are a projection of light intensities and spatial coordinates of 5,000,000 astronomical bodies obtained from the Sloan Digital Sky Survey. In 1-dimension, the data is often queried for light intensities and in 2 and 3-dimensions, for spatial coordinates. Data density varies significantly in the patch of sky that we have considered ranging from million of bodies in 1 arc second to tiny black holes with no objects. One of our motivations of using real data is that while the authors in [9] report high accuracy of their technique over synthetic data, as our experiments show, is not true over real data that have complex density functions.

Workload. We generate synthetic workloads (similar to [9]) consisting of random range queries in one or more dimensions. Our workload represents queries as seen in the real workload log of the SDSS. While the corner points of each selection range are generated independently, workloads exhibit locality of reference. The attribute values used for selection range corner points in these workloads are generated from piecewise uniform distributions in which there is an 80% probability of choosing a value from a locality range that is 20% of the domain. The locality ranges for the different dimensions are independently chosen at random according to a uniform distribution.

Histograms. We use 100 buckets for 1-d histograms and 50 buckets per dimension for 2-d and 3-d histograms, respectively. The one, two and three dimensional histograms occupy 1.2, 10.5 kilobytes of memory, respectively, which is much smaller than constructing histograms over the entire dataset. The histogram is first constructed using a training workload and its accuracy then tested over a test workload which is statistically similar. To measure accuracy over a test workload, we use the average relative estimation error i.e., (100 * abs(actual result size - estimated result size) / N * actual result size) to measure the accuracy.

Evaluation. Finally, we compare our approach (denoted here as least squares (LS)) with STGrid, which also uses minimum feedback to construct histograms in one and higher dimensions. STGrid uses the heuristic that buckets with higher frequencies contribute more to the estimation error than buckets with lower frequencies. Specifically, they assign the "blame" for the error to the buckets used for estimation in *proportion to their current frequencies* [9].

Table 1. Error Comparison with Increasing Training Workload

Dim	1		2		3	
Wkld Size	LS	STGrid	LS	STGrid	LS	STGrid
2000	3.04%	5.70%	13.29%	15.14%	23.15%	49.56%
2500	3.13%	3.23%	12.15%	19.29%	21.87%	49.13%
3000	2.19%	2.12%	11.25%	23.28%	20.35%	44.10%
4000	1.29%	2.20%	10.05%	21.16%	20.12%	52.39%

Accuracy. We evaluate the accuracy of our technique with increasing training size. For testing, we start with B samples and obtain an initial histogram. Then each new query refines the histograms. After the histogram is refined with the given workload, a test workload is used to estimate the overall error. As the training size increases, there is more feedback to learn the data distribution and so the accuracy of any workload-aware technique should improve with increase in training size. We see in Table 1 that this is true for LS but not for STGrid. In fact extra feedback often increases the error in case of 2 and 3-d range queries. Additionally queries which have higher cardinality amount to high relative estimation errors for the STGrid technique in the test workload. This

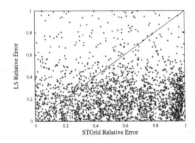

Fig. 2. Scatter Plot: Each point represents a query. Points below the diagonal line correspond to queries on which LS performs better than STGrid and vice-versa.

Model	Network Cost (GB)	Savings (GB)	% difference from optimal
No Caching	100.46		
Prescient	13.51	86.95	-0.00
LS	17.53	82.93	-4.02
STGrid	49.47	50.99	-35.96

Fig. 3. Impact of cardinality estimation on bypass-yield caching performance

is because STGrid looks at reducing the error of the current query but not the overall error. Thus it improves accuracy over a few queries but not all queries. This is clearly reflected in figure 2 in which we compare the error on a query-by-query basis. Figure 2 shows that for a few queries, STGrid has very low error, but as more queries come over the same region there is no guarantee that the error will be lower as the additional feedback produces inconsistent results.

Performance of BY caches. Our primary result (Figure 3) is to show the impact of an accurate workload-aware technique on the performance of BY caches. For this experiment we initialized a cache whose size varied from 1-5% of the database size. Our workload consists of range queries on several spatial and intensity attributes. Based on the yield of the incoming workload, the objective is to cache attributes such that network traffic is minimized. the workload consists of a mixture of single and multidimensional queries, each with equal probability of occurrence. In this experiment we compare the impact of the two approaches with a *prescient* estimator. A prescient estimator gives us an upper-bound on how well a caching algorithm could possibly perform when the cardinalities are known a-priori.

As a yield estimator for BY caching, LS outperforms STGrid dramatically and approaches the ideal performance of the prescient estimator. Even though LS has an accuracy range of 20-30% on the entire workload, for the caching system it performs close to the prescient estimator. This is because some of the queries on which LS performed poorly are sent to the server. The STGrid did not perform that poorly on those queries and thus severely affected cache performance. This experiment remarkably shows that a technique which improves the accuracy of a few queries can severely affect the performance of the BY cache.

The LS technique is computationally slower than STGrid. STGrid takes practically no time to update a histogram. In the worst case it has to refine the frequency value in B buckets. For the 1, 2, and 3 dimensional synthetic data sets LS takes an average of 5 secs, 65 and 115 secs over the entire training workload of 2000 queries. This is because with every feedback LS has to update an $O(B^2)$ matrix. More efficient approaches [14] for computing matrix transpose can be used. It is to be noted that the accuracy of LS comes at the expense of computational time. However, the considerable increase in accuracy is attractive, and the time to update is tolerable, making it a light-weight, accurate approach for remote applications.

6   Conclusions and Future Work

In this paper we have proposed a novel technique for learning workload-aware histograms that uses queries and their cardinality feedbacks only. The learned histograms are guaranteed to have the minimum sum of squared errors over the entire query sequence. Experiments show that minimizing the sum of squared errors provides far more accurate histograms than histograms learned by distributing error through heuristics. Further, our technique is neither tied to a query execution plan nor requires any artificial feedback. Therefore it is suitable for distributed applications that are remote from databases and need a light-weight, accurate self-learning cardinality estimation technique.

References

1. Olston, C., Widom, J.: Best-effort cache synchronization with source cooperation. In: ACM SIGMOD, pp. 73–84 (2002)
2. Malik, T., Burns, R.C., Chaudhary, A.: Bypass caching: Making scientific databases good network citizens. In: Intl' Conference on Data Engineering, pp. 94–105 (2005)
3. Ambite, J.L., Knoblock, C.A.: Flexible and scalable query planning in distributed and heterogeneous environments. In: Conference on Artificial Intelligence Planning Systems, pp. 3–10 (1998)
4. Poosala, V., Ioannidis, Y.E.: Selectivity estimation without the attribute value independence assumption. In: VLDB, 486–495 (1997)
5. Gibbons, P.B., Matias, Y., Poosala, V.: Fast incremental maintenance of approximate histograms. ACM Transactions on Database Systems 27, 261–298 (2002)
6. Malik, T., Burns, R., Chawla, N., Szalay, A.: Estimating query result sizes for proxy caching in scientific database federations. In: SuperComputing (2006)
7. Bruno, N., Chaudhuri, S., Gravano, L.: STHoles: A multidimensional workload-aware histogram. In: Proceedings of the ACM SIGMOD International Conference on Management of Data (2001)
8. Srivastava, U., Haas, P.J., Markl, V., Kutsch, M., Tran, T.M.: Isomer: Consistent histogram construction using query feedback. In: 22nd International Conference on Data Engineering, p. 39. IEEE Computer Society, Los Alamitos (2006)
9. Aboulnaga, A., Chaudhuri, S.: Self-tuning histograms: Building histograms without looking at data. In: Proceedings of the ACM SIGMOD International Conference on Management of Data, pp. 181–192 (1999)

10. Malik, T., Burns, R., Chawla, N.: A black-box approach to query cardinality estimation. In: Conference on Innovative Database System Research (2007)
11. Chen, C.M., Roussopoulos, N.: Adaptive selectivity estimation using query feedback. In: Proceedings of the ACM SIGMOD International Conference on Management of Data, pp. 161–172 (1994)
12. Fujiwara, H., Iwama, K.: Average-case competitive analyses for ski-rental problems. In: Intl. Symposium on Algorithms and Computation (2002)
13. Young, P.: Recursive Estimation and Time Series Analysis. Springer, New York (1984)
14. Ari, M.: On transposing large $2^n \times 2^n$ matrices. IEEE Trans. Computers 28, 72–75 (1979)

Is a Voting Approach Accurate for Opinion Mining?

Michel Plantié[1], Mathieu Roche[2], Gérard Dray[1], and Pascal Poncelet[1]

[1] Centre de Recherche LGI2P, Site EERIE Nîmes, École des Mines d'Alès - France
{michel.plantie,gerard.dray,pascal.poncelet}@ema.fr
[2] LIRMM, UMR 5506, Univ. Montpellier 2, CNRS - France
mathieu.roche@lirmm.fr

Abstract. In this paper, we focus on classifying documents according to opinion and value judgment they contain. The main originality of our approach is to combine linguistic pre-processing, classification and a voting system using several classification methods. In this context, the relevant representation of the documents allows to determine the features for storing textual data in data warehouses. The conducted experiments on very large corpora from a French challenge on text mining (DEFT) show the efficiency of our approach.

1 Introduction

The Web provides a large amount of documents available for the application of data-mining techniques. Recently, due to the growing development of Web 2.0, Web documents as blogs, newsgroups, or comments of movies/books are attractive data to analyze. For example, among the tackled issues addressed by the text mining community, the automatic determination of positive or negative sentiment in these opinion documents becomes a very challenging issue. Nevertheless, the storage of this kind of data in order to apply data-mining techniques is still an important issue and some research works have shown that a data warehouse approach could be particularly well adapted for storing textual data [9]. In this context, data warehouses approaches consider two dimensional tables with the rows representing features of the documents and the columns the set of document domains. For instance, if we consider opinion documents in the movie context, the domain could be the genre of the movie (e.g. fantastic, horror, etc). In this paper we focus on text-mining approaches to find the relevant features (i.e. the first dimension of the data warehouse) to represent a document. Then we deal with the data-mining algorithms in order to classify the opinion documents using these features, i.e. classifying documents according to opinion expressed such as positive or negative mood of a review, the favorable or unfavorable aspect given by an expert, the polarity of a document (positive, neutral, negative) and/or the intensity of each opinion (low, neutral, high), etc.

The rest of the paper is organized as follows. Firstly, we present previous works on opinion mining (section 2), followed by section 3 presenting our approach based on two main parts: the document representation techniques, and

I.-Y. Song, J. Eder, and T.M. Nguyen (Eds.): DaWaK 2008, LNCS 5182, pp. 413–422, 2008.
© Springer-Verlag Berlin Heidelberg 2008

the classification process. This process is based on machine learning and "text-mining" techniques paired with a vote technique. This vote technique (section 3.5) combines several classifiers in a voting system which substantially enhance the results of other techniques, and finally section 4 presents the obtained results.

2   Related Work

Classification of opinion documents as blogs or news is more and more addressed by the text mining community [1, 7, 23, 25].

Several methods exist for extracting the polarity of a document. Actually, the opinion polarities are often given by adjectives [7, 25]. The use of adverbs attached to adjectives (for instance, the adverb "very" attached to the adjective "interesting") allows to determine the intensity of phrases (group of words) [1].

For example P. Turney [23] proposes an approach based on the polarity of words in the document. The main idea is to compute correlations between both adjectives in the documents and adjectives coming from a seed set. Two seed sets are considered: positive (e.g. good, nice, ...) and negative (e.g. bad, poor, ...). The associations are calculated by statistical approaches based on the results of *(i)* search engine [22], *(ii)* LSA method [14]. Other approaches using supervised learning methods allow to define polarity degrees (positive, negative, objective) to the Wordnet lexical resource [16]. Besides many studies have shown that the grammatical knowledge are relevant for opinion mining approaches [6][10].

To calculate the polarity of words, supervised or unsupervised methods can be used to predict the polarity of a document. The supervised approaches have the advantage to automatically learn relevant features (words, phrases) to predict a domain opinion. It's important to extract domain dependent characteristics. The same word or group of words may be positive in a domain and negative for another domain: for example, the adjective "commercial" is positive for economic documents but expresses a negative sentiment to characterize a movie. Thus, these supervised methods are often used in national [8] and international [25] opinion mining challenges.

When we have well structured opinion corpora, machine learning techniques (based on training models on these corpora), outperform results. Methods based on individual word search cannot extract complete information on opinion texts and so produce less efficient classification results.

This paper proposes a new method called "COPIVOTE" (*classification of* OPI*nion documents by a* VOTE *system*) to classify document according to the expressed opinions. We thus define a new architecture based on coupling several techniques including a voting system adapted to each domain corpus in order to get better results. The main originality of our approach relies on associating several techniques: extracting more information bits via specific linguistic techniques, space reduction mechanisms, and moreover a voting system to aggregate the best classification results.

3   The Copivote Approach

For efficiency reasons our method does not try to search each opinion related word. Statistic techniques are able to produce a more comprehensive document representation. This characteristic allows us to manage the large complexity and the subtleties in opinion expression contained in the language as explained in subsection 3.2. The specificity of our approach lies on pre and post treatments adapted to the corpus types. However, the overall process presented in this paper may also be adapted to other kind of corpora.

3.1   Overall Process Presentation

Our method uses four main steps to classify documents according to opinion:

- **Linguistic treatments for vector space model representation:** In this step we use linguistic analysis adapted to opinion texts.
- **Vector space model reduction:** In order to get better performances and limited processing time we simplify the vector space model.
- **Classification:** This stage uses classifiers to compute model and to classify new texts
- **Classifier voting system:** This phase gather the classifiers results for one document and aggregate them in one answer for each document.

3.2   Linguistic Treatments for Vector Space Model Representation

Our first step is to apply several linguistic pre-treatments. The first one is based on the extraction of all linguistic units (lemmatised words or lemmas) used for document representation. For example the conjugated verb "presents" is replaced by its lemma: the infinitive verb "present".

We then eliminate words having grammar categories with a low discriminative power with regard to opinion mining: undefined articles and punctuation marks. In our approach, we keep lemmas associated with almost all grammatical categories (as adverbs) in order to specifically process opinion documents. Since we are on a machine learning approach based on corpora, we are able to use all information of documents. Each kind of word may contain opinion discriminative information even very slight. Further more we extract known expressions. extracting expressions and keeping almost all words will enhance the classification results.

For our purpose, we will call "index", the list of lemmas worked out for each corpus. Each corpus is represented by a matrix in compliance with the Salton vector space model representation [21]. In this representation, each row is associated to each document of the corpus and each column is associated with each lemma. Each matrix cell represents the number of occurrences for the considered lemma in the considered document.

In our approach, the whole set of documents of a corpus and therefore the associated vectors are used as training set.

3.3   Vector Space Model Reduction (Index Reduction)

The Vector space defined by the whole set of lemmas of the training corpus is very important in dimension. We thus perform an index reduction for each corpus. We use the method presented by Cover, which measures the mutual information between each vector space dimension and classes [5]. This method, explained in depth in [18], measures the interdependence between words and the document classifying categories by computing the entropy difference between the category entropy and the studied dimension (key word) entropy of the vector space. If the difference is high, then the discriminative information quantity of this word is high, and therefore the importance of this word is high in the categorization task.

Once the indexes are computed, we consider each computed key word in each index as the new dimensions of the new representation vectors space for each corpus of documents. The new vector spaces have a reduced number of dimensions. These new computed vectors are called: "reduced" vectors. As it is shown in [18], this reduction helps a lot to significantly improve the quality of results and drastically lower the computing times.

3.4   Use of Bigrams

In this approach we take into account words to compute the document vectors and we also add bigrams of the corpora (groups of two words). Only bigrams containing special characters are rejected (mathematical characters, punctuation, etc). This richer document representation allows us to extract information more adapted to opinion corpora. As an example, in the corpora we have used for experiments, bigrams like "not convincing", "better motivate", "not enough" are groups of words much more expressive of opinions than each word separately.

This enriched document representation using bigrams improve results, as we will see in section 4. In addition to the quality of document representation that improves the classification tasks, taking into account several classifiers (see next section) remains crucial to get good quality results.

3.5   Classifier Voting System

To improve the general method for classifying opinion documents, we have worked out a voting system based on several classifiers. Our vote method named COPIVOTEMONO (*classification of* OPI*nion documents by a* VOTE *system with* MONO*grams*) and COPIVOTEBI (*classification of* OPI*nion documents by a* VOTE *system with* BI*grams*) when monograms and bigrams are used, uses the specific data related to opinion documents presented in the previous subsections.

The voting system is based on different classification methods. We use three main classification methods presented afterwards. Several research works use voting of classifiers, Kittler and Kuncheva [12, 13] describe several ones. Rahman [20] shows that in many cases the quite simple technique of majority vote is the most efficient one to combine classifiers. Yaxin [2] compares vote techniques with summing ones. Since the probability results obtained by individual classifiers are not commensurate, vote techniques based on the final result of each classifiers is the most adequate to combine very different classifier systems.

In our approach we use four different procedures of vote:

- **Simple majority vote:** the class allocation is computed considering the majority of classifiers class allocation.
- **Maximum choice vote (respectively minimum):** the class allocation is computed as the classifier that gives the highest probability (respectively the lowest). In that situation, the probabilities expressed by each classifier must be comparable.
- **Weighted sum vote:** for each document d(i) and for each class c(j) the average of probabilities avg(i,j) is computed and the class allocated to the document i is based on the greatest average max(avg(i,j)).
- **Vote taking into account F-score, and/or recall and/or precision:** The classifier, for a given class, is elected if it produces the best result in F-score (and/or recall and/or precision) for this class. These evaluation measures (F-score, recall, precision) are defined below.

Precision for a given class i corresponds to the ratio between the documents rightly assigned to their class i and all the documents assigned to the class i. Recall for a given class i corresponds to the ratio between the documents rightly assigned to their class i and all the documents appertaining to the class i. Precision and recall may be computed for each of the classes. A trade-off between recall and precision is then computed: the F-score (F-score is the harmonic average of recall and precision).

3.6 Classification

We adapt the classification method for each training set. We have kept the most competitive classification method for a given corpus. The results are evaluated using the cross validation method on each corpus, based on the precision, recall and F-score measures.

Having described the vote system, we will now briefly present the different classification methods used by COPIVOTEMONO and COPIVOTEBI hereafter. A more detailed and precise description of these methods is given in [4].

- **Bayes Multinomial.** The Bayes Multinomial method [24] is a classical approach in text categorization; it combines the use of the probability well known Bayes law and the multinomial law.
- **Support Vector Machine (S.V.M.).** The SVM method [11, 19] draws the widest possible frontier between the different classes of samples (the documents) in the vector space representing the corpus (training set). The support vectors are those that mark off this frontier: the wider the frontier, the lower the classification error cases.
- **RBF networks (Radial Basis Function).** RBF networks are based on the use of neural networks with a radial basis function. This method uses a "k-means" type clustering algorithm [15] and the application of a linear regression method. This technique is presented in [17].

Our contribution relies on the association of all the techniques used in our method. First the small selection in grammatical categories and the use of bigrams enhance the information contained in the vector representation, then the space reduction allows to get more efficient and accurate computations, and then the voting system enhance the results of each classifiers. The overall process comes to be very competitive.

4   Results

4.1   Corpora Description

The third edition of the French DEFT'07 challenge (http://deft07.limsi.fr/) focused on specifying opinion categories from four corpora written in French and dealing with different domains.

- **Corpus 1:** Movie, books, theater and comic books reviews. Three categories: good, average, bad,
- **Corpus 2:** Video games critics. Three categories: good, average, bad,
- **Corpus 3:** Review remarks from scientific conference articles. Three categories: accepted, accepted with conditions, rejected,
- **Corpus 4:** Parliament and government members speeches during law project debates at the French Parliament. Two categories: favorable, not favorable,

These corpora are very different in size, syntax, grammar, vocabulary richness, opinion categories representation, etc. For example, table 1 presents the allocation of classes for each corpus. This table shows that corpus 4 is the largest, and corpus 3 is the smallest. On the other hand, we may find similarities between the corpora (for example, the first class is smaller for the 3 first corpora), Table 1 shows important differences with respect to the number of documents in each class.

Table 1. Allocation of the corpus classes for the DEFT'07 challenge

Classes	Corpus 1	Corpus 2	Corpus 3	Corpus 4
Class 1	309	497	227	10400
Class 2	615	1166	278	6899
Class 3	1150	874	376	$\emptyset$
Total	2074	2537	881	17299

Table 2 shows the vector space dimensions reduction associated to each corpus. This operation drastically decreases the vector spaces for all the DEFT07 challenge corpora with a reduction percentage of more than 90%.

4.2   Detailed Results

Table 4 shows that the vote procedures globally improve the results. Firstly, all the vote methods (see section 3) give rise to the same order of improvement

Table 2. Number of lemmas for each corpus before and after reduction

Corpus	Initial Number of linguistic units	Number of linguistic units after reduction	Reduction percentage
Corpus 1	36214	704	98.1%
Corpus 2	39364	2363	94.0%
Corpus 3	10157	156	98.5%
Corpus 4	35841	3193	91.1%

even if some results of the "weighted sum vote" are slightly better (also called "average vote"). Secondly, the bigram representations associated to vote methods (COPIVOTEBI) globally improved the results compared to those obtained without using bigrams (COPIVOTEMONO).

Table 3 shows the classification methods used in our vote system. We notice that the Bayes Multinomial classifier is very competitive with a very low computing time. Almost every time the SVM classifier gives the best results. The RBF Network classifier gives disappointing results.

Table 4 shows the results expressed with F-score measure (globally and for each class) obtained by the 10 fold cross validation process on each training set. These results point out the classes that may or may not be difficult to process for each corpus. For example, we notice in table 4 that the corpus 2 gives well balanced results according to the different classes. On the contrary, the neutral class (class 2) of corpus 1 leads to poor results meaning that this class is not very discriminative. This may be explained by the nearness of the vocabulary used to describe a film or a book in a neutral way comparatively to a more clear-cut opinion.

Table 5 shows the results associated with the test corpora given by the DEFT'07 challenge committee. Tables 4 and 5 give very close results, showing that the test corpus is a perfectly representative sample of the training data. Table 5 shows that only one corpus gives disappointing results: corpus 3 (reviews of conference articles). This may be explained by the low number of documents in the training set and by the noise contained in the data (for example, this corpus contains a lot of spelling errors). The vector representation of the documents is then poor and noise has a bad effect on the classification process. The bigram representation does not provide any improvement for this corpus. More effort should be made on linguistic pre-treatment on this corpus in order to improve results.

The outstanding results for corpus 4 (parliamentary debates) may be explained by its important size that significantly support the statistic methods used. With this corpus, the vote system improves a lot the results obtained by each of the classifiers (see table 3). We may notice that the F-score value exceeds for more than 4% the best score of the DEFT'07 challenge.

In table 5, we compare our results with the DEFT'07 challenge best results. It shows that our results was of the same order or even slightly better with COPIVOTEBI .

Table 3. F-score average for the different methods used in CopivoteMono on test corpus

Corpus	SVM	RBF-Network	Naive Bayes Mult.	CopivoteMono	CopivoteBi
Corpus 1	61.02%	47.15%	59.02%	60.79%	61.28%
Corpus 2	76.47%	54.75%	74.16%	77.73%	79.00%
Corpus 3	50.47%	X	50.07%	52.52%	52.38%
Corpus 4	69.07%	61.79%	68.60%	74.15%	75.33%

Table 4. F-score per class and global, on Learning corpus (10 fold cross validation)

Corpus	CopivoteMono				CopivoteBi			
	class 1	class 2	class 3	global	class 1	class 2	class 3	global
Corpus 1	64.6%	42.7%	75.2%	60.8%	64.8%	43.8%	75.3%	61.3%
Corpus 2	74.9%	76.9%	82.6%	78.1%	75.8%	79.1%	82.4%	79.1%
Corpus 3	52.3%	43.0%	62.7%	52.7%	47.9%	45.0%	64.48%	52.4%
Corpus 4	80.0%	68.5%		74.2%	81.2%	69.6%		74.2%

Table 5. F-score on DEFT07 Test corpus

Corpus	Vote type	CopivoteMono	CopivoteBi	Best submission of DEFT07
Corpus 1	Minimum	60.79%	61.28%	60.20%
Corpus 2	Average	77.73%	79.00%	78.24%
Corpus 3	Minimum	52.52%	52.38%	56.40%
Corpus 4	Average	74.15%	75.33%	70.96%
Total		**66.30%**	**67.00%**	**66.45%**

4.3   Discussion: The Use of Linguistic Knowledge

Before text classification, we also tried a method to improve linguistic treatments. Specific syntactic patterns may be used to extract nominal terms from tagged corpus [3] (*e.g.* Noun Noun, Adjective Noun, Noun Preposition Noun, etc). In addition to nominal terms, we extracted adjective and adverb terms well adapted to opinion data [1, 7, 25]. For instance the "Adverb Adjective" terms are particularly relevant in opinion corpora [1]. For example, *still insufficient, very significant, hardly understandable* extracted from the scientific reviews corpus (corpus 3) of the DEFT'07 challenge may be discriminative in order to classify opinion documents. We used the list of these extracted terms to compute a new index for vector representation. We obtained poor results.

Actually our approach CopivoteBi takes into account words and all the bigrams of the corpus to have a large index (before its reduction presented in section 3.3). Besides the number of bigrams is more important without the application of linguistic patterns. Then our CopivoteBi approach combining a voting system and an expanded index (words and all the bigrams of words) can explain the good experimental results presented in this paper.

5 Conclusion and Future Work

This paper lay down a new approach based on combining text representations using key-words associated with bigrams while combining a vote system of several classifiers. The results are very encouraging with a higher F-score measure than the best one of the DEFT'07 challenge. Besides, our results show that the relevant representation of documents for datawarehouses is based on words and bigrams after the application of linguistic and index reduction process.

In our future work, we will use enhanced text representations combining keywords, bigrams and trigrams which may still improve the obtained results. We also want to use vote systems based on more classifiers. Finally, a more general survey must be undertaken by using other kinds of corpora and moreover textual data in different languages.

References

1. Benamara, F., Cesarano, C., Picariello, A., Reforgiato, D., Subrahmanian, V.S.: Sentiment analysis: Adjectives and adverbs are better than adjectives alone. In: Proceedings of ICWSM conference (2007)
2. Bi, Y., McClean, S., Anderson, T.: Combining rough decisions for intelligent text mining using dempster's rule. Artificial Intelligence Review 26(3), 191–209 (2006)
3. Brill, E.: Some advances in transformation-based part of speech tagging. In: AAAI, Vol. 1, pp. 722–727 (1994)
4. Cornuéjols, A., Miclet, L.: Apprentissage artificiel, Concepts et algorithmes. Eyrolles (2002)
5. Cover, T.M., Thomas, J.A.: Elements of Information Theory. John Wiley, Chichester (1991)
6. Ding, X., Liu, B.: The utility of linguistic rules in opinion mining (poster paper). In: SIGIR 2007, Amsterdam, 23-27 July (2007)
7. Esuli, A., Sebastiani, F.: PageRanking wordnet synsets: An application to opinion mining. In: Proceedings of the 45th Annual Meeting of the Association for Computational Linguistics (ACL 2007), Prague, CZ, pp. 424–431 (2007)
8. Grouin, C., Berthelin, J.-B., El Ayari, S., Heitz, T., Hurault-Plantet, M., Jardino, M., Khalis, Z., Lastes, M.: Présentation de deft 2007 (défi fouille de textes). In: Proceedings of the DEFT 2007 workshop, Plate-forme AFIA, Grenoble, France (2007)
9. Gupta, H., Srivastava, D.: The data warehouse of newsgroups. In: Proceedings of the Seventh International Conference on Database Theory. LNCS, pp. 471–488. Springer, Heidelberg (1999)
10. Jindal, N., Liu, B.: Review spam detection (poster paper). In: WWW 2007, Banff, Canada, May 8-12 (2007)
11. Joachims, T.: Text categorisation with support vector machines: Learning with many relevant features. In: Nédellec, C., Rouveirol, C. (eds.) ECML 1998. LNCS, vol. 1398, pp. 137–142. Springer, Heidelberg (1998)
12. Kittler, J., Hatef, M., Duin, R.P.W., Matas, J.: On combining classifiers. IEEE Transactions on Pattern Analysis and Machine Intelligence 20(3), 226–239 (1998)
13. Kuncheva, L.I.: Combining Pattern Classifiers: Methods and Algorithms. John Wiley and Sons, Inc., Chichester (2004)

14. Landauer, T., Dumais, S.: A solution to plato's problem: The latent semantic analysis theory of acquisition, induction and representation of knowledge. Psychological Review 104(2), 211–240 (1997)

15. MacQueen, J.B.: Some methods for classification and analysis of multivariate observations. In: Proceedings of the 5th Berkeley Symposium on Mathematical Statistics and Probability (1967)

16. Miller, G.A., Beckwith, R., Fellbaum, C., Gross, D., Miller, K.J.: Introduction to WordNet: an on-line lexical database. International Journal of Lexicography 3(4), 235–244 (1990)

17. Parks, J., Sandberg, I.W.: Universal approximation using radial-basis function networks. Neural Computation 3, 246–257 (1991)

18. Plantié, M.: Extraction automatique de connaissances pour la décision multicritre. PhD thesis, École Nationale Supérieure des Mines de Saint Etienne et de l'Université Jean Monnet de Saint Etienne, Nîmes (2006)

19. Platt, J.: Machines using sequential minimal optimization. In: Schoelkopf, B., Burges, C., Smola, A. (eds.) Advances in Kernel Methods - Support Vector Learning (1998)

20. Rahman, A.F.R., Alam, H., Fairhurst, M.C.: Multiple Classifier Combination for Character Recognition: Revisiting the Majority Voting System and Its Variation, pp. 167–178 (2002)

21. Salton, G., Yang, C.S., Yu, C.T.: A theory of term importance in automatic text analysis. Journal of the American Society for Information Science 26, 33–44 (1975)

22. Turney, P.D.: Mining the Web for synonyms: PMI–IR versus LSA on TOEFL. In: Proceedings of ECML conference. LNCS, pp. 491–502. Springer, Heidelberg (2001)

23. Turney, P.D., Littman, M.: Measuring praise and criticism: Inference of semantic orientation from association. ACM Transactions on Information Systems 21(4), 315–346 (2003)

24. Wang, Y., Hodges, J., Tang, B.: Classification of web documents using a naive bayes method. In: Proceedings of the 15th IEEE International Conference on Tools with Artificial Intelligence, pp. 560–564 (2003)

25. Yang, H., Si, L., Callan, J.: Knowledge transfer and opinion detection in the trec 2006 blog track. In: Notebook of Text REtrieval Conference (2006)

Mining Sequential Patterns with Negative Conclusions

Przemysław Kazienko

Wrocław University of Technology, Institute of Applied Informatics
Wyb.Wyspiańskiego 27, 50-370 Wrocław, Poland
kazienko@pwr.wroc.pl

Abstract. The new type of patterns: sequential patterns with the negative conclusions is proposed in the paper. They denote that a certain set of items does not occur after a regular frequent sequence. Some experimental results and the SPAWN algorithm for mining sequential patterns with the negative conclusions are also presented.

1 Introduction

Sequential patterns are one of the typical methods within the data mining research domain. They have been studied in many scientific papers for over ten years [1, 3, 14, 15]. As a result, many algorithms to mine regular sequential patterns have been developed, including incremental and parallel ones, see sec. 2.1. The more unique solutions include mining sequential patterns in streams [12], documents [9] as well as discovery of hierarchical [16] or compressed [4] sequential patterns. All these patterns are positive frequent subsequences included and discovered within all sequences from the source multiset.

Another type of patterns are negative association rules that indicate the negative relationship between two sets of objects. If the first set occurs in source transactions, another set does not co-occur or co-occur very rarely in these transactions. Note that both the source transactions and output patterns operate on sets whereas sequential patterns refer sequences in time. There are some algorithms to mine either both positive and negative association rules together [2, 8, 12, 18].

In this paper, a new kind of patterns is proposed: positive patterns that possess negative conclusions. They combine frequent positive sequences (sequential patterns) and negative association rules. It can be shortly described in the following way: if there is a frequent sequence q then elements of set X usually do not occur after the sequence q. This type of patterns enables to verify the previously expected sequences in the negative way. For example, the management of the e-commerce web site supposes that their users usually terminate their visits with payments. Sequential patterns with negative conclusions extracted from web logs (navigational paths) can negatively verify this expectation. If users of the given e-commerce web site put certain products into their basket, afterwards they enter to the first step of the purchase – personal information delivery (sequence q), and next, they are not likely to finish their buy with none of the payments (set X – the negative conclusion), then such pattern debunks the beliefs of the e-commerce management.

I.-Y. Song, J. Eder, and T.M. Nguyen (Eds.): DaWaK 2008, LNCS 5182, pp. 423–432, 2008.

2 Regular Sequential Patterns

Discovery of regular sequential patterns is an important technique in the data mining domain.

Definition 1. Source sequences T in domain D is the multiset of sequences which items belong to set D.

For example, if $D=\{p_1,p_2,...,p_N\}$ is the set of N web pages existing in the single web site, then source sequences is the multiset of all navigational paths accumulated by the web server within the certain period. Hence, a navigational path is a single source sequence $t_i=<p_{i1},p_{i2},..., p_{im_i}>$ which all m_i items (pages) p_{ij} belong to the domain D of all web site pages, i.e. $\forall 1 \leq j \leq m_i$ $p_{ij} \in D$. The example navigational path $t=<p_4,p_6,p_1,p_6>$ means that first the user visited page p_4, next page p_6, p_1, and finished on page p_6.

The simple but quite effective path extraction from web logs consists in matching IP addresses and user agent fields from the HTTP requests gathered in the web logs Additionally, some time constraints are applied. A user navigational path is constituted by all request ordered by time that came from the same address IP_i and the same user agent with the idle time between two following requests of less than 30 minutes. Besides, the length of the path can be restricted to a few hundred.

In the multiset, repetitions are allowed, i.e. there may exist two separate source sequences with the same component elements equally ordered. For example, two different users a and b may have navigated through the web site in the same way: $t_a=t_b=<p_6,p_6,p_1,p_4,p_6>$. Moreover, every source sequence t may contain repetitions in any places. In t_a, we have $p_{a1}=p_{a2}=p_{a5}$.

2.1 Mining Regular Sequential Patterns

There are many algorithms to mine regular sequential patterns like AprioriSome, AprioriAll [1], PrefixSpan [14, 15] or SPAM [3]. There also exist some incremental approaches [5, 6, 13, 17] and parallel ones [7, 19].

2.2 Subsequences and Their Complements

Definition 2. A sequence $t_i=<p_{i1},p_{i2},..., p_{im_i}>$ contains another sequence $q_j=<p_{j1},p_{j2},...,p_{jn}>$, if there exist n integers $k_1<k_2<...<k_n$, called index K of q_j in t_i, such that $p_{ik_1}=p_{j1}$, $p_{ik_2}=p_{j2}$, ..., $p_{ik_n}=p_{jn}$. Sequence q_j is also called the subsequence of t_i. Overall, index K denotes positions of the subsequence q_j's items within the given sequence t_i. Item p_{ik_n} is called the end or the last item of q_j in t_i with respect to index K whereas its position in t_i, i.e. k_n, is called the end position and it is denoted with k^{end}.

Each sequence t_i may contain up to ($2^{m_i}-m_i-1$) different sequences q_j that consist of at least 2 items. For example, a navigational path $t=<p_4,p_6,p_1,p_6>$, i.e. the sequence, contains the following 2-, 3- and 4-item subsequences: $<p_4,p_6>$, $<p_4,p_1>$, $<p_4,p_6,p_1>$, $<p_4,p_1,p_6>$, $<p_4,p_6,p_6>$, $<p_4,p_6,p_1,p_6>$, $<p_6,p_1>$, $<p_6,p_6>$, $<p_6,p_1,p_6>$, $<p_1,p_6>$. Their number – 10 is less than the maximum (11) due to repetition of p_6. Note that subsequence $<p_4,p_6>$ have two separate indexes $K_1=(1,2)$ and $K_2=(1,4)$; the end positions are $k_1^{end}=2$ and $k_2^{end}=4$, respectively.

Definition 3. A complement $C(q,t,K)$ of subsequence q in sequence t with respect to index K is the set of all items of t that follows the last item of subsequence q in t. The complement of subsequence q in t, which has the largest number of items, is called the maximum complement of q in t and denoted with $C^{max}(q,t)$.

Obviously, q must be subsequence of t according to Def. 2. For the example navigational path $t=<p_4,p_6,p_1,p_6,p_1,p_6>$ and subsequence $q=<p_4,p_6>$, we have three end positions $k_1{}^{end}=2$, $k_2{}^{end}=4$, and $k_3{}^{end}=6$ for three corresponding indexes $K_1=(1,2)$, $K_2=(1,4)$, and $K_3=(1,6)$, respectively. Hence, complement $C_1(q,t,K_1)=C_2(q,t,K_2)=\{p_1,p_6\}$ and $C_3(q,t,K_3)=\varnothing$. The complement is not a multiset – repetitions are not allowed. The first two complements are the greatest so they are simultaneously the maximum complement $C^{max}(q,t)=C_1(q,t,K_1)=C_2(q,t,K_2)$.

Overall, the maximum complement corresponds to index K with the smallest value of end position $k_1{}^{end}=2$. Apparently, the number of different complements is less or equal than the length of sequence t minus the length of subsequence q.

Note that for the given sequence t and its subsequence q, the maximum complement contains all other complements of q in t: $C_1 \subseteq C^{max}$, $C_2 \subseteq C^{max}$, and $C_3 \subseteq C^{max}$. Based on this feature, we can easily prove that if item p does not belong to the maximum complement $C^{max}(q,t)$, then item p also does not belong to any its subsets, i.e. to any of the complements $C_i(q,t,K_i)$.

2.3 Support – A Measure for the Frequent Sequence

Definition 4. Support $sup(q)$ of sequence q in T is the number of all source sequences from T that contain q (Def. 2):

$$sup(q)=\frac{card(\{t \in T : q \text{ is a subsequence of } t\})}{card(T)}. \qquad (1)$$

Support can be expressed either as the regular number of source sequences or the percentage contribution in T's. If sequence q is contained frequently enough in the source sequences from T, i.e. $sup(q) \geq minsup$, where $minsup$ is the minimum threshold, then such sequence q is called a sequential pattern in T.

3 Sequential Patterns with Negative Conclusions

Definition 5. A sequential pattern q in T and set $X \subset D$ constitute a sequential pattern with the negative conclusion $s^-(q \rightarrow \sim X)$ if there are source sequences $t_i \in T$ that contain q, for which set X does not intersect their complements $C(q,t_i,K_j)$ for any K_j.

Note that the empty intersection of set X with any of the complements $C(q,t_i,K_j)$ is equivalent to $X \cap C^{max}(q,t_i)=\varnothing$. The expression "some source sequences $t_i \in T$" in Def. 5 really means that elements of set X either occur in the maximum complement $C^{max}(q,t_i)$ very rarely or they do not occur there at all.

A sequential pattern with the negative conclusion $s^-(q \rightarrow \sim X)$, which has 1-item sequence $q=<p>$ on its left side, is simply equivalent to the negative association rule $\{p\} \rightarrow \sim X$ [2, 8, 11, 12, 18].

For the case of web user navigational paths, a sequential pattern with the negative conclusion denotes: if users have visited sequence q of web pages, afterwards, they usually visit none of the pages from set X. In other words, we do not expect any of elements from X after sequence q. For example, the pattern $s^-(<p_4,p_6> \rightarrow \sim\{p_4,p_5,p_8\})$ means that if users visit page p_4 and then p_6, they visit neither p_4 nor p_5 nor p_8. The following source sequences support the above sequential pattern with the negative conclusion: $<p_1,p_4,p_3,p_6,p_1,p_2>$, $<p_1,p_4,p_5,p_4,p_5,p_8,p_6,p_1,p_3,p_1,p_6>$ whereas the sequences: $<p_4,p_3,p_6,p_4,p_2>$, $<p_4,p_6,p_4,p_6,p_2>$ (both due to the second p_4), and $<p_4,p_6,p_8>$ (due to p_8) do not match the pattern.

3.1 Measures of Sequential Patterns with Negative Conclusions

Similarly to association rules, each sequential pattern $s^-(q \rightarrow \sim X)$ with negative conclusion possesses two basic measures: support $sup^{s-}(q \rightarrow \sim X)$ and confidence $conf^{s-}(q \rightarrow \sim X)$. The former is expressed as follows:

$$sup^{s-}(q \rightarrow \sim X) = \frac{card(\{t \in T : q \text{ is a subsequence of } t \wedge C^{max}(q,t) \cap X = \varnothing\})}{card(T)}. \tag{2}$$

Confidence $conf^{s-}(q \rightarrow \sim X)$ is calculated in the following way:

$$conf^{s-}(q \rightarrow \sim X) = \frac{card(\{t \in T : q \text{ is a subsequence of } t \wedge C^{max}(q,t) \cap X = \varnothing\})}{card(\{t \in T : q \text{ is a subsequence of } t\})}. \tag{3}$$

Note that (see also Eq. (1)):

$$conf^{s-}(q \rightarrow \sim X) = sup^{s-}(q \rightarrow \sim X) / sup^q(q). \tag{4}$$

Only patterns $s^-(q \rightarrow \sim X)$ that exceed minimum thresholds are really considered, i.e. $sup^{s-}(q \rightarrow \sim X) \geq minsup^{s-}$ and $conf^{s-}(q \rightarrow \sim X) \geq minconf^{s-}$. Since the quantity of the domain – $card(D)$ is usually several orders of magnitude greater than the average length of source sequences, then typically $minconf^{s-}$ has the value close to 1.

3.2 SPAWN – Mining Sequential Patterns with Negative Conclusions

Sequential patterns with negative conclusion can be discovered using the previously obtained set of regular sequential patterns q – any algorithm can be used for this purpose, see sec. 2.1. Hence, having regular sequential patterns q mined, the frequent set of maximum complement $C^{max}(q,t_i)$ is extracted from source sequences t_i that contain each such regular sequential pattern q. Items from domain D (see Def. 1) that do not belong to any complement of these source sequences automatically become members of the negative pattern conclusion – set X. All other items that frequent maximum complement are treated as candidate members for conclusion set X. These candidates and their combinations X_i that exceed $minsup^{s-}$ and $minconf^{s-}$ thresholds form sequential patterns $s^-(q \rightarrow \sim X_i)$ with negative conclusions. The process is repeated separately for each positive sequential pattern q. The above concept is used in SPAWN – the algorithm to discover Sequential PAttern With Negative conclusions.

The SPAWN algorithm

Input: T – the multiset of source sequences
   D – the domain, e.g. the set of web pages existing in the site
   $minsup^{s\text{-}}$ – minimum support for sequential patterns with negative
      conclusions; expressed in number of sequences
   $minconf^{s\text{-}}$ – minimum confidence for sequential patterns with
      negative conclusions; expressed in %

Output: Q – the set of regular, positive sequential patterns
   $SUPQ$ – the set of support values for the appropriate patterns $q \in Q$
   $S^{\text{-}}$ – DB containing sequential patterns with negative conclusions

1. *extract regular sequential patterns q (with support); fill Q and SUPQ*
2. $S^{\text{-}}$ = empty
3. **for each** $q \in Q$ and $sup_q \in SUPQ$ {
4. $threshold_q$ = max $(minsup^{s\text{-}}, sup_q * minconf^{s\text{-}})$
5. CDB_q = empty
6.   **for each** $t \in T$
7.   **if** q is the subsequence of t **then**
8.   **if** $C^{max}(q,t) \neq \varnothing$ **then**
9.   append $C^{max}(q,t)$ to CDB_q
10. $CAND_1$ = distinct_p (CDB_q)
11. $M_q = D\,/\,CAND_1$
12.   check_and_add $(CAND_1, sup_q, threshold_q, CDB_q, S^{\text{-}}, R_1)$
13. $k = 1$
14.   **while** R_k is not empty {
15. $k = k+1$
16. $CAND_k$ = generate_candidates (R_{k-1})
17.   check_and_add $(CAND_k, sup_q, threshold_q, CDB_q, S^{\text{-}}, R_k)$
18.   }
19. $ABSENT_q$ = generate_candidates (M_q)
20.   **for each** $X \in ABSENT_q$ {
21.   append $s^{\text{-}}(q \rightarrow \sim X)$ to $S^{\text{-}}$
22. $sup^{s\text{-}}(q \rightarrow \sim X) = sup_q$
23. $conf^{s\text{-}}(q \rightarrow \sim X) = 100\%$
24.   }
25.   **for each** $X \in ABSENT_q$ and **each** $Y \in R_i$ {   /* for all i=1, 2, ..., k */
26.   append $s^{\text{-}}(q \rightarrow \sim(X \cup Y))$ to $S^{\text{-}}$
27. $sup^{s\text{-}}(q \rightarrow \sim(X \cup Y)) = sup^{s\text{-}}(q \rightarrow \sim Y)$
28. $conf^{s\text{-}}(q \rightarrow \sim(X \cup Y)) = conf^{s\text{-}}(q \rightarrow \sim Y)$
29. }}   /* **for each** $q \in Q$ and the entire algorithm*/
procedure check_and_add
Input: $CAND_k, sup_q, threshold_q, CDB_q, S^{\text{-}}$
Output: R_k – rare itemsets, $S^{\text{-}}$
30. **for each** $X \in CAND_k$ {   /* for $CAND_1$, X is a 1-itemset */
31. $sup^{s\text{-}}(q \rightarrow \sim X) = sup_q$
32.   **for each** C^{max} from CDB_q
33.   **if** $X \cap C^{max} \neq \varnothing$ **then**

34. $sup^{s^-}(q{\rightarrow}{\sim}X) = sup^{s^-}(q{\rightarrow}{\sim}X) - 1$
35.   **if** $sup^{s^-}(q{\rightarrow}{\sim}X) \geq threshold_q$ **then** {
36.   append X to R_k
37.   append $s^-(q{\rightarrow}{\sim}X)$ to S^-
38.   preserve $sup^{s^-}(q{\rightarrow}{\sim}X)$
39. $conf^{s^-}(q{\rightarrow}{\sim}X) = sup^{s^-}(q{\rightarrow}{\sim}X) / sup_q$
40.   }}   /* **for each** $X \in CAND_k$ and the entire procedure check_and_add */

Note that M_q is the set containing items from the domain D that do not occur in any complement of pattern q. For that reason, subsets of M_q can be utilized to create new negative conclusions and to supplement output patterns. It is achieved using either only subsets of M_q (lines 20-24) or by the extension of conclusions for the patterns that have been previously obtained in procedure *check_and_add* (lines 25-29).

The database scan (procedure *check_and_add*, lines 32-34) is performed only for the temporal CDB_q that contains maximum complements for the given regular pattern q, see Def. 4. Since the value of *minconf*$^{s^-}$ is usually closer to 1 rather than to 0, the concept of CDB_q scan consists in decreasing of initial, maximum support. It enables to interrupt the loop (lines 32-34) after the support value falls below the minimal thresholds.

The meaning of selected lines in the SPAWN algorithm is as follow:

- Line 1: Use any algorithm to discover regular sequential patterns q.
- Lines 3-29: One run of the loop creates all sequential patterns with negative conclusion for a single regular sequential pattern q extracted in line 1.
- Line 3: sup_q is the support in T for the corresponding q; it is expressed in number of sequences.
- Line 4: $threshold_q$ is the min. number of sequences $t \in T$ that must contain searched patterns $s^-(q{\rightarrow}{\sim}X)$; sup_q*$minconf^{s^-}$ is rounded up to integers.
- Line 5: CDB_q is the database (list) that contains sets with non-empty maximum complements $C^{max}(q,t)$ of subsequence q for all t in T that contain q.
- Lines 6-9 fill CDB_q. The database CDB_q is in a sense similar to the α-projected database used in the PrefixSpan algorithm [14].
- Line 10: find all distinct $p \in D$ that occur in CDB_q. $CAND_k$ is the set of candidates of length k.
- Line 11: M_q is the complement of $CAND_1$ in D. M_q contains items that never follows q in source sequences. For that reason, the elements of M_q can by default extend negative conclusions, see lines 19-29.
- Line 12: Procedure *check_and_add* (lines 30-40) tests the support for candidates from $CAND_1$ to find rare itemsets. From the rare enough sets, the new output patters $sup^{s^-}(q{\rightarrow}{\sim}X)$ are created together with their measures.
- Line 16: Function *generate_candidates* (R_{k-1}) generates new candidates with the length increased by 1 (k). Each new candidate is the sum of two sets from R_{k-1}.
- Line 17: Check k-item candidates. If they are rare create new sequential patterns with negative conclusions.
- Line 19: $ABSENT_q$ contains all possible subsets of M_q. For $M_q = \{p_1, p_3, p_4\}$, $ABSENT_q = \{\{p_1\}, \{p_3\}, \{p_4\}, \{p_1, p_3\}, \{p_1, p_4\}, \{p_3, p_4\}, \{p_1, p_3, p_4\}\}$.

- Lines 20-24: Generate new patterns using elements X from $ABSENT_q$ (i.e. negative conclusions $\sim X$) and the considered regular pattern q. Add the obtained patterns $s^{\cdot}(q \rightarrow \sim X)$ to the output $S^{\cdot}$. Elements of $ABSENT_q$ do not occur in any source sequence t after the regular pattern q. For that reason, $sup^{s^{\cdot}}(q \rightarrow \sim X) = sup_q$ and $conf^{s^{\cdot}}(q \rightarrow \sim X) = 100\%$.
- Lines 25-29: For each subset X of $ABSENT_q$ and each verified candidate Y from any set R_i (see line 36) create a new pattern for their unions $X \cup Y$, i.e. $s^{\cdot}(q \rightarrow \sim(X \cup Y))$. Since X does not occur in any t containing q ($\sim X$ "occurs" in all these t), the support and confidence of each $s^{\cdot}(q \rightarrow \sim(X \cup Y))$ are the same as for the appropriate $s^{\cdot}(q \rightarrow \sim Y)$, i.e. $sup^{s^{\cdot}}(q \rightarrow \sim(X \cup Y)) = sup^{s^{\cdot}}(q \rightarrow \sim Y)$ and $conf^{s^{\cdot}}(q \rightarrow \sim(X \cup Y)) = conf^{s^{\cdot}}(q \rightarrow \sim Y)$.
- Lines 30-40: One run of the loop checks the frequency of each candidate $X \in CAND_k$. If X is rare enough ($\sim Y$ is frequent enough) then a new pattern $s^{\cdot}(q \rightarrow \sim X)$ is created.
- Line 31: At first, we assume that X does not occur after q in any source sequence. Hence, the initial $sup^{s^{\cdot}}(q \rightarrow \sim X) = sup_q$ is the number of sequences $t \in T$ that contain q. In other words, X is maximally rare ($\sim X$ is maximally frequent). It also means that the initial $conf^{s^{\cdot}}(q \rightarrow \sim X)$ is 100%.
- Lines 32-34: Calculate support of candidate X using all maximum complements that contain q and are stored in CDB_q (see lines 5-9).
- Lines 33-34: If the intersection of X and maximum complement C^{max} is nonempty (at least one X's item belongs to C^{max}) then one source sequence does not support conclusion $\sim X$. It means that X is less rare and $\sim X$ is less frequent. Thus, initial support $sup^{s^{\cdot}}(q \rightarrow \sim X)$ has to be decremented. The processing could quit the loop *for each* when $sup^{s^{\cdot}}(q \rightarrow \sim X)$ falls below $threshold_q$.
- Lines 35-40: If X is rare enough ($\sim X$ is frequent enough) then add a new sequential pattern $s^{\cdot}(q \rightarrow \sim X)$ to the output set $S^{\cdot}$.
- Line 36: X is k-item set. Since X is rare enough $\Leftrightarrow \sim X$ is frequent enough X can be used to generate new $(k+1)$-item candidates by means of R_k.
- Line 37: Add a new sequential pattern with negative conclusion $s^{\cdot}(q \rightarrow \sim X)$ to the output collection $S^{\cdot}$.
- Line 38: $sup^{s^{\cdot}}(q \rightarrow \sim X)$ is expressed in number of source sequences.
- Line 39: Confidence $conf^{s^{\cdot}}(q \rightarrow \sim X)$ can be calculated using Eq. (4).

Procedure *check_and_add* tests the input candidates end returns only the rare ones, i.e. frequent negative. In the typical algorithms for association rule mining, the frequent itemsets are extracted whereas in the SPAWN algorithm, only the rare candidates X from $CAND_k$ are selected (line 35), used to create new patterns (lines 37-39) and to generate new candidates (added to R_k, line 36). The often X occurs in CDB_q, the worse. The occurrence means the non-empty intersection with maximum complements stored in CDB_q (line 33). For the negative conclusion the lower frequency of the candidate X, the better.

The final support $sup^{s^{\cdot}}(q \rightarrow \sim X)$ is expressed in the number of sequences and can be converted to percentage by means of division by $card(T)$. Confidence values $conf^{s^{\cdot}}(q \rightarrow \sim X)$ are percentages.

4 Experiments

The experiments have been performed on web logs collected by the main web server of Wrocław University of Technology. They contained 1000 user sessions (multiset T) with total 1421 HTTP requests for 67 distinct web pages (set D). 257 regular sequential patterns q of the length of up to 7-pages were discovered using $minsup$=0.2% (2 sequences). Next, sequential patterns with negative conclusions were extracted using the SPAWN algorithm (see sec. 3.2).

The number of sequential patterns with negative conclusions is the highest for conclusions of the length 11 (2,835,457 patterns) and 10 (2,834,) whereas the least quantity of patterns is for 22-pages conclusions (only 2 patterns), 21-pages (46), 20-pages (506) and 1-page conclusions (536), Fig. 1. The number of patterns strongly depends on minimum support $minsup^{s-}$ (Fig. 2) and less on minimum confidence $minconf^{s-}$ (Fig. 3).

The usability of sequential patterns with negative conclusions depends on their interpretations and often requires some kind of filtering.

After such manual scanning process, some interesting patterns were identified in the output set. Users who read information about student hostels and next about working possibilities (sequence q) do not visit Socrates/Erasmus Programme ($\sim X$);

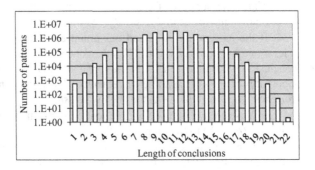

Fig. 1. No. of patterns with negative conclusions, $minsup^{s-}$=0.2%, $minconf^{s-}$=90%

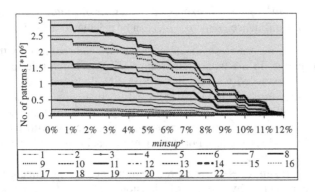

Fig. 2. No. of patterns in relation to $minsup^{s-}$ for different lengths of conclusions

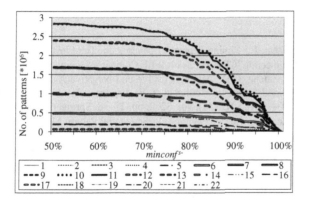

Fig. 3. No. of patterns in relation to $minconf^{s-}$ for different lengths of conclusions

$sup^{s-}(q\rightarrow\sim X)$=1.1%, $conf^{s-}(q\rightarrow\sim X)$=91%. In other words, those who consider working on the spot, are not likely to study abroad. Based on another pattern, users read general promotion information in English, next about study fees in English (q), do not navigate to any contact information in English ($\sim X$); $sup^{s-}(q\rightarrow\sim X)$=1.3%, $conf^{s-}(q\rightarrow\sim X)$=92%.

5 Conclusions and Future Work

Sequential patterns with negative conclusions are new patterns that describe frequent sequences not followed by some items. In the web environment, these patterns can reflect typical user behaviours, e.g. e-commerce users put some products into their basket, then start filling order form but rather do not finish their purchases, i.e. skip all payment pages.

To extract sequential patterns with negative conclusions the SPAWN algorithm can be used (see sec. 3.2). Note that the new patterns differ both from typical sequential patterns (new patterns have negative conclusions) and from negative association rules (the left side of the new patterns is a sequence not a set).

Due to the general profile of sequential patterns with negative conclusions, there are usually large amount of them. Hence, to make use of them it may be necessary to apply some filtering mechanisms in many application domains.

The new patterns, similarly to negative association rules, can be utilized to assess the previously existing connections between objects like hyperlinks binding web pages. This is especially useful for web content managers who can remove useless hyperlinks based on the knowledge provided by the negative patterns [11]. In this case, the new patterns are filtered according the set of hyperlinks. Besides, the new patterns can extend typical content-based recommendation systems by verification of similarities calculated by means of textual content analysis [10].

Further work will focus on the analysis of the usefulness of new patterns in re-commender systems [10] as well as the efficiency of their extraction.

Acknowledgments. The author is indebted to Janusz Wiśniowski for his help at experiments. The work was supported by The Polish Ministry of Science and Higher Education, grant no. N516 037 31/3708.

References

1. Agrawal, R., Srikant, R.: Mining Sequential patterns. In: The Eleventh International Conference on Data Engineering ICDE, pp. 3–14. IEEE Computer Society, Los Alamitos (1995)
2. Antonie, M.-L., Zaïane, O.R.: Mining Positive and Negative Association Rules: An Approach for Confined Rules. In: Boulicaut, J.-F., Esposito, F., Giannotti, F., Pedreschi, D. (eds.) PKDD 2004. LNCS (LNAI), vol. 3202, pp. 27–38. Springer, Heidelberg (2004)
3. Ayres, J., Gehrke, J.E., Yiu, T., Flannick, J.: Sequential Pattern Mining Using Bitmaps. In: KDD 2002, pp. 429–435. ACM Press, New York (2002)
4. Chang, L., Yang, D., Tang, S., Wang, T.: Mining Compressed Sequential Patterns. In: Li, X., Zaïane, O.R., Li, Z. (eds.) ADMA 2006. LNCS (LNAI), vol. 4093, pp. 761–768. Springer, Heidelberg (2006)
5. Chen, Y., Guo, J., Wang, Y., Xiong, Y., Zhu, Y.: Incremental Mining of Sequential Patterns Using Prefix Tree. In: Zhou, Z.-H., Li, H., Yang, Q. (eds.) PAKDD 2007. LNCS (LNAI), vol. 4426, pp. 433–440. Springer, Heidelberg (2007)
6. Cheng, H., Yan, X., Han, J.: IncSpan: incremental mining of sequential patterns in large database. In: KDD 2004, pp. 527–532. ACM Press, New York (2004)
7. Cong, S., Han, J., Padua, D.A.: Parallel mining of closed sequential patterns. In: KDD, pp. 562–567. ACM Press, New York (2005)
8. Cornelis, C., Yan, P., Zhang, X., Chen, G.: Mining Positive and Negative Association Rules from Large Databases. In: 2006 IEEE Conf. on Cybernetics & Intelligent Sys., pp. 613–618 (2006)
9. García-Hernández, R.A., Trinidad, J.F.M., Carrasco-Ochoa, J.A.: A New Algorithm for Fast Discovery of Maximal Sequential Patterns in a Document Collection. In: Gelbukh, A. (ed.) CICLing 2006. LNCS, vol. 3878, pp. 514–523. Springer, Heidelberg (2006)
10. Kazienko, P.: Filtering of Web Recommendation Lists Using Positive and Negative Usage Patterns. In: Apolloni, B., Howlett, R.J., Jain, L. (eds.) KES 2007, Part III. LNCS (LNAI), vol. 4694, pp. 1016–1023. Springer, Heidelberg (2007)
11. Kazienko, P.: Usage-Based Positive and Negative Verification of User Interface Structure. In: ICAS 2008, pp. 1–6. IEEE Computer Society, Los Alamitos (2008)
12. Li, H., Chen, H.: GraSeq: A Novel Approximate Mining Approach of Sequential Patterns over Data Stream. In: Alhajj, R., Gao, H., Li, X., Li, J., Zaïane, O.R. (eds.) ADMA 2007. LNCS (LNAI), vol. 4632, pp. 401–411. Springer, Heidelberg (2007)
13. Masseglia, F., Poncelet, P., Teisseire, M.: Incremental mining of sequential patterns in large databases. Data & Knowledge Engineering 46(1), 97–121 (2003)
14. Pei, J., Han, J., Mortazavi-Asl, B., Pinto, H., Chen, Q., Dayal, U., Hsu, M.: PrefixSpan: Mining Sequential Patterns by Prefix-Projected Growth. In: ICDE 2001, pp. 215–224. IEEE, Los Alamitos (2001)
15. Pei, J., Han, J., Mortazavi-Asl, B., Wang, J., Pinto, H., Chen, Q., Dayal, U., Hsu, M.: Mining Sequential Patterns by Pattern-Growth: The PrefixSpan Approach. IEEE Transaction on Knowledge and Data Engineering 16(11), 1424–1440 (2004)
16. Plantevit, M., Laurent, A., Teisseire, M.: HYPE: mining hierarchical sequential patterns. In: DOLAP 2006, pp. 19–26. ACM Press, New York (2006)
17. Ren, J.-D., Zhou, X.-L.: An Efficient Algorithm for Incremental Mining of Sequential Patterns. In: Yeung, D.S., Liu, Z.-Q., Wang, X.-Z., Yan, H. (eds.) ICMLC 2005. LNCS (LNAI), vol. 3930, pp. 179–188. Springer, Heidelberg (2006)
18. Wu, X., Zhang, C., Zhang, S.: Efficient Mining of Both Positive and Negative Association Rules. ACM Transaction on Information Systems 22(3), 381–405 (2004)
19. Zhu, T., Bai, S.: A Parallel Mining Algorithm for Closed Sequential Patterns. In: AINA 2007, vol. 1, pp. 392–395. IEEE Computer Society, Los Alamitos (2007)

Author Index

Lecture Notes in Computer Science

Sublibrary 3: Information Systems and Application, incl. Internet/Web and HCI

For information about Vols. 1– 4796
please contact your bookseller or Springer

Vol. 4956: C. Macdonald, I. Ounis, V. Plachouras, I. Ruthven, R.W. White (Eds.), Advances in Information Retrieval. XXI, 719 pages. 2008.

Vol. 4952: C. Floerkemeier, M. Langheinrich, E. Fleisch, F. Mattern, S.E. Sarma (Eds.), The Internet of Things. XIII, 378 pages. 2008.

Vol. 4950: A. Kerren, J.T. Stasko, J.-D. Fekete, C. North (Eds.), Information Visualization. IX, 177 pages. 2008.

Vol. 4947: J.R. Haritsa, R. Kotagiri, V. Pudi (Eds.), Database Systems for Advanced Applications. XXII, 713 pages. 2008.

Vol. 4936: W. Aiello, A. Broder, J. Janssen, E.E. Milios (Eds.), Algorithms and Models for the Web-Graph. X, 167 pages. 2008.

Vol. 4932: S. Hartmann, G. Kern-Isberner (Eds.), Foundations of Information and Knowledge Systems. XII, 397 pages. 2008.

Vol. 4928: A.H.M. ter Hofstede, B. Benatallah, H.-Y. Paik (Eds.), Business Process Management Workshops. XIII, 518 pages. 2008.

Vol. 4918: N. Boujemaa, M. Detyniecki, A. Nürnberger (Eds.), Adaptive Multimedia Retrieval: Retrieval, User, and Semantics. XI, 265 pages. 2008.

Vol. 4903: S. Satoh, F. Nack, M. Etoh (Eds.), Advances in Multimedia Modeling. XIX, 510 pages. 2008.

Vol. 4900: S. Spaccapietra (Ed.), Journal on Data Semantics X. XIII, 265 pages. 2008.

Vol. 4892: A. Popescu-Belis, S. Renals, H. Bourlard (Eds.), Machine Learning for Multimodal Interaction. XI, 308 pages. 2008.

Vol. 4882: T. Janowski, H. Mohanty (Eds.), Distributed Computing and Internet Technology. XIII, 346 pages. 2007.

Vol. 4881: H. Yin, P. Tino, E. Corchado, W. Byrne, X. Yao (Eds.), Intelligent Data Engineering and Automated Learning - IDEAL 2007. XX, 1174 pages. 2007.

Vol. 4877: C. Thanos, F. Borri, L. Candela (Eds.), Digital Libraries: Research and Development. XII, 350 pages. 2007.

Vol. 4872: D. Mery, L. Rueda (Eds.), Advances in Image and Video Technology. XXI, 961 pages. 2007.

Vol. 4871: M. Cavazza, S. Donikian (Eds.), Virtual Storytelling. XIII, 219 pages. 2007.

Vol. 4868: C. Peter, R. Beale (Eds.), Affect and Emotion in Human-Computer Interaction. X, 241 pages. 2008.

Vol. 4858: X. Deng, F.C. Graham (Eds.), Internet and Network Economics. XVI, 598 pages. 2007.

Vol. 4857: J.M. Ware, G.E. Taylor (Eds.), Web and Wireless Geographical Information Systems. XI, 293 pages. 2007.

Vol. 4853: F. Fonseca, M.A. Rodríguez, S. Levashkin (Eds.), GeoSpatial Semantics. X, 289 pages. 2007.

Vol. 4836: H. Ichikawa, W.-D. Cho, I. Satoh, H.Y. Youn (Eds.), Ubiquitous Computing Systems. XIII, 307 pages. 2007.

Vol. 4832: M. Weske, M.-S. Hacid, C. Godart (Eds.), Web Information Systems Engineering – WISE 2007 Workshops. XV, 518 pages. 2007.

Vol. 4831: B. Benatallah, F. Casati, D. Georgakopoulos, C. Bartolini, W. Sadiq, C. Godart (Eds.), Web Information Systems Engineering – WISE 2007. XVI, 675 pages. 2007.

Vol. 4825: K. Aberer, K.-S. Choi, N. Noy, D. Allemang, K.-I. Lee, L. Nixon, J. Golbeck, P. Mika, D. Maynard, R. Mizoguchi, G. Schreiber, P. Cudré-Mauroux (Eds.), The Semantic Web. XXVII, 973 pages. 2007.

Vol. 4823: H. Leung, F. Li, R. Lau, Q. Li (Eds.), Advances in Web Based Learning – ICWL 2007. XIV, 654 pages. 2008.

Vol. 4822: D.H.-L. Goh, T.H. Cao, I.T. Sølvberg, E. Rasmussen (Eds.), Asian Digital Libraries. XVII, 519 pages. 2007.

Vol. 4820: T.G. Wyeld, S. Kenderdine, M. Docherty (Eds.), Virtual Systems and Multimedia. XII, 215 pages. 2008.

Vol. 4816: B. Falcidieno, M. Spagnuolo, Y. Avrithis, I. Kompatsiaris, P. Buitelaar (Eds.), Semantic Multimedia. XII, 306 pages. 2007.

Vol. 4813: I. Oakley, S.A. Brewster (Eds.), Haptic and Audio Interaction Design. XIV, 145 pages. 2007.

Vol. 4810: H.H.-S. Ip, O.C. Au, H. Leung, M.-T. Sun, W.-Y. Ma, S.-M. Hu (Eds.), Advances in Multimedia Information Processing – PCM 2007. XXI, 834 pages. 2007.

Vol. 4809: M.K. Denko, C.-s. Shih, K.-C. Li, S.-L. Tsao, Q.-A. Zeng, S.H. Park, Y.-B. Ko, S.-H. Hung, J.-H. Park (Eds.), Emerging Directions in Embedded and Ubiquitous Computing. XXXV, 823 pages. 2007.

Vol. 4808: T.-W. Kuo, E. Sha, M. Guo, L.T. Yang, Z. Shao (Eds.), Embedded and Ubiquitous Computing. XXI, 769 pages. 2007.

Vol. 4806: R. Meersman, Z. Tari, P. Herrero (Eds.), On the Move to Meaningful Internet Systems 2007: OTM 2007 Workshops, Part II. XXXIV, 611 pages. 2007.

Vol. 4805: R. Meersman, Z. Tari, P. Herrero (Eds.), On the Move to Meaningful Internet Systems 2007: OTM 2007 Workshops, Part I. XXXIV, 757 pages. 2007.

Vol. 4804: R. Meersman, Z. Tari (Eds.), On the Move to Meaningful Internet Systems 2007: CoopIS, DOA, ODBASE, GADA, and IS, Part II. XXIX, 683 pages. 2007.

Vol. 4803: R. Meersman, Z. Tari (Eds.), On the Move to Meaningful Internet Systems 2007: CoopIS, DOA, ODBASE, GADA, and IS, Part I. XXIX, 1173 pages. 2007.

Vol. 4802: J.-L. Hainaut, E.A. Rundensteiner, M. Kirchberg, M. Bertolotto, M. Brochhausen, Y.-P.P. Chen, S.S.-S. Cherfi, M. Doerr, H. Han, S. Hartmann, J. Parsons, G. Poels, C. Rolland, J. Trujillo, E. Yu, E. Zimányie (Eds.), Advances in Conceptual Modeling – Foundations and Applications. XIX, 420 pages. 2007.

Vol. 4801: C. Parent, K.-D. Schewe, V.C. Storey, B. Thalheim (Eds.), Conceptual Modeling - ER 2007. XVI, 616 pages. 2007.

Vol. 4797: M. Arenas, M.I. Schwartzbach (Eds.), Database Programming Languages. VIII, 261 pages. 2007.